D1700914

# Lecture Notes in Earth System Sciences 143



For further volumes:
http://www.springer.com/series/10529

Andrej Ernst • Priska Schäfer • Joachim Scholz
Editors

# Bryozoan Studies 2010

*Editors*
Andrej Ernst
Institut für Geowissenschaften
Christian-Abstrechts-Universität zu Kiel
Kiel Schleswig-Holstein
Germany

Priska Schäfer
Institut für Geowissenschaften
Christian-Abstrechts-Universität zu Kiel
Kiel Schleswig-Holstein
Germany

Joachim Scholz
Forschungsinstitut und Naturmuseum
Senckenberg
Frankfurt
Germany

ISSN 2193-8571 ISSN 2193-858X (electronic)
ISBN 978-3-642-16410-1 ISBN 978-3-642-16411-8 (eBook)
DOI:10.1007/978-3-642-16411-8
Springer Heidelberg New York Dordrecht London

Library of Congress Control Number: 2012942188



Printed on acid-free paper

Springer is part of Springer Science+Business Media (www.springer.com)

# Preface

The International Bryozoology Association (IBA) held its 15th International Conference at Christian-Albrechts-University of Kiel, Germany, from August 2 to 6, 2010. The IBA records its sincere gratitude to the university for hosting the conference and giving logistic support to the organization. The IBA acknowledges the generous sponsorship provided by the Deutsche Forschungsgemeinschaft (German Research Foundation) and the Kiel University alumni association as well as a number of local and national companies, among which were EO-Electronen-Optik-Service GmbH, J. Bornhöft GmbH, Vollmer Privatrösterei GmbH, Citti Märkte GmbH & Co. KG, and Fielmann AG.

The IBA conference benefited from the enthusiasm, expertise, and logistic support provided by many colleagues in and outside Kiel: Heidi Blaschek, Mareike Ennels, Sebastian Meier, Sven Nielsen, Wolfgang Reimers, Ute Schuldt, Max Rademacher, Phillip Reuter, Arno Lettmann, Peter Appel, Ulrike Westernströer (Institute of Geosciences), Nils Andersen (Leibniz Laboratory for Age Determination and Isotope Research), Elena Nikulina, Ulrich Schmölke (Zentrum für Baltische und Skandinavische Archäologie, Schloss Gottorf), Michael Gruber (Seawater Aquarium), Frank Melzner (Leibniz Centre of Marine Sciences), Capt. Christoph and Crew (RV Littorina), Eckart Schrey (Multimar Wattforum, Tönning), John Jagst (Natuurhistorisch Museum Maastricht), Kirsten Grimm (Naturhistorisches Museum Mainz), and Brigitte Lotz, Sonja Wedmann (Forschungsinstitut und Naturmuseum Senckenberg).

When in 1968 a group of scientists especially engaged in the field of bryozoology met in order to build up the International Bryozoology Association, they held the first congress of the new association in Milan (Italy). Now more than 40 years have passed. The scientists of our field of interest have built a global community with vivid scientific and personal contacts among bryozoologists in many parts of the world. The 3-year spacing of the meetings and their varied venues, Milan (Italy) 1968, Durham (England) 1971, Lyon (France) 1974, Woods Hole – Massachusetts (USA) 1977, Durham (England) 1980, Wien (Austria) 1983, Bellingham – Washington (USA) 1986, Paris (France) 1989, Swansea (Wales) 1992, Wellington (New Zealand) 1995, Smithsonian Tropical Research Institute – Balboa (Republic of Panamá) 1998, Dublin (Ireland) 2001, Concepción (Chile) 2004, and Boone – North Carolina (USA) 2007, give a special rhythm to the conferences and make it possible for more people to attend. Ninety-five scientists and graduate students from 29 countries attended the conference in Kiel.

The conference in Kiel started with a 1-week field excursion to Denmark and Scania (Sweden) run by Eckart Håkansson. The field trip included 14 geological stops, spanning the stratigraphy from Silurian limestones in southern Scania to modern bryozoan communities dredged in the Kattegat (Kristineberg Marine Biological Station). Most outcrops visited the late Maastrichtian to middle Danian localities in Seeland and Jylland, Denmark. For many of the participants, the famous Stevns Klint K/T boundary locality, talked about at so many IBA conferences, was a highlight of the trip.

Following the tradition of the former conferences, a midweek excursion to visit the Hanseatic city of Lübeck was scheduled for Wednesday afternoon and evening. There, we toured the historic buildings and ate in a medieval guildhall-turned-restaurant. A special highlight of the conference was the visit to the seawater aquarium including its tanks for $CO_2$ experiments (ocean acidification) as part of the Cluster of Excellence "Future Ocean" at Kiel University. The conference dinner was held at the Maritim Hotel Bellevue, Kiel, spanning a fantastic view over the Kiel fjord.

Three 1-day field excursions followed the scientific sessions on Saturday, August 7, including a dredge tour with RV Littorina in the Kiel Bight (Elena Nikulina and Andrej Ostrovsky), a visit to the quarries near Kronsmoor and Lägerdorf (upper Campanian, lower Maastrichtian (Sven Nielsen)), and a visit to the coastal tidal-flat ecosystems near Westerhever, west coast of Schleswig-Holstein (Priska Schäfer). A 1-week field excursion to the Netherlands (Upper Maastrichtian and K/T boundary near Maastricht and western Germany), Middle Devonian in the Eifel Mountains, Middle Rhine valley, Tertiary of the Mainz Basin, the famous Messel site near Darmstadt (middle Eocene), and finally the public exhibits of the Senckenberg Naturmuseum Frankfurt, and Bryozoan collections donated by Professor Ehrhard Voigt to the Senckenberg Research Institute (Priska Schäfer and Andrej Ernst) ended the program of the conference.

Three workshops were held on Sunday, July 31, entitled "Bryozoa on the Web" (Scott Lidgard and colleagues), "Trepostomata: morphology, taxonomy, phylogeny and evolution" (Andrej Ernst and Caroline Buttler), and "Bryozoan geochemistry and carbonates: proxies for palaeoclimate and environment" (Marcus M. Key and Abigail Smith).

The scientific sessions held from August 2 to 6 included oral presentations as well as a full afternoon with poster presentations and marked the main program of the conference. Ninety-five scientists attended the conference, providing 98 abstracts for 62 oral presentations, of which 13 were invited or keynote lectures, and 36 posters. Jeremy B. C. Jackson presented a keynote lecture on the topic *The Future of the Ocean's Past*, which resulted in a very stimulating discussion. The abstracts covered the full breadth of actual scientific activities in the field of bryozoology. The scope of the oral and poster contributions covered thirteen thematic subjects including:

1. *Molecular Genetics and Phylogeny*
2. *Life History/Reproduction Biology/Anatomy*
3. *Climate Response*
4. *Research History*
5. *Scientific Collections*
6. *Faunistic Studies*
7. *Evolutionary Patterns*
8. *Faunistic Studies and Zoogeography*
9. *Bryozoan Ecology*
10. *Faunas and Evolutionary Patterns*
11. *Faunas and Taxonomic Revisions*
12. *Bryozoan Palaeoecology and Sediment Interactions*
13. *Bryozoan Taphonomy*

Of the 98 papers presented, 29, covering all above-mentioned topics, are included in the proceedings volume. The editors wish to thank the authors of the publications in this volume for submitting original and revised manuscripts in time to meet the publication schedule and also the colleagues who kindly refereed the papers. The editors also thank Helena Fortunato for kind help to non-English-speaking authors, especially as referee of the manuscripts.

Compliments and thanks go to all who organized and attended this wonderful meeting in Kiel. Hope to see you all soon in Catania.

Judy E. Winston, outgoing IBA president
Priska Schäfer, editor and conference host
Andrej Ernst, editor
Joachim Scholz, editor

# Contents

# Conference Photograph

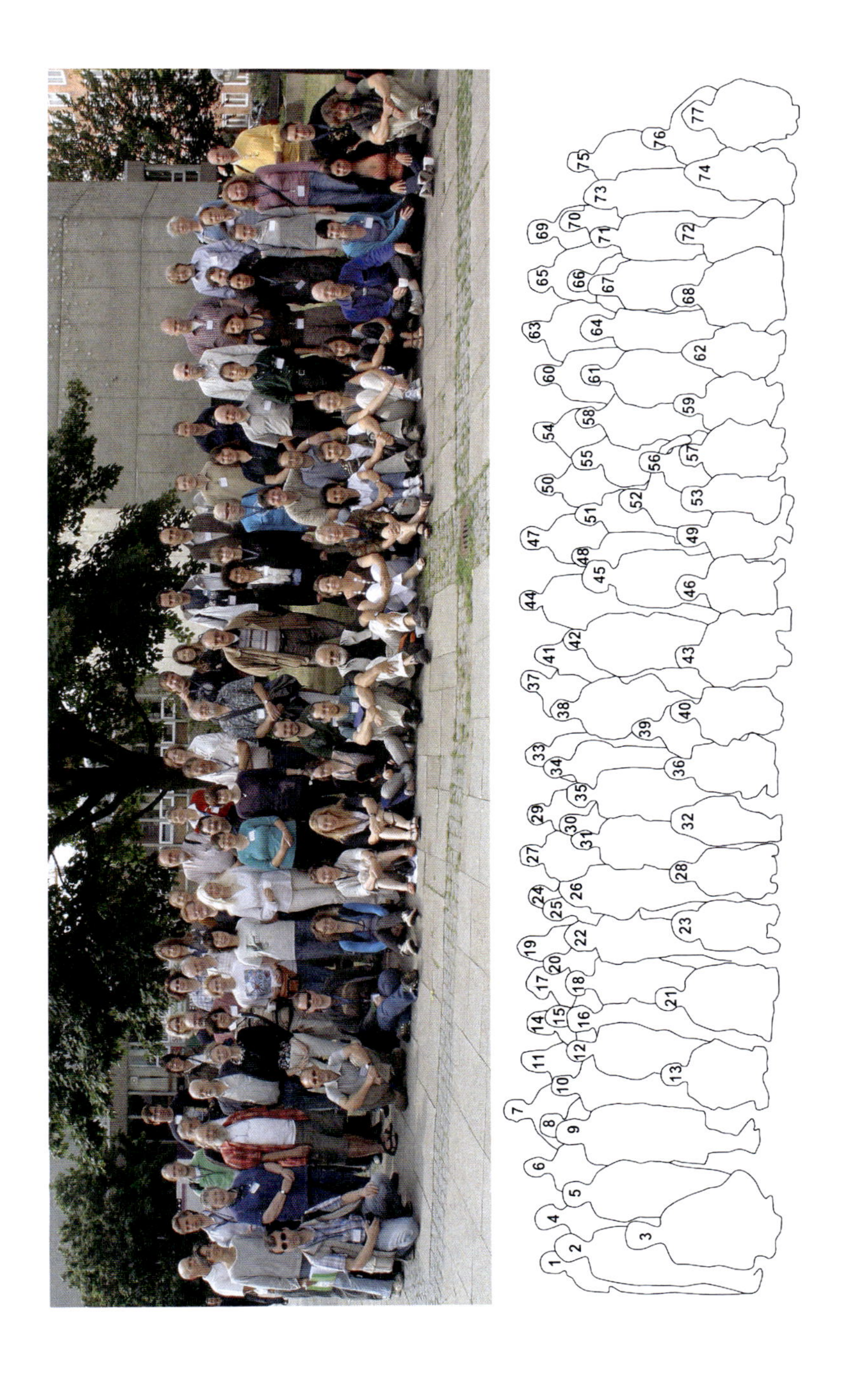

1 Timothy Wood, 2 Beth Okamura, 3 Joachim Scholz, 4 Matthias Obst, 5 Yvonne Bone, 6 Alexander Gruhl, 7 Piotr Kuklinski, 8 Norbert Vávra, 9 Eckart Hakansson, 10 Pierre Moissette, 11 Chiara Lombardi, 12 Urzula Hara, 13 Kamil Zágoršek, 14 Helena Fortunato, 15 Steve Hageman, 16 Amalia Herrera, 17 Mary Spencer Jones, 18 Maja Novosel, 19 Christine Davis, 20 Andrea Waeschenbach, 21 Michael Winson, 22 Ji Eun Seo, 23 Helen Jenkins, 24 Hanna Hartikainen, 25 Hans Arne Nakrem, 26 Emmy Wöss, 27 Ernest Gilmour, 28 Joanne Porter, 29 Priska Schäfer, 30 Michael Toma, 31 Abigail Smith, 32 Jennifer Loxton, 33 Nina Denisenko, 34 Mark Wilson, 35 Caroline Buttler, 36 Sally Rouse, 37 Javier Souto, 38 John Bartley, 39 Leandro Vieira, 40 Rolf Schmidt, 41 Consuelo Sendino, 42 Giampietro Braga, 43 Philip Bock, 44 Oscar Reverter-Gil, 45 Antonietta Rosso, 46 Facelucia Souza, 47 Carlos López-Fé, 48 Mary Sears, 49 Judith Winston, 50 Herwig Heidl, 51 Francoise Bigey, 52 Marcus Key, 53 Julia Cáceres, 54 Franziska Bitschofsky, 55 Catherine Reid, 56 Björn Berning, 57 Andrew Ostrovsky, 58 Kevin Tilbrook, 59 Karin Hoch Fehlauer-Ale, 60 Dennis Gordon, 61 Masato Hirose, 62 Blanca Figuerola, 63 Patrick Wyse Jackson, 64 Emanuela Di Martino, 65 Scott Lidgard, 66 Thomas Schwaha, 67 Lais Ramalho, 68 Matthew Dick, 69 John Ryland, 70 Jeremy Jackson, 71 Zoya Tolokonnikova, 72 Judith Fuchs, 73 Anna Koromyslova, 74 Loa Ramalho, 75 Paul Taylor, 76 Andrej Ernst, 77 Hans de Blauwe.

Not shown: Roger Cuffey, Aaron O'Dea

# In Memory of Richard Stanton Boardman (1923–2011)

It was with great sadness that I and many other members of the International Bryozoology Association (IBA) learned of the death of Richard S. Boardman, longtime colleague and founding member of the IBA. Any bryozoologist visiting the Smithsonian during the years Rich worked there was assured of his interest and hospitality. Rich and his wife Phyllis welcomed dozens of colleagues and students to their Bethesda home. Visitors were assured of good food, good wine, good conversation, and a 5:30 a.m. wake-up call (Rich was an early riser, who preferred to beat the DC traffic by arriving and leaving ahead of the rush hours). I first met Rich when I visited the Smithsonian as a graduate student. His friendship and encouragement helped greatly in dealing with the challenges of a research life, and my daughter and I enjoyed many visits to their retirement home in Sarasota, Florida.

Rich always said he had the great good fortune to work in paleontology at a time when American science was expanding rapidly, but Rich's intelligence, enthusiasm, and political skill played a large part as well. He received his Ph.D. from the University of Illinois and was recruited by the U.S. Geological Survey in 1951. In 1957, he became part of the National Museum of Natural History's Department of

Geology (later split into the Department of Paleobiology and the Department of Mineralogy). As curator-in-charge for the Department, Rich greatly expanded its staff to fill its space in the new east wing of the museum. He was responsible for hiring eight new invertebrate paleontologists, including a second bryozoan worker, Alan Cheetham.

His own research on bryozoans focused primarily on Paleozoic groups, especially trepostomes, but he was always seeking ways to connect fossil and living bryozoan biology. One of his biggest contributions was the method he developed for hard/soft thin-sectioning, a technique that makes it possible to study calcified bryozoans with both hard and soft tissue in place. Use of this technique in studying the internal morphology of both living and fossil species also yielded new information on the life histories of the fossils.

Rich was the lead editor for the revised Part G (Bryozoa) volume of the *Treatise on Invertebrate Paleontology* (1983) and the textbook *Fossil Invertebrates* (1987) as well as the author or coauthor of many scientific publications.

After his retirement in 1985, Rich and Phyllis enjoyed the garden and wildlife at their peaceful lakeside home in Sarasota, Florida, but he retained his interest in bryozoans, continuing work on the Trepostome volume of the Treatise in collaboration with other bryozoologists until the last few months of his life. Even in his 80s, Rich was never old – his enthusiasm for life was irrepressible. We will miss him.

Judy Winston

# In Memory of Frank K. (Ken) McKinney (1943–2011)

Ken McKinney contributed to the field of bryozoology with over 100 peer-reviewed publications and five books, including *Bryozoa Evolution* coauthored with Jeremy B.C. Jackson. Ken will be known for his many important scientific contributions, but he will be long remembered as a colleague, mentor, and friend because of the unique person that he was. One of the most remarkable things about Ken's career is that such productivity and reach came out of his appointment to a small geology department, at Appalachian State University, where he had no graduate students and, at that time, minimal support for research of any kind. Ken played an important role in leading his entire institution into a culture of original scholarship and research (see IBA Bulletin 7 (2), 2011, for complete publication list).

Ken McKinney grew up in Birmingham, Alabama, in the 1950s and met his future wife and lifelong collaborator, Marjorie Jackson, as an undergraduate student (his first professional paper was coauthored with Marge in 1963). Ken moved to the University of North Carolina, Chapel Hill, to study trilobites for his Ph.D., but, fortunately for us, he was "told" by his advisor, Joe St. Jean, to study middle Paleozoic bryozoans.

Ken never looked back. He studied bryozoans with Richard Boardman at the Smithsonian and soon began collaborations with researchers from around the world. The final and one of the most significant scholarly contributions from Ken McKinney will be the chapter on fenestrate Bryozoa in the *Treatise on Invertebrate Paleontology* to be published posthumously with Patrick Wyse Jackson.

A turning point in Ken's professional career occurred while he was on a field trip, organized for the 1983 Vienna IBA conference, and visited the marine lab in Rovinj, Yugoslavia (now Croatia). There he saw the living analogs of the creatures he had studied for so long only as fossils. Ken saw the bryozoan gardens brought to the surface through the eyes of a Paleozoic paleoecologist. This may be the greatest contribution that Ken made – not just to bryozoology but to the entire field of paleontology. Ken made connections from modern ecosystems, behavior and morphology, to the fossil record, connections that had relevance to broader evolutionary (paleo)biology but were not perceptible to those who study only fossils or only living organisms. Ken took this message to the paleontology community. Although he is not solely responsible, he was a leader in the 1980s up to the 2000s in what is now commonplace, the integration of paleo- and modern ecology.

Anyone who ever had the pleasure to collaborate with Ken McKinney can attest that he modeled responsible conduct in research. He was careful, meticulous, and thorough. No corners were cut, no attributions/citations were missed, and no result was merely "good enough." If a revision of a mature manuscript needed additional data, it was back to specimens for primary measurements or back to the outcrop for additional specimens.

For anyone junior to Ken McKinney who interacted with him, the term "mentor" is used frequently. So many of us benefitted from time spent with Ken, even if only for a brief period. He was never distracted or pretentious. He was genuinely interested in your bryozoan findings and problems. But most importantly, when Ken McKinney was talking with you, for that moment, you were the most important person in his life.

Steve Hageman

# Chapter 1
# Distribution over Space and Time in Epizoobiontic North Sea Bryozoans

## Distribution of North Sea Bryozoans

**Franziska Bitschofsky**

**Abstract** The epizoobiontic bryozoan-community on fronds of *Flustra foliacea* from 51 stations kept in four German museums was analysed. Sampling covers the entire North Sea and different time periods (1776–2008). Cluster analysis based on presence/absence shows a differentiation into a northern and a southern North Sea assemblage separated along the 50 m depth contour. The northern assemblage is characterized by *Amphiblestrum flemingii* (Busk), *Callopora dumerilii* (Audouin) and *Tricellaria ternata* (Ellis and Solander), while the southern North Sea is characterized by two cyclostome species – *Crisia eburnea* (Linnaeus), and *Plagioecia patina* (Lamarck) – and a cheilostome – *Electra pilosa* (Linnaeus). In the Helgoland Deep Trench (HTR), the occurrence of cyclostomes is more conspicuous than elsewhere, judging from the studied material. The occurrence of erect growth forms increases from the northern to the southern North Sea.

The results show a difference in spatial distribution of epizoobiontic bryozoans on *Flustra foliacea* in the North Sea. A temporal shift could not be detected in this investigation.

**Keywords** North Sea • Distribution pattern • Epizoobiontic bryozoans • *Flustra foliacea*

---

F. Bitschofsky (✉)
LOEWE – Biodiversity and Climate Research Centre (BiK-F), Senckenberganlage 25, D-60325 Frankfurt am Main, Germany
e-mail: bryozoa@bitschofsky.de

A. Ernst, P. Schäfer, and J. Scholz (eds.), *Bryozoan Studies 2010*,
Lecture Notes in Earth System Sciences 143, DOI 10.1007/978-3-642-16411-8_1,

## Introduction

The hornwrack bryozoan *Flustra foliacea* has been collected over decades at various depths throughout the entire North Sea basin. The very conspicuous colonies are abundantly present in many museum collections (Fig. 1.1). Individual colonies may grow for more than 12 years, forming bushy clumps up to 20 cm in diameter, thereby providing a reasonable stable and perennial substrate for various epizoobionts, many of which are bryozoans (Stebbing 1971; Menon 1975; Hayward and Ryland 1998).

Oceanographically, the North Sea is divided into a northern (on average 100–140 m depth) and a southern part (on average 10–30 m depth). The northern North Sea is mainly characterized by a strong Atlantic inflow with higher salinities and more stable temperatures, while the southern area is influenced by a weak inflow through the English Channel and the freshwater input of the rivers resulting in lower salinities and pronounced temperature variability (Rees 2007; OSPAR 2000). At the boundary between the stratified water masses in the northern and the well mixed water masses in the southern North Sea, frontal zones develop (for example the Helgoland-Flamborough-Front) correlating with some aspects of benthic communities' distribution (Glémarec 1973). The 50 m depth contour is deemed to be a boundary between different hydrographical conditions and biogeographical regions (Rees 2007).

Located in the southern North Sea, the Helgoland Deep Trench (HTR) is a region with unique hydrographic conditions. With an average depth of 50 m and a maximum depth of 59 m the HTR is a relatively deep depression southwest of the island of Helgoland. Its bottom temperature and salinity are more influenced by oceanic conditions and thus more stable than in the coastal areas (Blahudka and Türkay 2002). In contrast to the German Bight, which is characterized by sandy and muddy sediments, the HTR has secondary hard grounds, consisting of molluscan shells (Hertweck 1988; Schaller 2001; Blahudka and Türkay 2002).

Analyses of epibenthic communities in the North Sea showed a separation of northern and southern species assemblages approximately along the 50 m depth

**Fig. 1.1** Frond of *Flustra foliacea* sampled in 1776 as floatsam in Eckwarden/Oldenburg with *Scrupocellaria reptans* growing epizoobiontic at the basis of the colony; collection: "Museum für Naturkunde" of Humboldt University, Berlin; photo: Joachim Scholz

contour with increasing species richness from south to north (Frauenheim et al. 1989; Jennings et al. 1999; Zühlke et al. 2001; Callaway et al. 2002).

The epizoobiontic assemblage on *Flustra foliacea* fronds from the North Sea was analysed here in order to examine temporal and spatial changes in the bryozoan community. Special focus was laid on the distribution of cyclostomes and cheilostomes and different colony growth forms in the North Sea basin.

## Materials and Methods

*Flustra foliacea* specimens from 51 stations in the North Sea were analysed (Fig. 1.2). These samples are stored in four German museums: Research Institute and Natural History Museum Senckenberg, Frankfurt am Main (FIS); Zoological Museum of Christian-Albrechts-University, Kiel (CAU); "Museum für Naturkunde" of the Humboldt-University, Berlin (HU) and Zoological Museum of the University of Hamburg (UH). They were sampled between the years 1776 and 2008 (Table 1.1).

Due to the different sampling methods in the different time periods, samples varied widely in volume and quality. Therefore, I used the presence/absence data of epizoobiontic bryozoans for cluster analysis (PRIMER). The PRIMER subprogram SIMPER (similarity percentage) was used to examine the percentage contribution of each species to the similarity within the clusters respectively to the dissimilarity between two clusters (Bitschofsky et al. 2011).

Historical samples with exact coordinates for the sampling stations were available only from northern North Sea (CAU), therefore, temporal comparisons were done between the samples from the northern North Sea of the FIS collection (1980s) and the CAU collection (1904/1908).

The largest portion of studied samples came from the FIS collection. The material was sampled across the entire North Sea between 1980 and 2007, but mostly during the 1980's. For those samples it was possible to count and measure the epizoobiontic specimens (Table 1.1).

For better comparability between different samples the specimen numbers were divided by the dry-weight of the whole sample, to display them as the number of specimens per gram sample. These data were also used for the cluster analysis.

Using a micrometer scale specimens (colonies) were classified as small ($\leq$3 mm), medium (>3 mm) or erect. For this study, only the diameter of the specimens was considered, therefore it was not possible to differentiate between spot colonies and astogenetic early sheet colonies.

In general, specimen numbers correspond to colony numbers. The predominance of certain species such as *Crisia eburnea* in some localities may be due to the fact that colonies disintegrate and each internode passes as a full specimen. This is a well-known problem in quantitative analysis of bryozoan assemblages (Lagaaij and Gautier 1965). For the present study, this source of mistakes was avoided by counting only colony bases founded on *F. foliacea* fronds.

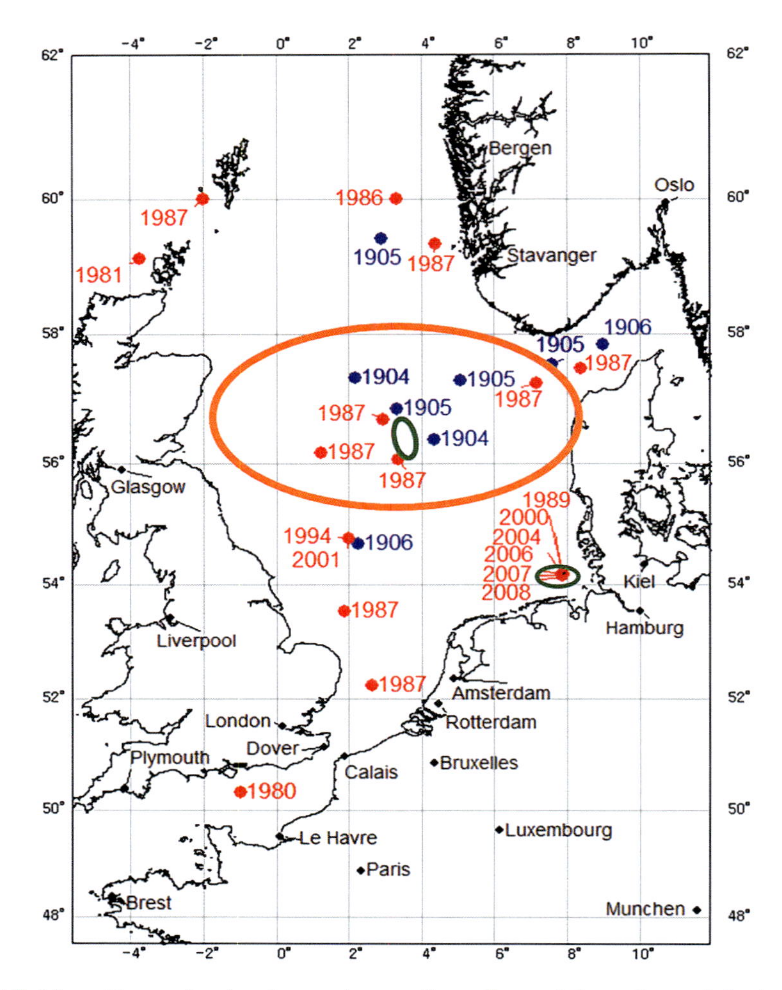


**Fig. 1.2** Map with sampling locations and years of sampling: red dots – Research Institute and Natural History Museum Senckenberg (FIS); *blue dots* – Zoological Museum of Christian-Albrechts-University Kiel (CAU); *orange circles* – Zoological Museum of University of Hamburg (UH); *green circles* – "Museum für Naturkunde", Berlin (HU)

## Results

A total of 45 bryozoan species were found in 51 samples. Thirty-nine were identified to species level (8 cyclostomes, 31 cheilostomes). Cluster analysis based on presence/absence data identifies a northern and a southern North Sea cluster. There is no significant separation in terms of species occurrence between historical and modern samples (Fig. 1.3) (Bitschofsky et al. 2011).

**Table 1.1** Collections, number of samples, sampling regions and periods, conservation and handling of the material analysed

| Collections | Total samples collected | Sampling period | Region | Conservation | Handling |
|---|---|---|---|---|---|
| Research Institute and Naturmuseum Senckenberg, Frankfurt (FIS) | 21 | 1980–2007 | Entire North Sea | Alcohol → air-dried | Quantitative |
| Zoological Museum, Christian-Albrechts-University, Kiel (CAU) | 10 | 1904–1908 | Northern North Sea | Alcohol | Qualitative |
| Natural History Museum, Humboldt-University, Berlin (HU) | 14 | 1776–1943 | Southern North Sea; Helgoland | Alcohol → partly dried herbarium | Qualitative |
| Zoological Museum, University of Hamburg (UH) | 5 | ca. 1926 | Central North Sea | Alcohol → partly dried | Qualitative/ quantitative |
| Fresh samples | 1 | 2008 | HTR (Helgoland Depth Trench) | Alcohol/dried | Quantitative |

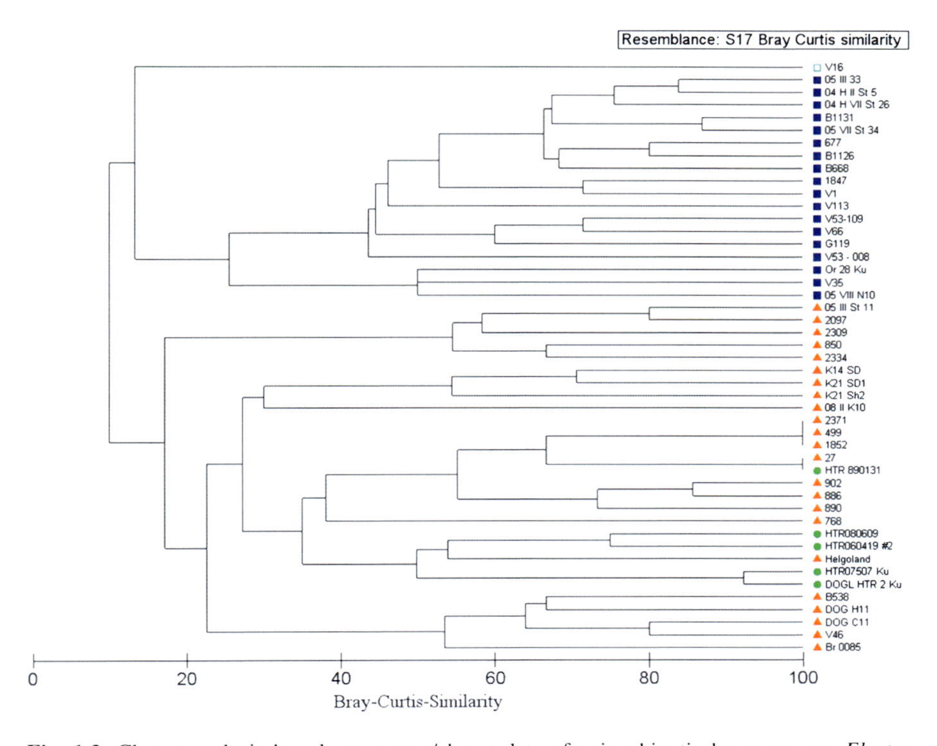


**Fig. 1.3** Cluster analysis based on present/absent data of epizoobiontic bryozoans on *Flustra foliacea; symbols* indicate the origin of samples *blue rectangles* – northern North Sea, *orange triangles* – southern North Sea, *green circles* – Helgoland Deep Trench (HTR)

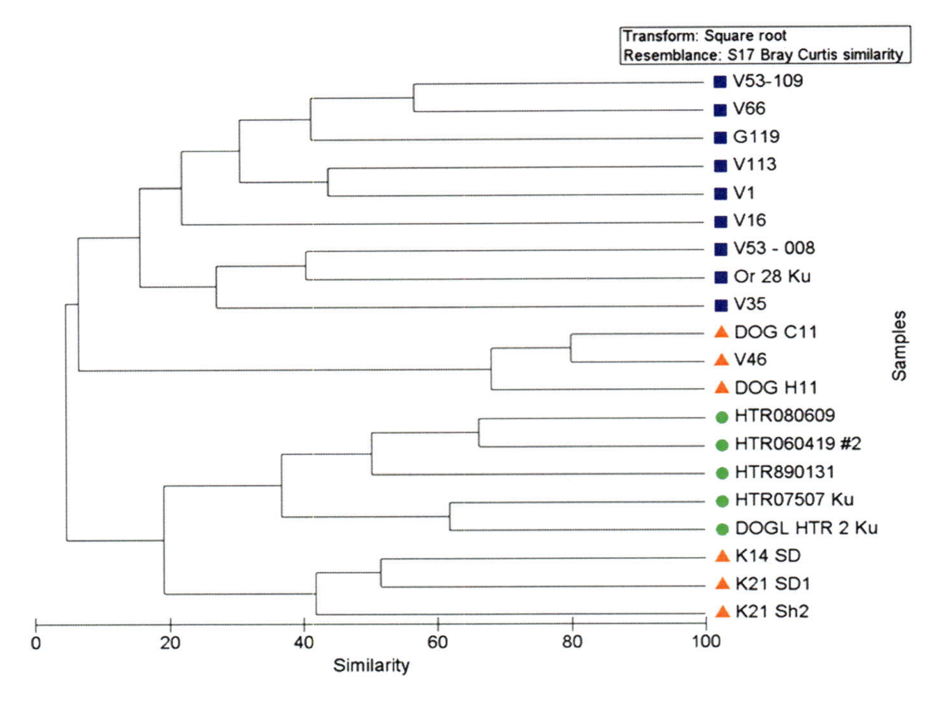


**Fig. 1.4** Cluster analysis based on specimen numbers of epizoobiontic bryozoans per gram of sample on *Flustra foliacea; symbols* indicate the origin of samples *blue rectangles* – northern North Sea, *orange triangles* – southern North Sea, *green circles* – Helgoland Deep Trench (HTR)

The typical species of the north cluster are the cheilostomes *Amphiblestrum flemingii*, *Tricellaria ternata* and *Callopora dumerilii*. Typical south cluster species are *Electra pilosa* and the cyclostomes *Crisia eburnea*, and *Plagioecia patina*.

Even though *C. eburnea* is defined as typical for the southern North Sea, specimens of *C. eburnea* also occur on four historical samples from the northern North Sea, whereas they are absent on modern samples from the northern North Sea.

The cluster analyses based on quantitative data of FIS samples show a similar grouping of northern and southern samples to the cluster analyses based on presence/absence data (Fig. 1.4), but there are some differences in detail. The samples from the Doggerbank, grouping together with the sample from the station V46 in the south of the Doggerbank, are added to the northern cluster. In the southern cluster, the samples from the Helgoland Deep Trench form a subcluster (Fig. 1.4).

The Helgoland Deep Trench (HTR) represented in the FIS material shows a high percentage of cyclostomes (79%) (Fig. 1.5). This is in contrast to the southern (without HTR) and northern North Sea areas, which show similar percentages of cyclostomes (22% and 33%, respectively) (Fig. 1.5). The percentage of erect species increases from north to south (Fig. 1.6). The encrusting to erect species ratio is 4.3 in the northern North Sea, 0.7 in the southern area, and 0.2 in the HTR. The high percentage of

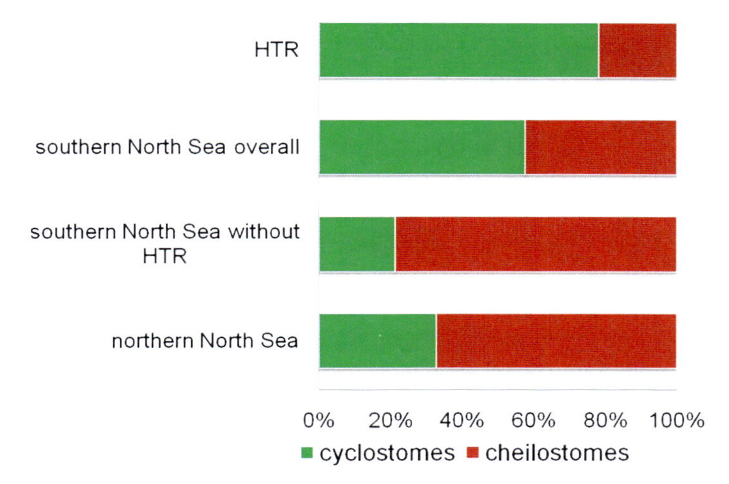


**Fig. 1.5** Bar chart showing percentages (n = 929 specimens on 22 samples) of cyclostomes and cheilostomes in the different North Sea regions: *green* = cyclostomes; *red* = cheilostomes

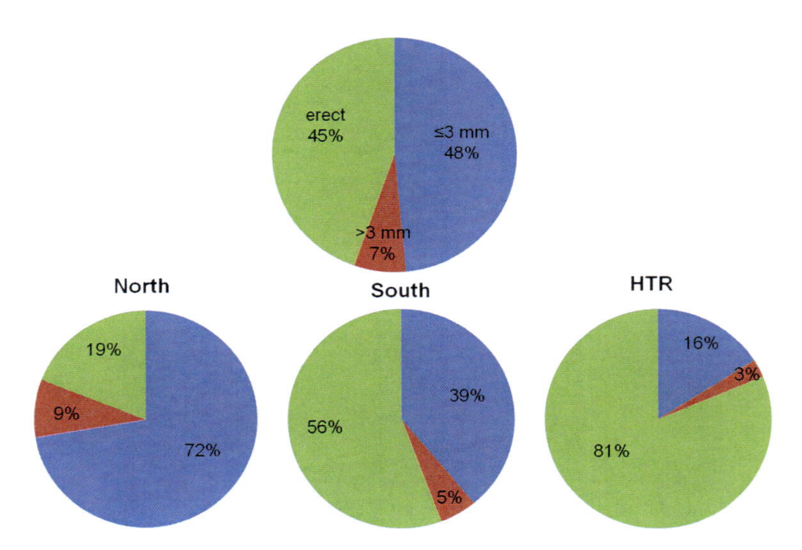


**Fig. 1.6** Pie charts showing percentages (n = 929 specimens on 22 samples) of the different colony diameters/growth forms over the entire North Sea (*at the top*) and the three regions of the North Sea

cyclostomes and of erect species in the HTR mainly contribute to a high amount of specimens of the genus *Crisia* including the species *Crisia eburnea* and *Crisia aculeata*. The species *Crisia aculeata, Bugula ciliata* and *Fenestrulina malusii* occur exclusively in the HTR.

# Discussion

## *Spatial Distribution of Zooepibiontic Bryozoans*

The two main clusters representing the spatial distribution of bryozoans attached on *Flustra foliacea* appear to correlate with the two main water masses in the North Sea characterizing the northern and southern regions and separated by the 50 m depth contour (Bitschofsky et al. 2011). The current system of the North Sea and thus larvae transport may directly reflect into the settlement pattern on *Flustra foliacea*. Plankton, as well as the benthic infauna, shows similar distribution schemes, outlining the separation into northern and southern communities (Rees 2007). Especially, hydrographic fronts appear as barriers for larval dispersal (James 1989). The regional scale heterogeneity of oceanographic conditions may, however, further control the settlement pattern of larvae on *Flustra foliacea* (Jablonsky and Lutz 1983; Fairweather and Quinn 1993). This is especially evident at the Doggerbank and in Helgoland Deep Trench. The Dogger bank is located in the central North Sea and the hydrographic regime is characterized by a complicated regime of currents and eddies (Kröncke and Knust 1995). Water masses from the central southern North Sea as well as water masses from the northern North Sea meet and mix at the Doggerbank region (Bo Pedersen 1994). This could be the reason for the different assignment of the Dogger bank samples to the two main clusters. Also, the Helgoland Deep Trench is a special ecological region in the southern North Sea.

## *Temporal Distribution of Zooepibiontic Bryozoans*

There is no separation of the historical samples in the cluster analyses. Both major time periods (1904/1908 and 1980s) show a positive North Atlantic Oscillation (NAO) Index and, therefore, similar climatic and oceanographic conditions. Furthermore, the warming of the North Sea is being noticed mainly during summer months (+1.4°C since the 1980s) (Mackenzie and Schiedek 2007). Due to a constant thermocline in the northern North Sea, where most historical samples originate from, no warming of the bottom water took place that could be noticed in changes in community patterns.

*Crisia eburnea*, however, characteristic for southern North Sea, especially for the HTR, occurs on some historical *Flustra* samples (1904 and 1905) (CAU material) from the northern North Sea but was not found on the modern *Flustra* specimens from the northern North Sea. This could be an evidence for the importance of the current pattern for larval transport and for the composition of the bryozoan assemblage. Even through the conditions in both time periods were similar, differences in the preceding time periods occurred. The years 1904–1908 were within a positive NAO-phase while the 1980s followed a preceding negative

NAO-phase (Osborn 2011). Years with positive NAO-index are characterized by a strong Atlantic inflow, and the entire North Sea is influenced by Atlantic water masses, whereas years with negative NAO-index are influenced by a weak Atlantic inflow in the North and an approximate outflow through the English Channel (Becker and Pauly 1996). One would expect, that in the ongoing positive NAO-phase since the 1980s *Crisia eburnea* again settled in the northern North Sea. Therefore, further investigations on North Sea bryozoans on historical as well as on Recent material are needed.

## *Cheilostomes Versus Cyclostomes*

While the characteristic species of the northern North Sea – *Amphiblestrum flemingii, Tricellaria ternata* and *Callopora dumerilii* – are all cheilostomes, the south-cluster species – *Crisia eburnea, Plagioecia patina* and *Electra pilosa* – contain two cyclostomes.

The southern North Sea is subject to a higher disturbance rate due to waves, tidal currents and trawling effort (Szekielda et al. 1988) than the northern region. McKinney (1992, 1995) suggests that the competition advantage of cheilostomes over cyclostomes causes a strong selective pressure for early reproduction in encrusting cyclostomes. This could be an advantage for the latter in the highly disturbed southern North Sea enabling them to rapidly colonize new substrates and grow to reproductive maturity before being destroyed. Especially the Helgoland Deep Trench seemed to be a refuge for cyclostomes.

## *The Helgoland Deep Trench (HTR): A Special Case*

The HTR has quite different hydrographical conditions from the rest of the southern North Sea. Lower bottom temperatures, higher salinities, and a reduced seasonality make the HTR more similar to the northern North Sea. Therefore, it should be expected that the species assemblage would resemble the northern pattern as well (Caspers 1939). This is not the case, however, as the species composition of the HTR is more similar to the southern region. It is cautiously speculated that this may show the dominating influence of current systems and consequent larvae transport on the epizoobiontic bryozoan community.

Results from specimen counts show a much higher percentage for cyclostomes in the HTR than in the rest of the southern North Sea. The genus *Crisia* including *C. eburnea* and *C. aculeata* accounts for nearly all cyclostomes specimens in the HTR.

## *Erect Species Versus Encrusters*

Epizoobiontic settlement patterns on *F. foliacea* are characterized either by small (<3 mm) or erect growth forms. This corresponds nicely with some growth form models established by Bishop (1989) and McKinney and Jackson (1989) as well as certain predictions for bryozoan growth forms in unstable or turbulent environments (Silén 1980). McKinney and Jackson (1989) pointed out, that calculated over all Recent taxa, encrusting growth forms are 1.7 times more common in the Atlantic Basin and the western Mediterranean than erect ones whereas erect and encrusting growth forms are equally abundant. In the present study, the overall ratio of encrusting to erect species shows a value of 1.2. Furthermore, the percentage of erect species increases from north to south (Fig. 1.6). The ratio of encrusting to erect species is 4.3 in the northern North Sea, 0.7 in the southern North Sea, and 0.2 in the HTR. The North Sea has high gradients of suspended matter ranging from more than 100 mg/l in the nearshore waters of Belgium and the Netherlands, to less than 0.2 mg/l in the northern part and the Norwegian Channel. Possibly, the siltation impact of in the southern North Sea is reflected in the higher occurrence of erect species (Lagaaij and Gautier 1965). More data is needed in order to know the reasons for such high percentages of erect cyclostomes in the HTR.

**Acknowledgement** This research was only possible with the help of researchers and curators who made their collection material available. The author is grateful to Dirk Brandis (Zoological Museum, Kiel), Carsten Lüter (Natural History Museum, Berlin), Andreas Schmidt-Rhesa (Zoological Museum, Hamburg), Joachim Scholz and Brigitte Lotz (Forschungsinstitut Senckenberg) and their staff for both discussions and providing the material. My research was supported by Stefan Forster and the University of Rostock (Institute for Biosciences) and by the Senckenberg Research Institute, Department of Aquatic Zoology (Frankfurt am Main). This manuscript benefited much from advice and criticism provided by the late Frank K. McKinney (Boone, NC U.S.A.). Further thanks go to the reviewers Priska Schäfer, Mary Spencer Jones and Steve Hageman for reading and giving further advice on the manuscript.

The present study took place at the Biodiversity and Climate Research Centre (BiK-F), Frankfurt a.M., and was financially supported by the research funding programme "LOEWE – Landes-Offensive zur Entwicklung Wissenschaftlich-ökonomischer Exzellenz" of Hessen's Ministry for Higher Education, Research, and the Arts.

# References

Becker GA, Pauly M (1996) Sea surface temperature changes in the North Sea and their causes. ICES J Mar Sci 53(6):887–898

Bishop JDD (1989) Colony form and the exploitation of spatial refuges by encrusting Bryozoa. Biol Rev 64:197–218

Bitschofsky F, Forster S, Scholz J (2011) Regional and temporal changes in epizoobiontic bryozoan-communities of *Flustra foliacea* (Linnaeus, 1758) and implications for North Sea ecology. Estuar Coast Shelf S 91(3):423–433

Blahudka S, Türkay M (2002) A population study of the shrimp *Crangon allmanni* in the German Bight. Helgoland Mar Res 56(3):190–197

Bo Pedersen F (1994) The oceanographic and biological tidal cycle succession in shallow sea fronts in the North Sea and the English Channel. Estuar Coast Shelf S 38(3):249–269

Callaway R, Alsvag I, de Boois J, Cotter J, Ford A, Hinz H, Jennings S, Kröncke I, Lancaster J, Piet G, Prince P, Ehrich S (2002) Diversity and community structure of epibenthic invertebrates and fish in the North Sea. ICES J Mar Sci 59:1199–1214

Caspers H (1939) Die Bodenfauna der Helgoländer Tiefen Rinne. Helgoland Mar Res 2(1):1–112

Fairweather PG, Quinn GP (1993) Seascape ecology: the importance of linkages. In: Battershill CB, Schiel DR, Jones GP, Cresse RG, MacDiarmid AB (eds) Proceedings of the Second International Temperate Reef Symposium, January 7–10, 1992, Auckland. NIWA Marine, Wellington, p 77–83

Frauenheim K, Neumann V, Thiel H, Türkay M (1989) The distribution of the larger epifauna during summer and winter in the North Sea and its suitability for environmental monitoring. Senckenberg Mar 20(3/4):101–118

Glémarec M (1973) The benthic communities of the European North Atlantic continental shelf. Oceanogr Mar Biol: Annu Rev 11:263–289

Hayward PJ, Ryland JS (1998) Cheilostomatous Bryozoa part I Aeteoidea – Cribrilinoidea, The Linnean Society of London. The Estuarine and Coastal Sciences Association by Field Studies Council, Shrewsbury

Hertweck C (1988) Zonierung der Faziesbereiche in der Heloländer Tiefen Rinne. Bochum Geol Geotec Arb 29:71–73

Jablonsky D, Lutz RA (1983) Larval ecology of marine benthic invertebrates: paleobiological implications. Biol Rev 58:21–89

James I (1989) A three-dimensional model of circulation in a frontal region of the North Sea. Ocean Dyn 42(3):231–247

Jennings S, Lancaster J, Woolmer A, Cotter J (1999) Distribution, diversity and abundance of epibenthic fauna in the North Sea. J Mar Biol Assoc UK 79:385–399

Kröncke I, Knust R (1995) The Dogger Bank: a special ecological region in the central North Sea. Helgoland Mar Res 49(1):335–353

Lagaaij R, Gautier YV (1965) Bryozoan assemblages from marine sediments of the Rhone delta, France. Micropaleont 11:37–58

Mackenzie BR, Schiedek D (2007) Daily ocean monitoring since the 1860s shows record warming of northern European seas. Glob Change Biol 13:1335–1347

McKinney FK (1992) Competitive interactions between related clades: evolutionary implications of overgrowth interactions between encrusting cyclostome and cheilostome bryozoans. Mar Biol 114(4):645–652

McKinney FK (1995) One hundred million years of competitive interactions between bryozoan clades: asymmetrical but not escalating. Biol J Linn Soc 56:465–481

McKinney FK, Jackson JBC (eds) (1989) Bryozoan evolution. University of Chicago Press, Chicago

Menon NR (1975) Observations on growth of *Flustra foliacea* (Bryozoa) from Helgoland waters. Helgoland Wiss Meer 27:263–267

Osborn TJ (2011) Winter 2009/2010 temperatures and a record-breaking North Atlantic Oscillation index. Weather 66(1):19–21

OSPAR (2000) Quality status report – region II: greater North Sea. Monitoring and Assessment, OSPAR

Rees HLE (ed) (2007) Structure and dynamics of the North Sea benthos, vol 259. ICES, Copenhagen

Schaller N (2001) Die Bryozoenfauna aus der Helgoländer Tiefen Rinne (Nordsee). Zool Inst. Universität Heidelberg, Heidelberg

Silén L (1980) Colony-substratum relations in Scrupocellariidae (Bryozoa, Cheilostomata). Zool Scr 9:211–217

Stebbing ARD (1971) The epizoic fauna of *Flustra foliacea* [Bryozoa]. J Mar Biol Assoc UK 51:283–300

Szekielda K-H, McGinnis D, Ambroziak B, McClain P, Clark D (1988) Satellite observations over the North Sea. Geol-Paläont Inst Univ Hamburg 65:1–33

Zühlke R, Alvsvag J, de Boois I, Cotter J, Ehrich S, Ford A, Hinz H, Jarre-Teichmann A, Jennings S, Kröncke I, Lancaster J, Pet G, Prince P (2001) Epibenthic diversity in the North Sea. Senckenberg Mar 31(2):269–281

# Chapter 2
# The World's Oldest-Known Bryozoan Reefs: Late Tremadocian, mid-Early Ordovician; Yichang, Central China

## Oldest Bryozoan Reefs

**Roger J. Cuffey, Xiao Chuantao, Zhongde Zhu (deceased), Nils Spjeldnaes (deceased), and Zhao-Xun Hu (deceased)**

**Abstract** The world's earliest-known bryozoan-built reef-mounds are in mid-Lower Ordovician (upper Tremadocian) strata near Yichang, Hubei Province, central China. Their framework, a globstone, was built by abundant rounded zoaria of the trepostome *Nekhorosheviella semisphaerica*. That framework further baffled micrite immediately around the colonies, more broadly surrounded regionally by bioclastic calcarenites (Fenxiang Formation). Fragments of delicate branching *Orbiramus normalis* occur rarely in the bryohermal micrite.

**Keywords** Bryozoan reefs • Early Ordovician • China • Yichang • *Nekhorosheviella* • *Orbiramus* • Trepostomes

---

R.J. Cuffey (✉)
Department of Geosciences, Pennsylvania State University, 412 Deike Building, University Park, PA 16802, USA
e-mail: rcuffey@psu.edu

X. Chuantao
Geosciences School, Yangtze University, No. 1 Nanhuan Road, Jingzhou, Hubei 434023, China
e-mail: ctxiao188@yahoo.com.cn

Z. Zhu (deceased)
Jianghan Petroleum Institute, Jingzhou/Jiangling, Hubei, China

N. Spjeldnaes (deceased)
University of Oslo, Oslo, Norway

Z.-X. Hu (deceased)
Academia Sinica, Nanjing (Jiangsu), China

A. Ernst, P. Schäfer, and J. Scholz (eds.), *Bryozoan Studies 2010*,
Lecture Notes in Earth System Sciences 143, DOI 10.1007/978-3-642-16411-8_2,

## Purpose

Bryozoan-built reef mounds, rare among bioherms and exotic for their phylum, occur episodically throughout bryozoans' geologic history (Cuffey 1977, 1985, 2006). The world's oldest-known bryozoan reefs were recently discovered in the Early Ordovician of China (Zhu et al. 1993). The purpose of this present paper is to elucidate their characteristics (Cuffey 1997a, 2006; Cuffey and Zhu 2010; Cuffey et al. 2010), to provide for comparison with all later bryoherms, and especially to the next-oldest one (Garden Island; Cuffey et al. 2002b).

As applied to these Chinese structures, several terms are interchangeable. A *mound* is a small or local elevation above the sea floor, a *bioherm* is a fossilized one with an organic-growth framework, a *reef* is one, which influenced adjacent sedimentation by resisting surrounding water movements (James 1983; Fagerstrom 1987; Wood 1999; Stanley 2001), and a *bryoherm* is one largely or entirely made of bryozoan colonies (Cuffey 1977: 185, 1997a, 2006: 35, 37).

### *Overview*

The Yichang bryozoan reefs in central China are the oldest-known such structures, of mid-Early Ordovician (late Tremadocian) age. Small mounds, they consist largely of globstone (Cuffey 1985) built by the recently described, round to domal trepostome *Nekhorosheviella semisphaerica* (Xia et al., 2007). The reef framework is in-filled with micrite, in which a few fragments of delicate branching *Orbiramus normalis* (Xia et al., 2007) are also found.

After completing and submitting the present paper, our attention was drawn by Hans Arne Nakrem to an abstract just published (Adachi et al. 2010) dealing with similar reef mounds nearby but as intergrowths of lithistid sponges and bryozoans, in contrast to the globular and columnar bryozoan frameworks documented herein.

## Location

The earliest bryozoan reefs are near Yichang city, in Hubei Province in central China, near the Yangtze River, 35 km SE or downstream from the Three-Gorges (Sandouping) Dam, and 95 km NW or upstream from Jianghan Petroleum Institute (now part of Yangtze University) at Jingzhou (formerly Jiangling). They are exposed at two localities (Fig. 2.1). Other maps showing these and related sites can be seen elsewhere (Zhu et al. 1993; Bingli et al. 1997; Rong et al. 2007; Xia et al. 2007; Wang et al. 2009).

One locality (YH) is a large limestone quarry immediately (1 km) W of and in the hill above Huanghuachang village, 20 km NNE of Yichang. Its coordinates are

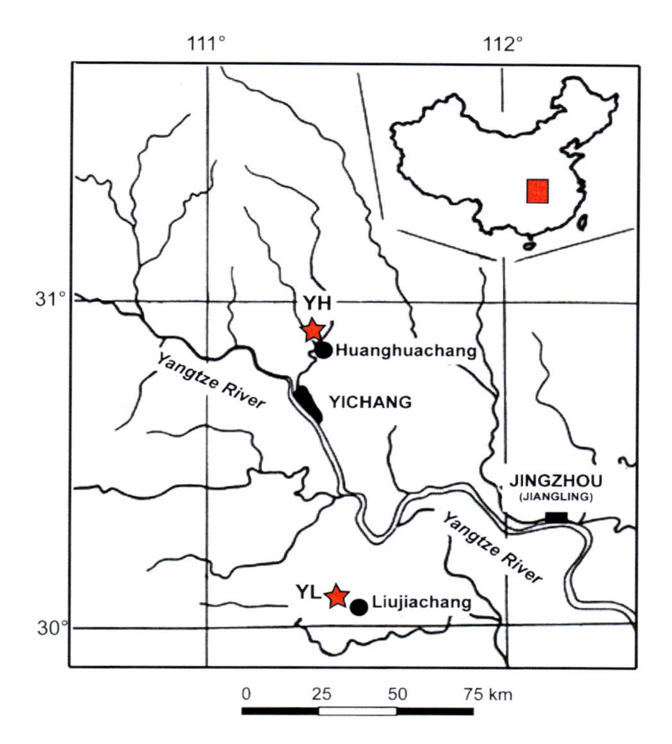


**Fig. 2.1** Bryozoan reef localities (YH quarry, YL pinnacle; see text) near Yichang in central China (*shaded square* in inset; N latitude, E longitude) (Modified from Bingli et al. (1997: 195))

30° 53′ 42″ N, 111° 21′ 12″ E. The Daping bryozoan site (Xia et al. 2007) is 5 km further NNE from YH. The other locality (YL) is a karst pinnacle in the woods WNW of Liujiachang village, 75 km SSE of Yichang. Its coordinates are 30° 02′ 48″ N, 111° 30′ 54″ E. The Guanzhuangping bryozoan site (Xia et al. 2007) is 12 km SE from YL.

After the present paper was completed, Chuantao found another, somewhat older, bryozoan reef near Songzi, 70 km SE of Yichang (Bingli et al. 1997: 195), on which investigation is beginning.

## Stratigraphy and Age

The Yichang bryoherms are developed in the middle part of the 68-m-thick Fenxiang Limestone Formation at both the YH and YL localities (Bingli et al. 1997: 195). Zhu et al. (1993, 1995) and Chen et al. (1995: 45, Chart 1) provide regional stratigraphic context. Xia et al. (2007: 1310–1311) mention similar bryozoan reefs but at the top of the Fenxiang, in their Daping section.

**Fig. 2.2** Bryozoan reefs exposed in YH quarry walls (people as scales). (**a**) (*above mid-level*): Low dome on path into eastern part of quarry. (**b**) (*lower left*): High mound in north wall in central part of quarry

The Fenxiang is of mid-Early Ordovician age, late Tremadocian, about 480 Ma old (Wang et al. 2009).

Chuantao's newly found bryoherm is below the Fenxiang, in the Nantsinkuan (Chen et al. 1995: 62) or Nanjinguan (Wang et al. 2009: 378) Formation, which is 170 m thick. This bryoherm is 45 m above the formation's base. Its age could thus be middle Tremadocian, or even the later part of the lower Tremadocian, which could possibly represent the oldest bryozoans in the world. However, precise analysis of its exact horizon, and comparison with other Early Ordovician bryozoan reports (see references under Species Identifications below) will be necessary to confirm that. And, note the recent report of a possible, even older (Late Cambrian), though non-reefal, arthrostylid-like bryozoan (Landing et al. 2010).

Interestingly, these reefs occur very early in the history of their phylum (Taylor and Ernst 2004). However, not enough is yet known to warrant speculation about possible connections between reef inhabitation and evolutionary differentiation.

The Lower Ordovician strata enclosing the Yichang bryozoan reefs also contain a number of small reef mounds built by other non-bryozoan, sessile taxa: (Zhu et al. 1993, 1995; Rigby et al. 1995; Bingli et al. 1997): sponges (*Archaeoscyphia, Jianghania*), dasyclads (*Calathium*), encrusting red algae (*Pulchrilamina*), encrusting cryptalgal laminae (*Girvanella, "Stromatactis"*), and stromatolites (blue-green cyanobacteria). Also preserved around these bioherms are typical Early Ordovician invertebrates: nautiloids, trilobites, brachiopods, pelmatozoans (crinoids? cystoids?), and gastropods (references above).

## Reef-Mound Form

The Yichang bryozoan reefs vary in shape and size, from low-domed (Fig. 2.2a) to higher-conical or higher-mounded (Fig. 2.2b), 1–3 m high by 3–10 m wide. They thus resemble other early bryoherms (Cuffey 2006). Where sufficient lateral exposure exists, the individual bryozoan mounds at a given horizon are spaced 3–20 m apart.

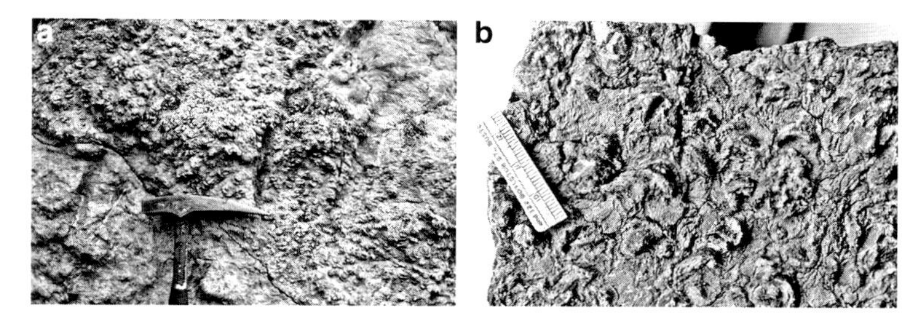

**Fig. 2.3** Weathered surfaces of bryozoan reef-rock (*Nekhorosheviella* globstone; difficult to see because zoaria and matrix are similar shade of gray; YH quarry). (**a**) Along path into eastern part of quarry (hammer for scale). (**b**) Slab from central part of quarry (scale in mm)

## Reef-Rock Characteristics

The Yichang bryozoan reefs consist of globular, domal, and crustose *Nekhorosheviella semisphaerica* colonies (Fig. 2.3), 1–2 cm across, mostly in-place and attached to one another, sometimes stacked up vertically in columnar fashion (Figs. 2.4a and 2.5d) as much as 4–5 cm high. Disoriented or broken colonies are also evident locally.

On average, the bryozoan zoaria make up 40–60% of the bryoherms' rock volume, although local extremes can range from as little as 10% up to as much as 90%.

In certain thin-sections cut vertically down the center of a columnar stack of domal colonies, individual zooecial tubes can be traced continuously from an earlier colony up through a later one (Fig. 2.4b, upper right quarter) and on further into a still-later colony, thereby demonstrating long-lived growth continuing through several successive zoaria. In other sectioned columns, the zooecia all end at the upper surface of the early colony, and a basal lamina covering them gives rise to new tubes which grow into the next zoarium (Fig. 2.4b, below center), thus suggesting death and re-encrustation in the formation of the single column. Caution must be applied here, however, because off-center vertical sections can artificially create an incorrect appearance of growth interruptions.

A few horizontal thin-sections show a round zoarium surrounded by thin concentric rings of similar skeletal material (Fig. 2.4d), apparently the lowermost-and-outermost edges of successively younger colonies encrusted on top of the central one, but above the plane of the section.

The Yichang bryozoan reef-rock is therefore mostly globstone (Cuffey 1985) made up of round *Nekhorosheviella* zoaria on top of one another, and locally cruststone (Cuffey 1985) where the colonies are thinner or flatter. Its structure is best seen on weathered rock surfaces (Fig. 2.3) or in thin-sections (Fig. 2.4), although a few freshly-broken pieces vaguely show the outlines of the rounded zoaria.

Between the *Nekhorosheviella* colonies and columns is micrite, lime-mud. Such fine-grained sediment presumably settled out (was baffled) around the

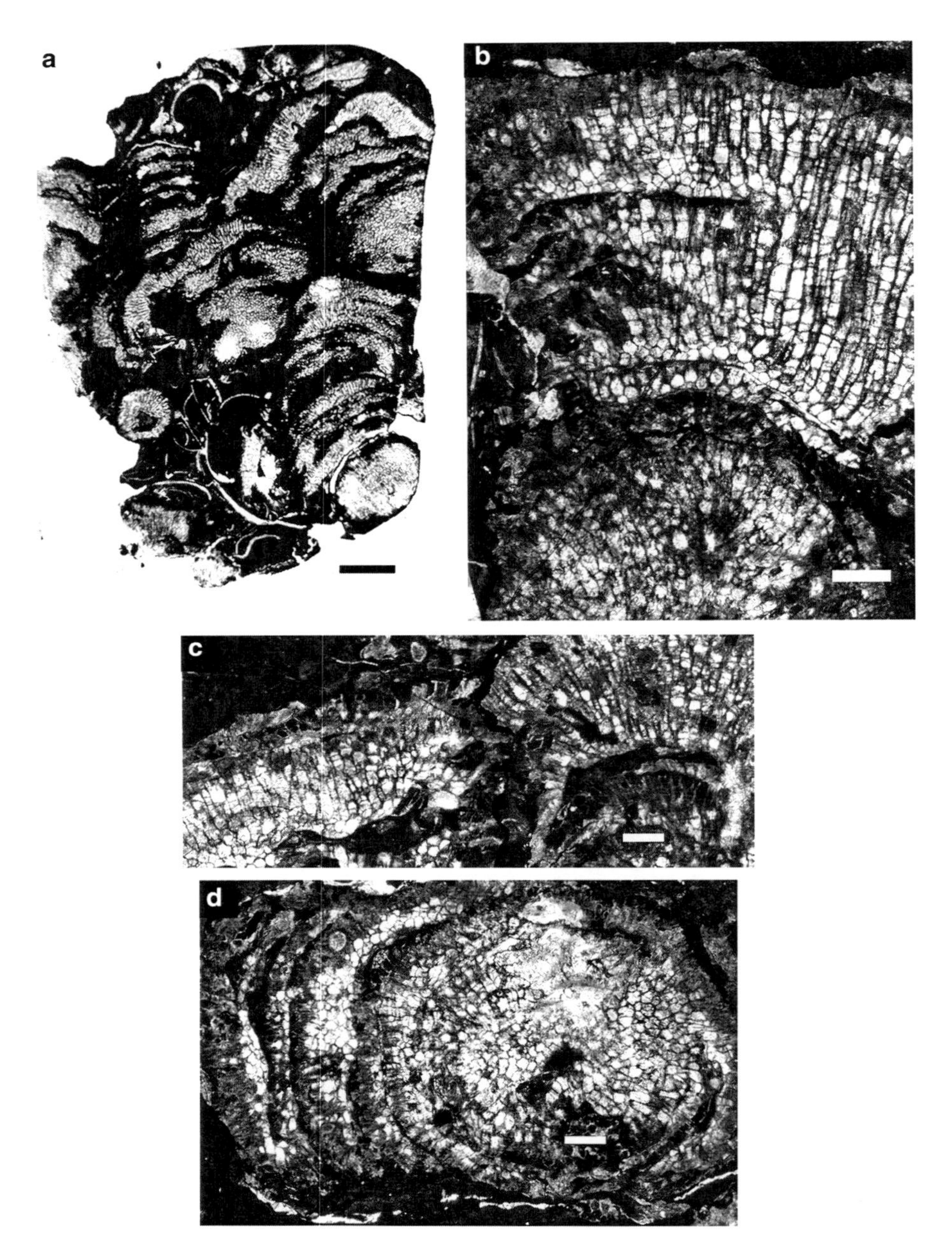

**Fig. 2.4** Thin-sections illustrating *Nekhorosheviella* globstone reef-rock. (YH quarry: *scale bars* (**a**) 10 mm, (**b–d**) 1 mm; here and in following figures, *scale bars* positioned to maximize their visibility). (**a**) Stacked columnar framework in vertical/longitudinal section; YH#31. (**b**) Globular and (**c**) Domal to *thick-crustose colonies* in vertical/longitudinal sections; YchO1. (**d**) Horizontal/transverse section; YchO1

*Nekhorosheviella* framework projecting up into the water flowing past. The reef-rock might therefore be termed a secondary or consequent bafflestone (Cuffey 1985; Zhu et al. 1993), but its most important component is clearly that skeletal framework.

In a few places, the micrite contains broken small bivalved shell fragments, probably recrystallized brachiopods (rather than pelecypods) and possibly also ostracods. Elsewhere, occasional sponges occur among the numerous bryozoans; in one section, a *Nekhorosheviella* can be seen encrusting the side of a large sponge, rather than the much more usual substrate provided by another *Nekhorosheviella*. The Fenxiang sponges were described by Bingli et al. (1997). Little or no echinoderm debris appears in the micrite in the bryoherms.

In a few Yichang reef-rock thin-sections, the micrite contains scattered broken thin branch fragments of a second bryozoan species, *Orbiramus normalis*. Where these do occur, they nowhere comprise more than 5–10% of the rock volume, and usually much less. They do not constitute part of the reef framework; none were seen in place.

## Species Identifications

When research on the Yichang reef mounds began, little was known of Chinese Ordovician bryozoan taxonomy, but progress has been made in recent years.

Initially, the dominant bryozoan species in the Yichang bryoherms was thought to be *Batostoma jinhongshanense* (Zhu et al. 1993, 1995; Ying and Xia 1986). Other Eurasian Early Ordovician species do not appear in these reef mounds (Bassler 1911; Dzik 1981; Pushkin and Popov 1999; Taylor and Cope 1987; Taylor and Curry 1985; Taylor and Rozhnov 1996; Taylor and Wilson 1999; Yang 1957). Later, connection was made to the Lower Ordovician genera *Hubei(o)pora* and *Yichangopora* (Hu and Spjeldnaes 1991: 181–182; Chen et al. 1995: 19, 91–92; Spjeldnaes 2001), but Hu and Spjeldnaes both died before formally describing those taxa and their included species, so that their names remain as *nomina nuda* and thus are not usable. Still later, similarities to *Dianulites* were noted (Cuffey 1997a, 2006: 37). Finally, species recently described from near-by localities (Xia et al. 2007) include those in the Yichang bryozoan reefs.

Preservation of the bryozoans varies from good to poor, irregularly or randomly, within the Yichang reef mounds. Skeletal microstructures are easily visible in many specimens, but obscurely recrystallized in others. In the former, spar cement fills all the originally open pore spaces, so that no porosity remains, which might store hydrocarbons.

Both Yichang species are common enough that their sizeable intraspecific variabilities are evident, with numerous intermediates grading between the morphologic extremes encountered in the thin-sections. In this respect, the Yichang species resemble branching *Tabulipora carbonaria* and hemispherical *Prasopora simulatrix* (Cuffey 1967, 1997b, respectively).

With only the one predominant *Nekhorosheviella* species and the other accessory *Orbiramus* species, the bryozoan diversity in the Yichang reefs is quite low, as might be expected so early in the phylum's evolutionary history. Later, as more bryoherms were built, reefal bryozoan diversity increased considerably (Cuffey 2003, 2006).

Specimens illustrated herein are from the collections of the Department of Geology at Jianghan Petroleum Institute, which has become the Geosciences School at Yangtze University.

## *Nekhorosheviella semisphaerica*

The abundant, round and domal Yichang bryozoans all represent one species, the orbiporid esthonioporine trepostome *Nekhorosheviella semisphaerica* (Xia et al. 2007: 1315–1321), the principal frame-builder of these reef mounds. Its zoarial form and more numerous diaphragms distinguish it from *Nekhorosheviella nodulifera*, also described by Xia et al. (2007: 1313–1315). It differs substantially in colony form from the robust branching *Batostoma jinhongshanense* (Ying and Xia, 1986), with which it had initially been identified (Zhu et al. 1993). *Hubeipora* or *Hubeiopora* (Hu and Spjeldnaes 1991: 181–182; Chen et al. 1995: 19; respectively) *simplex* (Spjeldnaes 2001) was the *nomen nudum* which would have been applied to *Nekhorosheviella semisphaerica*.

The Yichang *Nekhorosheviella* colonies grade from globular (spherical 10–17 mm diameter, to ovoid 10–18 mm by 8–12 mm), to domal (hemispherical 10–17 mm wide by 5–10 mm high), and down to thick crusts (3–4 mm high) (Figs. 2.3a, b, 2.4a–d, and 2.5c–e). Zooecial apertures are 0.20–0.30 mm in diameter and round or polygonal (Fig. 2.5a). Interzooecial walls appear microgranular and thin (Fig. 2.5b), varying 0.01–0.05 mm wide, due to tiny acanthostyles embedded within them, and are not separated into endo- and exozonal sections. Zooecial tubes are crossed by many, straight to gently curved diaphragms (Fig. 2.5b). Occasional smaller-diameter apertures may be either beginning zooecia or incompletely differentiated mesozooecia.

## *Orbiramus normalis*

Much less common are delicate cylindrical-ramose fragments (Fig. 2.6a), a second Yichang species, *Orbiramus normalis* (Xia et al. 2007: 1322), originally dwelling on or in the reef masses. Its affinities may lie with *Orbipora, Aisenvergia, Jordanopora*, or *Lamottopora*, but more study will be needed. Slight differences in aperture diameter and shape distinguish this species from *Orbiramus ovalis* and *O. minus* (Xia et al. 2007: 1322). *Yichangopora petaloformis* was to have been applied to this species (Hu and Spjeldnaes 1991: 181–182; Chen et al. 1995: 19; Spjeldnaes 2001), but remains a *nomen nudum*.

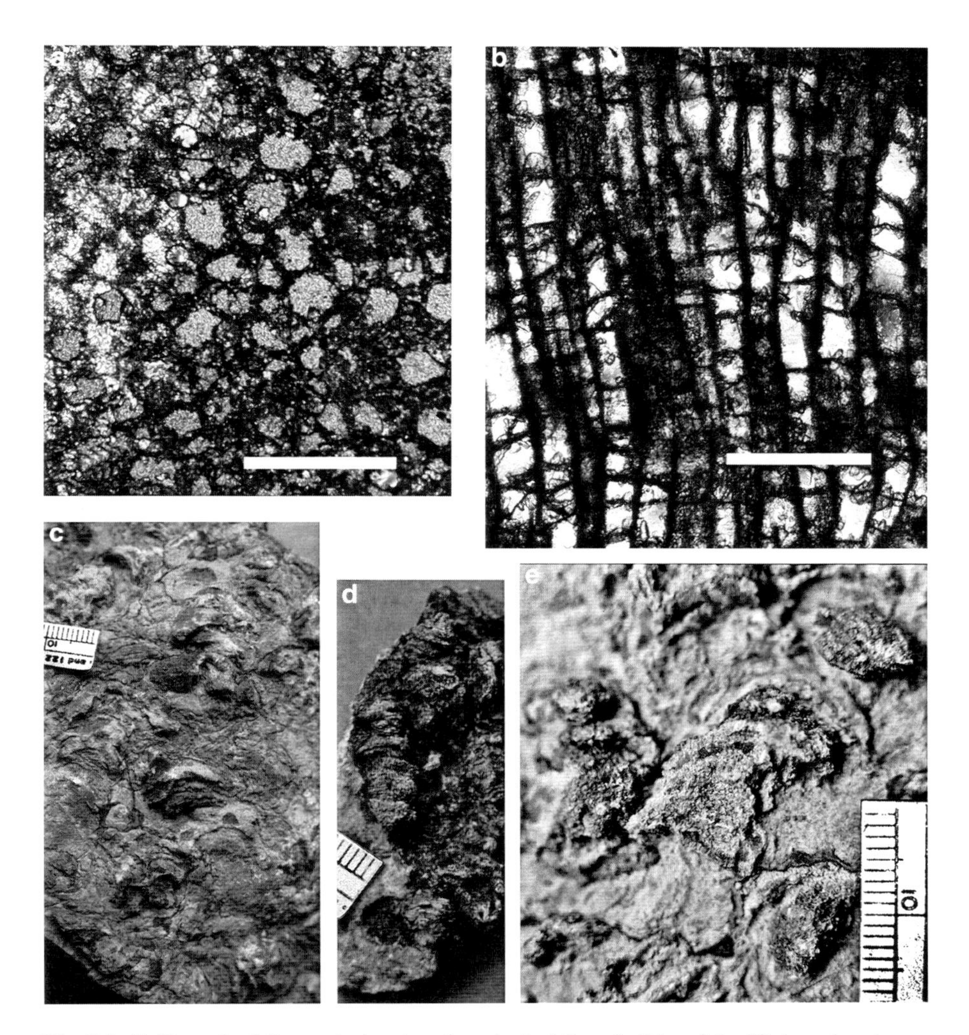


**Fig. 2.5** *Nekhorosheviella semisphaerica*, the principal framebuilder of the Yichang bryoherms (YH quarry (**a–c**, **e**), YL pinnacle (**d**) *scale bars* (**a–b**) 1 mm; (**c–e**) scales in mm). (**a**) Tangential section; YH #13. (**b**) Longitudinal section; YchO1. (**c–e**) Exteriors of colonies weathering out of reef-rock; (**c**, **e**) YH-bscq; (**d**) YL-ttp

The *Orbiramus normalis* branches in the Yichang bryoherms vary from 1–6 mm in diameter (averaging 3.2 mm), and display distinct endo- and exozones. Apertures are oval in deep tangential sections but grade up into petaloid or indented (due to acanthostyles) in shallow tangential sections (Fig. 2.6b); dimensions are 0.15–0.20 mm, but are difficult to measure due to their irregularities. Interzooecial walls are very thin (0.0 l mm) and possibly microgranular in the endozone, but become much thicker (as much as 0.20 mm) and laminated in the outer or shallow exozone. A few diaphragms are visible in the exozone (Fig. 2.6c).

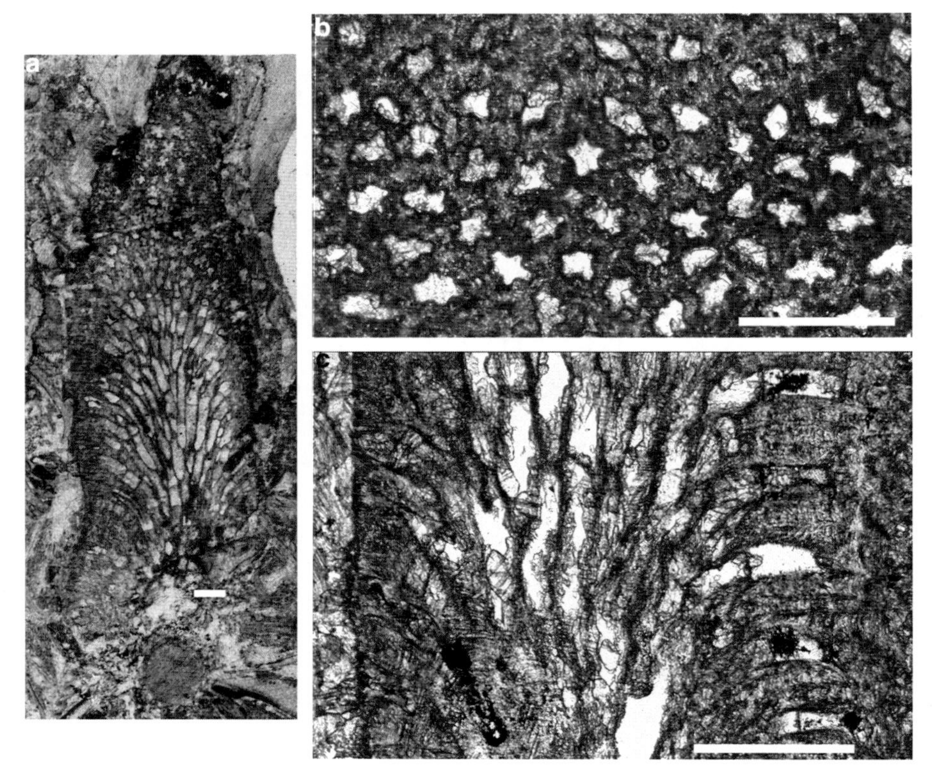

**Fig. 2.6** *Orbiramus normalis*, the accessory reef-dweller in the Yichang bryoherms (YH quarry; *scale bars*: 1 mm). (**a**) Inclined section grading from longitudinal below to tangential above; XLoh-5b. (**b**) Tangential section with star-shaped apertures; XLoh-5d. (**c**) Longitudinal section; XLoh-5a

## Bryoherm Zonation

The Yichang bryoherms do not exhibit internal reef zonation, either lateral (ecozonal) or vertical (seral or successional). In this respect, they resemble many other Ordovician bryozoan reefs, such as those in Virginia (Cuffey et al. 2002a) and Kentucky (Cuffey 1998). In contrast, a few other early bryoherms do show lateral flank ecozones (Cuffey and Cuffey 1995; Cuffey et al. 2002b; Weir et al. 2004), or vertical seral (ecologic-successional) zones (Cuffey and Taylor 1989).

Vertical successional changes extended over longer time intervals grade into community evolution like that documented among the Lake Champlain bryozoan reefs (Pitcher 1964). The Yichang bryoherms examined do not show such evolution, probably due to the much shorter time represented by the Chinese structures.

## *Surrounding Sediments*

The Yichang bryozoan reef mounds are founded on, surrounded by, and buried under medium- to thick-bedded bioclastic limestones, largely medium-sand-sized, relatively well-washed, spar-cemented calcarenites (grainstones), with varying proportions of pelmatozoan debris (Rigby et al. 1995; Zhu et al. 1995).

The reefs examined do not appear to be resting upon lithified hardground foundations (as does the Maysville, Kentucky, bryoherm; Cuffey 1998), but instead resemble the modern Joulters (Bahamas) bryozoan reef mounds, which are growing on weakly cemented carbonate sand (Cuffey et al. 1977).

# Bryozoan Constructional Roles

The orientation and abundance of the globular and domal colonies demonstrate that *Nekhorosheviella semisphaerica* functioned as the principal frame-builder (Cuffey 1977) in the Yichang bryoherms.

Even brief inspection of Yichang reef-rock thin-sections (Fig. 2.4) conveys an impression of fairly large zooecial tubes with substantial walls (in part due to many tiny embedded acanthostyles). Such sturdy skeletal construction may have - pre-adapted ("exapted") this *Nekhorosheviella* species for building these small reefs (Cuffey 1997a), much as similar characteristics of *Dianulites* did in later bryoherms like Chickasaw in Oklahoma (Cuffey and Cuffey 1995; Werts et al. 2001).

The fine-grained micrite surrounding the colonies indicates that the bryozoan framework also played a baffling or sediment-trapping role in these reefs' development, the lime-mud having settled out due to slowing of the currents flowing past the upward projecting framework.

The rarity and broken condition of the delicate ramose *Orbiramus normalis* indicate that it was a minor, accessory reef dweller, living in small patches scattered across the tops of the bryoherms, or down in crevices or fissures cut into them. In particular, their fossils are too rare for them to have formed reef-flank thickets, as *Champlainopora (Atactotoechus) chazyensis* did around the Mid-Ordovician Garden Island bryoherm (Cuffey et al. 2002b; Pitcher 1964).

# Palaeoenvironment

Characteristics of the calcarenites surrounding the Yichang bryozoan reefs indicate deposition on shallow (perhaps less than 10 m; Zhu et al. 1993: 88), turbulent, marine carbonate-sand bottoms. Incipient cementation of patches of those sands would have provided just enough firm substrate for *Nekhorosheviella* larvae to settle and grow into sizeable zoaria.

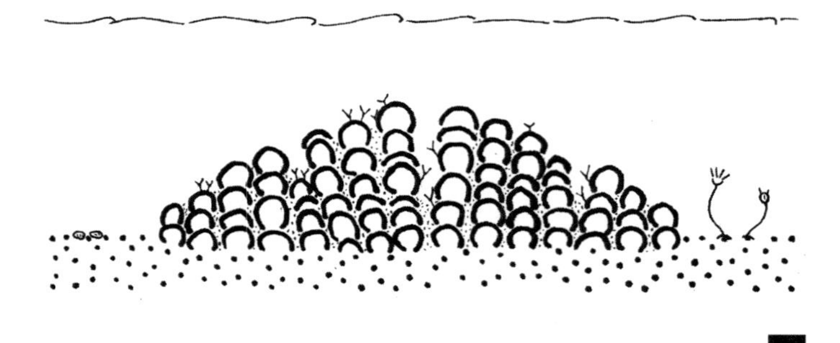

**Fig. 2.7** Diagrammatic reconstruction of a typical Yichang bryozoan reef mound, showing abundant *globular* and *domal Nekhorosheviella semisphaerica*, rare delicate branching *Orbiramus normalis*, micrite (*fine stipples*) trapped between the colonies, calcarenite (*coarse stipples*) surrounding the reef, pelmatozoans (*to right*) and brachiopods (*to left*) off–reef; *scale bar* for zoaria 2 cm, for reef-mound outline 30 cm

Previous paleogeographic and tectonic studies place the Yichang area at low paleolatitude (approximately 20° N; Smith et al. 1981: 90–91; Scotese 1997: 29–32), within the tropics, and near the edge of the Yangtze carbonate platform (Chen et al. 1995: 24–26, 30; Xu et al. 2010: 92).

The paleoenvironment described above strikingly imitates the shallow, high-energy, tropical, platform-edge, carbonate-sand environment in which the modern Joulters Cays (Bahamas) bryozoan reef mounds are found today (Cuffey et al. 1977).

## Comparison with Garden Island

Although earlier parts of this Yichang paper have mentioned specific points of comparison with other Ordovician bryozoan reefs, it is particularly instructive to note similarities and differences with Garden Island, the next-oldest bryoherm, out in Lake Champlain east of the New York shoreline.

In the sea-cliff at the southern tip of Garden Island (Pitcher 1964; Cuffey et al. 2002b; Mehrtens and Cuffey 2003), is a small (meter-sized) mound of massive reef-rock, surrounded by well-stratified limestone, much of which is calcarenitic, like the Yichang bryozoan reefs. Unlike Yichang, the Garden Island bryoherm consists of cruststone (Cuffey 1985), thin encrusting colonies of *Batostoma chazyensis* (Ross 1963: 859–860; Bolton and Cuffey 2005: 27–30), a trepostome species which, like *Nekhorosheviella*, seems pre-adapted to reef-building by structural strengthening, but by many mesopores containing numerous diaphragms rather than by thickened interzooecial walls (Cuffey 1997a). At Garden Island, ramose *Champlainopora (Atactotoechus) chazyensis* (Ross 1962: 734–737; Ross 1970: 374) is much more abundant than *Orbiramus*, and grew as branching thickets flanking around the *Batostoma* core. The Garden Island bryozoan mound is somewhat younger, roughly 15 Ma (Wang et al. 2009; early Mid–Ordovician, basal

Chazyan of classical North American usage; late Darriwilian of new global stages; in the Scott Limestone Member of the Day Point Formation). Not surprisingly, therefore, its bryozoan diversity, in the form of several additional rare species, is higher than Yichang's (Cuffey et al. 2002b; Cuffey 2003).

## Conclusions: Reconstruction

Rather than repeating the abstract and overview at the beginning of this paper, a visual summary reconstruction (Fig. 2.7; which greatly expands the initial drawing in Cuffey 2006: 37) of the Yichang *Nekhorosheviella* mounds will serve as a fitting conclusion to our examination of these, the world's earliest-known bryozoan reefs.

**Acknowledgments** Several individuals and organizations helped us complete this project, and we gratefully thank them all. J. Keith Rigby (Brigham Young University) originally brought Cuffey and Zhu together. Zhu's initial field and laboratory research was supported by the National Natural Science Foundation of China (Grant 49372116). Cuffey's visit to Jianghan Petroleum Institute (Hua Beizhuang, President; Zhou Yong-Mo, International Programs) was made possible by the China National Petroleum Corporation. While there, Liu Bingli, Jiang Yanwen, and Xiao Chuantao provided curatorial and field assistance. Pennsylvania State University's facilities enabled Cuffey to continue processing and examining the Yichang materials. Paul Taylor (Natural History Museum), Hans Arne Nakrem (Naturhistorisk Museum), and Steven Hageman (Appalachian State University) were instrumental in locating and coordinating Spjeldnaes' and Feng–Sheng Xia's (Nanjing Institute of Geology and Paleontology) manuscripts after the former's passing. Debra Lambert and William Ammerman prepared the present manuscript, which was improved by reviews from Hans Arne Nakrem and Marcus Key.

## References

Adachi N, Ezaki Y, Jianbo L (2010) The oldest bryozoan (Early Ordovician) reefs were constructed by an unusual mode of bryozoan growth. In: Third international palaeontology congress Program, p 61, Abstract, London, UK

Bassler RS (1911) The early Paleozoic Bryozoa of the Baltic provinces. US Natl Mus Bull 77:1–382

Bingli L, Rigby JK, Yanwen J, Zhu Z (1997) Lower Ordovician lithistid sponges from the eastern Yangtze Gorge area, Hubei, China. J Paleontol 71:194–207

Bolton TE, Cuffey RJ (2005) Bryozoa of the Romaine and Mingan formations (Lower and Middle Ordovician) of the Mingan Islands, Quebec, Canada. In: Moyano HI, Cancino JM, Wyse Jackson PN (eds) Bryozoan studies 2004. Balkema, Leiden, pp 25–41

Chen X, Rong JY, Wang XF, Wang ZH, Zhang YD, Zhan RB (1995) Correlation of the Ordovician rocks of China: charts and explanatory notes. Int Union Geol Sci Publ 31:1–104

Cuffey RJ (1967) Bryozoan *Tabulipora carbonaria* in Wreford Megacyclothem (Lower Permian) of Kansas, The University of Kansas Paleontological Contributions: Article 43 (Bryozoa 1), Kansas, pp 1–96

Cuffey RJ (1977) Bryozoan contributions to reefs and bioherms through geologic time. Am Assoc Petrol Geol Stud Geol 4:181–194

Cuffey RJ (1985) Expanded reef-rock textural classification and the geologic history of bryozoan reefs. Geology 13:307–310

Cuffey RJ (1997a) Functional-morphologic characteristics in relation to constructional contributions of the earliest bryozoan reef builders. In: Russian international conference on fossil and living Bryozoa. Bryozoa of the World. Abstract, St. Petersburg, pp 6–7, 47–48

Cuffey RJ (1997b) *Prasopora*-bearing event beds in the Coburn Limestone (Bryozoa; Ordovician; Pennsylvania). In: Brett CE, Baird GC (eds) Paleontological events: stratigraphic, ecological, and evolutionary implications. Columbia University Press, New York, pp 110–130

Cuffey RJ (1998) The Maysville bryozoan reef mounds in the Grant Lake Limestone (Upper Ordovician) of north-central Kentucky. Ohio Division of Geological Survey, Guidebook, vol 13. pp 38–44, 156, Columbus, OH

Cuffey RJ (2003) Bryozoan species, diversity, and constructional roles in early Mid-Ordovician bryozoan reefs of eastern and central North America. In: Viskova LA, Mezentseva OP, Morozova IP, Udodov VP (eds) Bryozoans of the Globe. Kuzbass State Pedagog Acad Russ Acad Sci Paleontol Inst, Novokuznetsk, vol. 1:20–26

Cuffey RJ (2006) Bryozoan-built reef mounds – the overview from integrating recent studies with previous investigations. Cour Forsch-Inst Senckenberg 257:35–47

Cuffey CA, Cuffey RJ (1995) The Chickasaw bryozoan reef in the Middle Ordovician of south-central Oklahoma. In: Short papers for the 7th international symposium on the Ordovician system. Pacific section of SEPM (Society Sedimentary Geology), vol 77. Las Vegas, pp 435–438

Cuffey RJ, Taylor JF (1989) Altoona bryozoan-coral-stromatoporoid reef, Uppermost Silurian, Pennsylvania. Can Soc Petrol Geol Mem 13:296–298

Cuffey RJ, Zhu Z (2010) Yichang (central China) – the oldest-known bryozoan reefs (mid-Early Ordovician), and comparison with the next-oldest (Garden Island, New York/Vermont, early Middle Ordovician). Geol Soc Am Abstr Program 42(1):64

Cuffey RJ, Gebelein CD, Fonda SS, Bliefnick DM, Kosich DF, Soroka LG (1977) Modern tidal-channel bryozoan reefs at Joulters Cays (Bahamas). In: Proceedings of the 3rd international coral reef symposium, vol 2. Miami, pp 339–345

Cuffey RJ, Cawley JC, Lane JA, Bernarsky-Remington SM, Ansari SL, McClain MD, Ross-Phillips TL, Savill AC (2002a) Bryozoan reefs and bryozoan-rich limestones in the Ordovician of Virginia. In: Proceedings of the 9th international coral reef symposium, vol 1. Bali, pp 205–210

Cuffey RJ, Robison MR, Mehrtens CJ (2002b) Garden Island – the earliest (and first-illustrated) bryozoan reef in North America (basal Chazyan, early Middle Ordovician; Lake Champlain, New York – Vermont). Geol Soc Am Abstr Program 34(1):A–72

Cuffey RJ, Zhu Z, Spjeldnaes N, Hu ZX (2010) The oldest-known bryozoan reefs (Yichang, central China; late Tremadocian, mid-Early Ordovician), and comparison with the next-oldest (Garden Island, New York/Vermont; basal Chazyan, early Middle Ordovician). Terra Nostra: Schr GeoUnion Alfred-Wegener-Stift 2010(4):36–37

Dzik J (1981) Evolutionary relationships of the early Palaeozoic ‘cyclostomatous’ Bryozoa. Palaeontology 24:117–120

Fagerstrom JA (1987) The evolution of reef communities. Wiley-Interscience, New York, pp 1–600

Hu ZX, Spjeldnaes N (1991) Early Ordovician bryozoans from China. Soc Sci Nat Ouest Fr Bull Mém HS 1:179–185

James NP (1983) Reef environment. Am Assoc Petrol Geol Mem 33:345–440

Landing E, English A, Keppie JD (2010) Cambrian origin of all skeletalized metazoan phyla – discovery of earth’s oldest bryozoans (Upper Cambrian, southern Mexico). Geology 38:547–550

Mehrtens C, Cuffey RJ (2003) Paleoecology of the Day Point Formation (lower Chazy Group, Middle – Upper Ordovician) and its bryozoan reef mounds, northwest Vermont and adjacent New York. Northeast Geol Environ Sci 25:313–329

Pitcher M (1964) Evolution of Chazyan (Ordovician) reefs of eastern United States and Canada. Bull Can Petrol Geol 12:632–691

Pushkin VI, Popov LE (1999) Early Ordovician bryozoans from North-Western Russia. Palaeontology 42:171–189

Rigby JK, Nitecki MH, Zhu Z, Bingli L, Yangwen J (1995) Lower Ordovician reefs of Hubei, China, and the western United States. In: Short papers for the 7th international symposium on the Ordovician system. Pacific section of SEPM (Society for Sedimentary Geology), vol 77. pp 423–426

Rong J, Jin J, Zhan R (2007) Early Silurian *Sulcipentamerus* and related pentamerid brachiopods from south China. Palaeontology 50:245–266

Ross JP (1962) Chazyan (Ordovician) leptotrypellid and atactotoechid Bryozoa. Palaeontology 5:727–739

Ross JP (1963) The bryozoan trepostome *Batostoma* in Chazyan (Ordovician) strata. J Paleontol 37:857–866

Ross JP (1970) Distribution, paleoecology and correlation of Champlainian Ectoprocta (Bryozoa), New York state, Part III. J Paleontol 44:346–382

Scotese CR (1997) Paleogeographic atlas. PALEOMAP – University of Texas, Arlington, pp 1–45

Smith AG, Hurley AM, Briden JC (1981) Phanerozoic paleocontinental world maps. Cambridge University Press, Cambridge, UK, pp 1–102

Spjeldnaes N (2001) Yichang bryozoan species. Unpublication manuscript with 8 pls. pp 1–17

Stanley GD (ed) (2001) The history and sedimentology of ancient reef systems. Kluwer Academic/ Plenum, New York, pp 1–458

Taylor PD, Cope JCW (1987) A trepostome bryozoan from the lower Arenig of South Wales: implications of the oldest described bryozoan. Geol Mag 124:367–371

Taylor PD, Curry GB (1985) The earliest known fenestrate bryozoan, with a short review of Lower Ordovician Bryozoa. Palaeontology 28:147–158

Taylor PD, Ernst A (2004) Bryozoans. In: Webby BD, Paris F, Droser ML, Percival IG (eds) The great Ordovician biodiversification event. Columbia University Press, New York, pp 147–156

Taylor PD, Rozhnov S (1996) A new early cyclostome bryozoan from the Lower Ordovician (Volkhov Stage) of Russia. Paläontol Z 70:171–180

Taylor PD, Wilson MA (1999) *Dianulites* Eichwald, 1829: an unusual Ordovician bryozoan with a high–magnesium calcite skeleton. J Paleontol 73:38–48

Wang X, Stouge S, Chen X, Li Z, Wang C (2009) Dapingian stage: standard name for the lowermost global stage of the Middle Ordovician series. Lethaia 42:377–380

Weir WK, Cuffey RJ, Ettensohn FR (2004) Bryozoan reef-mounds in the Lexington Limestone (Ordovician, Kentucky). Geol Soc Am Abstr Program 36(2):109

Werts SP, Cuffey RJ, Cuffey CA (2001) Bryozoan species in the Chickasaw bryozoan reef (Ordovician, Oklahoma). Geol Soc Am Abstr Program 33(5):A-54–A-55

Wood R (1999) Reef evolution. Oxford University Press, Oxford, pp 1–414

Xia FS, Zhang SG, Wang ZZ (2007) The oldest bryozoans: new evidence from the Late Tremadocian (Early Ordovician) of east Yangtze Gorges, China. J Paleontol 81:1308–1326

Xu C, Zhi-yi Z, Jun-xuan F (2010) Ordovician paleogeography and tectonics of the major paleoplates of China. Geol Soc Am Spec Pap 466:85–104

Yang KC (1957) Some Bryozoa from the upper part of the Lower Ordovician of Liangshan, southern Shensi (including a new genus). Acta Palaeontol Sin 5(1):1–12

Ying HM, Xia FS (1986) Discovery of genus *Batostoma* from the lower part of the Malieziken Group of Ruoqiang area, Xinjiang. Acta Micropalaeontol Sin 3(4):435–440

Zhu Z, Chengxian G, Bingli L, Mingyi H, Aimei H, Chuantao X, Xianfu M, Xiangming L (1993) Lower Ordovician reefs at Huanghuachang, Yichang, east of the Yangtze Gorge. Sci Geol Sin 2(1):79–90

Zhu Z, Yanwen J, Bingli L (1995) Paleoecology of Late Tremadocian reef-bearing strata in western Hubei Province of China. In: Short papers for the 7th international symposium on the Ordovician System. Pacific section of SEPM (Society for Sedimentary Geology), vol 77. Riverside, CA, pp 427–428

# Chapter 3
# Molecular Distance and Morphological Divergence in *Cauloramphus* (Cheilostomata: Calloporidae)

## Divergence in *Cauloramphus*

**Matthew H. Dick, Masato Hirose, and Shunsuke F. Mawatari**

**Abstract** Molecular phylogenetic analysis of cytochrome c oxidase I (COI) sequences in the calloporid genus *Cauloramphus* was used to examine (1) the correlation between COI genetic distance and morphological divergence in selected characters; (2) relative levels of intra- and inter-population COI genetic divergence; and (3) the utility of COI in discriminating species and species groups. The phylogeny includes representatives of 15 previously described morphospecies and five unidentified taxa. Kimura 2-parameter (K2P) distances within local populations of five morphospecies ranged from 0.00% to 3.01%. For three morphospecies, K2P distances ranged from 0.50% to 11.0% between populations separated geographically by 750–4,500 km, with no correlation between genetic distance and geographical separation; we identified at least one putative cryptic species. The phylogeny detected at least three undescribed morphospecies; two other specimens not identified to species prior to the analysis emerged as divergent populations of, or sister taxa to, previously described or newly detected morphospecies. Our results indicate that in *Cauloramphus*, and perhaps in many other cheilostomes, there is a necessary bias in application of the morphological species concept. Lack of detectable morphological differences between geographically separate populations says little about genetic distance between them, and operationally they must be considered as populations within a single morphospecies. On the other hand, geographically separate populations exhibiting overt differences in the form of, or non-overlapping differences in the ranges of, one or more characters indicate substantial genetic divergence and probable reproductive isolation, validating application of the morphological species concept as a proxy for the biological species concept in these cases.

**Keywords** Bryozoa • *Cauloramphus* • Phylogeny • Morphology • Genetic distance • Diversity

---

M.H. Dick (✉) • M. Hirose • S.F. Mawatari
Department of Natural History Sciences, Hokkaido University, N10 W8, Sapporo 060-0810, Japan
e-mail: mhdick@mail.sci.hokudai.ac.jp

A. Ernst, P. Schäfer, and J. Scholz (eds.), *Bryozoan Studies 2010*,
Lecture Notes in Earth System Sciences 143, DOI 10.1007/978-3-642-16411-8_3,

## Introduction

The taxonomy of cheilostomes, the dominant bryozoan group in modern seas, relies almost exclusively on characters of the calcified skeleton for identification and classification. Species delineated exclusively by skeletal characters are morphospecies rather than confirmed biological species. Jackson and Cheetham (1990), however, found for several species in each of one anascan and two ascophoran genera that morphospecies identity was heritable, and that skeletally identified morphospecies were genetically distinct in terms of allozyme allele frequencies. That study suggested that significant morphological divergence between cheilostome populations is indicative of both significant genetic divergence and interspecific differences. Although Jackson and Cheetham (1990) detected no morphologically cryptic species, the converse (that high genetic divergence correlates with detectable morphological differentiation) does not always hold; examples of cryptic or 'sibling' species are known among various marine invertebrates (Knowlton 1993, 2000), including cheilostome bryozoans (Gómez et al. 2007).

Morphospecies within cheilostome genera can be often delineated unambiguously using suites of characters, and detection of genetic divergence between seemingly morphologically similar populations often allows the subsequent detection of subtle character differences (e.g., Dick and Mawatari 2005), at which point the populations are no longer morphologically cryptic. Problems may arise, however, because the degree of morphological differentiation may not be correlated with the degree of genetic divergence between closely related but geographically separate populations. Furthermore, the degree of genetic distance indicative of distinct biological species can vary among taxa and among molecules, and there is no absolute scale (Harrison 1991; Avise 2000; Knowlton 2000).

Here we examine the extent of correlation between genetic divergence and morphological differentiation in *Cauloramphus* Norman, 1903, a member of the large paraphyletic cheilostome family Calloporidae. *Cauloramphus* contains roughly 30 nominal species and is well defined by a suite of characters including circum-opesial spines, uniporous septula, basally jointed avicularia, a vestigial kenozooidal ooecium, and internal brooding (Ostrovsky et al. 2007; Dick et al. 2009). Species are usually readily discriminated by a suite of characters including zooid size, spine number, pattern of intra-zooidal variation in spine morphology (Dick et al. 2011), spine color, extent of exposed gymnocyst, texture of the cryptocyst, form and placement of avicularia, and form of the ooecium. Borderline cases exist, however, where it has not been clear whether morphological differences represent intra- versus interspecific variation. For example, Dick et al. (2011) on the one hand used a non-overlapping difference in spine number to distinguish between *C. cheliferoides* and *C. oshurkovi* in the northwestern Pacific, but on the other hand detected scarcely overlapping variation in spine number and orifice size among geographically disjunct populations of nominal *C. ascofer* (these species as well as *C. ordinarius* and *C. peltatus* referred to herein were originally described by Dick et al. 2011).

Dick et al. (2009) used sequence data from the mitochondrial gene cytochrome c oxidase subunit I (COI) to reconstruct *Cauloramphus* phylogeny in order to investigate the relationship between divergent cribrimorph species and less-divergent, or stereotypical, species in the genus. Since that study, we obtained COI data for several additional species, and we also have limited data for geographical populations within several species. Here we use these data to assess (1) the correlation between COI genetic distance and morphological divergence in selected characters; (2) relative levels of intra- and inter-population COI genetic divergence; and (3) the utility of COI in discriminating species and species groups.

## Material and Methods

Table 3.1 lists localities or areas from which specimens were obtained; exact locality information is available under GenBank accession numbers. Specimens were identified to species based on morphological descriptions in Canu and Bassler (1930), Dick and Ross (1988), Seo (2001), Dick et al. (2005, 2011), and Grischenko et al. (2007); five specimens were not identified to species prior to analysis. Most of the values for the morphological characters presented in Table 3.1 were taken from the literature and refer to the same populations from which specimens were obtained for DNA analysis. In cases where we did not identify specimens to species, or specimens came from populations for which no data were available in the literature, we took measurements from voucher specimens. Length measurements were made either through a stereomicroscope fitted with an ocular micrometer or from SEM images by use of ImageJ v. 1.42 software (http://rsb.info.nih.gov/ij). For examination of each voucher by scanning electron microscopy (SEM), we first mounted the dried specimen on a stub, coated it with Pd-Pt in a Hitachi E-1030 sputter coater, and took electronic images with a Hitachi S-2380 scanning electron microscope at 15 kV accelerating voltage. We then removed the specimen from the stub, soaked it in a household bleach solution to remove non-skeletal material, rinsed and dried the specimen, remounted it on the stub, re-coated it, and took SEM images of the bleached specimen.

DNA was extracted from dried (82%), ethanol-preserved (11%), or fresh (7%) specimens; dried specimens from which DNA was successfully amplified were up to 25 years old, but generally within 1–5 years old. In all cases, part of each colony selected for DNA analysis was retained as a voucher specimen. Details of the methods used for DNA extraction, PCR amplification of the COI fragment with primers LCO1490 and HCO2198 (Folmer et al. 1994), cloning of amplicons into pGEM-T Easy vector, PCR screening of plasmid inserts, and sequencing are described or referenced in Dick et al. (2009).

We aligned the nucleotide sequences by using ClustalX (Thompson et al. 1997); there were no gaps in the alignment, which included 333 constant sites, 24 parsimony-uninformative sites, and 301 parsimony-informative sites among 658 bp. In addition, we constructed an alignment of deduced amino-acid

**Table 3.1** Values for selected characters, geographical range, and GenBank accession numbers for *Cauloramphus* specimens included in this study. In most cases, zooid length, avicularium position, spine counts, and cryptocyst condition were taken from the literature (R2) and are representative of the population from which the specimen came; zooid lengths are mean values. Values obtained in this study ("8" in R2 column) came from the DNA voucher specimens, in all cases a single colony from which DNA was extracted. The "Clone" column indicates the clone sequenced, as a cross reference to Dick et al. (2009)

| Specimen | Clone | R1[a] | R2[a] | Zooid length (mm) | Avic. position | Orificial spines[b] | Interm. spines | Opesial spines | Total spines | Dark spines[c] | CR[d] | Locality[e] | GenBank accession number |
|---|---|---|---|---|---|---|---|---|---|---|---|---|---|
| *C. ascofer* | CLC-1 | 4 | 4 | 0.71 | 2 | 4 [5] | 0 | 8–12 | 12–16 | - | A | W. Aleutians, AK | EU835953 |
| *C. brunea* | BRU-1 | 1 | 8 | 0.38 | 3/4 | 3 [4] | 2 | 8–11 | 13–16 | P | T | Galapagos Is. | HQ201940 |
| *C. cheliferoides* | DIS-1 | 4 | 4 | 0.56 | 3/4 | 2 | 2 | 10–15 | 14–19 | - | S | W. Aleutians, AK | EU835957 |
| *C. cryptoarmatus* 1 | CAR-1 | 5 | 5 | 0.60 | 2/3 | 3 [4] | 0 | 4–6 | 8–10 | - | T* | Akkeshi, JPN | HQ201929 |
| *C. cryptoarmatus* 3 | CAR-3 | 5 | 5 | 0.60 | 2/3 | 3 [4] | 0 | 4–6 | 8–10 | - | T* | Akkeshi, JPN | HQ201930 |
| *C. korensis* | KOR-1 | 7 | 7 | 0.52 | ? | 1–3 | 0 | 4–6 | 5–9 | P | T* | S. Korea | EU560977 |
| *C. magnus* KE | MAG-1 | 2 | 2 | 0.65 | 2/3, 3/4 | 4 [4–6] | 0 | 8–12 | 12–18 | - | T | Ketchikan, AK | EU835959 |
| *C. multiavicular.* KE | MUL-2 | 3 | 3 | 0.69 | Variable | 3 [–5] | 2 | 11–17 | 16–23 | - | T | Ketchikan, AK | EU835961 |
| *C. multiavicular.* WA | MUL-1 | 3 | 8 | 0.72 | Variable | 4 [3] | 2 | 13–16 | 18–21 | - | T | Anacortes, WA | HQ201931 |
| *C. multispinosus* AS | MSP-2 | 5 | 5 | 0.75 | No avic. | 4 | 2 | 14–20 | 20–26 | - | T | Akkeshi, JPN | EU835960 |
| *C. multispinosus* OS | MSP-1 | 5 | 8 | 0.75 | No avic. | 4 | 2 | 14–20 | 20–26 | - | T | Oshoro, JPN | EU560978 |
| *C. niger* 1 | CNI-1 | 5 | 5 | 0.56 | 2/3, 3/4 | 2–4 | 2 | 8–13 | 12–19 | P | T | Akkeshi, JPN | EU835954 |
| *C. niger* 4 | CNI-4 | 5 | 5 | 0.56 | 2/3, 3/4 | 2–4 | 2 | 8–13 | 12–19 | P | T | Akkeshi, JPN | HQ201948 |
| *C. ordinarius* | CAC-1 | 4 | 4 | 0.74 | 4/5 | 5 [4–6] | 0 | 9–13 | 14–18 | - | T | W. Aleutians, AK | EU835950 |
| *C. oshurkovi* | CDI-1 | 4 | 4 | 0.49 | 3/4 | 2 | 2 | 6–8 | 10–12 | - | S | Commander Is. | HQ201937 |

| | | | | | | | | | | | | | |
|---|---|---|---|---|---|---|---|---|---|---|---|---|---|
| *C. peltatus* | ALA-1 | 4 | 4 | 0.66 | 3, 3/4 | 2 | 2 | 15–19 | 19–23 | - | S | W. Aleutians, AK | EU835947 |
| *C. spectibilis* | SPE-1 | 2 | 2 | 0.58 | 5/6, 6 | 8 [9] | 0 | 11–16 | 19–24 | - | S | Kodiak, AK | HQ201939 |
| *C. spinifer*-2 KO | SPI-2 | 6 | 2 | 0.59 | 2/3 | 3 [4] | 0 | 5–10 | 8–13 | - | T | Kodiak, AK | HQ201925 |
| *C. spinifer*-3 KO | SPI-3 | 6 | 2 | 0.59 | 2/3 | 3 [4] | 0 | 5–10 | 8–13 | - | T | Kodiak, AK | HQ201926 |
| *C. spinifer*-8 AS | SPI-8 | 6 | 5 | 0.56 | 2/3 | 3 [2–5] | 0 | 5–10 | 8–13 | - | T | Akkeshi, JPN | HQ201927 |
| *C. tortilis* | TOR-2 | 3 | 3 | 0.63 | 3 | 4 [5] | 2 | 3–8 | 9–15 | - | T* | Ketchikan, AK | HQ201934 |
| Commanders sp. 1 | CIN-1 | - | 8 | 0.67 | ? | 6 [5] | 0 | 6–10 | 11–16 | - | T | Commander Is. | EU835952 |
| Korea Baeng. sp. 1 | KMU-1 | - | 8 | 0.52 | 2/3, 3/4 | 3 [4] | 2 | 5–8 | 10–13 | - | T | S. Korea | HQ201935 |
| Korea Baeng. sp. 2 | KVA-1 | - | 8 | 0.57 | 2/3 | 3 | 0 | 10–12 | 13–15 | P | T | S. Korea | EU835958 |
| Oshoro sp. 1 | COS-9 | - | 8 | 0.56 | 2/3 | 3 [4] | 0 | 7–9 | 9–14 | P | T | Oshoro, JPN | EU835956 |
| Oshoro sp. 2 | COS-5 | - | 8 | 0.61 | ? | 2 [–4] | 2 | 7–10 | 11–15 | - | T | Oshoro, JPN | HQ201950 |

[a] R1 is the reference of the original description; R2 is the source of values presented in the Table: 1, Canu and Bassler (1930); 2, Dick and Ross (1988); 3, Dick et al. (2005); 4, Dick et al. (2011); 5, Grischenko et al. (2007); 6, Johnston (1832); 7, Seo (2001); 8, this study

[b] Single values indicate an invariant number of spines; modal values are followed by an indication of the range in square brackets

[c] P, dark brown to black coloration of spines or spine bases; -, no dark coloration

[d] CR is cryptocyst texture: T, tuberculate; T*, entire mural rim cryptocysal and tuberculate, with little or no gymnocyst showing; S, smooth; A, absent

[e] Detailed locality information can be obtained through GenBank records; AK, Alaska, USA; JPN, Japan; WA, Washington State, USA

sequences by using DAMBE (Xia and Xie 2001); this alignment was 219 residues long (161 constant, 42 parsimony uninformative, 16 parsimony informative). We used four methods to reconstruct phylogeny: maximum parsimony (MP), neighbor-joining (NJ), and maximum likelihood (ML), as implemented in PAUP* (Swofford 2000), and Bayesian analysis (BA), as implemented in MrBayes (Ronquist and Huelsenbeck 2003). Plots of transitions and transversions against genetic distance, as implemented in DAMBE, showed that 3rd codon positions were highly saturated, so for the MP analyses we explored differential character weighting. MP analyses consisted of heuristic searches of 1,000 random-addition replicates; we estimated bootstrap support for nodes through analysis of 10 random-addition heuristic searches for each of 100 pseudoreplicates (parsimony-informative characters only), with replacement. We constructed NJ trees by using LogDet/paralinear distances for the DNA data set and mean character differences for the amino-acid data set. For ML analyses, the best-fit nucleotide substitution model determined with Modeltest (Posada and Crandall 2001) was GTR + $\Gamma$ + I. The search for the optimal ML tree started with a NJ tree based on LogDet distances, on which NNI branch swapping ensued. Nodal support for ML and NJ trees was assessed by bootstrap analyses of 200 and 1,000 pseudoreplicates, respectively. We conducted Bayesian analyses using the best-fit model (GTR + $\Gamma$ + I); analyses ran for six million generations, with trees sampled every 100 generations. From the resulting 60,001 trees, we discarded the first 12,500 as burn-in.

## Results and Discussion

### *Phylogenetic Analysis*

Analysis of the nucleotide (NT) data set generally produced incongruent trees among the four methods of analysis (MP, NJ, ML, BA), and among various character-weighting schemes (for MP, weighting transversions more than transitions, weighting codon positions $1 > 2 > 3$, or exclusion of 3rd codon positions; for ML and BA, exclusion of 3rd codon positions). Analyses of the amino acid (AA) data set by MP produced largely unresolved trees. Several methods and weighting schemes, however, produced similar trees that included a clade containing *C. ordinarius*, *C. cheliferoides*, *C. oshurkovi*, *C. peltatus*, and *C. ascofer* (Clade B, Fig. 3.1), though this clade received no nodal support except for (*C. peltatus* + *C. ascofer*). Morphological characters support the last four taxa as a monophyletic group (Dick et al. 2011): all have the bases of opesial spines calcified in mature zooids (loss of the opesial spine joints in *Cauloramphus* is known only among these four and several other related species) (Dick et al. 2011); *C. peltatus* and *C. ascofer* both have costal shields, and there is strong molecular support for them as sister taxa; and *C. peltatus*, *C. oshurkovi*, and *C. cheliferoides* all have a smooth cryptocyst (a rare character state in *Cauloramphus*), two orificial

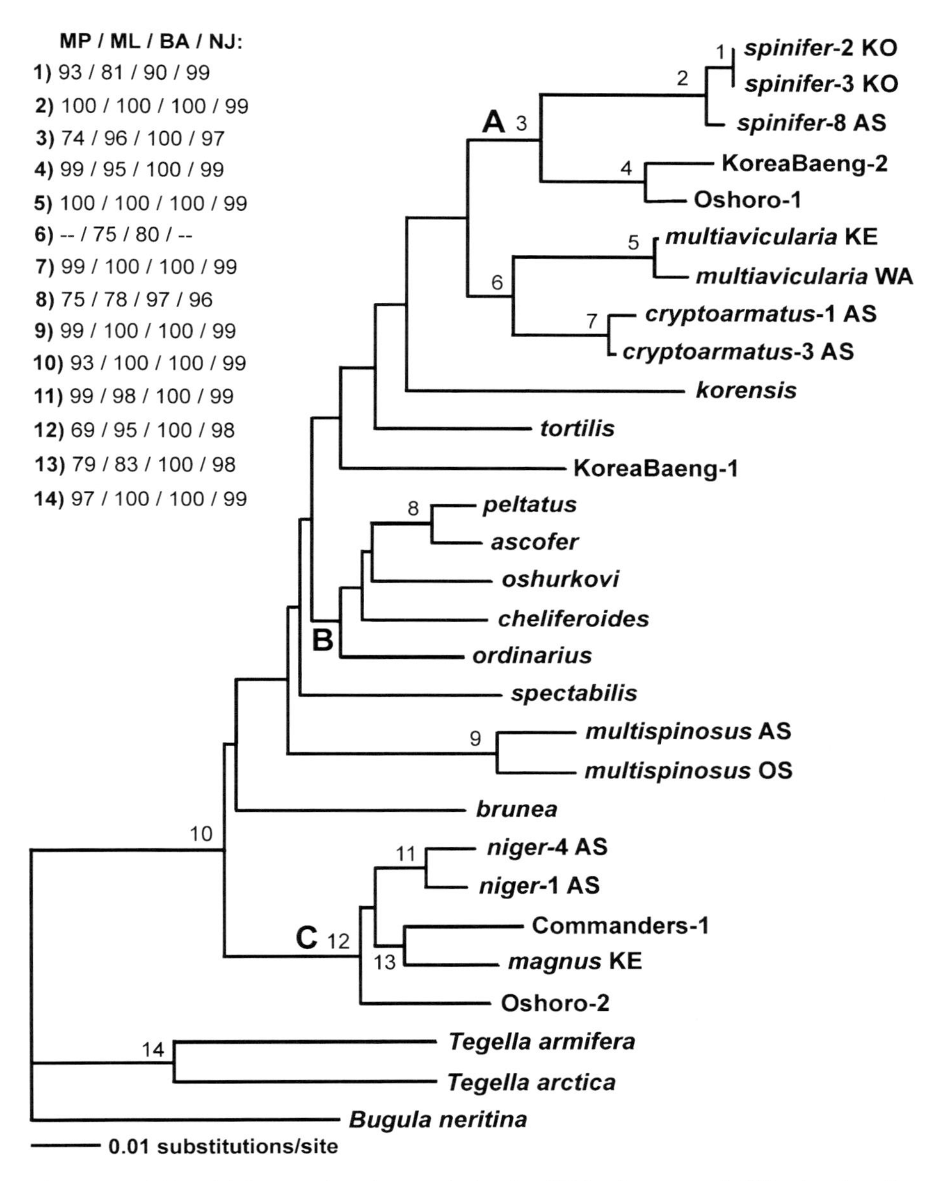


**Fig. 3.1** Neighbor-joining tree based on analysis of a 658 bp fragment of COI using Log-Det distances, with third codon positions omitted. Taxon names correspond to those in Table 3.1, with two-letter abbreviations following some names indicating the locality (AS, Akkeshi, Hokkaido, Japan; KE, Ketchikan, Alaska, USA; KO, Kodiak, Alaska, USA; OS, Oshoro, Hokkaido, Japan; WA, Anacortes, Washington, USA). Numbers near nodes refer to the key at the *upper left*, in which each set of four values includes bootstrap values in percent determined from analyses of the entire nucleotide data set by maximum parsimony (MP), maximum likelihood (ML), and neighbor-joining (NJ), and a posterior probability value in percent from a Bayesian analysis of the entire data set, respectively

spines, two intermediate spines (Table 3.1), and hypertrophied avicularia (Dick et al. 2011). Although we had no means of determining which of many incongruent trees was most reliable, we assumed that trees that retained Clade B were likely to be more reliable than those which did not, which contradicted strong morphological evidence. Three other multi-species clades evident in Fig. 3.1 were present in all trees and received high nodal support (support values greater than or equal to 95%) for one or more of the analyses of the full NT data set: Clade A, Clade C, and (*C. ascofer* + *C. peltatus*). In any case, conclusions in this study were based only on the well-supported intraspecific and multi-species clades, and were not dependent on the backbone topology.

The incongruence among trees produced by analysis of the NT data set by different methods very likely had to do with the high degree of saturation in 3rd codon positions, as all trees that included Clade B were from analyses that directly or indirectly differentially weighted 1st and 2nd codon over 3rd codon positions, or transversions over transitions, i.e., NJ analysis of the AA data set, NJ analysis of the NT data set with 3rd codon positions excluded, and MP analysis of the NT data set with transversions weighted 4:1 or 5:2 over transitions. It is not clear why COI did not give better resolution with *Cauloramphus*, as analyses of nominal *Celleporella hyalina* Linnaeus, 1767 by Gómez et al. (2007) and nominal *Membranipora membranacea* (Linnaeus, 1767) by Schwaninger (1999) using this gene placed most terminal taxa in well-supported clades separated by uncorrected distances of 8–20% and K2P distances of up to 16.9%, respectively. Backbone nodes were not resolved in either study, however, and maximum K2P distances among *Cauloramphus* taxa were even higher, up to 31.1%. The mitochondrial 16S rRNA gene may produce better resolution in intra-generic analyses (e.g., Dick et al. 2003; Nikulina and Schäfer 2008); 16S was our first choice in this study, but universal 16S primers that have worked well in other cheilostome genera (e.g., Dick et al. 2003; Dick and Mawatari 2005) did not amplify DNA at all in *Cauloramphus*, suggesting mutations in the primer-binding sites.

## Morphological Divergence and Genetic Distance Within Morphospecies

Five well-supported morphospecies clades (*C. spinifer* [Johnston, 1832]; *C. multiavicularia* Dick et al., 2005; *C. cryptoarmatus* Grischenko et al., 2007; *C. multispinosus* Grischenko et al., 2007; *C. niger* Grischenko et al., 2007) (Fig. 3.1) allowed us to assess genetic distances between or within populations, and the correlation between genetic distance and morphological divergence. In each case, specimens that were initially identified morphologically to a particular species grouped together (Fig. 3.1) with high nodal support. For the three species that involved inter-population comparisons (*C. multiavicularia*, *C. multispinosus*, *C. spinifer*), zooidal characters (Table 3.1) and the overall appearance of unbleached and bleached specimens (Figs. 3.2 and 3.3) were very similar across

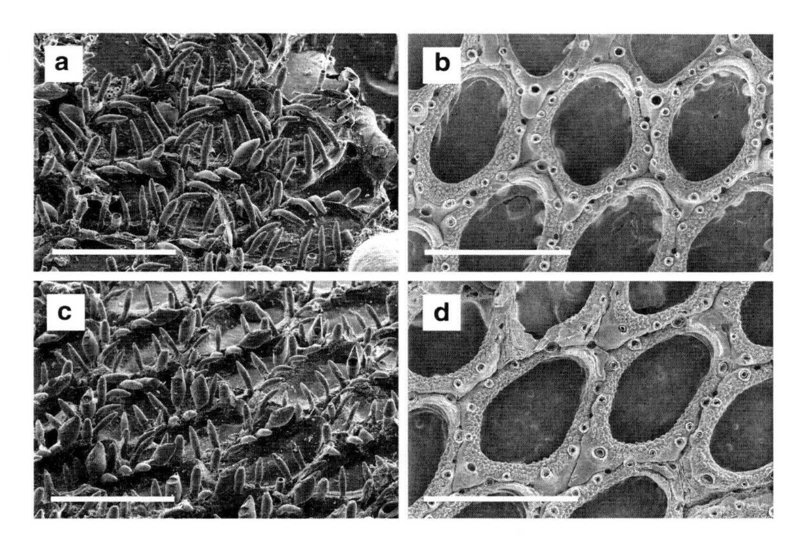


**Fig. 3.2** Comparison of morphology between DNA voucher specimens of *Cauloramphus spinifer* from two localities. (**a**, **b**) Kodiak, Alaska, USA (GenBank HQ201925). (**c**, **d**) Akkeshi, Hokkaido, Japan (GenBank HQ201927). (**a**, **c**) Dried, unbleached autozooids; (**b**, **d**) bleached autozooids. *Scale bars*: 0.50 mm

geographical distances of hundreds to thousands of kilometers (Table 3.2; Fig. 3.4). DNA voucher specimens of *C. spinifer*, for example, were very similar in the form of the spines, form and abundance of avicularia, extent and sculpturing of the cryptocyst, and form of the ooecium between Kodiak, Alaska (Fig. 3.2a, b) and Akkeshi, Japan (Fig. 3.2c, d). Likewise, *C. multispinosus* was very similar in overall appearance between Oshoro (Fig. 3.3a, c, e) and Akkeshi (Fig. 3.3b, d, f), Hokkaido, Japan.

These species with conserved morphology between populations showed both a wide range of K2P distances for COI and no correlation between genetic distance and geographical separation (Table 3.1; Fig. 3.4): the lowest genetic distance (0.5%) was between populations of *C. multiavicularia* 1,000 km apart on the Pacific Coast of North America; next highest was 2.26% between populations of *C. spinifer* in northeastern Japan and south-central Alaska 4,500 km apart; and highest was 11.0% between the Akkeshi and Oshoro populations of *C. multispinosus* 750 km apart on Hokkaido, northern Japan. The Akkeshi population, on the Pacific coast of Hokkaido, is separated by Tsugaru Strait from the Oshoro population on the Sea of Japan coast, and the large genetic distance suggests that Tsugaru Strait has long acted as a barrier to gene flow between the Pacific and the Sea of Japan.

Data for six *Cauloramphus* morphospecies showed distances within populations at single localities ranging from 0.00% to 3.01% (Table 3.2). These values are based on sample sizes of 2–4 individuals, and additional sampling would undoubtedly detect greater genetic variation. The level of variation detected within two populations (*C. niger*, 3.01%; *C. cryptoarmatus*, 2.03%) was roughly equivalent to

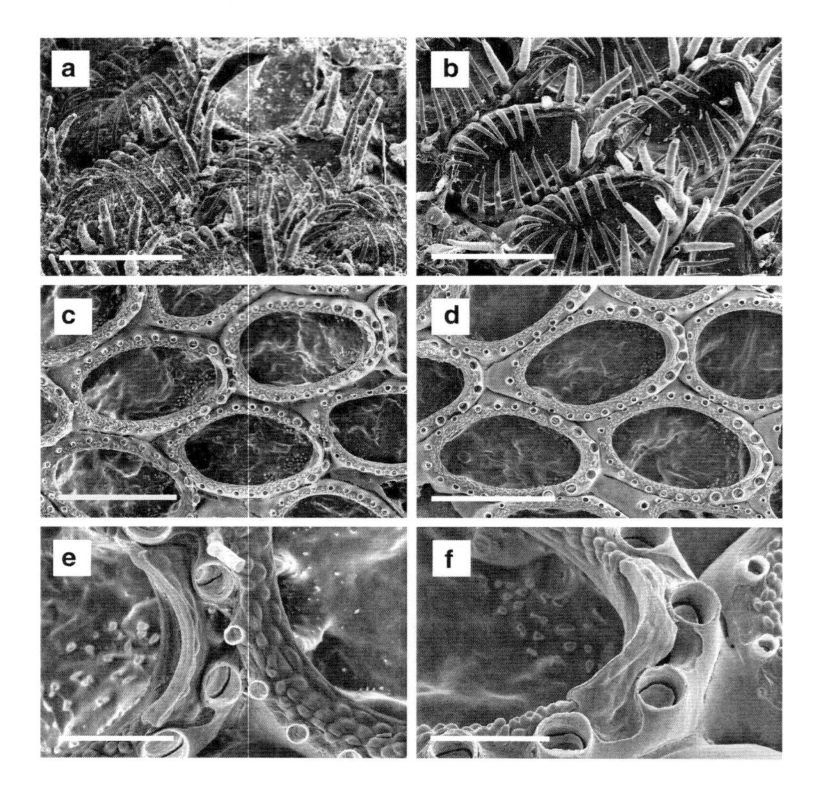


**Fig. 3.3** Comparison of morphology between DNA voucher specimens of *Cauloramphus multispinosus* from two localities. *Left column* (**a**, **c**, **e**), Oshoro, Hokkaido, Japan (GenBank EU560978). *Right column* (**b**, **d**, **f**), Akkeshi, Hokkaido, Japan (GenBank HQ201938). (**a**, **b**) Dried, unbleached autozooids; (**c**, **d**) bleached autozooids; (**e**, **f**) kenozooidal ooecia in bleached zooids. *Scale bars*: 0.50 mm

or exceeded that observed between populations thousands of kilometers apart for other species (*C. multiavicularia*, 0.5%, 1,000 km; *C. spinifer*, 2.26%, 4,500 km). These contrasting results likely reflect different population histories among the species examined.

## *Detection of Cryptic Species, Undescribed Morphospecies, and Multi-species Clades (Species Groups)*

### Cryptic Species

Although there were no detectable differences in morphological characters between specimens of *C. multispinosus* examined from Akkeshi and Oshoro, Hokkaido, Japan, the COI distance between the two populations was 11.0%. The molecular divergence

**Table 3.2** Average K2P distances for the 658-bp COI fragment within and between populations of *Cauloramphus* morphospecies, and geographical distances between the populations sampled

| Morphospecies | n | K2P distance within population (%) | K2P distance between populations (%) | Approximate geographical distance (km) |
|---|---|---|---|---|
| *C. spinifer* | | | 2.26 | 4,500 |
| Akkeshi, Japan | 4 | 0.16 | | |
| Kodiak, Alaska | 2 | 0.71 | | |
| Commanders-1 | 1 | – | 9.10 | 3,780 |
| *C. magnus* (Ketchikan) | 3 | 0.00 | | |
| KoreaBaeng-2 (Korea) | 1 | – | 10.0 | 1,360 |
| Oshoro-1 (Japan) | 1 | – | | |
| *C. multiavicularia* | | | 0.50 | 1,000 |
| Ketchikan, AK | 2 | 0.14 | | |
| Anacortes, WA | 1 | – | | |
| *C. multispinosus* | | | 11.0 | 750 |
| Akkeshi, Japan | 2 | 0.42 | | |
| Oshoro, Japan | 1 | – | | |
| *C. niger* | | | | |
| Akkeshi, Japan | 3 | 3.01 | – | – |
| *C. cryptoarmatus* | | | | |
| Akkeshi, Japan | 3 | 2.03 | – | – |

involves more than neutral variation, as there were five amino acid substitutions in the COI fragment between the two populations. This species was originally described from Akkeshi, and the genetically divergent Oshoro population is a good candidate for a cryptic species. Our results do not conclusively establish the Oshoro population as a cryptic species; this rests on whether there is reproductive compatibility and/or shared haplotypes indicative of gene flow between the two populations. Gómez et al. (2007), however, found reproductive isolation among major *Celleporella hyalina* lineages with COI divergences greater than 8% uncorrected p-distance, and found reproductive isolation among some populations within major lineages at lower COI divergences. Unfortunately, these authors did not indicate the minimum genetic distances at which they observed reproductive isolation.

Commanders-1 from the Commander Islands and *C. magnus* Dick and Ross, 1988 from Ketchikan, southeastern Alaska, grouped together with moderate to high nodal support (Fig. 3.1), and likewise are morphologically similar despite high molecular divergence (9.1%). DNA voucher specimens were virtually identical in zooid size (Table 3.1) and were similar in the overall form of the spines and bleached skeleton (compare Fig. 3.5a, b with Fig. 3.5c, d; note the orificial spine scars conspicuously larger than opesial spine scars in specimens from both populations). The Commanders specimen had a higher modal number of orificial spines and lower range of opesial spines (Table 3.1), though this was the only specimen examined from the Commanders. Characters of the *C. magnus* population

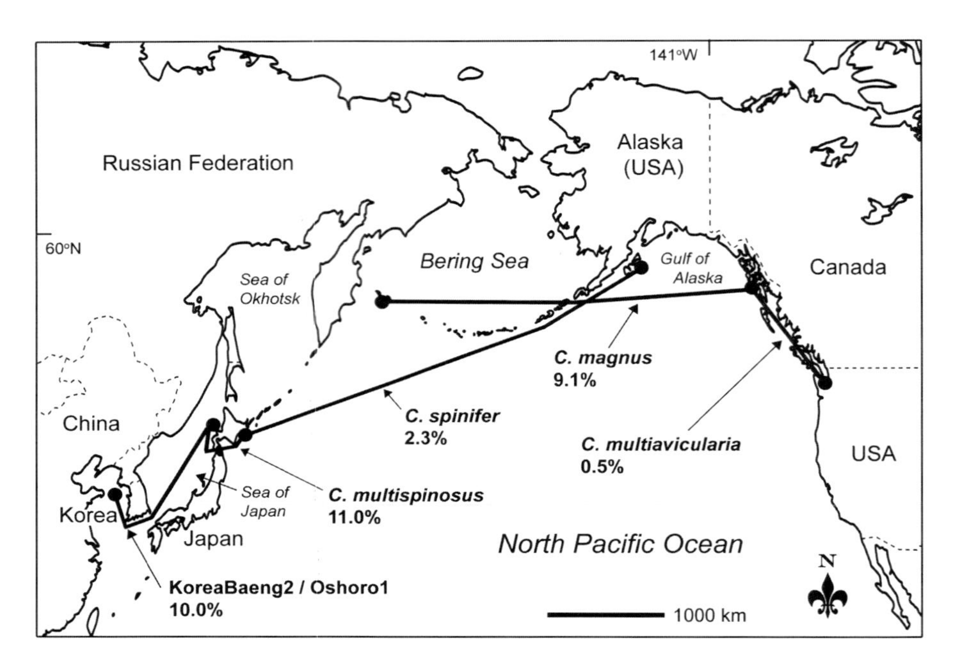


**Fig. 3.4** Map of the North Pacific illustrating the geographical distances separating populations of *Cauloramphus* morphospecies (*C. spinifer*, *C. multiavicularia*, *C. multispinosus*, *C. magnus*), and populations of a putative sister-species pair (KoreaBaeng-2 and Oshoro-1), examined in this study. Average COI distances (K2P, in percent) between the populations are also indicated

at Kodiak, the type locality (Dick and Ross 1988), overlap with the Commander and Ketchikan populations; unfortunately we were unable to amplify COI from dried material from Kodiak.

## Undescribed Morphospecies

This study detected three apparently undescribed morphospecies (KoreaBaeng-2 + Oshoro-1; KoreaBaeng-1; Oshoro-2); in these cases, specimens could not be identified morphologically with any known species prior to DNA extraction, and in trees (Fig. 3.1) failed to group with any previously described species included in the study. KoreaBaeng-2 from Korea and Oshoro-1 from Hokkaido, Japan, grouped together with very high nodal support in all analyses. Specimens from the two populations were similar in zooid size, attachment location of the avicularia (in the 2/3 position, but always close to the most distal opesial spine), number of orificial spines, and lack of intermediate spines. They did not overlap in number of opesial spines, though examination of more specimens from both populations will be necessary to confirm this difference. Zooids in both populations have black-colored spines in mature zooids. Bleached zooids are highly similar in the extent and texture of the cryptocyst, extent of marginal gymnocyst, and form of the

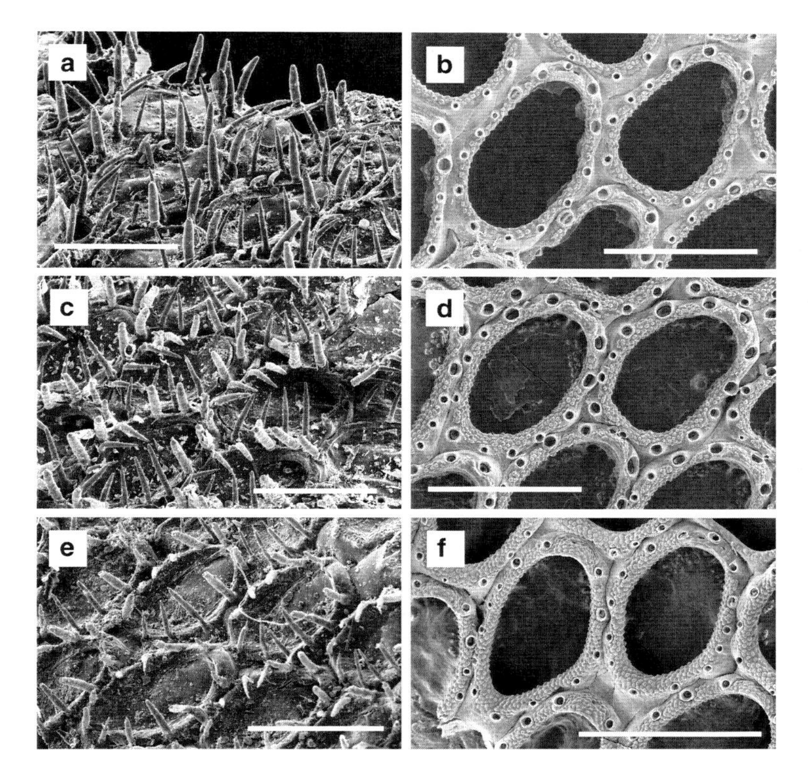


**Fig. 3.5** Comparison of morphology among DNA voucher specimens of (**a**, **b**) *Cauloramphus magnus* from Ketchikan, Alaska, USA (GenBank EU835959); (**c**, **d**) Commanders-1 from the Commander Islands, Russian Federation (GenBank EU835952); and (**e**, **f**) Oshoro-2 from Oshoro, Hokkaido, Japan (GenBank HQ201950). *Left column* (**a**, **c**, **e**), dried, unbleached autozooids; *right column* (**b**, **d**, **f**), bleached autozooids. *Scale bars*: 0.50 mm

ooecium, which is moderately large. The COI distance between KoreaBaeng-2 and Oshoro 1 was 10.0%; differences in opesial spine number (Table 3.1) and the form of the ooecium between the two suggests that they may be geographically disjunct sister species. Another species with dark spines, *C. korensis* Seo, 2001, occurs sympatrically with KoreaBaeng-2 in Korea, but *C. korensis* has markedly fewer spines, a more extensive cryptocyst, and a much larger ooecium than KoreaBaeng-2.

KoreaBaeng-1 did not group with any other species in the analysis, and COI distances between it and other *Cauloramphus* species ranged from 16.0% to 23.9%. This species lacks dark spine coloration and is similar in zooid size and some other characters to *C. spinifer*, with which it may have been confused in the past. It differs from *C. spinifer* in having intermediate spines and a more elongate avicularium with a distinct pedicel, and from all other species examined in having an extensive ooecium that is flush with the mural rim and sunken in the center, with lobes extending between the bases of the orificial spines.

Another orphan species was Oshoro-2 (Fig. 3.5e, f), which was basal to other taxa in Clade C (Fig. 3.1) and showed an average COI distance of 13.6% (range 11.5–17.2%) from them. Clade C itself received high nodal support, and clades comprising (*C. magnus* KE + Commanders-1) and two specimens of *C. niger* received moderate to high nodal support, to the exclusion of Oshoro-2. Oshoro-2 is thus genetically distinct from other taxa in Clade C and can be distinguished from them morphologically. For example, it lacks the black spine coloration and larger, heavily tuberculate ooecium of *C. niger*; it differs from *C. magnus* (Fig. 3.5a, b) and Commanders-1 (Fig. 3.5c, d) in having intermediate spines, in lacking the very large orifical spine bases (Fig. 3.5e, f) seen in the latter two species, and in having a more bulbous, conspicuously raised ooecium.

## Multi-species Clades (Species Groups)

Phylogenetic analyses detected at least three well-supported multi-species clades. One was Clade A, including *C. spinifer*, KoreaBaeng-2, and Oshoro-1 (the latter two either sister species or representing a deep intraspecific divergence). Another was Clade C, including *C. niger*, *C. magnus*, Commanders-1, and Oshoro-2, with *C. magnus* and Commanders-1 either trans-Pacific sister species or representing a deep intraspecific divergence, and Oshoro-2 likely a distinct species. The third comprised *C. peltatus* and *C. ascofer*, sister species separated by a COI distance of 12.1%; the former has been detected only in the western Aleutian Islands, Alaska, whereas the latter occurs in both Kamchatka and the western Aleutians (Dick et al. 2011).

Morphological characters have been used to delineate species groups in many genera of animals, with the implicit assumption (or hope) that these groups represent evolutionary units, or clades. In *Cauloramphus*, the only species group clearly recognizable on the basis of morphology comprises the species in Clade B, excluding *C. ordinarius* (Dick et al. 2011; also see Introduction). Clades A and C respectively comprise two or three, and three or four, species, yet we detected no single character or suite of characters delineating either of these clades: they appear to be cryptic clades within the genus, detectable only on the basis of molecular data. Conversely, our results ruled against some characters as indicating species groups. For example, the occurrence of dark-colored spines is an uncommon character that, according to the topology in Fig. 3.1, arose at least four times independently: certainly at least once each in Clades A (in Oshoro-1/KoreaBaeng-2) and C (in *C. niger*), and probably again twice in lineages leading to *C. brunea* Canu and Bassler, 1930 and *C. korensis* Seo, 2001. A relatively large, raised, globose ooecium is another uncommon but apparently convergent character in *Cauloramphus*, occurring in scattered species in the phylogeny: in *C. niger* (see Grischenko et al. 2007: Fig. 12, p. 1082) in Clade C, KoreaBaeng-2/Oshoro-1 (ooecium not illustrated) in Clade A, and *C. korensis* (see Seo 2001; Fig. 1b, c, p. 226).

## Concluding Remarks

Molecular phylogenetic and phylogeographic analyses of cheilostomes have detected deep genetic divergences within genera (e.g., Dick et al. 2003; Nikulina and Schäfer 2008) or within putatively cosmopolitan morphospecies (e.g., Schwaninger 1999; Gómez et al. 2007). In many cases, clades identified as genetically divergent also prove to be morphologically distinct (e.g., Dick and Mawatari 2005; Herrera-Cubilla et al. 2006, 2008), but in other cases (e.g., *C. multispinosus* in our study), morphological differences are not apparent. We found that morphospecies can show a broad range of genetic divergence in populations hundreds to thousands of kilometers apart, and even within local populations. In terms of practical taxonomy, this leads to an operationally skewed situation. It makes little sense to give new names to highly genetically divergent but morphologically identical populations within morphospecies, since there is no way to distinguish among members of component populations except through genetic analysis. Our results also suggest, however, that disjunct populations that are morphologically similar except for non-overlapping differences in some characters are also likely to be highly divergent genetically; we detected no cases where populations showing such differences also showed low genetic divergence. For example, *C. oshurkovi* in the Commander Islands differs from *C. cheliferoides* in the western Aleutians only in a non-overlapping range of opesial spine number (Dick et al. 2011), yet the genetic distance between the two is 18.8%. Ironically, despite the great resolving power of molecular phylogenetic analyses, for practical purposes we are not free operationally from the morphological species concept.

**Acknowledgments** We thank Andrei Grischenko and William Banta for providing some of the specimens for DNA extraction; Yoshinobu Nodasaka for making available SEM facilities; the Alaska Fisheries Science Center for facilitating collecting in the Aleutians; and Andrea Waeschenbach and Leandro Vieira for reviewing the manuscript. This study was supported in part by the COE Program 'Neo-Science of Natural History' at Hokkaido University, funded by MEXT, Japan, and by a grant-in-aid (KAKENHI C 17570070) from MEXT to MHD and SFM.

## References

Avise JC (2000) Phylogeography. Harvard University Press, Cambridge, MA

Canu F, Bassler RS (1930) The bryozoan fauna of the Galapagos Islands. Proc US Natl Mus 76(13):1–78

Dick MH, Grischenko AV, Mawatari SF (2005) Intertidal Bryozoa (Cheilostomata) of Ketchikan, Alaska. J Nat Hist 39:3687–3784

Dick MH, Herrera-Cubilla A, Jackson JBC (2003) Molecular phylogeny and phylogeography of free-living Bryozoa (Cupuladriidae) from both sides of the Isthmus of Panama. Mol Phylogenet Evol 27:355–371

Dick MH, Lidgard S, Gordon DP et al (2009) The origin of ascophoran bryozoans was historically contingent but likely. Proc R Soc B 276:3141–3148

Dick MH, Mawatari SF (2005) Morphological and molecular concordance of *Rhynchozoon* clades (Bryozoa, Cheilostomata) from Alaska. Invertebr Biol 124:344–354
Dick MH, Mawatari SF, Sanner J et al (2011) Cribrimorph and other *Cauloramphus* species (Bryozoa: Cheilostomata) from the northwestern Pacific. Zool Sci 28:134–147
Dick MH, Ross JRP (1988) Intertidal Bryozoa (Cheilostomata) of the Kodiak vicinity, Alaska. Cent Pac Northwest Stud Occas Pap 23:1–133
Folmer O, Black M, Lutz R et al (1994) DNA primers for amplification of mitochondrial cytochrome c oxidase subunit I from diverse metazoan invertebrates. Mol Mar Biol Biotechnol 3:294–299
Gómez A, Wright PJ, Lunt DH et al (2007) Mating trials validate the use of DNA barcoding to reveal cryptic speciation of a marine bryozoan taxon. Proc R Soc B 274:199–207
Grischenko AV, Dick MH, Mawatari SF (2007) Diversity and taxonomy of intertidal Bryozoa (Cheilostomata) at Akkeshi Bay, Hokkaido, Japan. J Nat Hist 41:1047–1161
Harrison RG (1991) Molecular changes at speciation. Annu Rev Ecol Syst 22:281–308
Herrera-Cubilla A, Dick MH, Sanner J et al (2006) Neogene Cupuladriidae of Tropical America. I: taxonomy of Recent *Cupuladria* from opposite sides of the Isthmus of Panama. J Paleontol 80:245–263
Herrera-Cubilla A, Dick MH, Sanner J et al (2008) Neogene Cupuladriidae of Tropical America. II: taxonomy of Recent *Discoporella* from opposite sides of the Isthmus of Panama. J Paleontol 82:279–298
Jackson JBC, Cheetham AH (1990) Evolutionary significance of morphospecies: a test with cheilostome Bryozoa. Science 248:579–583
Johnston G (1832) A descriptive catalogue of the recent zoophytes found on the coast of north Durham. Trans Nat Hist Soc Newcastle 2:239–272
Knowlton N (1993) Sibling species in the sea. Annu Rev Ecol Syst 24:189–216
Knowlton N (2000) Molecular genetic analyses of species boundaries in the sea. Hydrobiologia 420:73–90
Linnaeus C (1767) Systema Naturae. In: Regnum Animale, vol 1, 12th edn. Laurentii Salvii, Holmiae
Nikulina EA, Schäfer P (2008) An evaluation of the morphology of the genus *Electra* Lamouroux, 1816 (Bryozoa, Cheilostomata) with phylogenetic analyses of the ribosomal gene. In: Hageman SJ, Key MM Jr, Winston JE (eds) Bryozoan studies 2007, vol 15, Special Publication. Virginia Museum of Natural History, Martinsville
Ostrovsky AN, Dick MH, Mawatari SF (2007) The internal-brooding apparatus in the bryozoan genus *Cauloramphus* (Cheilostomata: Calloporidae) and its inferred homology to ovicells. Zool Sci 24:1187–1196
Posada D, Crandall KA (2001) Modeltest: testing the model of DNA substitution. Bioinformatics 14:817–818
Ronquist F, Huelsenbeck JP (2003) MRBAYES 3: Bayesian phylogenetic inference under mixed models. Bioinformatics 19:1572–1574
Schwaninger HR (1999) Population structure of the widely dispersing marine bryozoan *Membranipora membranacea* (Cheilostomata): implications for population history, biogeography, and taxonomy. Mar Biol 135:411–423
Seo JE (2001) A new species of the genus *Cauloramphus* (Bryozoa, Cheilostomata) from Korea. Korean J Syst Zool 17:223–228
Swofford DL (2000) *PAUP**: phylogenetic analysis using parsimony. (*and other methods), Ver. 4.0b10. Sinauer Associates, Sunderland
Thompson JD, Gibson TJ, Plewniak F et al (1997) The ClustalX windows interface: flexible strategies for multiple sequence alignment aided by quality analysis tools. Nucleic Acids Res 25:4876–4882
Xia X, Xie Z (2001) DAMBE: software package for data analysis in molecular biology and evolution. J Hered 92:371–373

# Chapter 4
# *Acanthoclema* (Rhabdomesina, Cryptostomata) from the Devonian of Europe

## *Acanthoclema* from Devonian of Europe

**Andrej Ernst**

**Abstract** The rhabdomesine genus *Acanthoclema* Hall, 1886 is known from the Devonian of North America and Eurasia. Many species referred to this genus were previously understood in a broad concept but do not belong to it. Six species of *Acanthoclema* are reported from the Devonian of Europe, which consistently show morphological characters shared with the type species. The generic diagnosis of *Acanthoclema* has been emended. The studied material was used for cluster analysis in order to reveal relationships between the six species of *Acanthoclema* and to compare them with *Nikiforovella gracilis*, a representative of the morphologically most similar genus. It is resumed that *Nikiforovella* derived from *Acanthoclema* in the Lower Devonian. These two genera show completely different trends in their evolution: a decrease of number of acanthostyles and metazooecia in *Acanthoclema*, and an increase of their number in *Nikiforovella*. A test using cluster analysis showed that Morisita's index is very useful for the heterogeneous data.

**Keywords** Rhabdomesines • Taxonomy • Devonian • Europe

## Introduction

Bryozoan taxonomy is very complicated and bears numerous problems in its current stand. One of the basic problems is to distinguish between species (Hageman 1991). The most promising method is the use of numerical statistics (Anstey and Perry 1970; Anstey and Pachut 2004; Snyder 1991a, b). This approach considers a comprehensive description of the available material including measurements of as many characters as possible.

A. Ernst (✉)
Institute of Geosciences, Kiel University, Ludewig-Meyn-Str. 10, Kiel D-24118, Germany
e-mail: ae@gpi.uni-kiel.de

A. Ernst, P. Schäfer, and J. Scholz (eds.), *Bryozoan Studies 2010*,
Lecture Notes in Earth System Sciences 143, DOI 10.1007/978-3-642-16411-8_4,

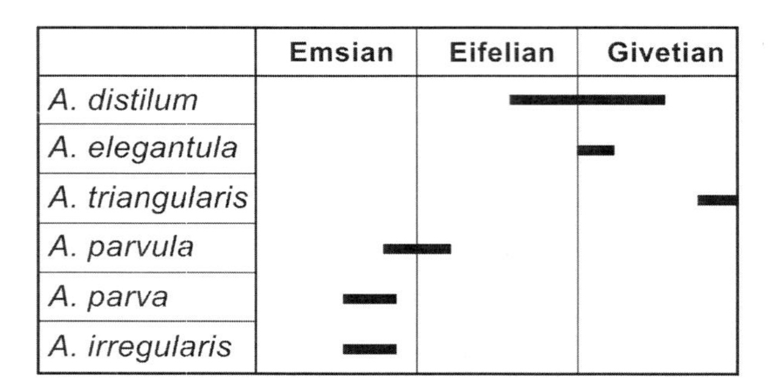


**Fig. 4.1** Distribution of *Acanthoclema* species in the Devonian of Europe

The main difficulties concerning descriptions of many bryozoan taxa are insufficient diagnoses, a deficiency of high-quality illustrations or the loss of type material. Furthermore, many taxa of Palaeozoic bryozoans were described without the use of thin sections, which is now the standard method for their study.

The genus *Acanthoclema* Hall, 1886 was established on the base of the type species *Trematopora alternata* Hall, 1883 from the Middle Devonian of New York, USA. The type material of this species was illustrated by Blake (1983, p. 584, Fig. 1a–c, 292). Besides the type species, several species were mentioned by various authors from the Silurian to the Late Carboniferous of North America, Europe, and China. However, many of these species were described without the use of thin sections and/or were badly illustrated, so that their assignment to *Acanthoclema* is questionable (see Diagnosis below).

During the study of the Devonian bryozoan faunas of Europe and adjacent areas (DFG project ER 278/4-1 and 2), several species of *Acanthoclema* were identified and comprehensively studied (Figs. 4.1, 4.2, and 4.3). These species share certain morphological characters, which distinctly outline the genus range. The aim of this paper is an emendation of the genus diagnosis and the morphological and numerical comparison of *Acanthoclema* species from the Devonian of Europe.

## Material and Methods

The morphology of the bryozoan species considered for the present paper was investigated in thin sections using a binocular microscope in transmitted light. The studied material is incorporated in different collections deposited at the Senckenberg Museum, Frankfurt (Main), Germany (SMF numbers), Geoscience Center Göttingen, Germany (GZG numbers), and at the Nationaal Natuurhistorisch Museum (Naturalis) in Leiden, Netherlands (RGM numbers).

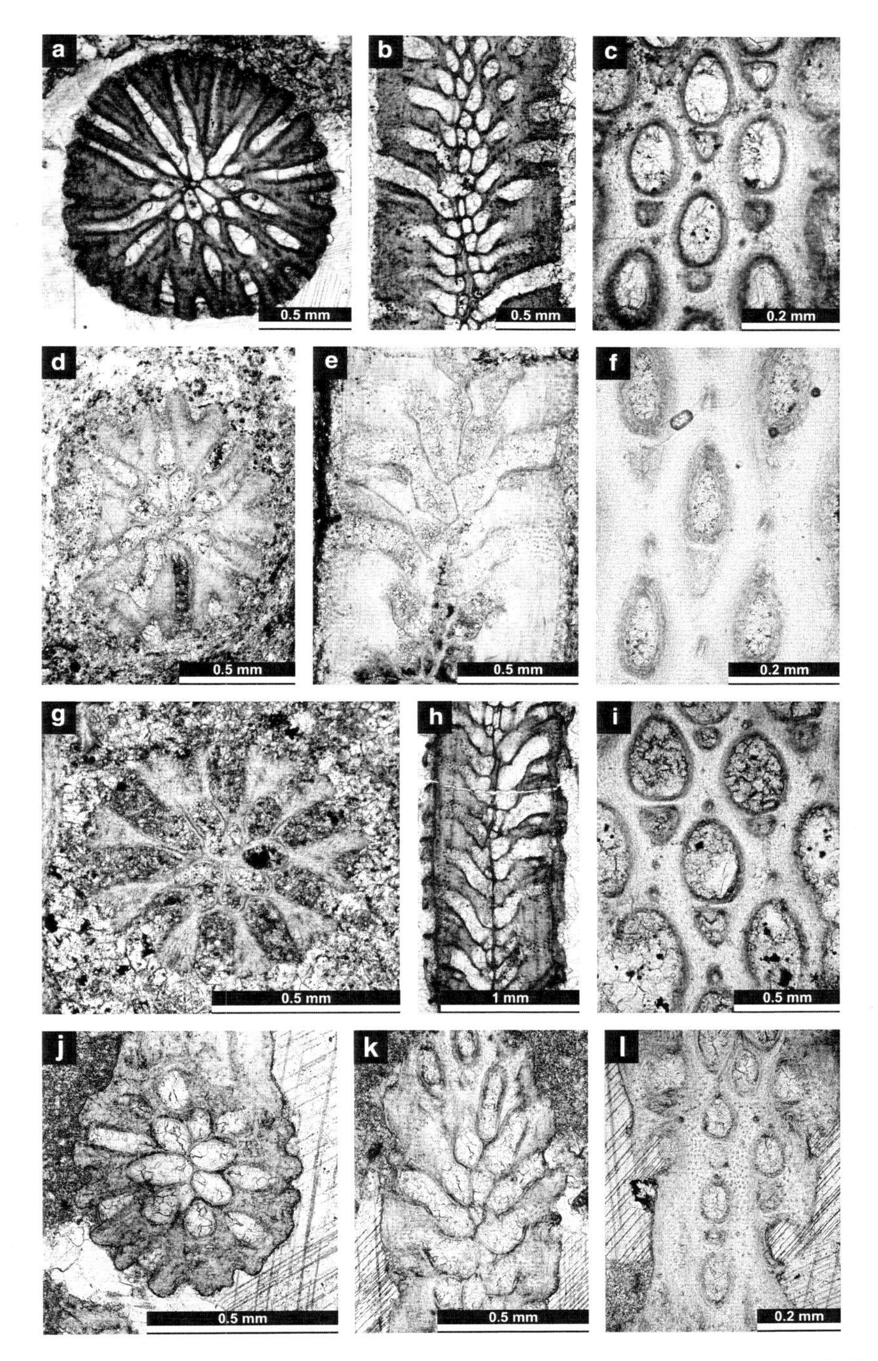


**Fig. 4.2** Species of *Acanthoclema* Hall, 1886 referred to in this study. *Acanthoclema distilum* Bigey, 1988. (**a**) transverse section, SMF 2216; (**b**) longitudinal section, SMF 2212; (**c**) tangential section, SMF 2220. *Acanthoclema elegantula* Ernst, 2008. (**d**) transverse section, SMF 20.142;

## Numerical Statistics

The aim of the statistic analysis was to estimate relations between *Acanthoclema* species from the Devonian of Europe. The following six species were included in the analysis (Table 4.1): *Acanthoclema distilum* Bigey, 1988 (Middle Devonian, Givetian; Germany), *A. elegantula* Ernst, 2008 (Middle Devonian, lowermost Givetian; Rhenish Massif, Germany), *A. triangularis* Ernst and Königshof, 2010 (Middle Devonian, uppermost Givetian; Western Sahara), *A. triangularis* Ernst and Königshof, 2010 (? Middle Devonian, Givetian; SW Spain), *Acanthoclema parvula* Ernst et al., 2011 (Santa Lucía Formation, Lower-Middle Devonian, Emsian-Eifelian; NW Spain), *A. parva* Ernst, 2011 (Lower Devonian, Emsian; NW Spain), *A. irregularis* Ernst, 2011 (Lower Devonian, Emsian; NW Spain), and *Nikiforovella gracilis* Ernst and Herbig, 2010 (Upper Devonian, uppermost Famennian; Rhenish Massif, Germany). The latter species was included in the matrix for testing of the consistence of morphological characters chosen for the analysis and for comparison between *Acanthoclema* and *Nikiforovella*. *Nikiforovella gracilis* occurs in the uppermost Velbert Formation (uppermost Famennian, Upper Devonian) of the Rhenish Massif, Germany, and is one of the earliest known species of this genus. The genus *Nikiforovella* Nekhoroshev, 1948 is proposed here as the closest to *Acanthoclema*, so the idea is to test overlapping between species of the two genera.

Data for the analysis were taken from the respective taxonomic descriptions. Measurements for *Acanthoclema distilum* Bigey, 1988 were taken from Ernst and Schroeder (2007). Statistics for *Acanthoclema triangularis* Ernst and Königshof, 2010 from SW Spain (Ernst and Rodríguez 2010) were measured from the sample SMF 20.802. The following 11 characters were involved in the analysis (presence/absence of mural spines was not considered here): branch width, exozone width, autozooecial aperture width, aperture spacing along branch, aperture spacing across branch (diagonally), acanthostyle diameter, acanthostyle number between successive apertures, metazooecia diameter, metazooecia number between successive apertures, autozooecial budding angle in endozone, autozooecial budding angle in exozone. The measurements are quite heterogeneous, depending on available material and the number of specimens regarded in the statistics. Also, the number of measurements for each character varies strongly. Basically, the arithmetic mean or range was considered here (Table 4.1). The cluster analysis was performed using PAST software (Hammer et al. 2001).

The numerical statistics was carried out by cluster analysis, using an unweighted pair-group average method (UPGMA), in which clusters are joined based on the average distance between all members in the two groups. Distances were measured

---

**Fig. 4.2** (continued) (**e**) longitudinal section, SMF 20.140; (**f**) tangential section, SMF 20.140. *Acanthoclema triangularis* Ernst and Königshof, 2010. (**g**) transverse section, SMF 20.545; (**h**) longitudinal section, SMF 20.511; (**i**) tangential section, 20.551. *Acanthoclema parvula* Ernst et al., 2011. SMF 20.856. (**j**) transverse section; (**k**) longitudinal section; (**l**) tangential section

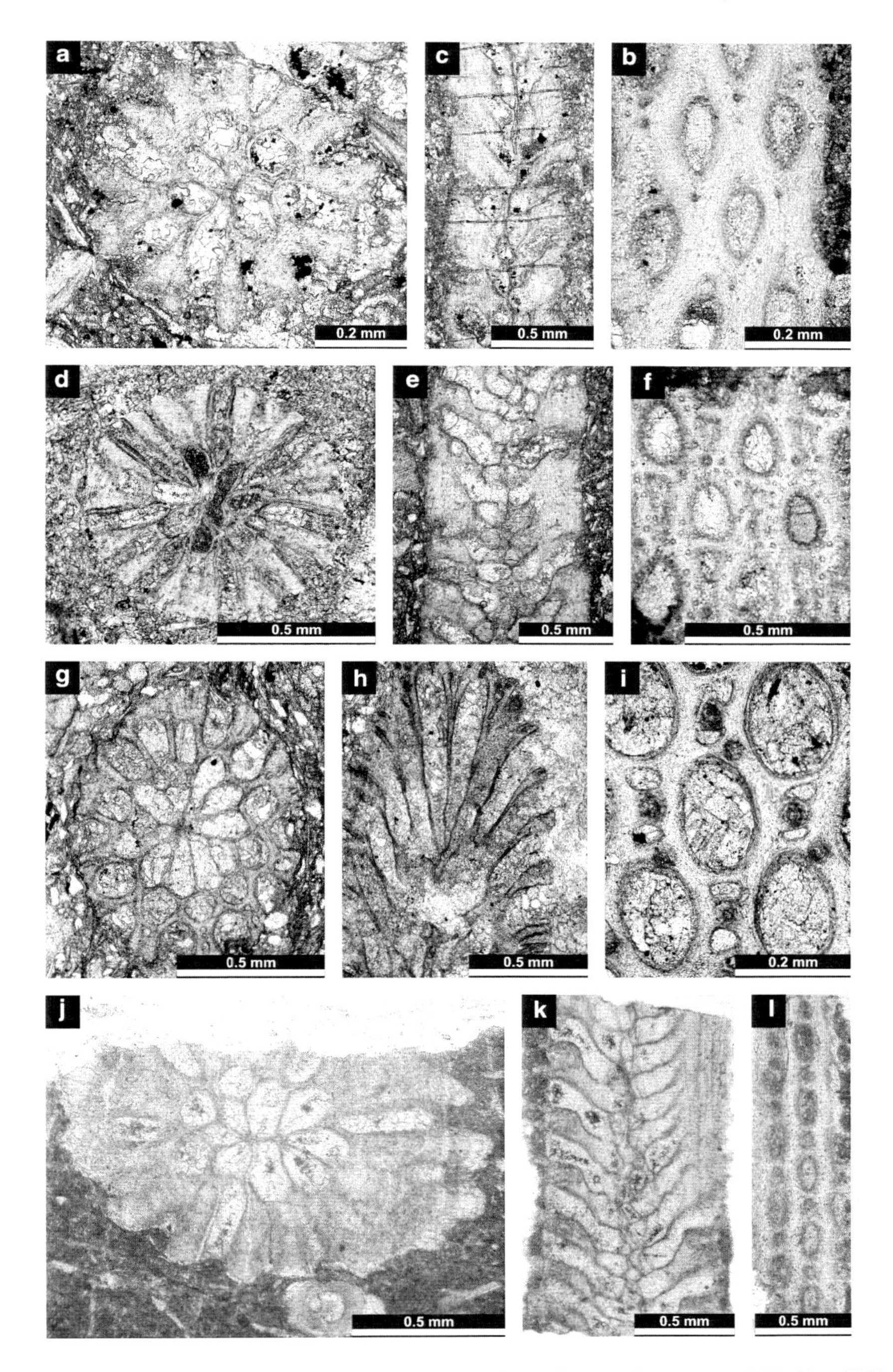


**Fig. 4.3** Species of *Acanthoclema* Hall, 1886, *Nikiforovella gracilis* Ernst and Herbig, 2010, and *Rozanovia unica* Gorjunova, 1992 referred to in this study. *Acanthoclema parva* Ernst, 2011; (**a**) transverse section, RGM 211 5452-b; (**b**) longitudinal section, RGM 211 545j; (**c**) tangential

using Euclidean distance metric and Morisita's index. Euclidean distance is the most common use of distance and examines the root of square differences between coordinates of a pair of objects. Morisita's overlap index is a statistical measure of dispersion used to compare overlap among samples (Morisita 1959). Among the non-binary indices the Morisita index has a major advantage in that it is independent of sample size and diversity (Morisita 1959; Wolda 1981).

## Systematic Palaeontology

Phylum Bryozoa Ehrenberg, 1831

Class Stenolaemata Borg, 1926

Order Cryptostomata Vine, 1884

Suborder Rhabdomesina Astrova and Morozova, 1956

Family Nikiforovellidae Gorjunova, 1975

Genus *Acanthoclema* Hall, 1886

[= *Rozanovia* Gorjunova, 1992]

*Type species*. *Trematopora alternata* Hall, 1883. Onondaga Limestones, Lower-Middle Devonian (upper Emsian to lower Eifelian); New York (USA).

*Emended Diagnosis*. Thin to moderately thick-branched colonies. Endozones and exozones distinctly separated. Autozooecia tubular, growing in spiral pattern from the distinct median axis at angles of 20–60° in endozones, abruptly bending in exozones and intersecting colony surface at angles of 44–90°; having a triangular to rhombic, tear-drop shape in transverse section of endozone. Autozooecial apertures oval, arranged in regular diagonal rows on branches. Basal diaphragms rare to absent. Hemisepta absent. Metazooecia short, positioned at proximal ends of the autozooecia; commonly single metazooecium (rarely 2) arranged between longitudinally successive autozooecial apertures. Acanthostyles usually present, small to moderately large, normally 1–2 arranged regularly between longitudinally successive autozooecial apertures; in some species distributed irregularly between apertures; sometimes absent. Autozooecial walls finely laminated, with dividing hyaline layer in endozone; laminated, without distinct boundaries, protruded by abundant mural spines in exozone. Autozooecial boundary commonly irregular,

---

**Fig. 4.3** (continued) section, holotype RGM 211 545-4b. *Acanthoclema irregularis* Ernst, 2011; (**d**) transverse section, GZG.IN.0.010.529c; (**e**) longitudinal section, RGM 211 536-1-4; (**f**) tangential section, RGM 211 536-1-4. *Nikiforovella gracilis* Ernst and Herbig, 2010; (**g**) transverse section, GIK 2267; (**h**) longitudinal section, GIK 2267; (**i**) tangential section, GIK 2269. *Rozanovia unica* Gorjunova, 1992, holotype PIN 1613/490; (**j**) transverse section; (**k**) longitudinal section; (**l**) tangential section

**Table 4.1** Overview of the measured characters used for analysis

| | A 1 | A 2 | A 3 | A 4 | A 5 | A 6 | A 7 | N 1 |
|---|---|---|---|---|---|---|---|---|
| **WB** | 0.8–2.0 | 0.75–1.17 | 0.63–1.35 | 0.48–0.66 | 0.58–0.80 | 0.53–0.66 | 0.72–1.05 | 0.75–1.05 |
| **ExW** | 0.33–0.72 | 0.23–0.35 | 0.17–0.60 | 0.10–0.24 | 0.14–0.22 | 0.11–0.24 | 0.20–0.21 | 0.40–0.63 |
| **AW** | 0.08 | 0.07 | 0.08 | 0.72 | 0.06 | 0.06 | 0.08 | 0.11 |
| **ADB** | 0.27 | 0.40 | 0.38 | 0.31 | 0.26 | 0.30 | 0.32 | 0.40 |
| **AAB** | 0.18 | 0.22 | 0.21 | 0.19 | 0.17 | 0.13 | 0.19 | 0.24 |
| **AD** | 0.03 | 0.03 | 0.03 | 0.028 | 0.02 | 0.02 | 0.04 | 0.05 |
| **AN** | 1–2 | 1 | 1–2 | 1–2 | 2–6 | 1 | 1–2 | 1–3 |
| **MD** | 0.04 | 0.05 | 0.04 | 0.04 | 0.03 | 0.03 | 0.03 | 0.03 |
| **MN** | 1 | 1 | 1 | 1 | 1 | 1 | 1–2 | 1–5 |
| **ABEND** | 47–60 | 35–60 | 32–42 | 38–39 | 25–38 | 33–39 | 20–35 | 24–36 |
| **ABEX** | 75–90 | 75–90 | 60–86 | 64–74 | 64–66 | 44–55 | 54–62 | 77–81 |

*Species*. *Acanthoclema distilum* (A 1), *A. elegantula* (A 2), *A. triangularis* I, Western Sahara (A 3), *A. triangularis* II, SW Spain (A 4), *A. parvula* (A 5), *A. parva* (A 6), *A. irregularis* (A 7), *Nikiforovella gracilis* (N1)

*Measured characters*. Branch width (WB, mm, range), exozone width (ExW, mm, range), autozooecial aperture width (AW, mm, mean), aperture spacing along branch (ADB, mm, mean), aperture spacing across branch (diagonally) (AAB, mm, mean), acanthostyle diameter (AD, mm, mean), acanthostyles between successive apertures (AN), metazooecia diameter (AD, mm, mean), metazooecia between successive apertures (AN), autozooecial budding angle, endozone (ABEND, range), autozooecial budding angle, exozone (ABEX, range) locally not visible. Mural spines regularly arranged in exozonal wall, opening into autozooecial cavities and on colony surface, varying in density and size.

*Comparison*. *Acanthoclema* Hall, 1886 is similar to *Nikiforovella* Nekhoroshev, 1948, but differs from it in the presence of abundant mural spines in exozonal wall. *Nikiforovella* commonly has more than one metazooecium between longitudinally successive autozooecial apertures. *Acanthoclema* differs from *Streblotrypella* Nikiforova, 1948 in the presence of mural spines in exozonal walls and the regular arrangement of 1–2 metazooecia between successive autozooecial apertures instead of few to abundant metazooecia in *Streblotrypella*.

Gorjunova (1992, p. 126, pl. 10, Fig. 4.3) established the genus *Rozanovia*, which differs from *Acanthoclema* in the absence of acanthostyles. However, the autozooecial morphology and the typical wall structure with mural spines are identical to *Acanthoclema* (Fig. 4.3j–l). These morphologies are consistent in *Acanthoclema*, whereas the number of acanthostyles is variable throughout the genus. Therefore, *Rozanovia* is regarded as the junior synonym of *Acanthoclema*, being at the bottom end of the latter's range in acanthostyle number.

*Stratigraphic and geographic range*. A number of species were previously included in *Acanthoclema* (Newton 1971; Gorjunova 1985). However, only the following species are placed with confidence in this genus: *Acanthoclema parva* Ernst, 2011, *A. irregularis* Ernst, 2011 (Lower Devonian, Emsian; Cantabrian Mountains, NW Spain), *A. parvula* Ernst et al., 2011 (Santa Lucia Formation, Lower-Middle Devonian, Emsian – Eifelian; Cantabrian Mountains, NW Spain), *A. alternatum* (Hall, 1883) (Middle Devonian, Eifelian-Givetian; New York, USA), *A. distilum* Bigey, 1988 (Middle Devonian, Eifelian-Givetian; France, Germany), *A. scutulatum* (Hall, 1883) (Ludlowville Formation, Middle Devonian, Givetian; New York, USA), *A. elegantula* Ernst, 2008 (Ahbach Formation, Middle Devonian, Givetian; Rhenish Massif, Germany), *A. triangularis* Ernst and Königshof, 2010 (Middle Devonian, uppermost Givetian; Southern Morocco, SW Spain), *A. unica* (Gorjunova, 1992) (Upper Devonian, Frasnian; Transcaucasia), *A. junggarensis* Xia, 1997 (Upper Devonian, Famennian; China).

Other species placed in *Acanthoclema* show morphologies different to the generic concept of this genus, or their assignment is doubtful:

*Acanthoclema asperum* (Hall, 1852) from the Silurian of USA and Canada possesses no metazooecia and has abundant acanthostyles as well as hemisepta.

*Acanthoclema divergens* Hall and Simpson, 1887, *A. ovatum* Hall and Simpson, 1887 and *A. sulcatum* Hall and Simpson, 1887 from the Middle Devonian (Eifelian-Givetian) of USA were described without thin sections, so that their assignment to this genus is questionable.

*Acanthoclema triseriale* (Hall, 1883) from the Lower Devonian (Emsian) of USA does not belong to *Acanthoclema* because of the absence of metazooecia and presence of longitudinal rows of tubercles. The internal morphology of this species is unknown.

*Acanthoclema ohioense* McNair, 1937 from the Traverse Group (Middle Devonian, Givetian) of Michigan, USA possesses no metazooecia and has longitudinal rows of tubercles on the colony surface. *Acanthoclema lineatum* McNair, 1937 from the same formation possesses inferior and superior hemisepta as well as longitudinal rows of tubercles and no metazooecia. These species do not belong to *Acanthoclema*, their assignment remains unclear because of insufficient illustrations.

*Acanthoclema bispinulatum* (Hall, 1883) from the Middle Devonian (Givetian) of USA has two acanthostyles between apertures, but no metazooecia. The internal morphology of this species is unknown.

*Acanthoclema cavernosa* Nekhoroshev, 1932 from the Upper Devonian (Famennian) of Germany possesses two large acanthostyles and 2–4 metazooecia, so that it does not belong to the genus *Acanthoclema*. This species belongs most probably to the genus *Nikiforovella*.

*Acanthoclema confluens* (Ulrich, 1888) from the Keokuk Formation (Mississippian) of USA was described only externally. This species seems to lack metazooecia according to available description and illustrations, but has a single acanthostyle between apertures and rows of granules (? = tubercles) on the colony surface. This species apparently does not belong to *Acanthoclema*.

*Acanthoclema carbonarium* Coryell *in* Morgan, 1924 from the Pennsylvanian of USA has no metazooecia, but possesses large single acanthostyle between apertures and tubules (? paurostyles) in the wall. This species belongs apparently to the genus *Rhombopora*.

## Discussion

### *Taxonomy*

The position of *Acanthoclema* within Rhabdomesina is not well understood. Bassler (1953, p. 131) placed *Acanthoclema* in the Family Rhabdomesonidae. He provided an extremely short diagnosis: "Zoarium with filiform axis; apertures in longitudinal series defined by ridges, with a megacanthopore between each pair of apertures". Figure G 91, 3 shows only external characters such as acanthostyles and autozooecial apertures. Bassler (1953) did not mention the presence of metazooecia. Blake (1983, p. 584) presented an extensive diagnosis restricting the genus to forms with a single metazooecium between longitudinally successive autozooecia, and with it placed *Acanthoclema* in the Family Nikiforovellidae Gorjunova, 1975. He illustrated thin sections of two species, *Acanthoclema alternatum* (Hall, 1883) and *A. scutulatum* (Hall and Simpson, 1887). Both species show characteristic mural spines in the skeleton. Gorjunova (1985, p. 115) generally followed the superficial diagnosis of Bassler (1953), and included in *Acanthoclema* species with single acanthostyles and no metazooecia. She placed *Acanthoclema* in the Family Rhomboporidae Simpson, 1895. Later, she followed Blake (Gorjunova, 1996) and placed *Acanthoclema* in the Family Nikiforovellidae Gorjunova, 1975.

Blake (1983) discussed the relations of *Acanthoclema* to *Nikiforovella*, *Streblotrypa*, *Streblotrypella*, and *Cuneatopora* on the basis of the presence of metazooecia, styles, and median axes. He mentioned only *Nikiforovella* occasionally having regularly arranged mural spines in exozonal walls. The species *Nikiforovella gracilis* Ernst and Herbig, 2010 from the uppermost Devonian of Germany possesses no such mural spines (Fig. 4.2g–i).

Investigation of the Devonian bryozoan faunas from Europe revealed several species, which share the following morphologic characters typical for *Acanthoclema* (Figs. 4.2 and 4.3): regularly arranged mural spines in exozonal wall, few metazooecia (normally one) and few acanthostyles, autozooecial shape and orientation (Table 4.1). Autozooecia in *Acanthoclema* have a tear-drop shape in transverse section at median axis, and they bend distinctly at the transition to exozone. One species (*A. irregularis* Ernst, 2011) possesses locally two metazooecia instead of one between autozooecial apertures. This is the only contradiction with the diagnosis of Blake (1983), and is regarded here as being within the normal range of the genus.

## *Numerical Comparison*

The computed dendrograms (Fig. 4.4) show that the Euclidean distance method is not able to separate *Nikiforovella gracilis* from the set of *Acanthoclema* species. Contrary, Morisita's index was able to recognize the dissimilarity between *Nikiforovella* and *Acanthoclema* species on the base of measured characters. Both methods could recognise a *triangularis* I + II cluster and *parvula-irregularis* cluster. *A. elegantula* and *A. distilum* are paired together in the Euclidean distance dendrogram, whereas in Morisita's index dendrogram *A. elegantula* is paired together with *A. triangularis* I + II cluster. *A. distilum* is paired with *A. parva* in Morisita's index dendrogram.

Interestingly, both methods place *Nikiforovella gracilis* near *Acanthoclema parva*, and in the Euclidean distance dendrogram this species is placed between *A. parva* and *A. irregularis*, both the apparently oldest species of the genus. These results show that early *Acanthoclema* species are similar to the early *Nikiforovella* species, even though there is a time span of about 35 Ma between them.

The analysis of the distribution of morphological characters within *Acanthoclema* reveals a distinct trend: The number of metazooecia and acanthostyles decreases during the evolution of the genus. Acanthostyles disappear completely in *A. unica*, one of the latest species of *Acanthoclema* (Frasnian). General trend in *Acanthoclema* is an increasing regularity of arrangement of autozooecia, metazooecia and acanthostyles. Mural spines are present in *Acanthoclema* species consistently.

To the contrary, the evolutionary trend in *Nikiforovella* was the increasing number of acanthostyles and metazooecia, as well as the increasing disorder in their arrangement and the loss (mainly) of mural spines. Earlier *Nikiforovella* species (Upper Devonian to Lower Carboniferous) have few and regularly arranged acanthostyles

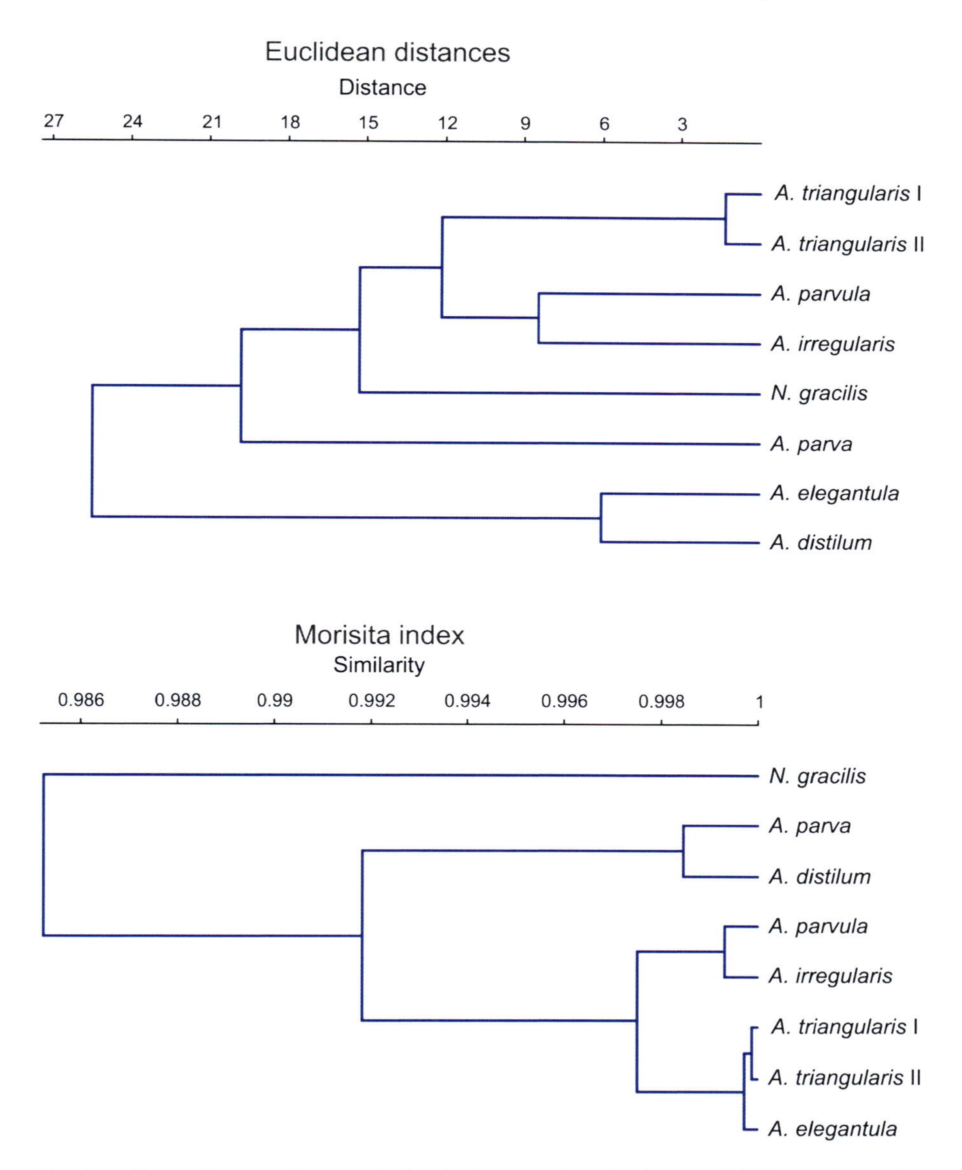


**Fig. 4.4** Cluster diagrams showing similarities between *Acanthoclema* and *Nikiforovella* species from the Devonian of Europe (UPGMA, Unweighted Pair Group Method with Arithmetic Mean, Euclidean distances metric and Morisita's index, PAST ver. 2.01, Hammer et al. 2001)

and metazooecia. Later species of this genus (Upper Carboniferous to Lower Permian) have abundant acanthostyles and metazooecia which are arranged irregularly.

It can be assumed that *Nikiforovella* diverged from *Acanthoclema* in the Lower Devonian, where one of the early *Acanthoclema* species was an ancestor of

*Nikiforovella* (? *A. irregularis*). The earliest *Nikiforovella* species are known from the Famennian. It should be expected that one or more species of *Nikiforovella* can be discovered in the time span between Emsian and Famennian (ca 35 Ma). Mural spines in *Nikiforovella* represent certainly an atavistic feature inherited from *Acanthoclema*, which appeared sporadically in some species.

## Conclusions

Considering the morphology of *Nikiforovella* and *Acanthoclema* and the results of the cluster analyses (Euclidean distance and Morisita's index) the following conclusions can be made:

1. Statistical and morphological analyses support the idea that *Nikiforovella* derived from *Acanthoclema*, apparently in the Lower Devonian.
2. Evolutionary trends in *Acanthoclema* were the decreasing number of metazooecia and acanthostyles and increasing regularity of arrangement of autozooecia, metazooecia and acanthostyles.
3. To the contrary, evolutionary trends in *Nikiforovella* were the increasing number of acanthostyles and metazooecia and increasing disorder in their arrangement, as well as the loss of mural spines.
4. The Morisita's index method delivers reliable results based on heterogeneous data available, whereas the Euclidean distance metric reveals some weaknesses in discriminating the species.

**Acknowledgments** The present study is part of the project ER 278/4-1 and 2 supported by the Deutsche Forschungsgemeinschaft (DFG), and is a contribution to IGCP 499 "Devonian land-sea interaction: evolution of ecosystems and climate". Priska Schäfer, Kiel, is thanked for reading of the manuscript and helpful comments. Roger Cuffey, Pennsylvania, and Patrick Wyse Jackson, Dublin, are thanked for their helpful reviews.

## References

Anstey RL, Pachut JF (2004) Cladistic and phenetic recognition of species in the Ordovician bryozoan genus *Peronopora*. J Paleontol 78(4):651–674

Anstey RL, Perry T (1970) Biometric procedures in taxonomic studies of Paleozoic bryozoans. J Paleontol 44(2):383–398

Astrova GG, Morozova IP (1956) About systematics of the order Cryptostomata. Dokl Akad Nauk 110(4):661–664, (In Russian)

Bassler RS (1953) Bryozoa. In: Moore RC (ed) Treatise on invertebrate paleontology, G1-G253. Geological Society of America/University of Kansas Press, Lawrence

Bigey FP (1988) Bryozoaires du Givetien et du Frasnien de Ferques (Boulonnais, France). In: Brice D (ed) Paleontologie, sedimentologie, stratigraphie, tectonique, vol 7, Biostratigraphy Paleoz. Le Devonien de Ferques, Bas-Boulonnais (N. France), pp 297–323

Blake DB (1983) Introduction to the suborder Rhabdomesina. Systematic descriptions for the Suborder Rhabdomesina. In: Robison RA (ed) Treatise on invertebrate paleontology, part G (1): Bryozoa (revised). Geological Society of America/University of Kansas, Lawrence, pp G530–G592

Borg F (1926) Studies on Recent cyclostomatous Bryozoa. Zool Bidr Uppsala 10:181–507

Coryell HN (1924) Bryozoa. In: Morgan GD (ed) Geology of the Stonewall Quadrangle, Oklahoma. Bureau of Geology, Norman, Oklahoma Bur Geol Bull 2:176–184

Ehrenberg CG (1831) Symbolae Physicae, seu Icones et descptiones Corporum Naturalium novorum aut minus cognitorum, quae ex itineribus per Libyam, Aegiptum, Nubiam, Dongalaam, Syriam, Arabiam et Habessiniam, studia annis 1820–25, redirent. Pars Zoologica 4, Animalia Evertebrata exclusis Insectis. Berolini, 10 pls Mittler

Ernst A (2008) Non-fenestrate bryozoans from the Middle Devonian of the Eifel (western Rhenish Massif, Germany). Neues Jb Geol Paläontol Abh 250(3):313–379

Ernst A (2011) Cryptostome (ptilodictyine and rhabdomesine) Bryozoa from the Lower Devonian of NW Spain. Palaeontol A 293(4–6):1–37

Ernst A, Herbig H-G (2010) Stenolaemate bryozoans from the Late Devonian (Famennian) of SW Germany. Geol Belg 13(3):173–182

Ernst A, Königshof P (2010) Bryozoan fauna and microfacies from a Middle Devonian reef complex (Western Sahara, Morocco). Abh Senckenberg Nat Ges 568:1–91

Ernst A, Rodríguez S (2010) Bryozoan fauna from the oolitic limestone near Pajarejos, SW Spain. Rev Esp Paleontol 25(2):81–86

Ernst A, Schroeder S (2007) Stenolaemate bryozoans from the Middle Devonian of the Rhenish Slate Massif (Eifel, Germany). Neues Jb Geol Paläontol Abh 246(2):205–233

Ernst A, Dorsch T, Keller M (2011) A bryozoan fauna from the Santa Lucia Formation (Lower – Middle Devonian) of Abelgas, Cantabrian Mountains, NW-Spain. Fazies 57:301–329

Gorjunova RV (1975) Permskie mshanki Pamira [Permian bryozoans of the Pamir]. Tr Paleontol Inst Akad Nauk SSSR 148:1–127, (In Russian)

Gorjunova RV (1985) Morphology, system und phylogeny of Bryozoa (Order Rhabdomesida). Tr Paleontol Inst Akad Nauk SSSR 208:1–152, (In Russian)

Gorjunova RV (1992) Morphology and system of the Paleozoic bryozoans. Tr Paleontol Inst Akad Nauk SSSR 251:1–168, (In Russian)

Gorjunova RV (1996) Phylogeny of the Paleozoic Bryozoa. Tr Paleontol Inst Ross Akad Nauk 267:1–161

Hageman SJ (1991) Discrete morphotaxa from a Mississippian fenestrate faunule: presence and implications. In: Bigey FP (ed) Bryozoaires actuels et fossil: Bryozoa living and fossil, Bull Soc Sci Nat l'Quest Fr Mèm HS 1:147–150

Hall J (1852) Organic remains of the lower middle division of the New York system. Natural history of New York, Part 6. Palaeontol New York 11:viii + 1–363

Hall J (1883) Bryozoans of the Upper Heldelberg and Hamilton Groups. Trans Albany Inst 10:145–197

Hall J (1886) Bryozoa of the Upper Heldelberg groups; plates and explanations. State Geologist of New York, 5th annual report, 1885, pp 25–53

Hall J, Simpson GB (1887) Corals and Bryozoa from the Lower Heldelberg, Upper Heldelberg and Hamilton Groups. New York State Geological Survey, vol 6. pp 98, 265–288

Hammer Ø, Harper DAT, Ryan PD (2001) PAST: paleontological statistics software package for education and data analysis. Palaeontol Electron 4(1):1–9

McNair AH (1937) Cryptostomatous Bryozoa from the Middle Devonian Traverse Group of Michigan, vol 5. Museum Paleontology, University of Michigan, Ann Arbor pp 103–170

Morisita M (1959) Measuring of the dispersion and analysis of distribution patterns. Mem Fac Sci Kyushu Univ Ser E Biol 2:215–235

Nekhoroshev VP (1932) Die Bryozoen des deutschen Unterkarbons. Abh Preuss Geol Landesanst (NF) 141:1–74

Nekhoroshev VP (1948) Devonian Bryozoa of the Altai. Paleontol USSR 3(2):1–172, (In Russian)

Newton GB (1971) Rhabdomesid bryozoans of the Wreford megacyclothem (Wolfcampian, Permian) of Nebraska, Kansas, and Oklahoma. Univ Kansas Paleontol Contr 56:1–71

Nikiforova AI (1948) Lower Carboniferous bryozoans of Karatau. Izd Akad Nauk Kazakhskoi SSR, Alma-Ata, p 53, (In Russian)

Simpson GB (1895) A handbook of the genera of the North American Paleozoic Bryozoa. Fourteenth annual report of the State Geologist of New York for the year 1894, pp 407–608

Snyder EM (1991a) Revised taxonomic procedures and paleoecological applications for some North American Mississippian Fenestellidae and Polyporidae (Bryozoa). Palaeontol Am 57:1–272

Snyder EM (1991b) Revised taxonomic approach to acanthocladiid Bryozoa. In: Bigey FP (ed) Bryozoaires actuels et fossil: Bryozoa living and fossil. Bull Soc Sci Nat l´Quest Fr Mèm HS 1:431–445

Ulrich EO (1888) Waverly Bryozoa. Bull Lab Denison Univ 4:62–96

Vine GR (1884) Fourth report of the Committee consisting of Dr. H. R. Sorby and Mr. G. R. Vine, appointed for the purpose of reporting on fossil Polyzoa. Report of the 53rd meeting of the British association for the advancement of sciene, pp 161–209

Wolda H (1981) Similarity indices, sample size and diversity. Oecology 50:296–302

Xia F (1997) Marine microfaunas (bryozoans, conodonts, and microvertebrate remains) from the Frasnian-Famennian interval in Northwestern Junggar Basin of Xinjiang in China. Beitr Paläontol 22:91–207

# Chapter 5
# Growth Rates, Age Determination, and Calcification Levels in *Flustra foliacea* (L.) (Bryozoa: Cheilostomata): Preliminary Assessment

## Morphology, Growth and Calcification Levels *in Flustra foliacea* (L.)

**Helena Fortunato, Priska Schäfer, and Heidi Blaschek**

**Abstract** Potential consequences for species distribution, abundances and diversity and their imprint in food chains and ecosystems call for more studies of the short and long term impacts of ocean acidification. Bryozoans have been overlooked in this respect even though they play an important role in benthic temperate ecosystems. *Flustra foliacea* colonies from the North and Baltic Seas were used to assess morphology, growth rates, wall structure and preservation aiming to build up a baseline to use this species as a 'sentinel' of acidification levels. Though no significant differences in mean zooid size among the studied basins were found, North Sea colonies show periodic oscillations across generations in mean frontal area index and zooid density. Preliminary geochemistry analyzes show: (1) similar carbon contents (TC, TIC, TOC) in both basins; (2) skeletal walls composed of IMC; (3) over 50% weight loss in dissolution experiments during the first hour. A winter growth stop marked by growth-check lines is postulated. In order to obtain calibrated results, we need experimental data, and moreover, access to collections done over the past 200 years.

**Keywords** Climate change • North Sea • Baltic Sea • *Flustra foliacea* • Morphology • Geochemistry

H. Fortunato (✉) • P. Schäfer • H. Blaschek
Institut für Geowissenschaften, Christian-Albrechts-Universität, Ludewig-Meyn-Str. 10, D-24118 Kiel, Germany
e-mail: fortunatomh@hotmail.com; hf@gpi.uni-kiel.de

A. Ernst, P. Schäfer, and J. Scholz (eds.), *Bryozoan Studies 2010*,
Lecture Notes in Earth System Sciences 143, DOI 10.1007/978-3-642-16411-8_5,

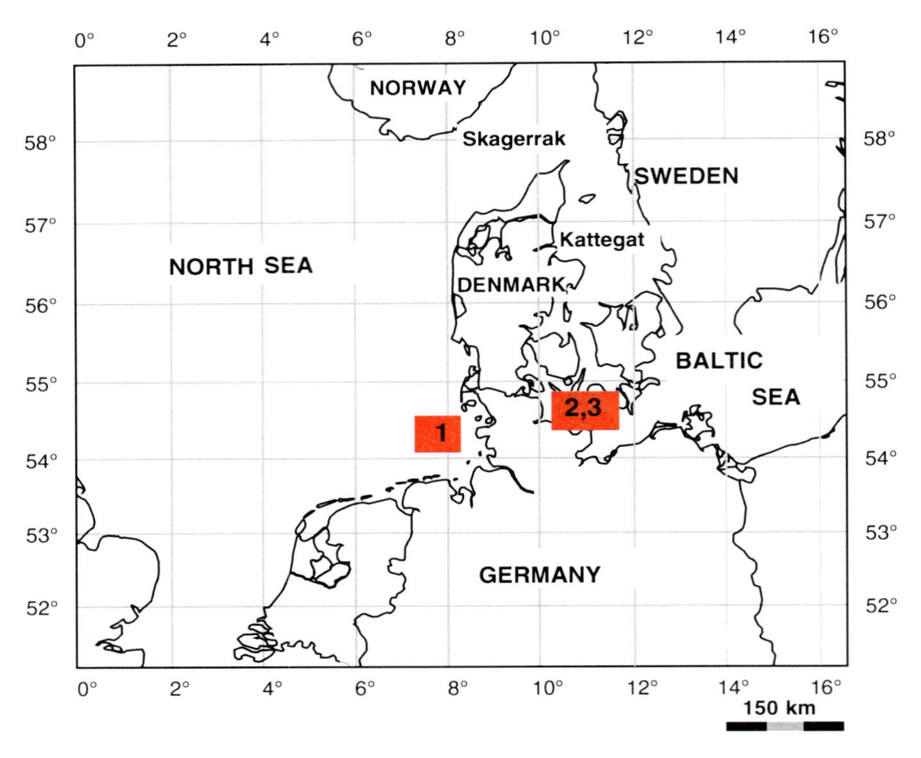


**Fig. 5.1** Map showing the study area. 1 – Helgoland (North Sea, Germany); 2 – Bredgrund (Baltic Sea Denmark); 3 – Langeland (Baltic Sea, Denmark)

## Introduction

The current reduction of oceanic pH, already out of the range values known for the past several thousand years and most probably for the past few million years, is affecting the saturation state of seawater with respect to calcite (Ω calcite) and aragonite (Ω aragonite) with severe consequences for marine ecosystems in general (Royal Society 2005). Evidence from the southern ocean shows the shoaling of the lysocline and its imprint in the skeletal elements while travelling up the slope and the shelf (Royal Society 2005; McNeil and Matear 2008). Although the same can be expected in the temperate northern seas, we still lack the appropriate data to properly assess it (Orr et al. 2005; Royal Society 2005; Howard and Tillbrook 2008), thus the need to collect information from these areas to assess the present day situation and plan for the future.

Though a small sea on a global scale the Baltic Sea (Fig. 5.1) is ecologically unique and highly sensitive to environmental impacts. With a very limited water exchange with the North Sea through the Belts and the Kattegat, its water can remain in the basin for up to 30 years. The infrequent inflows of North Sea water are extremely important to the Baltic ecosystem because of the oxygen they transport into the Baltic deeps. They used to happen on average every 4–5 years until the 1980s but became less frequent in recent decades. The latest three occurred in 1983, 1993 and 2003 suggesting a new inter-inflow period of about 10 years. Baltic surface salinity varies from about 20 psu (parts per 1,000) in the Kattegat to 1–2 psu in the northernmost Bothnian Bay and the northernmost Gulf of Finland. Salinity also varies with depth and the halocline lies at around 60–80 m. Bottom oxygen contents are often at a deficit. Carbonate chemistry measurements show that $CaCO_3$ under saturation already occurs in the central Baltic especially during winter with respect to both calcite and aragonite. The situation is even worse in the Bothnian Sea and bay. This under saturation state is documented by the scarcity of coccolithophores in the area (Tyrell et al. 2008). Low and highly fluctuating pH values in bottom waters are characteristic of several other areas as well (M. Wahl, 2010).

The North Sea (Fig. 5.1) is mostly a shallow basin lying along the European continental shelf with a mean depth of 90 m. The exception is the Norwegian trench parallel to the Norwegian shoreline where depths are slightly over 700 m. It is also the drainage basin of most European continental watersheds including the Baltic Sea. Temperature gradients display important seasonal patterns varying between 17°C in summer and 6°C in winter. Salinity ranges from 28 to 33 psu near the coast, and averages 34–35 psu in open waters. Both parameters are often distributed in a gradient perpendicular to the coast. Though more saturated, a decrease of 0.1 in pH level over the next 50 years is expected for the southern North Sea, which would account for a total acidification of 0.5pH units below pre-industrial levels at atmospheric $CO_2$ concentrations of 1,000 ppm (Blackford and Gilbert 2007; Blackford et al. 2008).

Experiments on marine organisms that build carbonate skeletons have shown the impact of acidification (Kleypas et al. 2006; Langdon and Atkinson 2005; Orr et al. 2005; Riebesell et al. 2000, 2007; IPCC 2007). However, most laboratory studies used short-term exposures raising the question of what will happen when the same organisms are exposed to the same conditions during years and decades. Besides, the focus has been on plankton organisms and tropical corals (acidification is expected to happen mostly in surface waters; corals are valuable habitat providers). Data concerning the shoaling of the lysocline call for more studies of benthic systems. The few long-term studies at community level show shifts in benthic communities without any indication of adaptation or replacement of indicator species (Riebesell 2008; Hall-Spencer et al. 2008).

Although underused in this context, the phylum Bryozoa may have high potential in monitoring the effects of ocean acidification. In northern temperate latitudes, bryozoans have been studied mostly under the point of view of skeletal mineralogy, growth rates, and carbonate production as well as monitoring of seasonal changes in temperature and primary productivity (Bader 2001; Bader and Schäfer 2005; Schäfer and Bader 2008). Bryozoans can grow skeletons composed exclusively of calcite, aragonite or a combination of both. They exhibit a relatively broad range of $MgCO_3$ (from 0 to 14 wt%), which is partly phylogenetically controlled

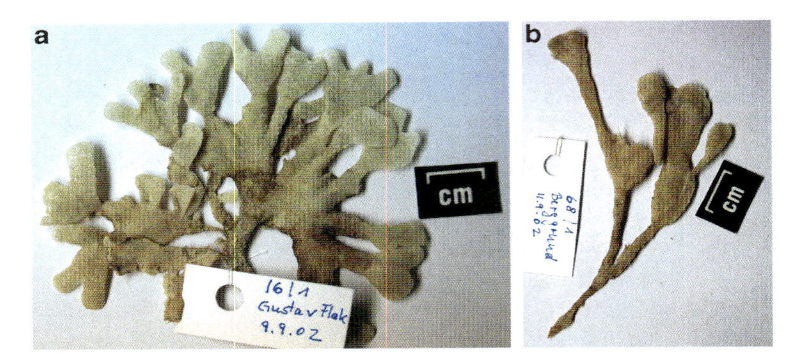


**Fig. 5.2** *Flustra foliacea* (L.) pictures. (**a**) colony from Baltic Sea (Bredgrund). (**b**) colony from North Sea (Helgoland). Scale = 1 cm

(Smith et al. 2006; Taylor et al. 2009). Studies on dead skeletons show that low pH effects depend on their mineralic composition (LMG-/HMG-calcite, aragonite). Different species will manifest the impact of acidification at different times and in different ways depending, among others of their mineralic composition. Thus, some particularly vulnerable species may act as "sentinels" providing an early warning for shelf communities affected by acidification (Smith 2009). This seems most likely for bryozoan species with either intermediate or bi-mineralic skeletons many of which form thickets on the seafloor and are important ecologic constituents and carbonate producers in mid-latitude north- and south-hemisphere shelf communities.

The bryozoan *Flustra foliacea* (L.) (Fig. 5.2) forms dense thickets ecologically structuring the sea floor and providing habitat to a rich epifauna (Stebbing 1971a; Bitschofsky 2012, this volume). It is widespread on gravel grounds in the North Sea but rare in many areas of the Baltic Sea. North Sea and Baltic populations show considerable morphologic variability (i.e. predation by nudibranchs in the North Sea colonies leads to the development of well-formed spines, which are mostly absent from Baltic colonies) (Harvell 1984a, b, 1992). In spite of this, preliminary data shows that only one base pair separates North Sea and Baltic populations (Schäfer P.S. and Nikulina E.N., 2004, unpublished data). This species is also an important producer of biologically active metabolites such as Flustramine E from North Sea colonies, which interferes with the development of *Botrytis cinerea* (de Bary) and *Rhizotonia solani* (J.G. Kühn) thus being a potential subject for medical studies (Lysek et al. 2002; Liebezeit 2005; Sala et al. 2005; Sharp et al. 2007).

Growth bands can be easily discriminated in many organisms providing a reliable tool to calculate growth rates. These data have been used to infer relationships between growth and environmental parameters such as temperature and productivity levels (Bader and Schäfer 2005; Okamura 1987; Amui-Vedel et al. 2007). *Flustra foliacea* colonies form growth-check lines more pronounced in the colony younger (upper) parts and easily observable with naked eye and under an optical microscope. Using this, Stebbing (1971b) studied monthly growth rates in

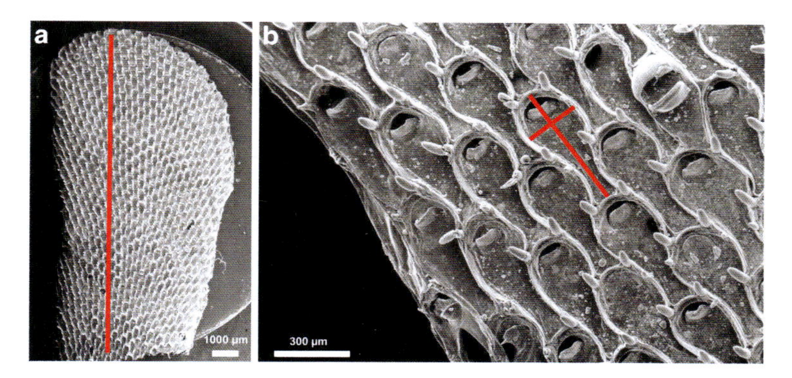


**Fig. 5.3** SEM pictures of studied colonies. (**a**) Profile across generations from growing tip to base; (**b**) Close up of zooid showing max width and length measurements

material collected from South Wales finding a linear growth pattern in height and similar growth rates at all ages. On the other hand, Menon (1975) found a direct relationship between temperature and growth rates and an inverse relationship with zooecia average size. Such findings on seasonal growth cycles were later used to infer environmental regimes, especially temperature changes, as documented by zooid size profiling in perennial bryozoan skeletons (O'Dea and Okamura 1999, 2000a, 2000b; O'Dea 2005).

This study is aimed to calibrate growth rates and calcification levels to further assess effects of increasing ocean pH on mineralization rates in the Baltic and North Sea basins. Colonies from the North and Baltic Seas were studied to assess growth rates, wall structure, and wall preservation in order to build up a baseline to use this species as an environmental indicator of acidification levels.

## Material and Methods

### *Morphology*

Material for the study was collected in two consecutive years approximately in the same season – May 2002 (North Sea) and July 2003 (Baltic Sea). Several branches from five colonies from each basin were used (Fig. 5.2a, b). In branches used for SEM, and prior to preparation, all smaller side branches were removed in such a way that the resulting branch represents a continuous growing unit. These resulting branches were then cut, the pieces mounted in numbered stubs, sputter coated with gold-palladium and examined using a CAMSCAN Series-2-CS-44. All biometric data was collected in mounted SEM photographs with the same scale (300 µm). A profile was then traced from the growing tip of the branch and moving downwards towards its base following the sequential generations of budding zooids (Fig. 5.3a). Maximum zooid length and width were measured for three zooids in each sequential generation along the profile (Fig. 5.3b). Length and width were used to obtain an index of zooid frontal area (length × width) and zooid shape (length/width) (O'Dea and Jackson 2002). Zooid density, collected every 4.6 mm intervals along profiles starting from the growing tip, was measured as the number of zooids in an area of 0.88 $mm^2$.

The map shown in Fig. 5.1 is drawn using the free program http://www.pangaea.de.Software/PanMap/.

## *Geochemistry*

Material from eight dried colonies (six from the North Sea and two from the Baltic Sea) were used for geochemistry analyzes. Total, organic and inorganic carbon as well as carbonate contents was analyzed with a Coulomat 702 (Carlon-Erba method) using dried material. In addition, muffle oven burning of dried colonies was used to analyze the relation between organics and carbonate material as a proxy for calcification levels.

X-Ray diffraction analysis (XRD) of the skeleton geochemical composition ($MgCO_3$ vs. $CaCO_3$) was done on individual segments of *F. foliacea* and compared with that of *Hornera lichenoides* (L.) from the Greenland Sea, and *Cellaria sinuosa* (Hassall) from the English Channel using a Phillips PW1710 (Cu K$\alpha$ 40 kV, 0.002° 2$\theta$/s) in order to observe Mg peaks. Quartz (d = 3.343Å) served as internal standard. Contents were measured as weight% $MgCO_3$ and weight% $CaCO_3$. The comparative material (i.e. *C. sinuosa* and *H. lichenoides*) was chosen arbitrarily mostly because the first is a bi-mineralic species (Schäfer and Bader 2008) and the second a mono-mineralic one (Kuklinski and Taylor 2009, own data).

For dissolution experiments, dried colony fragments from *F. foliacea*, *C. sinuosa* and *H. lichenoides* were weighted and then bathed in 1% acetic acid for 5, 10, 15, 20, 30, 45, 60, 75, 90, 120, 150 and 180 min. After each bath, specimens were dried and weighted. Acetic acid is commonly used in this type of studies (Smith et al. 1992). Further work will include several other acids in order to have some proxies of dissolution levels related to actual ocean acidification.

# Results

## *Morphology*

Preliminary data show no significant differences in mean zooid size (maximum length and width) as well as in zooid frontal area in both basins (Table 5.1). Zooid density is also very similar in both areas. Figures 5.4 and 5.5 show plots of the mean zooecia frontal index and mean zooid density along the profiles. A periodic

**Table 5.1** Descriptive statistics for *Flustra foliacea* for North and Baltic seas

| | Max length (mm) | | | Max width (mm) | | | Zooid frontal area (mm) | | | Zooid density (zooid/mm$^2$) | | |
|---|---|---|---|---|---|---|---|---|---|---|---|---|
| | Mean | Stdev | Median | Mean | Stdev | Median | Mean | Stdev | Median | Mean | Stdev | Median |
| Baltic Sea | 0.470 | 0.032 | 0.468 | 0.266 | 0.026 | 0.266 | 0.125 | 0.016 | 0.125 | 11.03 | 1.074 | 11 |
| North Sea | 0.498 | 0.044 | 0.493 | 0.255 | 0.026 | 0.261 | 0.127 | 0.017 | 0.127 | 11.46 | 1.818 | 12 |

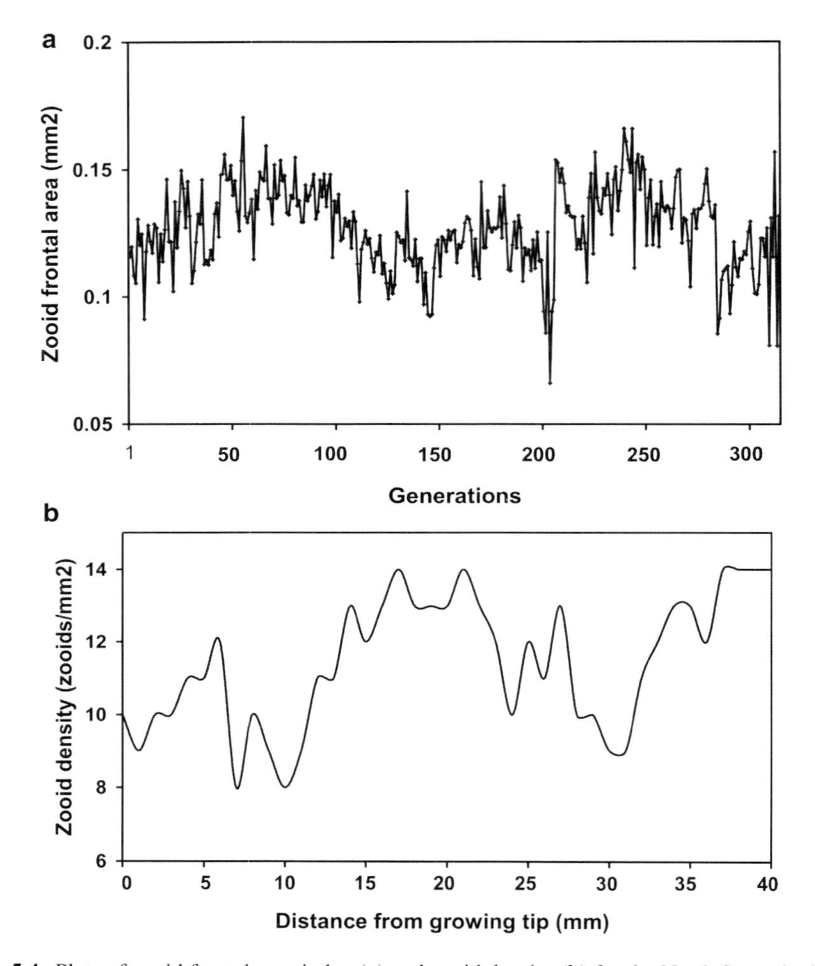


**Fig. 5.4** Plots of zooid frontal area index (**a**) and zooid density (**b**) for the North Sea colonies

oscillation across generations can be noticed especially in the North Sea material (Fig. 5.4a). This oscillation in zooecia frontal index in the North Sea material is somewhat mirrored in the plot of zooid density (Fig. 5.4b). Although also present in the Baltic material (Fig. 5.5a, b), these oscillations are less clear and lack the regularity of ones observed in the North Sea. Mean size is 5.7 cm for the Baltic Sea colonies and 6.3 cm for the North Sea ones. Due to high rates of sediment accumulation and epizootic coverage, measurements of zooid density and zooecia width and length could not be done in the basal region of the studied branches (about 1–1.5 cm length). Thus, data plotted in Figs. 5.4 and 5.5 cover only the top 300–350 generations (about 5–6 cm) where zooids were well visible and could be measured.

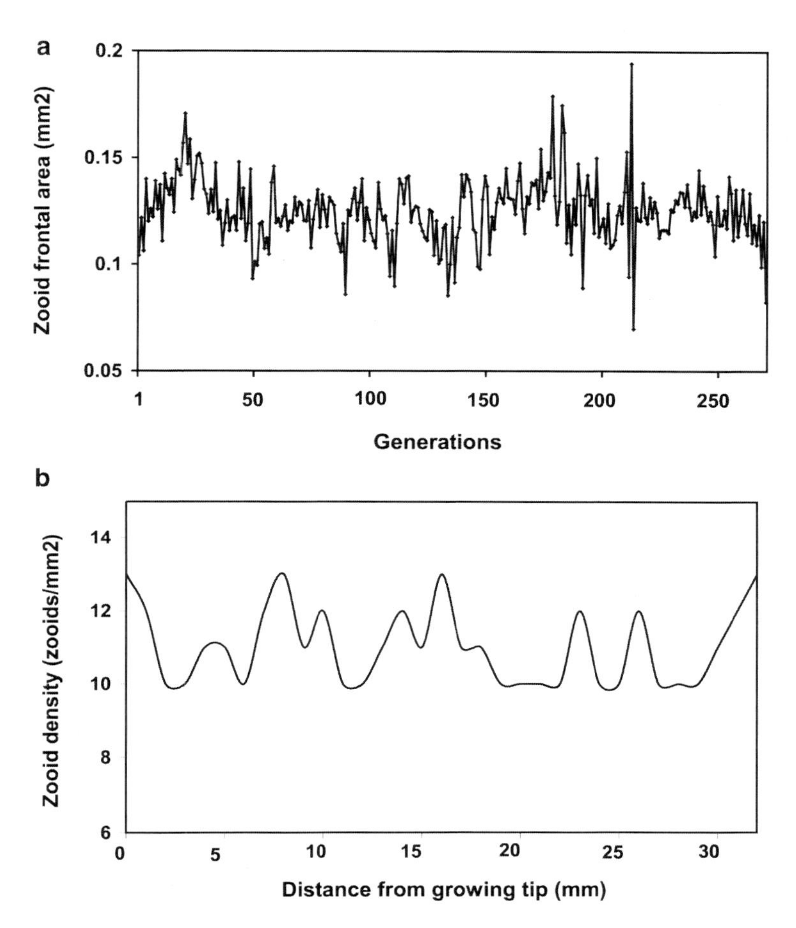


**Fig. 5.5** Plots of zooid frontal area index (**a**) and zooid density (**b**) for the Baltic Sea colonies

## *Geochemistry*

Results from chromatography experiments are shown in Fig. 5.6. Data show no significant differences in total (TC), organic (TOC) and inorganic (TIC) carbon contents between Baltic and North seas. Carbonate contents (wt% of $CaCO_3$) are also similar although a higher dispersal among the samples can be noticed with an exceptional low value for one of the two Baltic samples (sample 7).

Muffle oven experiments show a smaller weight loss in *Flustra* samples from the Baltic (38.3%) than in those from the North Sea (49.8%).

X-ray diffraction indicates that the skeletal walls of *Flustra* are formed of IMC (intermediate magnesium calcite) whereas *C. sinuosa* shows both intermediate and high magnesium calcite (HMC) and *H. lichenoides* has only low magnesium calcite (LMC).

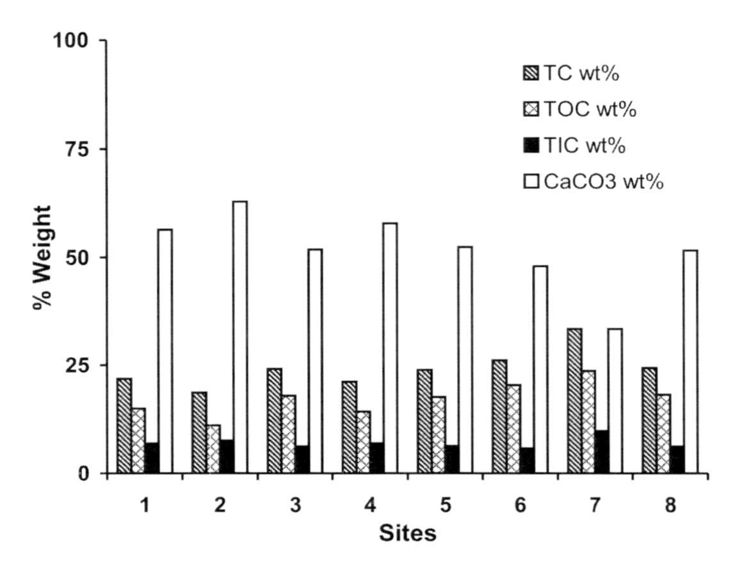


**Fig. 5.6** Plot showing total, organic and inorganic carbon and carbonate contents in North and Baltic seas' colonies. Sites 1–6 are located in the North Sea; 7–8 are in the Baltic Sea

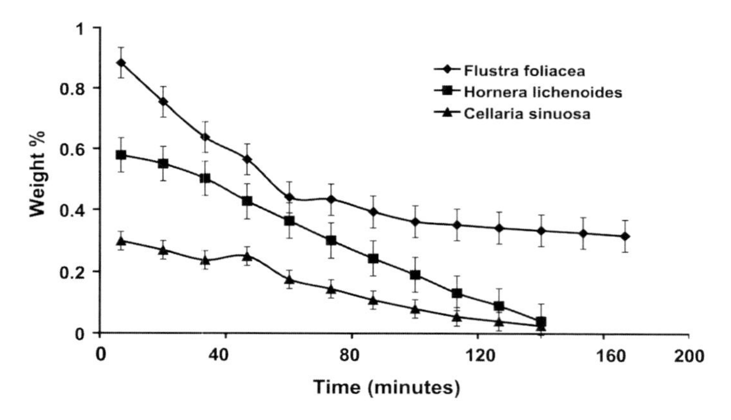


**Fig. 5.7** Plot showing carbonate dissolution rates in three bryozoan species: *Flustra foliacea* (L.), *Hornera lichenoides* (L.), *Cellaria sinuosa* (Hassall). Small bars represent error bars

In dissolution experiments, *Flustra* shows a rapid loss of calcite (Fig. 5.7). Indeed, 59.1% of the original weight is lost during the first 60 min. After 180 min, most carbonate is dissolved. The remaining 0.317 g (38.8% of the original weight) are mostly organics and are left as an insoluble residue. In both *H. lichenoides* and *C. sinuosa* dissolution takes place more gradually. Both species reach the dissolution peak only after 120 min. The remaining insoluble residue is 0.040 g (*C. sinuosa*) and 0.025 g (*H. lichenoides*).

## Discussion and Conclusions

This study confirms the existence of a relationship between morphological parameters and environmental conditions such as temperature. It is difficult to compare obtained growth rates with data reported in the literature because almost every worker uses its own method and different parameters. For example, Eggleston (1972) reports that growth-check lines in colonies from the Isle of Man were 2–3 cm apart, whereas Menon (1975) used heights, having found an average height of 2.12 cm in Helgoland colonies after 2 years, and 7.93 cm after 8 years. Such data show that colonies grow faster during the first couple of years (about 1.05 cm/year), slowing down afterwards, which could be due to the lateral growth of the fronds. Stebbing (1971b) reported a yearly increment of 1.68 cm in frond growth for the Gower Peninsula (South Wales) while O'Dea and Okamura (2000a) used incremental intervals (distance between two growth-check lines) and found similar yearly growth rates for *Flustra* from the Menai Straits (Wales) and the Skagerrak (Baltic) – 2.65 and 2.28 cm respectively, but only 1.38 cm for material from Minas Basin (Scotland). On the other hand, Kahle et al. (2003) obtained growth rates of 0.3% of the initial weight per day for colonies grown in vitro under controlled conditions. Note that the later results represent really the potential growth of this species under optimal conditions. Our colonies from the North Sea had a mean height of 6.3 cm and were at least 4–5 years old (according to the number of growth-check lines that could be discriminated) which gives a growth rate between 1.26 and 1.57 cm/year. Colonies from the Baltic Sea had 5.7 cm mean height and were approximately of the same age, which gives a slower growth rate of 1.42–1.14 cm/year.

Earlier studies of seasonal variation in *Flustra* colonies from several localities have shown that zooecia sizes vary with ambient temperature (O'Dea and Okamura 2000a). These studies were based on observations concerning growth-check lines and yearly growth rates done by Stebbing (1971b) in field material and some experimental work showing an inverse relationship between temperature increase and zooecia size (Menon 1975). Following this we postulated that if zooid size differences are indeed substantial across growth lines these should be tracked in SEM photographs.

Nevertheless, growth-check lines traditionally used to age *Flustra* colonies cannot be discriminated in SEM photos (although they are visible at both naked eye and under an optical microscope). This probably means that there are no abrupt size changes across these lines, but rather a small and constant increment in zooecia size following the changes in outside conditions as the year progresses from spring through winter. This may explain the absence of significant differences in mean zooecia size (maximal width and length) and mean zooecia density (number of zooecia per area unit) in the studied basins. Such also agrees with remarks by Stebbing (1971b) concerning similar zooecia size both in different fronds of one colony and in different parts of one frond within the years.

Following several other authors (Eggleston 1963; Stebbing 1971b), we postulate that *Flustra* stops growing in winter and re-starts adding new zooids in early spring. Growth-check lines represent most probably the position of the growing tip edge during this winter growth stop. Such lines are easily discriminated in the colonies' younger (several top centimeters) region, whereas increased sediment and epifaunal deposition in older (lower) parts difficult its recognition.

In spite of the lack of significant differences in mean zooecia size and density, zooecia frontal area index shows a distinct oscillation pattern across generations along a multiyear profile. Such oscillation is most noticeable in the North Sea material (Fig. 5.4a). Colonies were collected in late spring/early summer (2002) when growth re-starts after winter arrest. Zooecia have smaller sizes increasing backwards towards the previous winter with a decrease in temperature. Assuming that growth is almost absent during winter (or very slow), only a couple of generations at most should be formed. Sizes then decrease again towards summer/spring as the temperature increases. Interesting to notice that there is a slight increase in the area size during summer also noticed by Stebbing (1971b) who attributed it to a likely decrease in food resources at the peak of the season. Note that such food resource variability agrees well with reported peaks in plankton productivity in the North Sea (Leterme et al. 2005, 2008; Wiltshire et al. 2010). Size increases once more towards the winter. Such oscillation cycle is repeated across the profile. These oscillations in zooecia frontal index are somewhat mirrored in the plot of zooid density (Fig. 5.4b) where lower densities match areas with higher frontal indexes (winter) and higher densities match areas with lower frontal indexes (spring/summer). Although also present in the Baltic material (Fig. 5.5a, b), these oscillations lack the regularity observed in the North Sea. This lack of regularity noticed in the Baltic colonies is most probably due to a much higher variability of this basin (Smetacek 1985; Smetacek et al. 1987). Indeed, many factors that affect the growth of benthic organisms (food resources, salinity, temperature, dissolved oxygen, etc.) show a high degree of unpredictability as shown in other studies. For example, Trutschler and Samtleben (1988) noticed the lack of a distinct growth band pattern in *Astarte elliptica* (Brown) from the Kiel Bight (Baltic Sea) attributed by them to the high variable environmental conditions of the area. Similarly, Dunca et al. (2009) report a very weak signal in the growth, structure and chemical composition of *Arctica islandica* (L.) from the Baltic Sea (Kattegat and Skagerrak) used to monitor local climate. The authors attribute this to environmental interferences such as that from salinity.

The oscillation noticed in our material could be related to environmental regimes such as temperature changes. These data agree with findings from cupuladriid bryozoans (O'Dea and Jackson 2002) from the Panama region. The clear relation between temperature regimes and zooecia area rather than linear size could reinforce the idea that sizes may be more dependent of their position in the colony due to their importance in zooid tessellation (O'Dea and Okamura 1999). On the other hand, Lombardi et al. (2006) found that zooid sizes in *Pentapora fascialis* (Pallas) were more sensitive to temperature regimes than zooid area whereas Amui-Vedel et al. (2007) found that *Cryptosula pallasiana* Moll colonies collected in July had longer and wider zooids than those collected in January while colonies grown in laboratory under different temperature regimes presented significantly longer and wider zooids at lower temperature values. More data are needed in order to calibrate these parameters as well as growth rates in order to further use them in ocean acidification studies.

One of the questions is whether the possible reductions in calcification levels can be compensated by an increase in the skeletons' organic contents. Preliminary results show quite similar carbon contents in both the Baltic and the North seas. Carbonate contents are also similar. The fact that *Flustra* from the Baltic shows less weight loss than that from the North Sea was somewhat unexpected and does not fit our assumption that Baltic *Flustra* should be more affected by carbonate deficit thus having lower mineralization levels. More data is needed in order to validate these results.

The skeletal walls of *F. foliacea* are formed of IMC (intermediate magnesium calcite) as shown in the X-ray diffraction analyses in contrast to the two Mg peaks identifiable in the bi-mineralic *C. sinuosa*. The resolution of the XRD technique is not very high so further measurements will be necessary to test the geochemical variability occurring in individual walls and wall layers.

Dissolution experiments have been widely adopted to test the taphonomic resistance of bryozoan skeletons (Smith et al. 1992; Smith and Nelson 2003; Smith 2009). For dead parts of otherwise living colonies or dead colonies exposed to either low-pH bottom waters or shallow burial pore waters, certainly, one would expect different dissolution profiles. In our experiments, we compared the dissolution behavior of dead but still fresh colonies of *F. foliacea* (North Sea) with that of the low magnesium calcite *H. lichenoides* (Greenland Sea) and the bi-mineralic (low and intermediate magnesium calcite) *C. sinuosa* (English Channel). In general, *F. foliacea* loses carbonate much faster than the two other species thus showing a higher sensitivity of its skeleton to dissolution. On the other hand, the relatively high percentage of material remaining after 180 min of exposure indicates the presence of a high amount of organics (including cystid and polypide) in *Flustra*. Though such taphonomic behavior of dead *Flustra* colonies with respect to dissolution cannot be compared with the calcification success in living colonies, it may give an idea concerning the calcification potential of the species in low pH environments (Smith 2009).

## Conclusions

With changing ocean chemistry adding to factors such as increased pollution and global warming, stress in marine organisms may lead to heightened extinction and dramatic shifts in marine ecosystems in general. More data on processes such as biomineralization are of utmost importance if we want to ameliorate present day and predicted effects of ocean acidification on biodiversity.

Ongoing experiments include monthly measurements of *Flustra* colonies grown under laboratory conditions to refine results presented here in order to age colonies and calculate growth rates. Calibrated growth rates and age patterns will then be used to assess effects of changing pH/$pCO_2$ in laboratory conditions.

These data will be compared with data obtained from collections from pre-industrial times in order to evaluate possible effects of changing ocean chemistry in morphology and geochemistry of *F. foliacea*. Results will complement efforts towards a better knowledge of ocean acidification impact in marine organisms and systems.

**Acknowledgements** We thank all that helped making this paper a better one. P. Fiedler (IfG, Institute of Geosciences, Kiel University) for the XRD analyses, I. Dold (IfG) for the carbon analyses (Coulomat). U. Schuldt provided invaluable support with the SEM. A special word of thanks is due to W. Blaschek (Pharmaceutical Institute, University Kiel) in whose laboratory the muffle oven analyzes were conducted. This paper benefited from discussions with several colleagues (A. O'Dea, B. Okamura, A. Smith) as well as the suggestions of F. Bitschofsky and D. Barnes who kindly served as reviewers. This work is part of the Excellence Cluster "Future Ocean" project 2009/1 CP0924.

## References

Amui-Vedel A-M, Peter JH, Porter JS (2007) Zooid size and growth rate of the bryozoan *Cryptosula pallasiana* Moll in relation to temperature, in culture and in its natural environment. J Exp Mar Bio and Ecol 353:1–12

Bader B (2001) Modern Bryomol-sediments in cool-water, high-energy setting: the inner shelf off Northern Brittany. Facies 44:81–104

Bader B, Schäfer P (2005) Impact of environmental seasonality on stable isotope composition of skeletons of the temperate bryozoan *Cellaria sinuosa*. PPP 226:58–71

Bitschofsky F (2012) Distribution over space and time in some selected epizoobiontic North Sea bryozoans. In: Ernst A, Schäfer P, Scholz J (eds) Bryozoan studies 2010. Springer, Heidelberg/ Berlin/New York, this volume

Blackford JC, Gilbert FJ (2007) PH variability and $CO_2$ induced acidification in the North Sea. J Mar Syst 64:229–241

Blackford JC, Jones N, Proctor R, Holt J (2008) Regional scale impacts of distinct $CO_2$ additions into the North Sea. Mar Pollut Bull 56:1461–1408

Dunca E, Mutvei H, Göransson P, Mörth C-M, Schöne BR, Whitehouse MJ, Elfman M, Baden SP (2009) Using ocean quahog (*Arctica islandica*) shells to reconstruct palaeoenvironment in Öresund, Kattegat and Skagerrak, Sweden. Int J Earth Sci (Geol Rundsch) 98:3–17

Eggleston D (1963) The marine Polyzoa of the Isle of Man. Thesis, University of Liverpool, p 297

Eggleston D (1972) Patterns of reproduction in marine Ectoprota off the Isle of Man. J Nat Hist 6:31–38

Hall-Spencer JM, Rodolfo-Metalpa R, Martin S, Ransome E, Fine M, Turner SM, Rowley SJ, Tedesco D, Buia M-C (2008) Volcanic carbon dioxide vents show ecosystem effects of ocean acidification. Nature 454:96–99

Harvell CD (1984a) Predator-induced defense in a marine bryozoan. Science 224:1357–1359

Harvell CD (1984b) Why nudibranchs are partial predators: intracolonial variation in bryozoan palatability. Ecology 65(3):716–724

Harvell CD (1992) Inducible defenses and allocation shifts in a marine bryozoan. Ecology 73 (5):1567–1576

Howard W, Tillbrook B (2008) Ocean acidification: Australian impacts in the global context (Australian Department of Climate Change and Antarctic Climate and Ecosystems Cooperative Research Centre, Hobart). http://www.pangaea.de.Software/PanMap/

IPCC Fourth Assessment Report: Climate change (2007) http://www.ipcc.ch/publicationsanddata/publicationsanddatareports.htm#1. Accessed 19 Sept 2010

Kahle J, Liebezeit G, Gerdes G (2003) Growth aspects of *Flustra foliacea* (Bryozoa, Cheilostomata) in laboratory culture. Hydrobiologia 503:237–244

Kleypas JA, Feely RA, Fabry VJ, Langdon C, Sabine CL, Robbins LL (2006) Impacts of ocean acidification on coral reefs and other marine calcifiers: a guide for future research. In: Report of a workshop, Sponsored by the NSF NOAA and USGS, St. Petersburg, 18–20 Apr 2005, p 89

Kuklinski P, Taylor PD (2009) Mineralogy of Arctic bryozoan skeletons in a global context. Facies 55:489–500

Langdon C, Atkinson MJ (2005) Effect of elevated $pCO_2$ on photosynthesis and calcification of corals and interactions with seasonal change in temperature/irradiance and nutrient enrichment. J Geophys Res. doi:10.1029/2004JC002576

Leterme SC, Edwards M, Seuront L, Attrill MJ, Reid PC, John AWG (2005) Decadal basin-scale changes in diatoms, dinoflagelates, and phytoplankton color across the North Atlantic. Limnol Oceanogr 50(4):1244–1253

Leterme SC, Pingree RD, Skogen MD, Seuront L, Reid PC, Attrill M (2008) Decadal fluctuations in North Atlantic water inflow in the North Sea between 1958–2003: impacts on temperature and phytoplankton populations. Oceanologia 50(1):59–72

Liebezeit G (2005) Aquaculture of "non-food organisms" for natural substance production. Adv Biochem Engg/Biotechnol 97:1–28

Lombardi C, Cocito S, Occhipinti-Ambrogi A, Hiscock K (2006) The influence of seawater temperature on zooid size and growth rate in *Pentapora fascialis* (Bryozoa: Cheilostomata). Mar Biol 149:1103–1109

Lysek N, Rachor E, Lindel T (2002) Isolation and structure elucidation of deformylflustrabromine from the North Sea Bryozoan *Flustra foliacea*. Z Naturforsch 57c:1056–1061

McNeil BI, Matear RJ (2008) Southern Ocean acidification: a tipping point at 450-ppm atmospheric $CO_2$. PNAS 105:18860–18864

Menon NR (1975) Observations on growth of *Flustra foliacea* (Bryozoa) from Helgoland waters. Helgoländer wiss Meeresunters 27:263–267

O'Dea A, Jackson JBC (2002) Bryozoan growth mirrors contrasting seasonal regimes across the Isthmus of Panama. PPP 185:77–94

O'Dea A, Okamura B (1999) Influence of seasonal variation in temperature, salinity and food availability on module size and colony growth of the estuarine bryozoan *Conopeum seurati*. Mar Biol 9:267–273

O'Dea A, Okamura B (2000a) Life history and environmental interference through retrospective morphometric analysis of bryozoans: a preliminary study. J Mar Biol Ass UK 80:1127–1128

O'Dea A, Okamura B (2000b) Intracolony variation in zooid size in cheilostome bryozoans as a new technique for investigating palaeoseasonality. PPP 162:319–332

O'Dea A (2005) Zooid size parallels contemporaneous oxygen isotopes in a large colony of *Pentapora foliacea* (Bryozoa). Mar Biol 146:1075–1081

Okamura B (1987) Seasonal changes in zooid size and feeding activity in epifaunal colonies of *Electra pilosa*. In: Ross JRP (ed) Bryozoa: present and past. Western Washington University, Bellingham, pp 197–203

Orr JC, Fabry JA, Aumont O, Bopp L, Doney SC, Feely RA, Gnanadesikan A, Gruber N, Ishida A, Joos F, Key RM, Lindsay K, Maier-Reimer E, Matear R, Monfray P, Mouchet A, Najjar RG, Plattner G-K, Rodgers KB, Sabine CL, Sarmiento JL, Schlitzer R, Slater RD, Totterdell IJ, Weirig M-F, Yamanaka Y, Yoo A (2005) Anthropogenic ocean acidification over the twenty-first century and its impact on calcifying organisms. Nature 437:681–686

Riebesell U (2008) Acid test for marine biodiversity. Nature 454:46–47

Riebesell U, Zondervan I, Rost B, Tortell PD, Zeebe E, Morel FMM (2000) Reduced acidification of marine plankton in response to increased atmospheric $CO_2$. Nature 407:364–367

Riebesell U, Schulz KG, Bellerby RGJ, Botros M, Fritsche P, Meyerhöfer M, Neill C, Nondal G, Oschlies A, Wohlers J, Zöllner E (2007) Enhanced biological carbon consumption in a high $CO_2$ ocean. Nature 450:545–548

Sala F, Mulet J, Reddy KP, Bernal JA, Wikman P, Valor L, Peters L, König GM, Criado M, Sala S (2005) Potentiation of human _4_2 neuronal nicotinic receptors by a *Flustra foliacea* metabolite. Neurosci Lett 373:144–149

Schäfer P, Bader B (2008) Geochemical composition and variability in the skeleton of the bryozoan *Cellaria sinuosa* (Hassall): biological vs. environmental control. Virginia Mus Nat Hist Spec Pub 15:269–279

Sharp JH, Winston MK, Porter JS (2007) Bryozoan metabolites: an ecological perspective. Nat Prod Rep 24:659–673

Smetacek V (1985) The annual cycle of Kiel Bight plankton: a long term analysis. Estuaries 8:145–157

Smetacek V, Bodungen B, Bolter M, Bröckel K, Dawson R, Knoppers B, Liebezeit G, Martens P, Peinert R, Pollehne F, Stegmann P, Wolter K, Zeitzschel B (1987) The pelagic system. In: Rumohr J, Walger E, Zeitzschel E (eds) Seawater-sediment interactions in coastal waters, Lecture notes on coastal and estuarine studies. Springer, Heidelberg, pp 32–68

Smith AM (2009) Bryozoans as southern sentinels of ocean acidification: a major role for a minor phylum. Mar Freshw Res 60:475–482

Smith AM, Nelson CS (2003) Effects of early sea-floor processes on the taphonomy of temperate shelf skeletal carbonate deposits. Earth Sci Rev 63:1–31

Smith AM, Nelson C, Danaher PJ (1992) Dissolution behaviour of bryozoan sediments: taphonomic implications for non-tropical carbonates. PPP 93:213–226

Smith AM, Key MM Jr, Gordon DP (2006) Skeletal mineralogy of bryozoans: taxonomic and temporal patterns. Earth Sci Rev 78:287–306

Society R (2005) Ocean acidification due to increased atmospheric carbon dioxide, Royal society policy document 12/05. The Royal Society, London, p 60

Stebbing ARD (1971a) The epizoic fauna of *Flustra foliacea* (Bryozoa). J Mar Biol Assoc UK 51:283–300

Stebbing ARD (1971b) Growth of *Flustra foliacea* (Bryozoa). Mar Biol 9:267–273

Taylor PD, James NP, Bone Y, Kuklinski P, Kyser TK (2009) Evolving mineralogy of cheilostome bryozoans. Palaios 24:440–452

Trutscher K, Samtleben C (1988) Shell growth of *Astarte elliptica* (Bivalvia) from Kiel Bay (Western Baltic Sea). Mar Ecol Prog Ser 42:155–162

Tyrell T, Schneider B, Charalampopoulou A, Riebesell U (2008) Coccolithophores and calcite saturation state in the Baltic and Black seas. Biogeosciences 5:485–494

Wiltshire KH, Kraberg A, Bartsch I, Boersma M, Franke H-D, Freund J, Gebühr C, Gerdts G, Stockmann K, Wichels A (2010) Helgoland roads, North Sea: 45 years of change. Estuar Coast 33:295–310

# Chapter 6
# Life on the Edge: *Parachnoidea* (Ctenostomata) and *Barentsia* (Kamptozoa) on Bathymodiolin Mussels from an Active Submarine Volcano in the Kermadec Volcanic Arc

## Vent-Faunal Ctenostome and Kamptozoan

**Dennis P. Gordon**

**Abstract** Very few bryozoans have been reported from the vicinity of hot vents. *Parachnoidea rowdeni* n. sp. (Ctenostomata: Arachnidiidae) has been found associated with the bathymodiolin mussel *Gigantidas gladius* (Bivalvia: Mytilidae) from Rumble V seamount in the southern part of the Kermadec volcanic arc. The mussels exhibit high population densities at sulphur-rich hydrothermal springs on active submarine volcanoes at depths of 216–755 m. Seawater temperatures above the mussel beds are above ambient and dredge hauls containing the mussel include elemental sulphur and smell strongly of it. The mussel's nutrition appears to involve chemosynthesis by the sulphur-oxidising symbiotic bacteria that are concentrated in its extremely enlarged gills. The colonies of *Parachnoidea* are partly encrusting, partly erect, forming tangles of stoloniform zooids above the shell surface. Newly formed zooids are transparent; older zooids have a dark coating of inorganic particles that include, inter alia, sulphur, iron and manganese. Growth and morphological attributes of *P. rowdeni* suggest that it may be a genuine vent-endemic bryozoan. An unidentified species of *Barentsia* is the first record of the phylum Kamptozoa from the vicinity of a hydrothermal vent.

**Keywords** Hydrothermal vent • Bryozoa • Stroke • *Parachnoidea* • Kamptozoa • *Barentsia* • *Gigantidas gladius* • Kermadec Ridge • New Zealand

D.P. Gordon (✉)
National Institute of Water & Atmospheric Research (NIWA), Private Bag 14901, Kilbirnie, Wellington, New Zealand
e-mail: dennis.gordon@niwa.co.nz

A. Ernst, P. Schäfer, and J. Scholz (eds.), *Bryozoan Studies 2010*,
Lecture Notes in Earth System Sciences 143, DOI 10.1007/978-3-642-16411-8_6,

## Introduction

Three decades of investigation on deepwater chemosynthetic habitats have yielded much information about the biota that is associated with hydrothermal vents and cold seeps. Deep-sea hydrothermal vents associated with volcanic arcs and mid-ocean ridge systems depend on reduced gases like hydrogen sulphide and methane to support an array of chemoautotrophic microbes. In turn, these microbes directly or indirectly support diverse assemblages of suspension-feeding and grazing invertebrates and their attendant scavengers and predators including fish. A broad variety of taxa have been reported from deep-sea hydrothermal vents but bryozoans have not generally been among them (Grassle 1985; Desbruyères et al. 2006). Recently, however, ten bryozoan species, comprising at least four species of Cyclostomata and the rest apparently cheilostomes, were collected from a low-temperature hydrothermal vent field on the Mohn Ridge north of Iceland (Schander et al. 2010). Additionally, two species of Cheilostomata have been reported encrusting disarticulated valves of *Calyptogena gallardoi* Sellanes and Krylova, 2005 (Bivalvia, Vesicomyidae) at a cold-seep locality off Concepción, Chile, where methanotrophic bacteria thrive (Moyano 2008).

Here I report on the first records of the Ctenostomata and Kamptozoa from a hydrothermal vent field on an active submarine volcano. Both were found encrusting live *Gigantidas gladius* Cosel and Marshall, 2003 (Bivalvia, Mytilidae, Bathymodiolinae), a giant vent-mussel species that harbours sulphur-oxidising symbiotic bacteria in its enlarged gills. The mussels occur in high population densities at sulphur-rich hydrothermal springs on Rumble V, a basaltic-andesite stratovolcano northeast of the Bay of Plenty, New Zealand. Dredge hauls containing the mussel included elemental sulphur and smelled strongly of it.

First discovered in 1992, when a large plume of bubbles was acoustically detected rising from its summit, Rumble V is the southernmost of the Quaternary Rumbles group of seamounts at the southern end of the Kermadec Ridge. The ridge is part of the Kermadec–Tonga arc, itself part of the ~2,000 km-long, southward-propagating and actively widening Lau–Havre–Taupo arc-back-arc complex (Fig. 6.1a) associated with the Pacific-Australian plate convergence (Wright 1994; Parson and Wright 1996; Clark and O'Shea 2001; de Ronde et al. 2001). Rumble V rises 2,233 m from the seafloor at 2,600 to 367 m below the sea surface and is one of seven cone- and caldera-forming volcanoes in the area discovered to have hydrothermal-vent fields. It has an area of 48 km$^2$. Apart from *G. gladius*, the conspicuous macrofauna comprises mostly Brachyura, Galatheidae and Ophiuroidea (Clark and O'Shea 2001; Ahyong 2008), in association with hexactinellid sponges, octocorals, gastropods, isopods, asteroids, echinoids and three sympatric species of *Alvinocaris* (Crustacea, Alvinocarididae) (Ahyong 2009).

The new ctenostome and kamptozoan, and the habitat in which they occur, are described below and discussed in relation to what is known about these taxa and the hydrothermal vent environment.

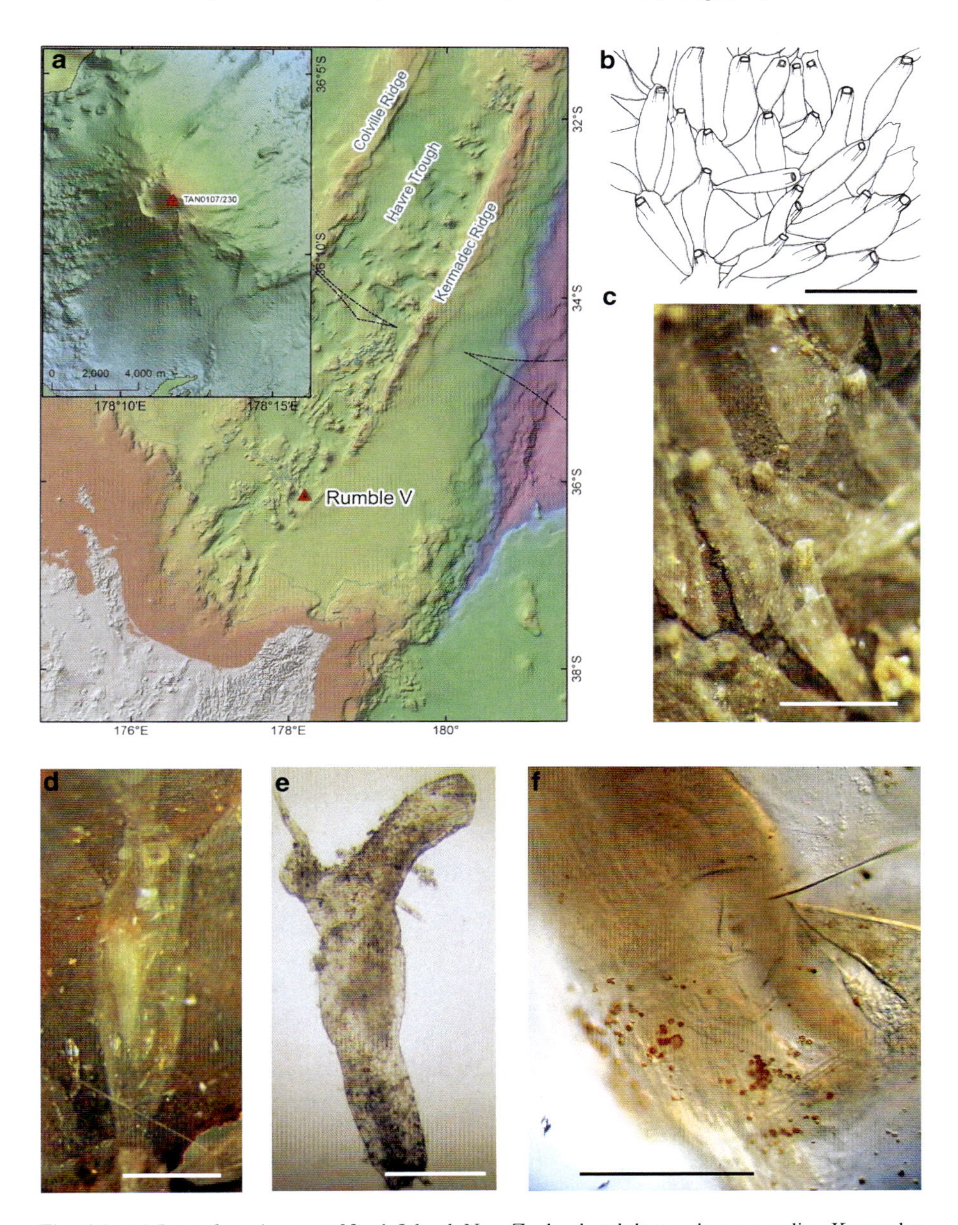


**Fig. 6.1** (**a**) Part of northeastern North Island, New Zealand and the northeast-trending Kermadec Ridge, Havre Trough and Colville Ridge, with the position of Rumble V seamount indicated. Inset: Swath-bathymetric image of Rumble V, with the location of NIWA Station TAN0107/230 indicated on the caldera rim. (**b**) General form of encrusting neanic zooids of *Parachnoidea rowdeni* sp. nov., traced from a photograph. *Scale bar*: 2 mm. (**c**) Some neanic encrusting zooids showing their general form, branching of zooid rows, and squared retracted zooidal peristomes. *Scale bar*: 1 mm. (**d**) Individual zooid on mussel substratum. *Scale bar*: 0.5 mm. (**e**) Individual zooid in transparency. *Scale bar*: 0.3 mm. (**f**) Base of lophophore, pharynx and oesophagus. *Scale bar*: 0.1 mm

## Materials and Methods

Specimens of the vent-faunal mussel *Gigantidas gladius* were collected by RV Tangaroa (belonging to the National Institute of Water & Atmospheric Research) from Rumble V seamount on 24 May 2001. An epibenthic sled approximately 1 m wide by 1.5 m long was towed upslope from 755 m depth to the summit at 367 m and some of the catch preserved onboard ship in 70% isopropanol. Later examination of the mussels at NIWA's Wellington laboratory led to the discovery of the ctenostome, which comprised a colony (or colonies) easily visible to the eye on one of the valves. Some zooids of a kamptozoan were found growing among the basal parts of the ctenostome on the mussel surface. Part of the mussel valve was removed and preserved in two glass tubes. The alcohol subsequently evaporated from one of them, causing the specimen to dry; it was later reconstituted in aqueous trisodium phosphate.

Both taxa were measured using optical microscopy and an eyepiece micrometer. Selected zooids of the ctenostome were air-dried from alcohol, placed on a Jeol 12 mm stub on carbon tape and coated with 8 nm platinum followed by a 4 nm coat of carbon. The elemental composition of both the mussel and bryozoan surfaces were determined using energy-dispersive X-ray analysis (EDS) in a Jeol JSM 6500F equipped with a backscatter detector. The SEM was operated at 15 kV accelerating voltage and a probe current of 18.

A later NZ-US cruise (KOK-05-05) by RV Ka-imikai-o-Kanaloa in April-May 2005 carried out measurements of environmental parameters in *Gigantidas* vent fields, with videography from Pisces V submersible dives, as part of a wider survey of the Kermadec arc submarine volcanoes (Merle et al. 2005).

## Results

### *Environmental Setting*

Rumble V is dominated by basaltic lavas, and shows a typical progression from effusive to explosive eruption products with decreasing water depth. *Gigantidas gladius* occurs in dense beds at several locations, usually in areas of diffuse venting, as is typical for bathymodiolins (Colaço et al. 2002), or adjacent to small rifts where the difference in ambient water temperature could be detected by a shimmering of the water as seen in videography. Ambient temperature in the vicinity of the vent field subsequent to the collection of the bryozoan and kamptozoan was measured as 10.2°C, with temperatures at three locations just above the mussel beds measured at 11.2°C, 12.6°C and 39.7°C; pH at two locations measured 4.75 and 5.06 (background seawater 7.9) (Massoth et al. 2005). Adjacent rock on the Rumble V caldera rim included bare pillow lavas, white/yellow rock from vents, rock with red/yellow, black or brown crusts and rock with bacterial mats.

A relatively large single colony (possibly a compound colony) of the bryozoan was discovered, attached to the surface of a live mussel which was itself already coated with precipitate from seawater. The kamptozoan occurred sparsely, ramifying among the encrusting zooids of the bryozoan. Two species of asteroids, *Rumbleaster eructans* and *Coronaster reticulatus*, and gastropods including *Hirtomurex tangaroa* and *H. taranui*, were seen crawling over the mussels in video images. Numerous other invertebrates occurred in the general vicinity and on adjacent slopes (see appendix).

## Systematic Part

Phylum Bryozoa

Class Gymnolaemata

Order Ctenostomata

Suborder Euctenostomata

Superfamily Arachnidioidea

Family Arachnidiidae

Genus *Parachnoidea* d'Hondt, 1978

Type species: *Parachnoidea rylandi* d'Hondt, 1978

*Parachnoidea rowdeni* sp. nov.

Figures 6.1b–d, and 6.2a–d

Material examined: Holotype NIWA 60511, from NIWA station TAN0107/230 (Z10811), 36°08.48′ S, 178°11.70′ W – 36°08.79′ S, 178°11.53′ E, 755–367 m, 24 May 2001, Rumble V seamount, southern Kermadec Ridge, New Zealand. Paratype NIWA 60512, station details as for holotype.

Etymology: The species is named for Dr Ashley A. Rowden, NIWA, one of the biological observers on the KOK-05-05 cruise.

Description: Colony initially encrusting, spreading over an area of ca. 70 mm$^2$. Autozooids in uniserial rows that tend to bifurcate or trifurcate, individual zooids having an elongate-subclavate form, 0.92 to 2.25 mm long (mean 1.45 mm), 0.25–0.57 mm wide (mean 0.48 mm), with the lateral margins generally somewhat parallel before the zooid tapers somewhat abruptly to its origination point on the parent zooid. Lateral walls with visibly thicker cuticularisation than the somewhat flattened frontal wall. Peristome subterminal, somewhat elongate, its normal retracted height around one fifth of zooid length in neanic zooids, the orifice then almost squared in outline but generally just a little wider (~0.12 mm) than long (~0.11 mm). Non-feeding ephebic or dead zooids with much longer fully everted peristome that is more or less cylindrical, often ~ two-thirds of zooid length and

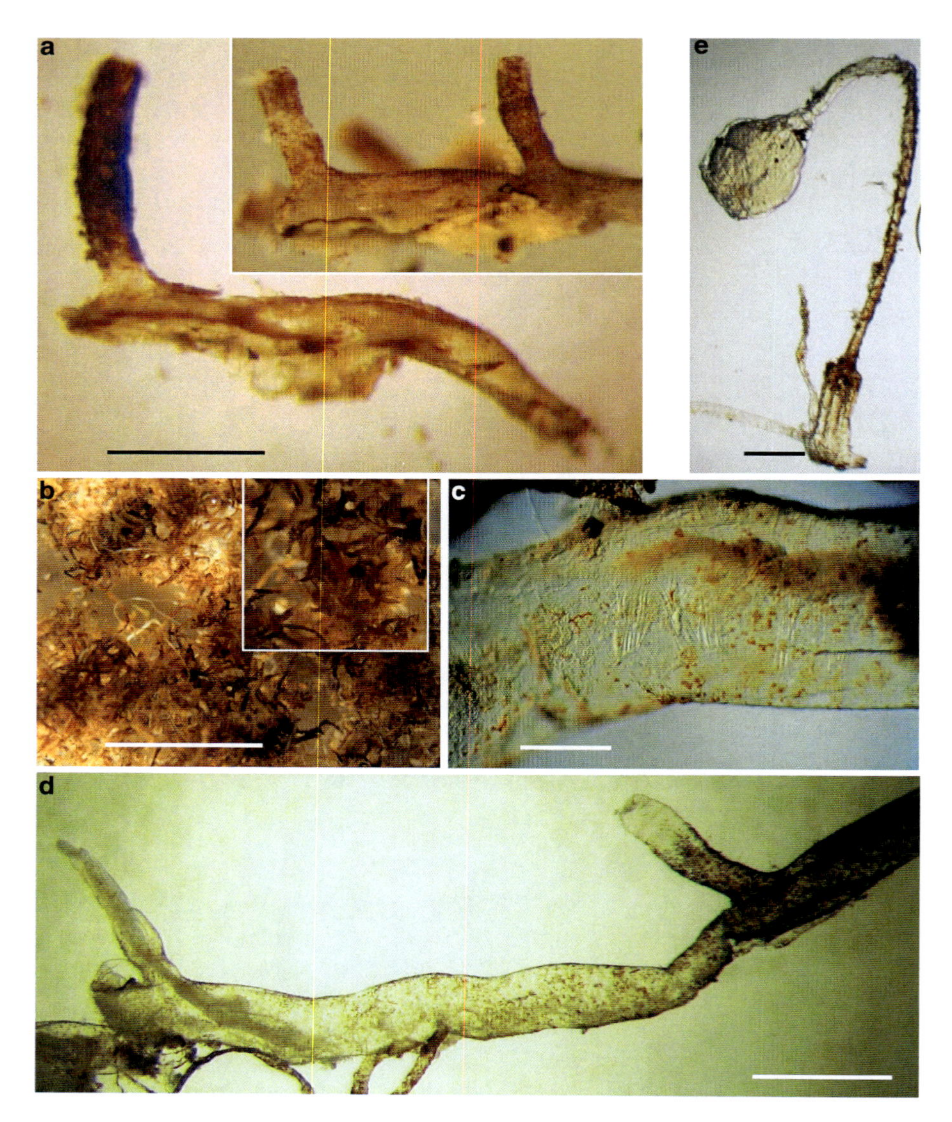


**Fig. 6.2** (**a**) (*scale bar*: 0.5 mm) and inset. Ephebic zooids of *Parachnoidea rowdeni* sp. nov., discoloured by inorganic precipitate. (**b**) Tangle of non-encrusting stoloniform zooids (*scale bar*: 1 cm) inset: close-up of part of the tangle. (**c**) Part of stoloniform zooid as seen in transparency, showing groups of parietal muscles. *Scale bar*: 0.1 mm. (**d**) Stoloniform zooids as seen in transparency. *Scale bar*: 0.5 mm. (**e**) *Barentsia* sp. *Scale bar*: 0.1 mm

darkened with a surface crust. Each zooid typically having one mid-distal budding locus and frequently also two distolateral budding loci, the latter variably placed; budding can also take place from a site on a lateral wall. In crowded parts of neanic colonies with many budding sites, adjacent zooids can become laterally appressed and overgrow one another.

Zooid growth can continue above the substratum, in which case the zooids become elongated and stoloniform. Stoloniform zooids bud further stoloniform zooids, which can attach to other such zooids, eventually producing an erect tangled mass of such zooids up to 10 mm or more above the substratum.

Retracted lophophore in stoloniform zooids 0.59–0.63 mm long, pharynx 0.15 mm long. No stomach gizzard. Parietal muscles in a longitudinal series of short, separated linear groups on either side of the zooid, stretching from the basal to the frontal wall. Embryos and ancestrula not seen.

Remarks: *Parachnoidea* is a little-known genus. D'Hondt distinguished it from *Arachnidium* on the basis of the tubular peristome with discrete rim (i.e. non-mamillate) in the retracted state and also noted the absence of the stoloniform cross-connections and filiform processes found in some species ascribed to *Arachnidium*. The type and sole species *Parachnoidea rylandi* was discovered at 320–1,050 m in the Bay of Biscay [Gulf of Gascony, Golfo de Vizcaya] on shell debris. The present material accords with d'Hondt's (1978) generic diagnosis and the two species share the same kind of form, zooidal budding and peristome. The species overlap in depth range and zooidal dimensions but d'Hondt (1978) indicated that the zooids in *P. rylandi* are a little more elongate-hexagonal and he described the peristome rim as pentagonal whereas it is squared, when retracted, in *P. rowdeni* and apparently thicker-walled. Both species also share linear rows of grouped parietal muscles.

Phylum Kamptozoa

Order Coloniales

Suborder Stolonata

Family Barentsiidae

Genus *Barentsia* Hincks, 1880

Type species. *Barentsia bulbosa* Hincks, 1880

*Barentsia* sp.

Figure 6.2e.

Material examined: Figured material from NIWA station TAN0107/230 (Z10811), 36°08.48′ S, 178°11.70′ W – 36°08.79′ S, 178°11.53′ E, 755–367 m, 24 May 2001, Rumble V seamount, southern Kermadec Ridge, New Zealand.

Description: Colony ramifying, comprising narrow thread-like stolons bearing sparse, minute, erect zooids. Muscular bulb 0.078–0.090 mm diameter, 0.17–0.25 mm high, with vertical pleats; stalk (pedicle) 0.022–0.034 mm diameter, 0.75–1.07 mm high, calyx 0.101–0.134 mm wide, 0.190–0.224 mm high, hence the combined total height of zooids in the colony 1.11–1.54 mm. Stalk without articulation or geniculation. Basal stolon 0.022–0.034 mm diameter, i.e. same as stalk diameter.

Remarks: Most zooids in the colony had lost the calyx, hence comprised only a series of stalks. Only five complete zooids were obtained. These had been dried but were reconstituted in trisodium phosphate. It is possible that bulb, stalk and calyx widths were affected by this treatment but not overall height.

It is likely that this species is new to science but the few available zooids are inadequate for proper characterisation of a new taxon. In its diminutive size it resembles *Barentsia gracilis* (Sars, 1835) (0.5–1.5 mm height), *B. stiria* (Jullien *in* Jullien and Calvet, 1903) (0.85 mm height), *Barentsia parva* O'Donoghue and O'Donoghue, 1923 [see also Wasson 1997] (0.45–2.7 mm height), and *B. minuta* Winston and Håkansson, 1986 (0.7–1.18 mm height), but differs from all of these in details. *Barentsia gracilis* has nodes in the pedicel; *B. stiria* has a disproportionately small muscular bulb; *B. parva* can also have nodes and the basal stolon has the same diameter as the pedicel; and *B. minuta* has a proportionately longer pedicel and wider calyx. The vent-faunal species also has vertical pleating in the muscular bulb but it is possible that this feature (not illustrated in other *Barentsia* species) could be a consequence of drying and subsequent rehydration and reconstitution. Either way, the present material definitely constitutes the first known vent-faunal kamptozoan.

## *Elemental Analysis*

The shell surface of *Gigantidas gladius* was covered by a dark-brown coating that also occurred on epizoic organisms including *Parachnoidea rowdeni* zooids. Whereas neanic zooids are largely semi-transparent, ephebic zooids become covered by the same dark-brown or dark orange-brown coating as the substratum, especially the peristomes, which become long and cylindrical with a rounded orifice and evidently incapable of retraction.

It is well known that metalliferous sulphides precipitate out of hot-vent fluids (Styrt et al. 1981), the emitted sulphur deriving largely (up to 90%) from leaching as primary sulphide from basalts (Jannasch 1989). To test the hypothesis that the crust on the bryozoan mostly comprises vent-fluid precipitate, the surfaces of a neanic and an ephebic zooid of *Parachnoidea rowdeni*, and of the mussel substratum, were subjected to elemental analysis.

The mussel surface showed the presence of sulphur, iron and manganese plus some calcium, magnesium and silicon. [High peaks of sodium and phosphorus may have reflected the residue of these elements following treatment of the bryozoan-encrusted mussel surface with trisodium phosphate.] Analysis of the surface of a neanic zooid showed a near-identical result, while the black-brown crust on an ephebic zooidal peristome showed a little more manganese.

## Discussion

Samples of the bryozoan *Parachnoidea rowdeni* sp. nov. and the kamptozoan *Barentsia* sp. were obtained only from a single vent mussel, hence there is no information about their distribution and abundance on either Rumble V or adjacent submarine volcanoes in the Kermadec arc. Seawater temperature above the mussel beds on Rumble V was measured as 1.0–29.5°C higher than ambient, suggesting that *P. rowdeni* and *Barentsia* sp. are capable of surviving in at last the lower part of this range. The occurrence of a surface coating that includes metalliferous sulphides is direct evidence that these suspension-feeding organisms survive in relative proximity to vent fluids containing them.

Elemental analyses of the surfaces of the host mussel and the frontal walls of neanic and ephebic *P. rowdeni* showed results consistent with previous analyses of vent-fluid characteristics from Rumble V. Out of seven seamounts analysed, it had the highest concentrations of $H_2S$ and volatile sulphur in the plume, and very high concentrations of total dissolvable manganese (de Ronde et al. 2001).

Neither coccoliths nor diatoms were found on the surfaces examined by SEM and it is likely that bacteria, ingested either separately or on inorganic precipitate, are an important source of nutrition for mussel epizoites. Similar vents in a Southern Mariana back-arc vent field are associated with epsilonproteobacteria and hyperthermophilic archaebacteria (Kato et al. 2010) but a variety of prokaryotes have been identified from a range of hydrothermal vents and cold seeps (Reed et al. 2009) and it appears that, by differentially affecting microbial populations, variation in vent-fluid characteristics is a key factor in the variation of local sources at vents (De Busserolles et al. 2009). Comita et al. (1984) noted that bacteria-rich particles emanating from some measured vents are much richer in nitrogen and lipid compounds than particles collected in peripheral areas. Studies of trophic structure in Atlantic vent systems show short food chains, with free-living and symbiotic bacteria comprising the base. A number of taxa (especially bathymodiolin mussels and some shrimps) host symbiotic bacteria while also carrying out filtration; the mussels have functional guts (Page et al. 1990). Bathymodiolin mussels transplanted away from vents show minimal growth and loss of body weight, and stable-isotope analysis of vent-peripheral suspension and deposit feeders indicate a local (chemosynthetic organisms) source rather than pelagic (Grassle 1985). *Gigantidas gladius* is a large species (up to 316 mm shell length) and an assemblage of these mussels will collectively draw particle-laden water to the whole bed, thereby benefiting epizoitic suspension-feeders. Bacterivory has been demonstrated to sustain survival and growth in both marine (Gosselin and Qian 2000) and freshwater (Richelle et al. 1994) bryozoans. Faecal particles and flocculent material are likely to serve as supplementary food sources for vent-faunal bryozoans and kamptozoans. Gut contents were non-determinable in the present material.

Vent-faunal assemblages are frequently associated with high levels of endemism (Ramirez-Llodra et al. 2007). Nevertheless, even taxa with lecithotrophic larvae – the dominant type among vent species (Tyler and Young 1999; Young 2003) can disperse long distances. The lower temperatures of bathyal and abyssal waters are associated with long metabolic life spans of larvae and/or developmental arrest of embryos of vent polychaetes such as *Riftia pachyptila* and *Alvinella pompejana* (Marsh et al. 2001; Pradillon et al. 2001), useful attributes given that individual vents are ephemeral at geological time scales, lasting only on the order of years to decades (Micheli et al. 2002), while vent fields may last for hundreds of thousands of years (Boetius 2005). The mussel host of *Parachnoidea rowdeni* and *Barentsia* sp. is so far known only from Rumble V and nearby Rumble III less than 50 km to the northeast on the same arc. Nevertheless it shows significant heterogeneity in allozyme gene frequencies (Smith et al. 2004), suggesting limited inter-seamount dispersal even though it is likely to have a planktonic larva. Further, preliminary results of a multivariate analysis of presence-absence data for macro-invertebrates from epibenthic-sled samples from the two seamounts showed faunal dissimilarity, with an average Bray-Curtis dissimilarity of >90% (A. Rowden, pers. comm. in Smith et al. 2004). It is unlikely that the kamptozoan, which would have a feeding larva with a possible long planktonic stage (Nielsen 1971), or even the bryozoan, with an assumed non-feeding larva, is an obligate associate of *Gigantidas gladius*, but the mussel beds would provide a favourable habitat for these organisms should they be able to traverse the distances between seamounts.

Schander et al. (2010) recorded 180 species-level taxa, including 10 bryozoans, from a hydrothermal vent field on the Mohn Ridge in the North Atlantic. They noted that very few of the 180 taxa were vent-endemic, though some may be vent-associated. The most speciose higher taxa were polychaetes (71 taxa) and crustaceans (64 taxa), with 11 species of echinoderms. An interim macrofaunal list from Rumble V includes just 84 species. Polychaetes have not yet been identified, and only a few crustacean groups have been examined in detail. In contrast, the best-studied group, Echinodermata, is represented by 48 species on Rumble V, of which 41 are ophiuroids.

Of the five bryozoans in the Mohn Ridge vent field identified to species, all are known from non-vent habitats elsewhere in the northeastern Atlantic. In contrast, *Parachnoidea rowdeni* has morphological and growth attributes that give evidence that it may be a genuine vent endemic. It is striking for its plasticity of colony form. It appears that larvae first settle on the hard substratum of the mussel shell and relatively quickly encrust into available space. When space is no longer available for peripheral spread, budding takes place into free space, causing the zooids to become more or less cylindrical in cross section and much more elongate, even stoloniform, not that actual stolons are formed. Both encrusting and non-encrusting zooids become coated with vent precipitate that contains metal sulphides. This is most marked on the emergent peristomes of ephebic zooids. It appears that the peristomes become so encrusted that they can no longer retract. However, zooids thus affected, while no longer able to feed, appear able to bud daughter zooids. Notably, most of the zooids in a colony are stoloniform, forming a tangle above the mussel substratum. The rate of production of new zooids exceeds the rate at which zooids become coated by precipitate.

Erect growth via stoloniform zooids may be an adaptation to:

1. Eventual lack of substratum space after initial colonisation;
2. Enhanced access to food particles by adopting a three-dimensional form;
3. Compensating for the rate of coating by metal-sulphide precipitates; growth into three-dimensional space potentially exploits micro-variations in precipitation.

Vent-associated bryozoans have been reported from the fossil record. Von Bitter et al. (1992) and Morris et al. (2002) described 10-mm spheroidal multilamellar encrustations composed of alternating trepostome-bryozoan and microbial layers from a Lower Carboniferous chemosynthetic hydrothermal-vent system.

## Appendix A. Interim List of Taxa from Rumble V

**Porifera**
*Farrea occa* Bowerbank, 1862
Lyssanacida gen. et sp. indet. 1
Lyssanacida gen. et sp. indet. 2

**Cnidaria**
Acanthogorgiidae sp.
Actiniaria indet.
*Anthomastus robustus* Versluys, 1906
*Paragorgia whero* Sánchez, 2005
Pennatulacea indet.
*Placogorgia* sp.

**Kamptozoa**
*Barentsia* sp.

**Bryozoa**
*Parachnoidea rowdeni* sp. nov.

**Mollusca**
*Gigantidas gladius* Cosel and Marshall, 2003
*Hirtomurex tangaroa* Marshall and Oliverio, 2009
*Hirtomurex taranui* Marshall and Oliverio, 2009

**Annelida**
Polychaeta indet.

**Asteroidea**
*Allostichaster* sp.
Asteriidae sp.
*Coronaster reticulatus* (H.L. Clark, 1916)
*Rumbleaster eructans* McKnight, 2006

**Ophiuroidea**
*Amphiophiura laudata* (Koehler, 1904)
*Amphioplus ctenacantha* Baker, 1977
*Amphioplus longirima* Fell, 1952
*Amphioplus* sp.
*Amphiura magellanica* Ljungman, 1867
*Amphiura* sp.

*Aspidophiura* sp.
*Asteroschema bidwillae* McKnight, 2000
*Ophiacantha brachygnatha* H.L. Clark, 1928
*Ophiacantha duplex* Koehler, 1897
*Ophiacantha* sp.
*Ophiactis abyssicola* Sars, 1861
*Ophiactis hirta* Lyman, 1879
*Ophiactis profundi* Lütken and Mortensen, 1899
*Ophiochiton lentus* Lyman, 1879
*Ophiocopa spatula* Lyman, 1883
*Ophiocten cryptum* McKnight, 2003
*Ophiogeron* sp.
*Ophiolepis biscalata* McKnight, 2003
*Ophioleuce brevispinum* (H.L. Clark, 1911)
*Ophioleuce regulare* (Koehler, 1901)
*Ophiologimus prolifer* (Studer, 1882)
*Ophiolycus farquhari* McKnight, 2003
*Ophiomedea liodisca* (H.L. Clark, 1911)
*Ophiomusium relictum* Koehler, 1904
*Ophiomusium scalare* Lyman, 1878
*Ophiomyxa brevirima* (H.L. Clark, 1915)
*Ophiomyxa* sp.
*Ophiophrixus confinis* Koehler, 1922
*Ophiophyllum teplium* McKnight, 2003
*Ophioplax lamellosa* Matsumoto, 1915
*Ophioplinthaca miranda* Koehler, 1904
*Ophioplinthus inornata* Lyman, 1878
*Ophiopristis dissidens* (Koehler, 1905)
*Ophioscolex* sp.
*Ophiotreta valenciennesi* (Lyman, 1879)
*Ophiotreta* sp.
*Ophiura irrorata* (Lyman, 1878)
*Ophiura rugosa* (Lyman, 1878)
*Ophiura* sp.
*Ophiurothamnus clausa* (Lyman, 1878)

**Echinoidea**
*Caenopedina* sp.
Cidaroida indet.
*Dermechinus horridus* (Agassiz, 1879)
*Gracilechinus multidentatus* (H.L. Clark, 1915)

**Chordata**
*Beryx splendens* Lowe, 1834
*Caelorinchus aspercephalus* Waite, 1911
*Caelorinchus supernasutus* McMillan and Paulin, 1993

*Centriscops humerosus* (Richardson, 1846)
*Hyperoglyphe antarctica* (Carmichael, 1819)
*Oxynotus bruniensis* (Ogilby, 1893)
*Polyprion oxygeneios* (Schneider and Forster, 1801)
Scorpaenidae indet.
*Tripterophycis gilchristi* Boulenger, 1902

**Crustacea**
*Aegapheles mahana* Bruce, 2009
*Aegiochus gordoni* Bruce, 2009
*Aegiochus nohinohi* Bruce, 2009
*Alvinocaris alexander* Ahyong, 2009
*Alvinocaris longirostis* Kikuchi and Ohta, 1995
*Alvinocaris niwa* Webber, 2004
*Dorhynchus* sp.
*Mathildella mclayi* Ahyong, 2008
*Munida endeavourae* Ahyong and Poore, 2004
*Munida magniantennulata* Baba and Türkay, 1992
*Projasus parkeri* (Stebbing, 1902)
*Rochinia ahyongi* McLay, 2009

## References


Ahyong ST (2008) Deepwater crabs from seamounts and chemosynthetic habitats off eastern New Zealand (Crustacea: Decapoda: Brachyura). Zootaxa 1708:1–72

Ahyong ST (2009) New species and new records of hydrothermal vent shrimps from New Zealand (Caridea: Alvinocarididae, Hippolytidae). Crustaceana 82:775–794

Boetius A (2005) Lost city life. Science 307:1420–1422

Clark R, O'Shea S (2001) Hydrothermal vent and seamount fauna from the southern Kermadec Ridge, New Zealand. InterRidge News 10(2):14–17

Colaço A, Dehairs F, Desbruyères D (2002) Nutritional relations of deep-sea hydrothermal fields at the Mid-Atlantic Ridge: a stable isotope approach. Deep-Sea Res I 49:395–412

Comita PB, Gagosian RB, Williams PM (1984) Suspended particulate organic material from hydrothermal vent waters at 21° N. Nature 307:450–453

d'Hondt J-L (1978) Nouveaux Bryozoaires Cténostomes bathyaux et abyssaux. Bull Soc Zool Fr 103:325–333

De Busserolles F, Sarrazin J, Gauthier O, Gélinas Y, Fabri MC, Sarradin PM, Desbryuères D (2009) Are spatial variations in the diets of hydrothermal fauna linked to local environmental conditions? Deep-Sea Res II 56:1649–1664

de Ronde CEJ, Baker ET, Massoth GJ, Lupton JE, Wright IC, Feely RA, Greene RR (2001) Intra-oceanic subduction-related hydrothermal venting, Kermadec volcanic arc, New Zealand. Earth Plan Sci Lett 193:359–369

Desbruyères D, Segonzac M, Bright M (eds) (2006) Handbook of deep-sea hydrothermal vent fauna, vol 18, 2nd edn, Dennisia. Biologiezentrum der Oberösterreichisches Landesmuseen, Linz, pp 1–544

Gosselin LA, Qian P-Y (2000) Can bacterivory sustain survival and growth in early juveniles of the bryozoan *Bugula neritina*, the polychaete *Hydroides elegans* and the barnacle *Balanus amphitrite*? Mar Ecol Progr Ser 192:163–172

Grassle JF (1985) Hydrothermal vent animals: distribution and biology. Science 965:713–717
Hincks T (1880) On new Hydroida and Polyzoa from Barents Sea. Ann Mag Nat. Hist s.5, 6:277–286
Jannasch HW (1989) Sulphur emission and transformations at deep sea hydrothermal vents. In: Brimblecombe P, Lein AYu (eds) Evolution of the global biogeochemical sulphur cycle. Wiley, New York
Jullien J, Calvet L (1903) Bryozoaires provenant des campagnes de l'Hirondelle, vol 23, Résult camp sci accomp prince Albert I 23:1–188, 18 pls
Kato S, Takano Y, Kakegawa T, Oba H, Inoue K, Kobayashi C, Utsumi M, Marumo K, Kobayashi K, Ito Y, Ishibashi J, Yamagishi A (2010) Biogeography and biodiversity in sulfide structures of active and inactive vents at deep-sea hydrothermal fields of the Southern Mariana trough. Appl Environ Microbiol 76:2968–2979
Marsh AG, Mullineaux LS, Young CM, Manahan D (2001) Larval dispersal potential of the tubeworm *Riftia pachyptila* at deep-sea hydrothermal vents. Nature 411:77–80
Massoth G, Butterfield D, Lupton J (2005) Volcano chemistry: vent fluids. In: Merle S, Embley R, Chadwick W (eds) New Zealand American submarine ring of fire 2005 (NZASRoF'05): Kermadec arc submarine volcanoes. R/V Ka-imikai-o-Kanaloa, cruises KOK05-05 and KOK05-06, Pisces V dives PV-612–PV628, RCV-150 dives D310–D312. NOAA (National Oceanographic & Atmospheric Administration), Washington, DC
Merle S, Embley R, Chadwick W (2005) New Zealand American submarine ring of fire 2005 (NZASRoF'05): Kermadec arc submarine volcanoes. R/V *Ka-imikai-o-Kanaloa*, cruises KOK05-05 and KOK05-06, *Pisces V* dives PV-612–PV628, *RCV-150* dives D310–D312. NOAA, Washington DC. http://oceanexplorer.noaa.gov/explorations/05fire/logs/leg2_summary/media/srof05_cruisereport_final.pdf
Micheli F, Peterson CH, Mullineaux LS, Fisher CR, Mills SW, Sancho G, Johnson GA, Lenihan HS (2002) Predation structures communities at deep-sea hydrothermal vents. Ecol Monogr 72:365–382
Morris PA, von Bitter PH, Schenk PE, Wentworth SJ (2002) Interactions of bryozoans and microbes in a chemosynthetic hydrothermal vent system: Big cove formation (Lower Codroy Group, Lower Carboniferous, Middle Viséan/Arundian), Port au Port Peninsula, western Newfoundland, Canada. In: Wyse Jackson PN, Buttler CJ, Spencer Jones ME (eds) Bryozoan studies 2001. Swets & Zeitlinger, Lisse, pp 181–186
Moyano HI (2008) Bryozoa on *Calyptogena* sp. shells: a scientific note. IBA Bull 4(2):18
Nielsen C (1971) Entoproct life-cycles and the entoproct/ectoproct relationship. Ophelia 9:209–341
O'Donoghue CH, O'Donoghue E (1923) A preliminary list of Bryozoa (Polyzoa) from the Vancouver Island region. Contrib Can Biol Fish 1:143–201
Page HM, Fisher CR, Childress JJ (1990) Role of filter-feeding in the nutritional biology of a deep-sea mussel with methanotrophic symbionts. Mar Biol 104:251–257
Parson LM, Wright IC (1996) The Lau–Havre–Taupo back-arc basin: a southward-propagating, multi-stage evolution from rifting to spreading. Tectonophysics 263:1–22
Pradillon F, Shillito B, Young CM, Gaill F (2001) Developmental arrest in vent worm embryos. Nature 413:698
Ramirez-Llodra E, Shank TM, German CR (2007) Biodiversity and biogeography of hydrothermal vent species – thirty years of discovery and investigations. Oceanography 20:30–41
Reed AJ, Dorn R, Van Dover CL, Lutz RA, Vetriani C (2009) Phylogenetic diversity of methanogenic, sulfate-reducing and methanotrophic prokaryotes from deep-sea hydrothermal vents and cold seeps. Deep-Sea Res II 56:1665–1674
Richelle E, Moureau Z, Van de Vyver G (1994) Bacterial feeding by the freshwater bryozoan *Plumatella fungosa* (Pallas, 1768). Hydrobiologia 291:193–199
Sars M (1835) Beskrivelser og Iagttagelser over nogler maerkelige eller nye i Havet ved den Bergenske Kyst levende Dyr. T Hallager, Bergen

Schander C, Rapp HT, Kongsrud JA, Bakken T, Berger J, Cochrane S, Oug E, Byrkjedal I, Todt C, Cedhagen T, Fosshagen A, Gebruk A, Larsen K, Levin L, Obst M, Pleijel F, Stöhr S, Warén A, Mikkelsen NT, Hadler-Jacobsen S, Keuning R, Petersn KH, Thorseth IH, Pedersen RB (2010) The fauna of hydrothermal vents on the Mohn Ridge (North Atlantic). Mar Biol Res 6:155–171

Sellanes J, Krylova E (2005) A new species of *Calyptogena* (Bivalvia, Vesicomyidae) from a recently discovered methane seepage area off Concepción Bay, Chile (~36°S). J Mar Biol Assoc UK 85:969–976

Smith PJ, McVeagh SM, Won Y, Vrijenhoek RC (2004) Genetic heterogeneity among New Zealand species of hydrothermal vent mussels (Mytilidae: *Bathymodiolus*). Mar Biol 144:537–545

Styrt MM, Brackmann AJ, Holland HD, Clark BC, Pisutha-Arnond V, Eldridge CS, Ohmoto H (1981) The mineralogy and the isotopic composition of sulphur in hydrothermal sulphide/sulphate deposits on the East Pacific Rise, 21°N latitude. Earth Planet Sci Lett 53:382–390

Tyler PA, Young CM (1999) Reproduction and dispersal at vents and cold seeps. J Mar Biol Assoc UK 79:193–208

von Bitter PH, Scott SD, Schenk PE (1992) Chemosynthesis: an alternate hypothesis for Carboniferous biotas in bryozoan/microbial mounds, Newfoundland, Canada. Palaios 7:466–484

von Cosel R, Marshall BA (2003) Two new species of large mussels (Bivalvia: Mytilidae) from active submarine volcanoes and a cold seep off the eastern North Island of New Zealand, with a description of a new genus. Nautilus 117:31–46

Wasson K (1997) Systematic revision of colonial kamptozoans (entoprocts) of the Pacific coast of North America. Zool J Linn Soc 121:1–63

Winston JE, Håkansson E (1986) The interstitial bryozoan fauna from Capron Shoal, Florida. Am Mus Novit 2865:1–50

Wright IC (1994) Nature and tectonic setting of the southern Kermadec submarine arc volcanoes: an overview. Mar Geol 118:217–236

Young CM (2003) Reproduction, development and life history traits. In: Tyler PA (ed) Ecosystems of the deep oceans. Elsevier, London, pp 381–426

# Chapter 7
# Occurrence and Identity of "White Spots" in Phylactolaemata

## White Spots in Phylactolaemata

**Alexander Gruhl**

**Abstract** Localised epidermal gland complexes of unknown function have previously been recognised as "white spots" in *Pectinatella magnifica* and *Lophopodella carteri*, but not in other phylactolaemate species. In this study a similar glandular organ is described for *Lophopus crystallinus*. It is a complex epidermal gland that consists of two types of gland cells, one of which contains light-refracting, possibly lipidic, vesicles. This gland is situated at the anal side of the duplicature, distal to a pore, which most likely resembles the vestibular or statoblast pore. Such a pore had always been postulated, but histological evidence was lacking so far. The most likely functions of the glandular organs could either be connected to statoblast expulsion or to chemical defence.

**Keywords** Bryozoa • Ultrastructure • Confocal microscopy • Epidermal glands • Statoblast pore • Toxicity

## Introduction

The epidermis of phylactolaemate bryozoans is rich in gland cells producing a variety of secretory substances (Gasser 1962; Mukai and Oda 1980; Mukai et al. 1997). Most of these cells are scattered throughout the cystid epidermis and produce the ectocyst, the chitinous or gelatinous organic layer that forms the outer surface of the colony. In the gelatinous species *Lophopodella carteri* (Hyatt, 1866) and *Pectinatella magnifica* (Leidy, 1851) additional discrete complexes of epidermal gland cells occur, which have been described as "white bodies" or "white spots" (Mukai and Oda 1980). The functions of these organs are still unknown and their isolated occurrence in two phylactolaemate species only poses several questions.

A. Gruhl (✉)
Department of Zoology, Natural History Museum, Cromwell Road, SW7 5BD London, UK
e-mail: a.gruhl@nhm.ac.uk

A. Ernst, P. Schäfer, and J. Scholz (eds.), *Bryozoan Studies 2010*,
Lecture Notes in Earth System Sciences 143, DOI 10.1007/978-3-642-16411-8_7,

However, in a third, also gelatinous, species *Lophopus crystallinus* (Pallas, 1768) accumulations of light-refracting particles have been noticed at the anal side of the duplicature (Dumortier and van Beneden 1850; Allman 1856; Nitsche 1868; Marcus 1934). Allman named these "brilliant corpuscules" and Marcus referred to them as "fuchsinophile Zellen" according to their histochemical characteristics, but no plausible explanation as to their structure and function has been found so far. The aim of this paper is to provide a detailed morphological description of these structures and to compare the morphology and histology with the "white spots" of *L. carteri* and *P. magnifica*. My results show that these structures are glandular, thus, I propose the general term "vestibular glands" for all of them. I provide the first histological evidence of a vestibular pore in Phylactolaemata. I will also consider functional aspects of these specific structures and evaluate their potential homology based on currently available phylogenetic hypotheses.

## Material and Methods

Colonies of *Lophopus crystallinus* were sampled from the following localities: Hücker Moor, Spenge, Germany (2005), Schiedersee, Schieder-Schwalenberg, Germany (2009); Barton Blow Wells, North Lincolnshire, UK (2010). Animals were kept in aquaria on the undersides of plastic petri dishes. *Pectinatella magnifica* was collected in Lake Huffman, Dayton, Ohio, USA in June 2010. Light micrographs of live animals were taken under fibre-optics illumination with Leica DM 6 and Nikon dissecting microscopes equipped with CCD camera.

For histology and transmission electron microscopy, colonies were fixed with 2.5% glutaraldehyde in 0.01 M phosphate buffer (PB) for 4 h at 4°C, subsequently rinsed in PB, post-fixed with 2% $OsO_4$ for 30 min on ice, rinsed again and stored in storage buffer (PB + 0.05% $NaN_3$). Specimens were dehydrated in a graded ethanol series and embedded into Araldite via acetone. Semithin (0.5 μm) and ultrathin (65 nm) sections were produced on a Reichert Ultracut S microtome (Leica). Semithin sections were mounted on glass slides, stained with toluidine blue (1% toluidine, 1% $Na_2B_4O_7$, 20% sucrose) for 1 min at 60°C and sealed using Araldite and cover slips. Ultrathin sections were collected on formvar-coated single-slot copper grids and stained automatically with uranyl acetate and lead citrate. Histological examination was done with an Olympus BX 51 compound microscope, equipped with SIS CCD camera. Ultra-structural observation was accomplished on Philips CM 120 (60 kV) and Hitachi 7100 (100 kV) transmission electron microscopes.

For confocal microscopy, specimens were narcotized with 7% $MgCl_2$ and subsequently fixed with 4% paraformaldehyde in PB for 2 h at room temperature. Specimens were rinsed in PB, permeabilised with PB containing 0.25% Triton X-100 for 2 h and stained with AlexaFlour-488 conjugated phalloidin (Molecular Probes) overnight at room temperature at a final concentration of 2U/ml. Specimens were rinsed several times with PB, counterstained with propidium iodide (1:250) for

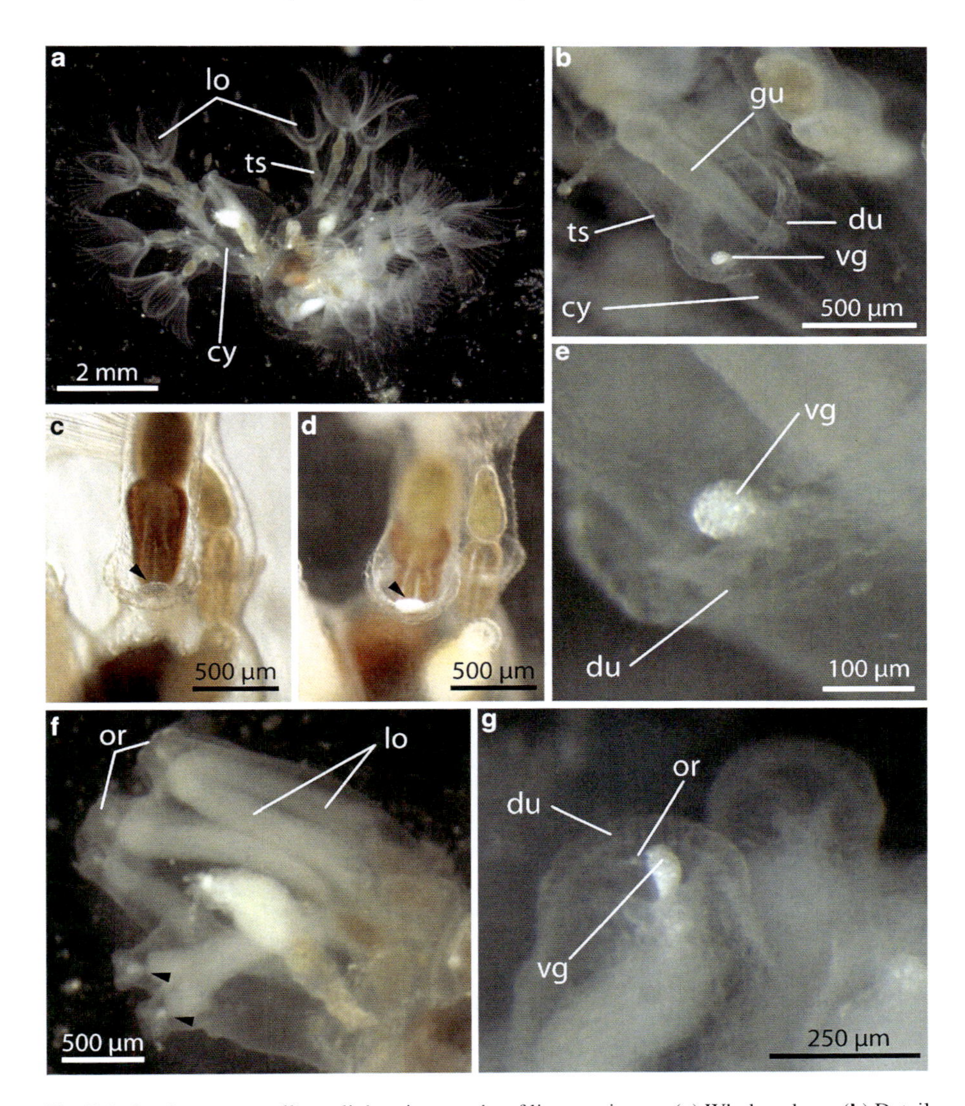


**Fig. 7.1** *Lophopus crystallinus*, light micrographs of live specimens. (**a**) Whole colony. (**b**) Detail caption of the duplicature region. The vestibular gland is situated at the anal side, just between duplicature and tentacle sheath. (**c**, **d**) Same zooid under direct illumination (**c**) and illumination from the side (**d**), demonstrating the differences in appearance of the vestibular gland (*arrowheads*). (**e**) The vestibular gland contains small light refracting granules. (**f**) Part of colony with retracted lophophores. (**g**) Detail of lophophore retraction. The duplicature forms the rim of the orifice. The vestibular gland is just inside the orifice. cy – cystid, du – duplicature, gu – gut, lo – lophophore, or – orifice, ts – tentacle sheath, vg – vestibular gland

20 min, rinsed again and mounted on glass slides in 90% glycerol, 10% PB, 0.25% DABCO. Confocal image stacks were taken on a Leica TCS SP microscope. Image data were viewed and processed using MBF ImageJ (McMaster Biophotonics Facility, Ontario, Canada).

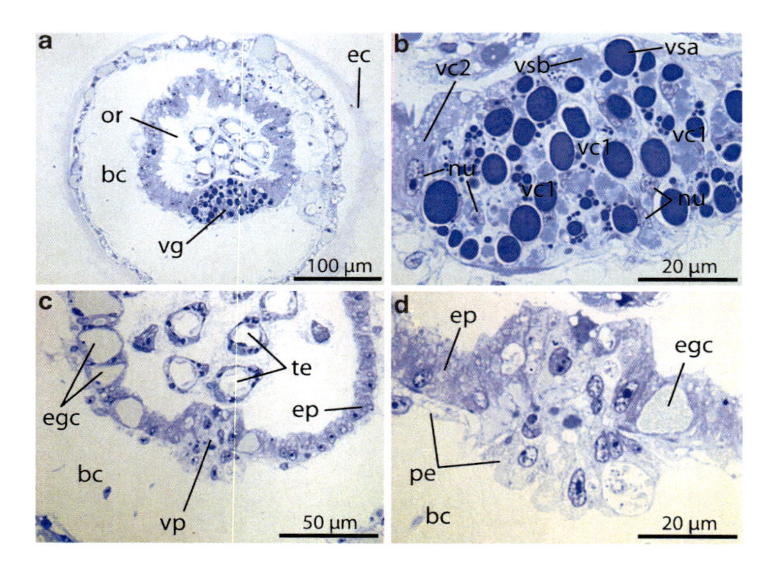


**Fig. 7.2** *Lophopus crystallinus*, toluidine blue stained semithin cross sections through zooid with retracted lophophore. (**a**) The vestibular gland is situated at the distal rim of the orifice, facing into the orifice. (**b**) The centrally locate type 1 gland cells (vc1) contain large secretory vesicles (vsa, vsb). A second type of glandular cells (vc2) surrounds the type 1 cells. (**c**) This cross section shows the vestibular pore and is 15 µm proximal to the cross section shown in (**a**). (**d**) The pore is recognisable as swelling of both the epidermis and the peritoneum. bc – body cavity, ec – ectocyst, egc – epidermal gland cell, ep – epidermis, nu – nucleus, or – orifice, pe – peritoneum, te – tentacle, vc1 – type 1 gland cell, vc2 – type 2 gland cell, vg – vestibular gland, vp – vestibular pore, vsa – type A vesicles, vsb – type B vesicles

# Results

## Lophopus crystallinus *(Pallas, 1768)*

The vestibular gland of *Lophopus crystallinus* is situated on the anal side of the duplicature (Figs. 7.1a, b and 7.4a, b). The duplicature is an annular bulge of the body wall that forms the boundary between the tentacle sheath and the apical part of the cystid. When viewed under direct illumination or transmitted light the vestibular gland is transparent, appearing only as inconspicuous hillock with a granular structure (Fig. 7.1c). Under illumination from the side and against a dark background it becomes visible due to its strong light-refracting characteristics (Fig. 7.1d). When the lophophore is everted, the gland appears either flattened (Fig. 7.1d) or mushroom-like (Fig. 7.1e), however, transitions between these two appearances in the same individual have not been observed, although a well-developed musculature is present near the gland (see below). The size can be quite variable, generally increasing with the age of the zooid. The largest vestibular

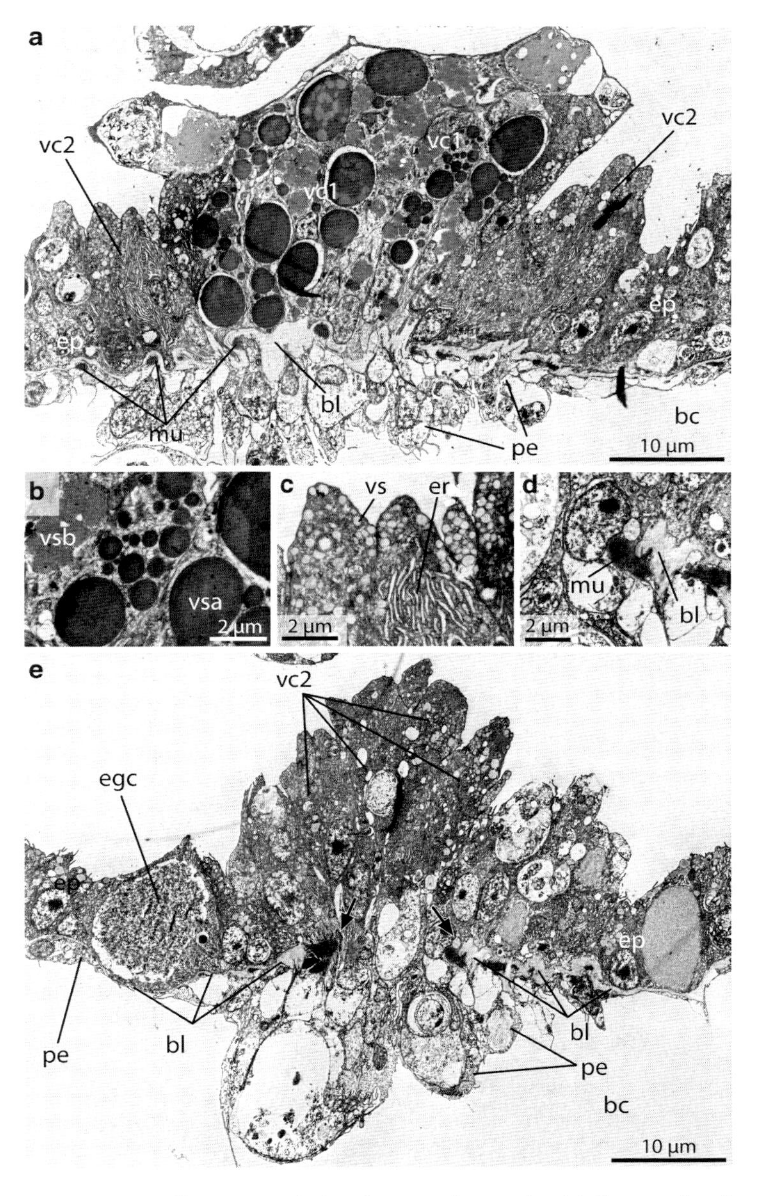


**Fig. 7.3** *Lophopus crystallinus*, transmission electron micrographs. (**a**) Cross section through the vestibular gland showing distribution of the two gland cell types. (**b**) Type 1 gland cells contain homogeneously electron-dense, roundish vesicles (vsa) as well as irregularly shaped, lobate vesicles (vsb) with electron-lucent flocculent content. (**c**) Type 2 cells bear small electron-lucent vesicles in their apical regions and show a pronounced endoplasmic reticulum. (**d**) Detail of muscles inserting at the point where the basal lamina ends. (**e**) Cross section through the vestibular pore. Both the ectodermal and the peritoneal layer are considerably thickened. The cells in the ectoderm resemble the type 2 glandular cells of the vestibular gland. The basal lamina that lies

glands found were ~200 μm in width and ~100 μm in thickness. There is also considerable variation across colonies. The gland is densely packed with translucent refracting vesicles (Fig. 7.1e). When the lophophore is retracted, the duplicature will form the rim of the orifice. This way the gland becomes situated directly in the orifice, sometimes even somewhat exposed to the exterior (Figs. 7.1f, g and 7.2a).

The vestibular gland comprises two types of glandular cells. Type1 cells form the central region of the vestibular gland (Figs. 7.2b and 7.3a). They are inversely flask-shaped settling with their narrower part on the basal lamina. Nuclei are found either in the basal or in the apical part of the cytoplasm. Apart from the usual cytoplasmic contents, two types of vesicles occur (Figs. 7.2b and 7.3b). Type A vesicles are almost homogeneously electron-dense in ultrastructure and appear dark blue in toluidine-stained sections. These vesicles occur in different sizes, the smallest ones are spherical with a diameter of ~100 nm, the largest are oval, measuring ~5 μm in length. In the largest vesicles, the membrane is often detached from the vesicular content, leaving a bright halo, however, this might be a fixation artefact. The histological and ultrastructural characteristics suggest lipid-like contents. Vesicles of type B are of similar size range and also membrane-bound, but they appear lobate and their contents are spongy and more electron-lucent. Toluidine staining results in a light blue color. Type 2 gland cells form a ring around the type 1 cells in the periphery of the vestibular gland (Figs. 7.2b and 7.3a). They are smaller than the type 1 cells, only slightly higher than the other epidermal cells. They have basal nuclei and a very pronounced endoplasmic reticulum. In their apical regions they bear numerous small electron-lucent vesicles (Fig. 7.3c). Some of these cells form thin apical extensions that form a layer covering the central type 1 cells (Fig. 7.3a). The basal lamina underneath the vestibular gland is rather thick (up to 3 μm) and considerably wrinkled (Fig. 7.3a). Several muscle strands are found attaching to the inner side of the basal lamina. The cells of the vestibular gland are clearly distinguishable from other epidermal gland cells in their histological and ultrastructural characteristics as well as by fluorescence microscopy (Figs. 7.2b, c, 7.3a, e, and 7.4b–d).

The vestibular pore is situated directly adjacent to the proximal side of the vestibular gland (Figs. 7.2c and 7.4c–e). In cross section it appears as a bulge of both the endodermal and ectodermal layer (Figs. 7.2d and 7.3e). This bulging is most likely caused by musculature that is associated with the pore and can close it like a sphincter (Figs. 7.3d, e and 7.4c–e). But even in a closed state, the existence of the pore can be proven by the fact that the basal lamina, which is situated between the epidermis and peritoneum throughout the entire body wall, is

**Fig. 7.3** (continued) between epidermis and peritoneum is interrupted in the centre of the pore (*arrows*). Here peritoneal and epidermal cells are directly adjacent. The gap is surrounded by strong musculature. bc – body cavity, bl – basal lamina, egc – epidermal gland cell, ep – epidermis, mu – muscles, pe – peritoneum, vc1 – type 1 gland cell, vc2 – type 2 gland cell, vsa – type A vesicles, vsb – type B vesicles

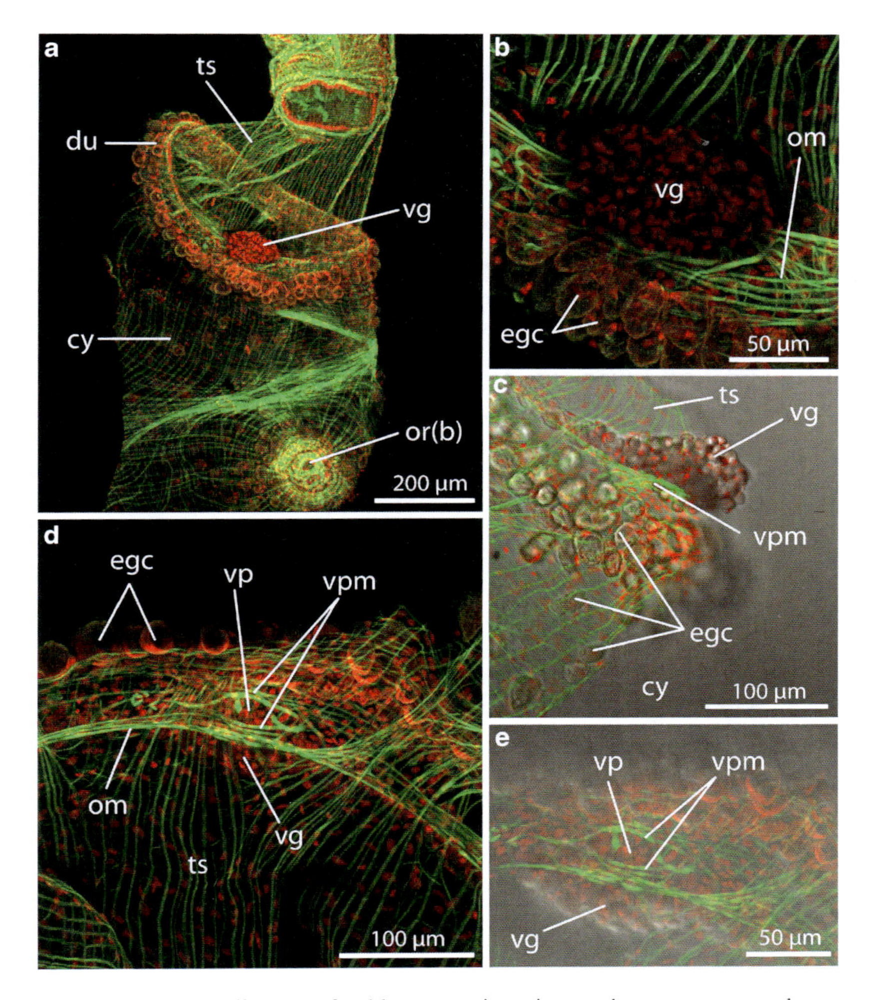


**Fig. 7.4** *Lophopus crystallinus*, confocal laser-scanning micrographs, *green* = musculature, *red* = nuclei. Maximum intensity projections of 3D stacks. (**a**) Duplicature region. (**b**) Detail of the vestibular gland. Due to the light-refracting nature of the gland, structures beneath appear shaded. (**c**) A lateral view shows the pronounced musculature of the vestibular pore being located proximally adjacent to the vestibular gland. (**d**) Dissected piece of the body wall oriented with the peritoneal layer to the *top*, epidermis with vestibular gland to the *bottom*. Short curved muscles surround the vestibular pore. (**e**) Detail of d, merged with transmitted light micrograph to demonstrate special relationship of vestibular pore and its musculature to the vestibular gland. cy – cystid, du – duplicature, egc – epidermal gland cell, om – orifice sphincter muscle, or – orifice, ts – tentacle sheath, vg – vestibular gland, vp – vestibular pore, vpm – vestibular pore musculature

interrupted at the pore (Fig. 7.3d, e), leaving a gap of ~12 µm. The cells of the epidermis and those of the peritoneal layer are directly adjacent at the rim of the gap.

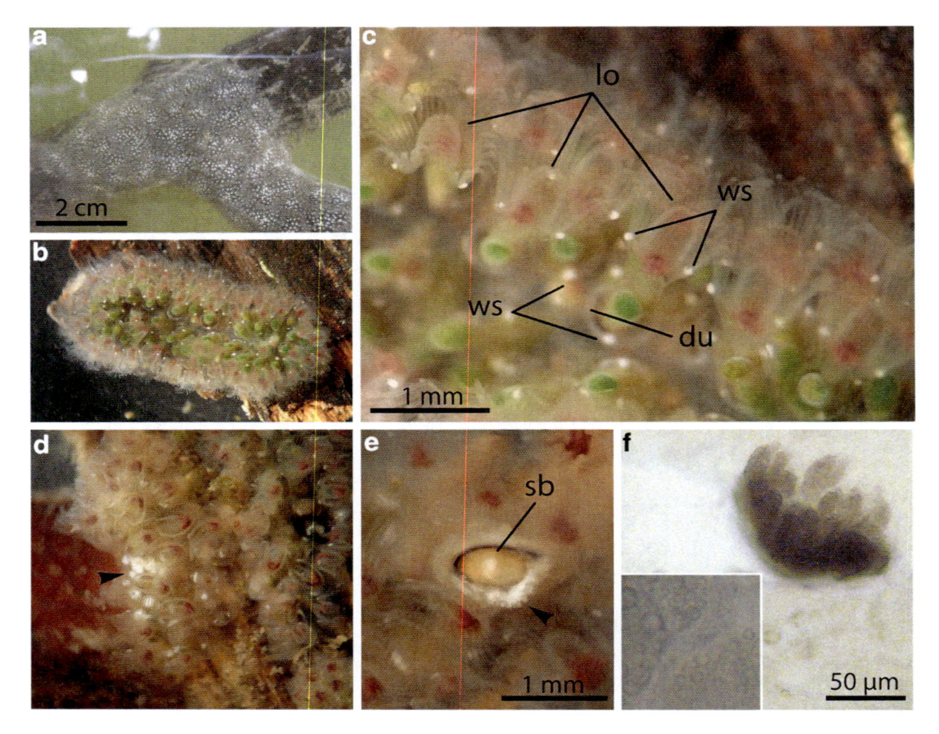


**Fig. 7.5** *Pectinatella magnifica* (**a**) Photo of medium-sized colony on piece of wood. (**b–f**) Light micrographs. (**b**) Small colony. (**c**) Detail of colony showing the white spots. One white spot is at the end of each lophophore arm. One, sometimes also two spots are located on the anal side of the duplicature. (**d**) White spots appear pronounced in regions of injury (*arrowhead*). (**e**) A statoblast was observed to be released through a passage that must be very close to the white spot, because similar looking tissue (*arrowhead*) is seen close to the statoblast. (**f**) Light micrographs of the lophophoral white spot at 400× and 1,000× (inset) magnification. du – duplicature, lo – lophophore, sb – statoblast, ws – white spot

## Pectinatella magnifica *(Leidy, 1851)*

White spots in *P. magnifica* occur at two different locations: at the end of each lophophore arm and at the anal side of the duplicature (Fig. 7.5b, c). In the latter location mostly one spot occurs, however, it is not situated in the midline of the zooid, but slightly offsets to the right or left hand side. In some cases also two white spots are present at the duplicature. In larger colonies the spots form a patchy pattern, which reflects the arrangement of the zooids in clusters with the zooids' anal sides facing to the centre of the clusters (Fig. 7.5a). In areas where a colony has faced a major injury, accumulations of white spots occur (Fig. 7.5d). Probably they remain in the colony surface, when a zooid degenerates. The white spots of the duplicature must be located closely to a putative statoblast pore in this species as well. In one case the expulsion of a statoblast was witnessed, and white spot-like tissue was seen near the emerging statoblast (Fig. 7.5e). With a compound microscope it becomes visible that the white spot is made up of cells, which contain very small, clear, droplet-shaped vesicles (Fig. 7.5f).

## Discussion

Although the structure that is described in the present paper as a vestibular gland did attract the attention of previous workers, its glandular nature was not recognized. Allman (1856) suspected a ring-like canal to be situated in the duplicature that was filled with "peculiar, spherical or oval brilliant corpuscules". He had also observed that these developed in the entire ectocyst in specimens kept in aquaria for a few days. Marcus (1934) described the structures as residing in the coelom and suspected an excretory function. The present data show clearly that the structures are epidermal glands, very similar to the "white spots" of *L. carteri* and *P. magnifica*. However, Mukai and Oda (1980) found differences concerning the histochemical properties between the latter two species, and a comparison of the ultrastructure of the glandular vesicles of *P. magnifica* (Mukai et al. 1997) to those of *L. crystallinus* shows most similarity to the electron-dense vesicles of type 1 cells in the latter species. The vesicles in *P. magnifica*, however, are smaller, only up to 1 μm, and in addition have an electron lucent core. These differences could eventually cause the different optical properties.

Although the content of the vesicles (proteins, lipids) are potentially of high nutritive value, a major function in energy storage for primary metabolism is unlikely as the organs are very small in relation to the body size. Given the size and position of the structures, a significant contribution to the formation of the gelatinous ectocyst appears unlikely. Histochemical studies by Mukai and Oda (1980) do not support this either.

Expulsion of statoblasts through a putative pore on the anal side of a zooid has been reported in *Stolella evelinae*, *Stolella agilis*, and *Hyalinella carvalhoi* by Marcus (1941, 1942) and *Plumatella emarginata*, *P. repens* and *P. fruticosa* by Wiebach (1952, 1953). Although histological studies had always failed to document such a pore it has repeatedly been incorporated into later general anatomical accounts of phylactolaemates. The present results demonstrate for the first time that at least in *L. crystallinus* a vestibular pore is definitively present. The presence of musculature at the pore suggests that it can be actively opened or closed by the animals. The position of the glandular cells close to this pore could point to a possible function in secreting, e.g., lubricating, protective or adhesive substances for the statoblasts. However, this explanation does not apply for the white spots in *L. carteri* and the lophophoral white spots in *P. magnifica*.

Excretory organs have not been shown convincingly so far and it appears likely that these are generally lacking in Bryozoa (see Gruhl et al. 2009). The only exception so far could be a bladder described on the anal side of the forked canal in *Cristatella mucedo* (e.g., Marcus 1934; Schwaha et al. 2011). A simple pore as demonstrated in the present study, however, would not be sufficient as an excretory organ alone. In all animals excretion is a selective process that involves separation of waste substances from essential ions and metabolites. This is mostly accomplished either by ultrafiltration of body cavity fluid and subsequent modification of the primary urine or by selective transcytosis. However, other mechanisms might exist: in brachiopods, for example, waste metabolites are thought to be taken up by specialized coelomocytes, which are subsequently discharged via nephridial canals (James 1997). A large variety of coelomocytes is also found in bryozoans (Mano 1964). Although release of coelomic fluid including its coelomocytes has been observed, it is still unclear whether this is occasional or a regular phenomenon (Marcus 1934; Oda 1958; Mano 1964).

Freshwater bryozoans constitute a common prey of fishes and various invertebrates like, e.g., insect larvae, isopods, mites, and flatworms (Bushnell 1966; Wood and Okamura 2005). Lacking physical defense mechanisms like strong skeletal reinforcement or avicularia, phylactolaemates are very likely to employ chemical weapons against predators, as well as against parasites or epibionts. A toxic effect of *Lophopodella carteri* to fish was firstly reported by Rogick (1957). Experimental exposition of fish to whole colony homogenates demonstrated fatal effects predominantly by severe histological damage to gill tissue (Tenney and Woolacott 1964; Collins et al. 1966). Meacham and Woolacott (1968) could show that the toxin is not specific to fish and located in the bryozoan tissue rather than the coelomic fluid. This makes it possible that it is located in the white spots. Peterson (2002) found evidence that the substance has neurotoxic effects and is probably an alkaloid. Lauer et al. (1999) have shown that *L. carteri* can inhibit recruitment of zebra mussels (*Dreissena polymorpha*), possibly by inhibitory substances. About 200 bioactive substances have so far been found in marine bryozoans or their bacterial symbionts (Sharp et al. 2007). The white spots in *L. crystallinus* and *P. magnifica* are in a good position to defend the uncovered orifice when the lophophore is retracted as bite attacks by macropredators would very likely cause discharge of the vesicle contents. In *P. magnifica* the lophophoral white spots would further help protecting the exposed lophophore. In the latter species, the peculiarity of the white spots could additionally be an aposematic adaptation against visually oriented predators like fish.

Although the phylogeny of phylactolaemates is not yet fully resolved, recent phylogenetic analyses provide consistent support for the monophyly of a family Lophopodidae comprising the genera *Lophopus*, *Lophopodella* and *Asajirella* (Wood and Lore 2005; Okuyama et al. 2006; Hirose et al. 2008). The position of *Pectinatella magnifica*, however, is elusive, with affinities either to Plumatellidae, Fredericellidae or Cristatellidae. Thus, there is no common ancestor from which only those species bearing vestibular glands would have derived. If assuming such a structure as ancestral for Lophopodidae it would have to be reduced (or not yet discovered) in *Asajirella gelatinosa*. However, the white spots could also be derived from ancestral bryozoan features. Two types of specialized glands have been reported from cheilostome bryozoans: autozooidal vestibular glands and avicularian glands. The former are epidermal invaginations, either paired at the tentacle sheath or unpaired near the orifice (Waters 1892; Calvet 1900; Lutaud 1964). They are thought to excrete mucous substances or contain toxins or bacteria (Lutaud 1965; Winston and Bernheimer 1986). Avicularian glands are situated closely to the vestigial polypide and regarded homologous to the vestibular glands (Waters 1892; Marcus 1939; Carter et al. 2010). However, more ultrastructural data are needed for a detailed comparison of gymnolaemate and phylactolaemate epidermal glands. As statoblasts are an apomorphic character of phylactolaemates, a vestibular opening is also likely to have originated in a phylactolaemate ancestor. However, not all phylactolaemates possess floatoblasts, and expulsion has only been observed in few species, thus more data on the presence of a vestibular pore in further phylactolaemate species are still needed.

The present study demonstrates that there are still many "white spots" in our knowledge on bryozoan morphology. In future studies, it would be desirable to survey further phylactolaemate species for the occurrence of vestibular glands and pores. A next step in order to unravel the function of these organs might be a detailed chemical analysis of the secretions produced by the vestibular glands. Further studies could also focus on the variation of this character across individuals or populations and correlations to external biotic and abiotic factors like, e.g., presence of predators, parasites, or pathogens. A closer examination of the process of statoblast expulsion including the state of the vestibular gland before and after this can yield valuable insights.

**Acknowledgments** I wish to thank Hanna Hartikainen (EAWAG, Zürich, Switzerland), Beth Okamura (Natural History Museum, London, UK) and Tim Wood (Wright State University, Dayton, OH, USA) for fruitful discussions on the topic. I am also most grateful to Tim for his hospitality during a stay in Dayton. Many thanks to Lauren Howard, Gabrielle Kennaway and Alex Ball (Natural History Museum, London, UK) for microscopy support.

## References

Allman GJ (1856) A monograph of the fresh-water Polyzoa. Ray Soc Lond 18:1–119

Bushnell JH (1966) Environmental relations of Michigan Ectoprocta, and dynamics of natural populations of *Plumatella repens*. Ecol Monogr 36:95–123

Calvet L (1900) Contribution à l'histoire naturelle des bryozoaires ectoproctes marins. Trav Inst Zool Univ Montpellier 8:1–488

Carter MC, Gordon DP, Gardner JPA (2010) Polymorphism and vestigiality: comparative anatomy and morphology of bryozoan avicularia. Zoomorphology 129:195–211

Collins EJ, Tenney WR, Woolcott WS (1966) Histological effects of the poison of *Lophopodella carteri* (Hyatt) on the gills of *Carassius auratus* (Linnaeus) and larval *Ambystoma opacum* (Gravenhorst). Va J Sci 17:155–163

Dumortier BC, van Beneden PJ (1850) Histoire naturelle des polypiers composés d'eau douce des bryozoaires fluviatiles. Hayez, Bruxelles

Gasser F (1962) L'épiderme du cystide de *Plumatella repens* (Linné) (bryozaire ectoprocte phylactolème) precisions histologiques, cytologiques et histochimiques. Arch Zool Exp Gén 101:1–13

Gruhl A, Wegener I, Bartolomaeus T (2009) Ultrastructure of the body cavities in Phylactolaemata (Bryozoa). J Morphol 270:306–318

Hirose M, Dick MH, Mawatari SF (2008) Molecular phylogenetic analysis of phylactolaemate bryozoans based on mitochondrial gene sequences. In: Hageman GS, Key MM Jr, Winston JE (eds) Bryozoan studies 2007. Virginia Museum of Natural History, Martinsville, pp 65–74

Hyatt A (1866) Observations on polyzoan order Phylactolaemata. Communications of the Essex Institute 4:197–228

James MA (1997) Brachiopoda: internal anatomy, embryology, and development. In: Harrison FW, Woollacott RM (eds) Microscopic anatomy of invertebrates, vol 13, Lophophorates, entoprocta and cycliophora. Wiley-Liss, New York, pp 298–407

Lauer TE, Barnes DK, Ricciardi A, Spacie A (1999) Evidence of recruitment inhibition of zebra mussels (*Dreissena polymorpha*) by a freshwater bryozoan (*Lophopodella carteri*). J N Am Benthol Soc 18:406–413

Leidy J (1851) On Cristatella magnifica. Proceed. Acad. Nat. sc. Philadelphia 5:265–266

Lutaud G (1964) Sur la structure et le rôle des glandes vestibulaires et sur la nature de certains organes de la cavité cystidienne chez les bryozoaires chilostomes. Cah Biol Mar 5:201–231

Lutaud G (1965) Sur la présence de microorganismes spécifiques dans les glandes vestibulaires et dans l'aviculaire de *Palmicellaria skenei* (Ellis et Solander), bryozoaire chilostome. Cah Biol Mar 6:181–190

Mano R (1964) The coelomic corpuscles and their origin in the freshwater bryozoan, *Lophopodella carteri*. Sci Rep Tokyo Kyoiku Dai (sec B) 11:211–235

Marcus E (1934) Über *Lophopus crystallinus* (Pall.). Zool Jb Anat Ontog Tiere 58:501–606

Marcus E (1939) Briozoários marinhos brasileiros III. Bol Fac Fil, Cienc Letr Univ Sao Paulo, Zool 3:111–353

Marcus E (1941) Sôbre Bryozoa do Brasil. Bol Fac Fil, Cienc Letr Univ Sao Paulo, Zool 5:3–208

Marcus E (1942) Sôbre Bryozoa do Brasil II. Bol Fac Fil, Cienc Letr Univ Sao Paulo, Zool 6:57–105

Meacham RH, Woolacott WS (1968) Studies of the coelomic fluid and isotonic homogenates of the freshwater bryozoan *Lophopodella carteri* (Hyatt) on fish tissues. Va J Sci N S 19:143–146

Mukai H, Oda S (1980) Histological and histochemical studies on the epidermal system of higher phylactolaemate bryozoans. Annot Zool Japan 53:1–17

Mukai H, Terakado K, Reed CG (1997) Bryozoa. In: Harrison FW, Woollacott RM (eds) Microscopic anatomy of invertebrates, vol 13, Lophophorates, Entoprocta and Cycliophora. Wiley-Liss, New York, pp 45–206

Nitsche H (1868) Beiträge zur Anatomie und Entwickelungsgeschichte der phylactolaemen Süsswasserbryozoen insbesondere von *Alcyonella fungosa*. Dissertation, Friedrich-Wilhelms-Universität zu Berlin

Oda S (1958) On the outflow of the blood in colonies of freshwater Bryozoa. Kagaku (Tokyo) 28:37

Okuyama M, Wada H, Ishii T (2006) Phylogenetic relationships of freshwater bryozoans (Ectoprocta, Phylactolaemata) inferred from mitochondrial ribosomal DNA sequences. Zool Scr 35:243–249

Pallas PS (1768) Descripto Tubulariae fugosa propae Volodemirum mense Julio 1768 observatae. Novi Commentarii academiae scientiarum imperialis Petropolitanae 12:565–572

Peterson NL (2002) Neurotoxin found in the freshwater bryozoan *Lophopodella carteri*. In: Wyse Jackson PN, Spencer Jones ME (eds) Bryozoan Studies 2001. Sets and Zeilinger, Lisse, pp 257–260

Rogick MD (1957) Studies on freshwater Bryozoa, XVIII *Lophopodella carteri* in Kentucky. Trans Ky Acad Sci 18:85–87

Schwaha T, Handschuh S, Redl E, Walzl M (2011) Organogenesis in the budding process of the freshwater bryozoan *Cristatella mucedo* Cuvier, 1798 (Bryozoa, Phylactolaemata). J Morphol 272:320–341

Sharp JH, Winson MK, Porter JS (2007) Bryozoan metabolites: an ecological perspective. Nat Prod Rep 24:659–673

Tenney WR, Woolacott WS (1964) A comparison of the responses of some species of fishes to the toxic effect of the bryozoan, *Lophopodella carteri* (Hyatt). Va J Sci 15:16–20

Waters AW (1892) Observations on the gland-like bodies in the Bryozoa. Zool J Linn Soc 24:272–278

Wiebach F (1952) Über den Ausstoss von Flottoblasten bei *Plumatella fruticosa* (Allman). Zool Anz 149:181–185

Wiebach F (1953) Über den Ausstoss von Flottoblasten bei Plumatellen. Zool Anz 151:266–272

Winston JE, Bernheimer AW (1986) Haemolytic activity in an Antarctic bryozoan. J Nat Hist 20:369–374

Wood TS, Lore M (2005) The higher phylogeny of phylactolaemate bryozoans inferred from 18S ribosomal DNA sequences. In: Moyano HI, Cancino JM, Wyse Jackson PN (eds) Bryozoan studies 2004. Balkema, London, pp 361–367

Wood TS, Okamura B (2005) A new key to the freshwater bryozoans of Britain, Ireland and Continental Europe. Freshwater Biological Association, Cumbria, UK

# Chapter 8
# Testing Habitat Complexity as a Control over Bryozoan Colonial Growth Form and Species Distribution

## Colonial Growth Form Ecology

**Steven J. Hageman, Frank K. McKinney (deceased), and Andrej Jaklin**

**Abstract** The aim of this study is to test the effects of fine scale (microhabitat) environmental variation on the distribution of bryozoan species and potential variation in growth habit diversity and disparity. Data are derived from six microhabitats in replicate, on designed apparatuses, providing surfaces of varied complexity and orientation. The apparatuses were deployed on a sediment substrate at 24 m depth offshore of Rovinj, Croatia and recovered 14 months later. Species distributions were documented for each microhabitat and indexed for relative abundance. Twenty-five bryozoan species were recorded in multiple $0.5 \times 0.5$ cm cells in multiple patches on each microhabitat. Species richness was relatively uniform in each microhabitat, but most individual species and several growth habit attributes differed in abundance or presence among microhabitats.

**Keywords** Recruitment • Species richness • Growth habit • Ecology • Adriatic Sea

## Introduction

This study arose from a long-term search for a better understanding of the environmental factors that control the distribution and abundance of bryozoan colony growth forms and species richness in space and time. The broad correlation of sedimentary environments with the dominance and diversity of colony growth forms (Stach 1936) has been used in reverse in many publications in which growth forms of ancient bryozoans have been used to infer paleoenvironments.

S.J. Hageman (✉) • F.K. McKinney (deceased)
Department of Geology, Appalachian State University, Boone, NC, USA
e-mail: hagemansj@appstate.edu

A. Jaklin
Ruđer Bošković Institute, Center for Marine Research, Rovinj, Croatia
e-mail: jaklin@cim.irb.hr

A. Ernst, P. Schäfer, and J. Scholz (eds.), *Bryozoan Studies 2010*,
Lecture Notes in Earth System Sciences 143, DOI 10.1007/978-3-642-16411-8_8,

Although the ultimate goal of this project is to understand controls on growth habits, this paper deals only with microhabitats and the bryozoan fauna found in them. The question addressed is: Given a range of microhabitats expressed by textural complexity and orientation, is there an apparent effect on species richness and difference in microhabitat specificity of taxa and growth habit?

## *Previous Work*

There is a rich literature reporting results of in situ and of laboratory studies of bryozoans' behaviour and/or survivorship in response to specific environmental influences on the entire life cycle from embryo to sexually mature colony. Citing only single examples from the literature base of representative topics, successful completion of a life cycle involves seasonality of larval release (Mariani et al. 2005), larval swimming duration and dispersal distance (Pemberton et al. 2007), larval behaviour before (Ryland 1977) and after (Burgess et al. 2009) contact with a potential substrate, substrate pre-emption (Sutherland and Karlson 1977), substrate quality (Dobretsov and Qian 2006), and post-settlement interactions with competitors (Barnes and Dick 2000) and predators (Lidgard 2008).

All of these topics – and others – are critically important to whether or not a given species is present or absent on a given substrate. For this study, however, they comprise a contextual background for an abundant pool of larvae from which a species-rich (Hayward and McKinney 2002) bryozoan fauna could potentially be recruited onto various microhabitats provided within experimental apparatuses.

# Material and Methods

Apparatuses consisting of multiple substrates (e.g. panels) with varied orientations and textures replicated in each apparatus were constructed for this study (Fig. 8.1a). The goals of the design were to provide equal access for bryozoan larvae to varied microhabitats (treatments) within a local environment, but with minimal interaction among microhabitats. All substrates were composed of plastic and the surfaces were mildly abraded with fine sandpaper. Treatment panels were attached to the frame by plastic cable ties via holes drilled in the PVC pipes. Each of the microhabitats provided a minimum of 600 $cm^2$ available for settlement.

## *Apparatuses and Microhabitat*

The microhabitats created for this study (Fig. 8.1a) included: (1) vertical smooth panels (Fig. 8.1f), (2) strings of nylon netting (0.5 mm diameter) composed of

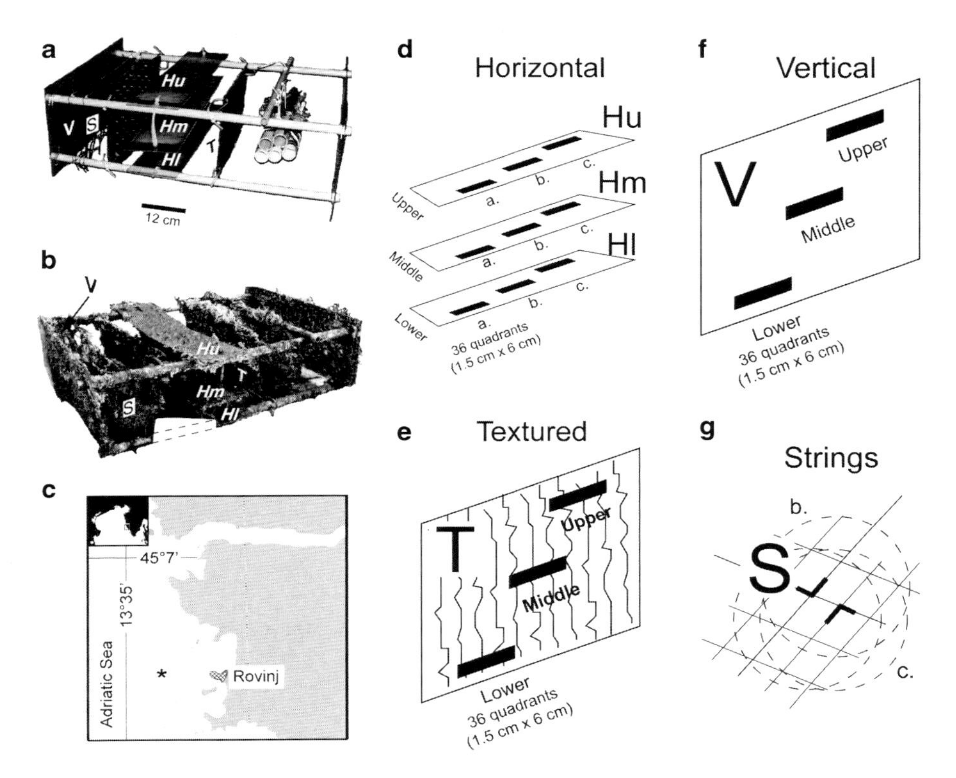


**Fig. 8.1** Experimental apparatus and deployment location. (**a**) PCV-supported microhabitat panel prior to deployment; (**b**) Microhabitat panel after deployment. V = vertical, S = Strings, H = Horizontal, u = upper, m = middle, l = lower, T = textured; (**c**) Location of deployment in the northeastern Adriatic Sea (*), approximately 2.0 km WSW of Rovinj peninsula, beneath oceanographic observation buoy at 45° 04.960′ N, 13° 36.233′ E, (**d–g**), Microhabitat panels

multiple twisted strands (Fig. 8.1g), (3) the under side of horizontal panels mounted separately on the upper, middle and lower part of the device (Fig. 8.1d), (4) vertical panels with an irregular corrugated surface with ~2.0 mm of relief (Fig. 8.1e). Other microhabitat panels were used in the apparatuses, but are not included in this study.

Two replicate apparatuses (A and B) were deployed beneath the moored oceanographic observation buoy 2.0 km WSW of the peninsula of the town of Rovinj (Fig. 8.1c). The apparatuses were deployed by SCUBA in March of 2007 in 24 m water depth and secured to the sandy sea floor approximately 2 m apart. The apparatuses were recovered by SCUBA in May 2008, with no signs of significant natural or artificial disruption to their placement.

Apparatuses were returned to the laboratory in sea water and photographed whole (Fig. 8.1b) immediately upon exposure to the air. Apparatuses were then disassembled and each microhabitat surface photographed in its entirety in the original wet state with 7.1 Mb digital images. Each treatment panel was then thoroughly rinsed in fresh water and allowed to air dry in a natural setting. Large soft-bodied organisms such as sponges and ascidians and macro-invertebrates such as ophiuroids and polychaetes were noted and removed. Dried treatment panels were individually wrapped and transported to Appalachian State University for detailed study. Due to their fragile state, bryozoan specimens were studied with dried cuticle and soft tissue intact.

## Experimental Design

Data for this study are hierarchically arranged. In descending order, levels are:

5. *Replicate* apparatuses: A and B at the one locality (observation buoy)
4. *Microhabitat*: six within each replicate, Horizontal Upper, Horizontal Middle, Horizontal Lower, Vertical, Textured, Strings (Fig. 8.2).
3. *Patch*: three (1.5 × 6 cm) within each microhabitat.
2. *Grid*: 36 (0.5 × 0.5 cm) grids within each Patch (3 × 12)
1. *Occurrence* of species: scored as 1 for present and 0 for absent.

The fundamental unit of observation in this study is presence or absence (occurrence) of each bryozoan species in a (0.25 $cm^2$) grid cell. Within five of the microhabitats each 1.5 × 6 cm patch, represented by black strips on panels in Fig. 8.1, was divided into 36 grid squares each 0.5 × 0.5 cm. Thus, the maximum score for each species for each patch is 36 and each microhabitat was 108 for each replicate apparatus (total pooled 216 per treatment). For the String microhabitat, the netting was cut between each knot, resulting in an "X" (Fig. 8.1g). The surface area of one half of this unit, a "V" is approximately equivalent to one grid (0.25 $cm^2$) and thus was used as a unit of observation. The string "Vs" were randomly placed in three groups of 36 (n = 108) for each Apparatus Replicate (position within the string treatment was not identified).

The experimental design of this study was exploratory and does not allow for a full factorial partitioning of sources of variation. Other potential sources of variation (distribution of colonies of bryozoan species) include: (1) positional (edge) effects within the apparatus and among microhabitats (effects presumed to be minimal, but unknown), (2) positional (edge) effects within microhabitats (effects apparently significant, and partially accounted for in this study), (3) compositional effects of different plastics among the microhabitat substrates (effects presumed to be minimal). In spite of these limitations, results from this study can provide valuable insight into the distribution of bryozoan species at the microhabitat level.

## Data Collection

Panels (microhabitats) were examined with an Olympus SZX 12 microscope with a field of view of approximately 1.5 × 4 cm, and digital photos were taken. A 3 × 4

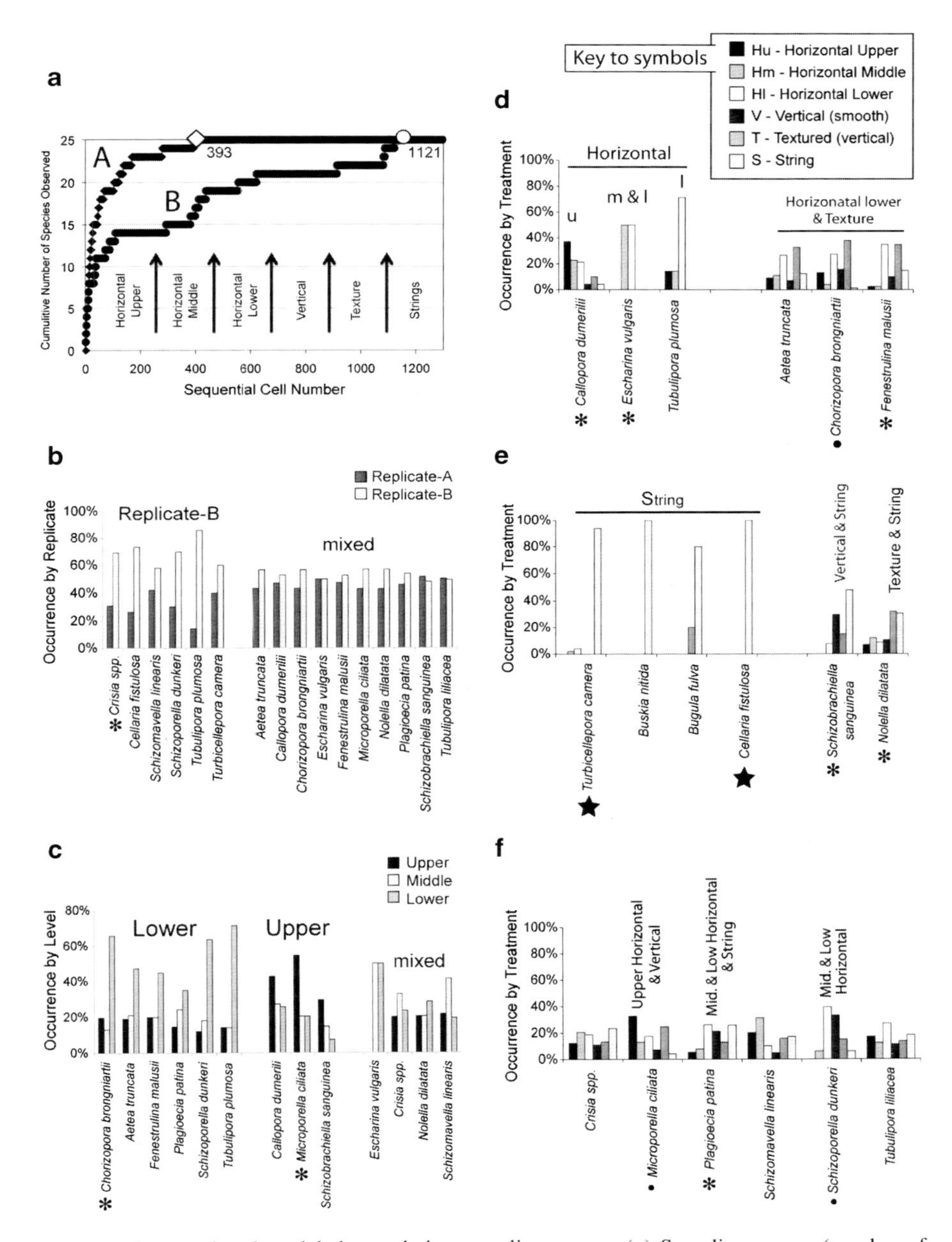

**Fig. 8.2** Observed and modeled cumulative sampling curves. (**a**) Sampling *curves* (number of species observed per successive observation) for A = data randomized by collecting sequence and B = observed by collected through successive microhabitats. All 25 species were encountered by the 393rd cell in the random distribution, but the same 25 species were not observed until the 1,121st cell was censused in the actual sequence collected sequentially by microhabitat; (**b–f**) Relative abundance of common species (present in 14 or more cells) calculated within taxa among

grid (0.25 $cm^2$ each cell) was digitally burned into the image and printed as a reference map. Using higher magnification, each grid cell was evaluated using the reference map and all bryozoan species were identified using Hayward and McKinney (2002) and scored 1 if present and 0 if absent. Criteria for occurrence counting was based on presence/absence of a species in a cell, not frequency or dominance, thus: (1) species with multiple colonies in a cell, whatever their origin, were counted as a single occurrence; (2) a single colony overlapping into two adjacent cells was counted as occurring once in each cell; (3) colonies represented only by ancestrulae or the primary zone of astogenetic variation were not counted.

## Data Analysis

### *Sampling Curve to Evaluate Species Heterogeneity Among Microhabitats*

If bryozoan species are randomly distributed across all microhabitats within an apparatus, then a plot of the number of successive observations, here grid cells, against the cumulative number of species observed should plot as a relatively smooth curve (Sanders 1968). The curve is expected to initially increase rapidly and then asymptotically approach the maximum number of observed species for the assemblage. A sampling curve was modeled subsequent to data collection by generating a curve of species richness from data randomized by the order of their collection (Fig. 8.2a). In contrast, if species are not randomly distributed (i.e. heterogeneity related to microhabitat) and a cumulative species sampling curve is generated by systematically collecting across the apparatus one microhabitat at a time, the curve should stair step as the sampling progresses.

### *Comparing Replicate Among and Within Microhabitat Variation*

Non-parametric Kruskal-Wallace tests were performed using PAST v. 2.01 (Hammer 2010) in order to determine the relative importance of differences among Replicates, Microhabitats and Level above substrate. Data were pooled for patches based on their

**Fig. 8.2** (continued) treatments, i.e. total for each species = 100%. Asterisk indicate significance a $p \leq 0.05$ for Kruskal-Wallace test, stars indicate a high degree of significance at $p \leq 0.001$ and a *dot* indicates those approaching significance $0.1 \leq p \leq 0.05$; (**b**) Species distribution between replicate apparatuses; (**c**) distribution of species based on relative height above the primary substrate; (**d**) distribution of species showing preference for horizontal and textured microhabitats; (**e**) distribution of species showing preference for strings, vertical and textured microhabitats; (**f**) distribution of species showing less pronounced preference of individual microhabitats

relative position as measured on vertical panels (lower ~5 cm, middle ~15 cm and upper ~25 cm above the primary substrate) and the three patches each from the respective horizontal panels. For each replicate, microhabitat and level, the relative abundance of species with 14 or more occurrences in the entire study was plotted in histograms.

### *Microhabitat Preference of Individual Species*

Typically, occurrence data are reported by treatment and species are indexed by which species is most common/abundant at a given treatment or site. This obscures the potential recognition of microhabitats or locations that are of particular importance for a given species, which may otherwise not dominate at any habitat or location. In this study, species occurrences are evaluated as to the distribution (percent occurrence) of one species among all microhabitats. (number of grid cells in which "species X" is present for a microhabitat, divided by the total number of all occurrences for "species X" in all microhabitats × 100) from raw data in Table 8.1. This value expressed as a percentage allows for comparison of the most or least "important" microhabitat for each species as in Fig. 8.2.

## Results

### *Sampling Curve to Evaluate Species Heterogeneity Among Microhabitats*

In hypothetical random data collection design (or species distribution), the total number (25) should be observed before the 400th cell observed (curve "A" in Fig. 8.2a). In the actual curve (curve "B" in Fig. 8.2a), the total species abundance was not observed until after the 1,100th cell, after data from the sixth microhabitat was included. This suggests that microhabitat heterogeneity is at least partially responsible for the distribution of the species. It is possible that factors other than microhabitat differences could be responsible for the deviation from the expected curve (randomized across apparatus in Fig. 8.2a, line a) but none are in evidence.

### *Frequency and Distribution of Species*

The absolute occurrence totals of the 25 species are given in Table 8.1. The number of occurrences (sum of species presences in cells) organized by Replicate and Microhabitat ranged from 112 (Vertical–Replicate-A) to 476 (Strings–Replicate B). The total number of occurrences was greater for Replicate-B relative to Replicate-A

**Table 8.1** Columns are 36-cell patches within a microhabitat treatment for a given replicate apparatus. The value for each species is the number of grid cells for which the species was present. Label abbreviations are: A-B for replicate, H-V-S for Horizontal, Vertical or String orientation of the treatment, S-T for smooth or textured surfaces, U-M-L for upper, middle or lower level above the primary substrate, and a-b-c for random replicates patches within horizontal and string treatments

| Species | A_H_S_U_a | A_H_S_U_b | A_H_S_U_c | A_H_S_M_a | A_H_S_M_b | A_H_S_M_c | A_H_S_L_a | A_H_S_L_b | A_H_S_L_c | A_V_S_L | A_V_S_M | A_V_S_U | A_V_T_L | A_V_T_M | A_V_T_U | A_S_a | A_S_b | A_S_c | Total |
|---|---|---|---|---|---|---|---|---|---|---|---|---|---|---|---|---|---|---|---|
| *Aetea sica* | 0 | 0 | 0 | 0 | 0 | 0 | 2 | 1 | 0 | 0 | 0 | 0 | 0 | 0 | 0 | 0 | 0 | 0 | 3 |
| *Aetea truncata* | 0 | 0 | 0 | 3 | 6 | 2 | 18 | 3 | 1 | 0 | 0 | 3 | 19 | 0 | 0 | 4 | 5 | 2 | 66 |
| *Bugula fulva* | 0 | 0 | 0 | 0 | 0 | 0 | 0 | 0 | 0 | 0 | 0 | 0 | 0 | 1 | 0 | 0 | 1 | 0 | 2 |
| *Buskia nitida* | 0 | 0 | 0 | 0 | 0 | 0 | 0 | 0 | 0 | 0 | 0 | 0 | 0 | 0 | 0 | 0 | 0 | 0 | 0 |
| *Callopora dumerilii* | 6 | 2 | 4 | 2 | 4 | 0 | 4 | 2 | 1 | 0 | 0 | 1 | 2 | 2 | 1 | 1 | 0 | 1 | 33 |
| *Cellaria fistulosa* | 0 | 0 | 0 | 0 | 0 | 0 | 0 | 0 | 0 | 0 | 0 | 0 | 0 | 0 | 0 | 2 | 0 | 2 | 4 |
| *Celleporina caminata* | 0 | 0 | 2 | 0 | 0 | 0 | 1 | 0 | 0 | 0 | 0 | 1 | 0 | 0 | 3 | 0 | 0 | 0 | 7 |
| *Chartella tenella* | 0 | 0 | 0 | 0 | 1 | 0 | 3 | 0 | 0 | 0 | 0 | 0 | 0 | 0 | 0 | 0 | 0 | 0 | 4 |
| *Chorizopora brongniartii* | 0 | 2 | 2 | 0 | 1 | 1 | 0 | 1 | 10 | 4 | 0 | 0 | 9 | 3 | 0 | 0 | 0 | 0 | 33 |
| *Crisia* spp. | 4 | 2 | 5 | 12 | 7 | 3 | 6 | 11 | 3 | 5 | 13 | 2 | 0 | 8 | 4 | 6 | 3 | 6 | 100 |
| *Cryptosula pallasiana* | 0 | 0 | 0 | 0 | 0 | 0 | 0 | 0 | 0 | 0 | 0 | 0 | 0 | 0 | 0 | 0 | 0 | 1 | 1 |
| *Escharina vulgaris* | 0 | 0 | 0 | 0 | 4 | 0 | 2 | 2 | 0 | 0 | 0 | 0 | 0 | 0 | 0 | 0 | 0 | 0 | 8 |
| *Fenestrulina malusii* | 1 | 0 | 0 | 0 | 0 | 0 | 0 | 1 | 3 | 0 | 1 | 3 | 3 | 5 | 1 | 0 | 0 | 1 | 19 |
| *Hippothoa flagellum* | 0 | 0 | 0 | 0 | 0 | 0 | 0 | 0 | 0 | 0 | 0 | 0 | 0 | 0 | 0 | 0 | 0 | 0 | 0 |
| *Microporella ciliata* | 5 | 6 | 15 | 9 | 0 | 0 | 2 | 0 | 0 | 1 | 2 | 7 | 3 | 3 | 5 | 0 | 1 | 0 | 59 |
| *Nolella dilatata* | 2 | 5 | 2 | 8 | 3 | 9 | 8 | 6 | 0 | 0 | 2 | 0 | 9 | 8 | 21 | 14 | 13 | 6 | 116 |
| *Plagioecia patina* | 1 | 3 | 6 | 3 | 7 | 1 | 20 | 0 | 3 | 6 | 11 | 10 | 1 | 12 | 3 | 10 | 5 | 8 | 110 |
| *Schizobrachiella sanguinea* | 0 | 0 | 0 | 0 | 0 | 0 | 2 | 0 | 0 | 0 | 0 | 0 | 0 | 4 | 0 | 2 | 5 | 1 | 14 |
| *Schizomavella linearis* | 0 | 14 | 12 | 0 | 11 | 8 | 4 | 1 | 9 | 5 | 6 | 0 | 11 | 0 | 1 | 2 | 3 | 8 | 95 |
| *Schizomavella subsolana* | 0 | 0 | 0 | 0 | 0 | 0 | 0 | 0 | 0 | 1 | 0 | 0 | 0 | 0 | 0 | 0 | 0 | 0 | 1 |
| *Schizoporella dunkeri* | 0 | 0 | 0 | 0 | 1 | 0 | 0 | 0 | 0 | 8 | 0 | 0 | 0 | 0 | 1 | 0 | 0 | 0 | 10 |
| *Schizoporella magnifica* | 0 | 0 | 0 | 0 | 0 | 0 | 0 | 0 | 0 | 0 | 0 | 0 | 0 | 7 | 0 | 0 | 0 | 0 | 7 |

| | | | | | | | | | | | | | | | | | | |
|---|---|---|---|---|---|---|---|---|---|---|---|---|---|---|---|---|---|---|
| *Tubulipora liliacea* | 18 | 6 | 9 | 10 | 9 | 6 | 21 | 14 | 14 | 4 | 11 | 5 | 5 | 8 | 9 | 5 | 7 | 9 | 170 |
| *Tubulipora plumosa* | 0 | 0 | 0 | 2 | 0 | 0 | 0 | 0 | 0 | 0 | 0 | 0 | 0 | 0 | 0 | 0 | 0 | 0 | 2 |
| *Turbicellepora camera* | 0 | 0 | 0 | 0 | 0 | 1 | 0 | 0 | 0 | 0 | 0 | 0 | 0 | 0 | 0 | 2 | 9 | 8 | 20 |
| Patch total | 37 | 40 | 57 | 49 | 54 | 31 | 93 | 42 | 44 | 34 | 46 | 32 | 62 | 61 | 49 | 48 | 52 | 53 | |
| Treatment total | 134 | | | 134 | | | 179 | | | 112 | | | 172 | | | 153 | | | 884 |
| Number of species | 10 | | | 14 | | | 15 | | | 13 | | | 15 | | | 16 | | | 23 |

| Species | B_H_S_U_a | B_H_S_U_b | B_H_S_U_c | B_H_S_M_a | B_H_S_M_b | B_H_S_M_c | B_H_S_L_a | B_H_S_L_b | B_H_S_L_c | B_V_S_L | B_V_S_M | B_V_S_U | B_V_T_L | B_V_T_M | B_V_T_U | B_S_a | B_S_b | B_S_c | Total |
|---|---|---|---|---|---|---|---|---|---|---|---|---|---|---|---|---|---|---|---|
| *Aetea sica* | 1 | 0 | 0 | 0 | 0 | 0 | 0 | 0 | 0 | 0 | 0 | 0 | 1 | 0 | 1 | 1 | 0 | 1 | 5 |
| *Aetea truncata* | 13 | 1 | 0 | 0 | 6 | 0 | 10 | 9 | 0 | 0 | 0 | 8 | 12 | 15 | 4 | 0 | 6 | 2 | 86 |
| *Bugula fulva* | 0 | 0 | 0 | 0 | 0 | 0 | 0 | 0 | 0 | 0 | 0 | 0 | 0 | 0 | 0 | 0 | 2 | 1 | 3 |
| *Buskia nitida* | 0 | 0 | 0 | 0 | 0 | 0 | 0 | 0 | 0 | 0 | 0 | 0 | 0 | 0 | 0 | 1 | 1 | 2 | 4 |
| *Callopora dumerilii* | 5 | 9 | 0 | 6 | 4 | 0 | 2 | 1 | 5 | 0 | 1 | 1 | 1 | 0 | 1 | 0 | 1 | 0 | 37 |
| *Cellaria fistulosa* | 0 | 0 | 0 | 0 | 0 | 0 | 0 | 0 | 0 | 0 | 0 | 0 | 0 | 0 | 0 | 4 | 4 | 3 | 11 |
| *Celleporina caminata* | 0 | 0 | 0 | 0 | 0 | 0 | 0 | 2 | 0 | 0 | 0 | 0 | 0 | 0 | 0 | 1 | 1 | 1 | 5 |
| *Chartella tenella* | 0 | 0 | 0 | 0 | 1 | 0 | 0 | 0 | 0 | 0 | 0 | 1 | 0 | 0 | 0 | 0 | 0 | 0 | 2 |
| *Chorizopora brongniartii* | 0 | 5 | 1 | 0 | 1 | 0 | 0 | 4 | 6 | 8 | 0 | 0 | 8 | 4 | 5 | 0 | 0 | 1 | 43 |
| *Crisia* spp. | 21 | 5 | 3 | 11 | 7 | 27 | 1 | 23 | 17 | 2 | 12 | 2 | 8 | 6 | 17 | 18 | 20 | 23 | 223 |

(continued)

(continued)

| Species | B_H_S_U_a | B_H_S_U_b | B_H_S_U_c | B_H_S_M_a | B_H_S_M_b | B_H_S_M_c | B_H_S_L_a | B_H_S_L_b | B_H_S_L_c | B_V_S_L | B_V_S_M | B_V_S_U | B_V_T_L | B_V_T_M | B_V_T_U | B_S_a | B_S_b | B_S_c | Total |
|---|---|---|---|---|---|---|---|---|---|---|---|---|---|---|---|---|---|---|---|
| *Cryptosula pallasiana* | 0 | 0 | 0 | 0 | 0 | 0 | 0 | 0 | 0 | 0 | 0 | 0 | 0 | 0 | 0 | 0 | 1 | 0 | 1 |
| *Escharina vulgaris* | 0 | 0 | 0 | 0 | 4 | 0 | 0 | 4 | 0 | 0 | 0 | 0 | 0 | 0 | 0 | 0 | 0 | 0 | 8 |
| *Fenestrulina malusii* | 0 | 0 | 0 | 1 | 0 | 0 | 10 | 0 | 0 | 0 | 0 | 0 | 1 | 1 | 3 | 3 | 1 | 1 | 21 |
| *Hippothoa flagellum* | 0 | 0 | 0 | 0 | 0 | 0 | 0 | 0 | 6 | 0 | 0 | 0 | 0 | 0 | 0 | 0 | 0 | 0 | 6 |
| *Microporella ciliata* | 6 | 8 | 5 | 9 | 0 | 0 | 14 | 7 | 1 | 0 | 0 | 0 | 0 | 5 | 18 | 0 | 3 | 2 | 78 |
| *Nolella dilatata* | 1 | 5 | 3 | 9 | 3 | 0 | 5 | 4 | 0 | 25 | 0 | 1 | 20 | 13 | 15 | 19 | 15 | 15 | 153 |
| *Plagioecia patina* | 1 | 0 | 2 | 0 | 7 | 1 | 6 | 14 | 20 | 10 | 10 | 4 | 4 | 6 | 5 | 15 | 12 | 12 | 129 |
| *Schizobrachiella sanguinea* | 0 | 0 | 0 | 0 | 0 | 0 | 0 | 0 | 0 | 0 | 0 | 8 | 0 | 0 | 0 | 3 | 0 | 2 | 13 |
| *Schizomavella linearis* | 0 | 4 | 16 | 20 | 11 | 21 | 0 | 0 | 9 | 0 | 0 | 0 | 5 | 17 | 2 | 3 | 12 | 11 | 131 |
| *Schizomavella subsolana* | 1 | 0 | 0 | 0 | 0 | 0 | 0 | 0 | 0 | 2 | 0 | 0 | 1 | 0 | 0 | 0 | 0 | 0 | 4 |
| *Schizoporella dunkeri* | 0 | 0 | 0 | 0 | 1 | 0 | 7 | 6 | 0 | 0 | 0 | 3 | 0 | 4 | 0 | 1 | 0 | 1 | 23 |
| *Schizoporella magnifica* | 3 | 0 | 0 | 0 | 0 | 0 | 0 | 0 | 0 | 0 | 0 | 0 | 0 | 0 | 0 | 0 | 0 | 0 | 3 |
| *Tubulipora liliacea* | 6 | 9 | 10 | 4 | 9 | 4 | 14 | 13 | 15 | 7 | 4 | 7 | 6 | 8 | 10 | 14 | 13 | 14 | 167 |
| *Tubulipora plumosa* | 2 | 0 | 0 | 0 | 0 | 0 | 7 | 1 | 2 | 0 | 0 | 0 | 0 | 0 | 0 | 0 | 0 | 0 | 12 |
| *Turbicellepora camera* | 0 | 0 | 0 | 0 | 0 | 0 | 0 | 0 | 2 | 0 | 0 | 0 | 0 | 0 | 0 | 11 | 7 | 10 | 30 |
| Patch total | 60 | 46 | 40 | 60 | 54 | 53 | 76 | 88 | 83 | 54 | 27 | 35 | 67 | 79 | 81 | 94 | 99 | 102 | |
| Treatment total | 146 | | | 167 | | | 247 | | | 116 | | | 227 | | | 295 | | | 1,198 |
| Number of species | 13 | | | 13 | | | 15 | | | 11 | | | 13 | | | 18 | | | 25 |
| Treatment total | H_U | | | H_M | | | H_L | | | V | | | T | | | S | | | |
| Study total | 280 | | | 301 | | | 426 | | | 228 | | | 399 | | | 448 | | | 2,082 |

**Table 8.2** Results from Kruskal-Wallis Tests for significant differences between ranked Raw occurrence data for equivalence among factors for each species (p-values reported from 1-way test Chi Square approximation). Factors and their cases are (1) Replicate, A and B are replicate devices for all treatments at one locality, (2) Level, is the height above the primary substrate, U = upper (~25 cm), M = Middle (~15 cm), L = lower (~5 cm), (3) Microhabitat, is an incomplete decomposition of orientation of the panel (H = Horizontal and V = vertical, both smooth surfaces), Surface texture (T = Texture) and substrate type (S = string). Only species with $n \geq 14$ occurrences are reported. Significant effects are highlighted in bold and [a] indicates a result approaching significance (potentially significant with larger n)

| | | Replicate | Level | Treatment |
|---|---|---|---|---|
| Species | n | A vs. B | U vs. M vs. L | Hu vs. Hm vs. Hl vs. V vs. T vs. S |
| *Aetea truncata* | 152 | 0.52 | 0.33 | 0.30 |
| *Callopora dumerilii* | 70 | 0.64 | 0.60 | **0.020** |
| *Cellaria fistulosa* | 15 | 0.53 | – | <**0.0001** |
| *Chorizopora brongniartii* | 76 | 0.58 | **0.032** | 0.07[a] |
| *Crisia* spp. | 323 | **0.022** | 0.11 | 0.45 |
| *Escharina vulgaris* | 16 | 0.71 | 0.22 | **0.040** |
| *Fenestrulina malusii* | 40 | 0.72 | 0.76 | **0.027** |
| *Microporella ciliata* | 137 | 0.76 | **0.040** | 0.07[a] |
| *Nolella dilatata* | 269 | 0.61 | 0.83 | **0.002** |
| *Plagioecia patina* | 239 | 0.60 | 0.25 | **0.014** |
| *Schizobrachiella sanguinea* | 27 | 0.50 | 0.99 | **0.012** |
| *Schizomavella linearis* | 226 | 0.65 | 0.37 | 0.17 |
| *Schizoporella dunkeri* | 33 | 0.14 | 0.71 | 0.72 |
| *Tubulipora liliacea* | 337 | 0.90 | 0.23 | **0.007** |
| *Tubulipora plumosa* | 14 | 0.16 | 0.40 | 0.10[a] |
| *Turbicellepora camera* | 50 | 0.88 | – | <**0.0001** |

(1,198 vs. 884, Table 8.2). All 25 observed species were found in Replicate-B, while the two rarest species (*Cryptosula pallasiana*, *Hippothoa flagellum*) were not present on Replicate-A (Table 8.1). The number of species present on any given microhabitat/replicate combination ranged from 10 to 19 with a mean of 13.8 (Table 8.1). The number of species present on any microhabitat (pooled replicates) ranged from 15 to 19, with an average of 16.7.

Large, sessile macroinvertebrates (polychaete worms, *Anomia* pelecypods, ascidians, sponges) recruited most abundantly onto the Middle and Upper Horizontal microhabitats, resulting in low numbers of observations (though more than for the Vertical microhabitat) and number of species (Table 8.1).

## *Kruskal-Wallis Test for Apparatus, Level, and Microhabitat Effects*

Kruskal-Wallis tests for each bryozoan species with 14 or more occurrences result in a progressively higher number of significant ($p \leq 0.05$) distributional patterns from Apparatus to Level to Microenvironment (Table 8.2). Only *Crisia* spp. is significant at the $p \leq 0.05$ level between Apparatus A vs. B. That is, there is no difference between the replicate apparatuses when examined species by species ($n \geq 14$), except that *Crisia* spp. is more abundant on Apparatus-B, which has more occurrences overall (n = 1,198 vs. n = 884, Table 8.1). Occurrences of only two species are significant ($p \leq 0.05$) in tests for difference among Levels above the substrate (lower, middle, upper). *Chorizopora brongniartii* is more common in the lower region and *Microporella ciliata* is more abundant in the upper region (Table 8.2).

Nine species have significantly different occurrences among Microhabitats (Table 8.2), and two of the nine species (*Cellaria fistulosa*, *Turbicellepora camera*) are highly significant ($p \leq 0.0001$). Three additional species approach significance ($0.10 > p > 0.05$) and could potentially achieve significance with a larger sample size. Four species displayed no recognizable difference among microhabitat treatments.

## *Microhabitat Preference of Individual Species*

When interpreting the preference of species by microhabitat, affinities are complex and best treated case by case (Fig. 8.2b–f). First considering species most important on the Horizontal treatment, *Callopora dumerilii* is common throughout, but is most important on the Upper Level of the Horizontal microhabitat (Fig. 8.2d). *Escharina vulgaris* is restricted to the Middle and Lower Levels of the Horizontal microhabitat (Fig. 8.2d). *Tubulipora plumosa* is restricted to the Horizontal microhabitat and is most abundant on the Lower Level. *Aetea truncata, Chorizopora brongniartii*, and *Fenestrulina malusii* are present in all microhabitats, but are most abundant on the Texture and Lower Horizontal microhabitats (Fig. 8.2d). The strongest preference was shown for several species limited or nearly limited to the Strings (Fig. 8.2e). *Turbicellepora camera* and *Cellaria fistulosa* were observed in other microhabitats outside of the study grids but nowhere as abundant as on Strings. *Bugula fulva* was rare (n = 5), but was observed exclusively in the String and Texture microhabitats. *Buskia nitida* was also rare (n = 4), and all observed colonies were well developed and found exclusively on Strings. No specimens of *Buskia nitida* were observed in natural microhabitats beyond the scope of the study grid either.

Two species that were preferentially abundant in the String microhabitat were also associated with a second treatment (Fig. 8.2e). *Nolella dilatata* formed extensive mats of runners in both String and Texture microhabitats. *Schizobrachiella sanguinea* formed large hollow conical colonies centered on Strings but in contrast formed large encrusting disks in Vertical (smooth) microhabitat.

The most evenly distributed species among microhabitats were *Crisia* spp., *Schizomavella linearis*, and *Tubulipora liliacea* (Fig. 8.2f). Other widely distributed species that showed slight preferences for a microhabitat(s) included: *Microporella ciliata* (Vertical and Upper Horizontal), *Plagioecia patina* (Strings and Middle & Lower Horizontal), and *Schizoporella dunkeri* (Middle & Lower Horizontal).

## Discussion

The question addressed in this study was: Given a range of textural complexities (flat plates, corrugated plates, woven strings) and orientations (horizontal or vertical plates, omnidirectional string surfaces), is there an apparent difference in microhabitat specificity among the bryozoans recruited onto the different microhabitats? Results suggest (but do not directly test) the idea that an increase in the number of kinds of microhabitats within a local environment (in this study $<1$ m$^3$), will result in an increase in species richness and individual occurrences relative to an equal area of uniform microhabitat.

The fact that the species list and general abundance did not differ significantly between apparatuses (Fig. 8.2b), suggests that there was no difference in bryozoan recruitment between A and B, i.e. equivalent sampling from a common larval pool. The slightly greater number of occurrences on Apparatus B (Fig. 8.2b and Table 8.1) suggests more favourable conditions for colony growth (size) rather than absolute colony number.

### *Species Richness Among Microhabitats*

Species richness was relatively constant among the microhabitats (average of 16.7); only about two-thirds of the total number of species observed (25) was present in any given microhabitat. Thus, although diversity is relatively constant, the combination of species present within microhabitats varies. This suggests some level of selection (preferences or differential survival) at the microhabitat level. The presence of many species within and among many microhabitats suggests that larvae had an equal chance of encountering any of the microhabitat treatments. That is, no "hot zones" or "dead zones" of overall settlement frequency were observed within or between apparatuses.

Clearly some but not all species have better success in certain microhabitats than others. That is, not every species has a preferred microhabitat. In addition, some 36-cell patches within one microhabitat are more similar in species composition to patches in other microhabitats than other regions of the same microhabitat. This means that either unobserved environmental conditions control these distributions, or more likely, that controls and preferences are loose enough at this scale that strict boundaries between species compositions are not established.

Regardless of the controlling ecological factors, it is evident that microhabitat variation does have *some* influence in the distributions and abundance of bryozoan species in a community in an early stage of development ($<1.2$ year).

In this study, the total species richness is not a simple function of the area of substrate sampled. The number of kinds of substrate sampled (microhabitats) plays an important role in determining the over all species richness (Fig. 8.2a, Table 8.1).

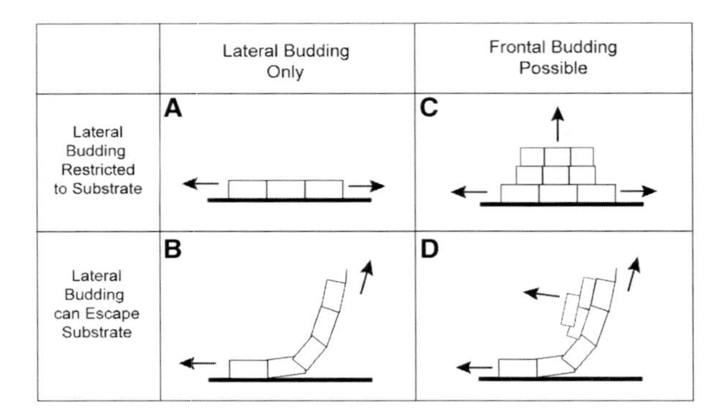


**Fig. 8.3** Four primary growth habit characteristics associated with microhabitat preference defined by combinations of restricted and non-restricted lateral and frontal budding. Type-A was found only rarely in the String microhabitat, but was common in the horizontal and vertical microhabitats. Type-C was rare in the horizontal and vertical habitats, but common to abundant in the String microhabitat. Types-B and D were common to abundant in all microhabitats

## *Distribution of Growth Habit Characters Among Microhabitats*

The most abundant species in the study were "weedy" cyclostomes (*Crisia* spp., *Tubulipora liliacea*, *Plagioecia patina)* and gynmolaemates (*Aetea truncata*, *Nolella dilata*) and were widely distributed. However, many taxa exhibited growth habit attributes that were relatively informative about microhabitat preferences. Some of the more obvious patterns were:

1. Frontal budding (*Celleporina caminata*, *Turbicellepora camera*) – much better represented on Strings; present but not abundant elsewhere.
2. Rhizoidal attachments (*Bugula fulva, Cellaria fistulosa,* but not *Crisia* spp.) – much more abundant on Strings; present but less common elsewhere.
3. Encrusting sheets with multizooidal budding zone that can escape substrate (*Schizomavella linearis, Plagioecia patina*, *Schizobrachiella sanguinea*) – prominent on Strings; common on other microhabitats. *S. sanguinea* is the only species that exhibited two growth habits, encrusting on flat substrates but forming large, hollow, conical colonies on Strings.
4. Encrusters with growing edge restricted to substrate (*Callopora dumerilii, Chorizopora brongniartii*, *Escharina vulgaris, Fenestrulina malusii, Hippothoa flagellum, Microporella ciliata, Schizomavella subsolana, Schizoporella dunkeri*, *Schizoporella magnifica*) – rare on Strings.
5. Runners (*Aetea sica, Aetea truncata, Nolella dilatata)* – most abundant on String and Texture microenvironmets and on Lower levels, but present throughout.

In summary, the greatest contrast in growth habit attributes among microhabitats is between Strings, favorable to bryozoan species with some attribute that allows some degree of escape from the substrate (Fig. 8.3 Types-B, C and D), and the less flexible broader microhabitats, favored by encrusting species with the growing edge restricted to the substrate (Fig. 8.3 Type-A). Thus, morphological characters related to interaction with the substrate, such as multizooidal budding (Lidgard and Jackson 1989), rhizoid attachments and those that allow for upward growth, such as frontal budding and budding zones raised from the substrate (Lidgard and Jackson 1989), may be of more ecological importance for determining growth habit disparity than selection on other characters (Fig. 8.3).

Our understanding of broader environmental and phylogenetic controls on the distribution of bryozoan colonial growth habits remains incomplete. However, the methods employed in this study, potentially with more simplified designs, performed in a wide range of settings, promise insights into controls over local species richness and morphological disparity.

**Acknowledgements** We thank M. Novosel and A. Novosel for their help and advice in acquiring materials for constructing the experimental apparatuses. This research was supported by a Fulbright Research Fellowship to SJH, and the University Research Council of Appalachian State University.

## References

Barnes DKA, Dick MH (2000) Overgrowth competition in encrusting bryozoan assemblages of the intertidal and infralittoral zones of Alaska. Mar Biol 136:813–822

Burgess SC, Hart SP, Marshall DJ (2009) Pre-settlement behavior in larval bryozoans: the roles of larval age and size. Biol Bull 216:344–354

Dobretsov S, Qian P-Y (2006) Facilitation and inhibition of larval attachment of the bryozoan *Bugula neritina* in association with mono-species and multi-species biofilms. J Exp Mar Biol Ecol 333:263–274

Hammer O (2010) PAST, PAleontological STatistics version 2.01 reference manual. Natural History Museum, University of Oslo

Hayward PJ, McKinney FK (2002) Northern Adriatic Bryozoa from the vicinity of Rovinj, Croatia. Bull Am Mus Nat Hist 270:1–139

Lidgard S (2008) Predation on marine bryozoan colonies: taxa, traits and trophic groups. Mar Ecol Prog Ser 359:117–131

Lidgard S, Jackson JBC (1989) Growth in encrusting cheilostome bryozoans. Paleobiology 15:255–282

Mariani S, Alcoverro T et al (2005) Early life histories in the bryozoan *Schizobrachiella sanguinea*: a case study. Mar Biol 147:735–745

Pemberton AJ, Hansson LJ et al (2007) Microscale genetic differentiation in a sessile invertebrate with cloned larvae: investigating the role of polyembryony. Mar Biol 153:71–82

Ryland J (1977) Taxes and tropisms of bryozoans. In: Woollacott RM, Zimmer RL (eds) Biology of bryozoans. Academic, New York

Sanders H (1968) Marine benthic diversity: a comparative study. Am Nat 102:243–282

Stach LW (1936) Correlation of zoarial form with habitat. J Geol 44:60–65

Sutherland J, Karlson R (1977) Development and stability of the fouling community at Beaufort, North Carolina. Ecol Monogr 47:425–446

# Chapter 9
# Distribution and Diversity of Erect Bryozoan Assemblages Along the Pacific Coast of Japan

## Erect Bryozoan Assemblages of Japan

**Masato Hirose, Shunsuke F. Mawatari, and Joachim Scholz**

**Abstract** To assess factors involved in the high diversity of benthic fauna in Sagami Bay, we examined the species composition of bryozoans forming rigidly erect colonies, possibly occuring in dense assemblages called bryozoan thickets. We identified erect bryozoans from collections made in the bay in four time intervals over ~125 years. To examine latitudinal effects on diversity, we also identified specimens in collections from near Otsuchi Bay to the north and the Nansei Islands to the south. In addition, we compared the composition and diversity of erect bryozoans in Sagami Bay with those in bryozoan thickets at Otago Shelf, New Zealand. We categorized erect bryozoans into five form categories based on the colony morphologies; detected 17 species in representing the five forms in Sagami Bay, five species representing four forms in the Nansei Islands, and three species representing two forms at Otsuchi Bay. Erect bryozoan diversity thus did not show a latitudinal gradient; it was higher in Sagami Bay than farther north or farther south, though we cannot rule out sampling effects. We speculate that the high diversity in Sagami Bay is due to greater environmental complexity than the other areas, including warming and cooling influences from the Kuroshio and

---

M. Hirose (✉)
Department of Natural History Sciences, Faculty of Science, Hokkaido University, N10 W8, Sapporo 060-0810, Japan

Department of Zoology, National Museum of Nature and Science Tokyo, Tsukuba, Ibaraki 305-0005, Japan
e-mail: m-hirose@sci.hokudai.ac.jp; mhirose@kahaku.go.jp

S.F. Mawatari
Department of Natural History Sciences, Faculty of Science, Hokkaido University, N10 W8, Sapporo 060-0810, Japan

J. Scholz
Sektion Marine Evertebraten 3, Forschungsinstitut und Naturmuseum Senckenberg, Senckenberganlage 25, Frankfurt D-60325, Germany
e-mail: joachim.scholz@senckenberg.de

A. Ernst, P. Schäfer, and J. Scholz (eds.), *Bryozoan Studies 2010*,
Lecture Notes in Earth System Sciences 143, DOI 10.1007/978-3-642-16411-8_9,

Oyashio Currents, respectively, in different parts of the bay. We detected no clear differences in species composition between eastern and western Sagami Bay, but did detect an apparent loss of diversity of four species overall and nine species in western Sagami Bay between the 1928–1988 and the 2001–2005 intervals, suggesting differential environmental changes in different parts of the bay. Sagami Bay was richer in rigidly erect species than the bryozoan thickets at Otago Shelf, though the same colony morphologies were represented.

**Keywords** Bryozoan assemblage • Bryozoan diversity • Bryozoan thickets • Colony form • Kuroshio Current • Oyashio Current • Phidoloporid • Sagami Bay

## Introduction

### *Bryozoan Assemblages of Japan*

Sagami Bay on the central Pacific coast of Honshu Island, Japan, is perceived as having an exceptionally diverse benthic fauna, both within Japan and among comparable sites at similar latitude worldwide (Isono 1988; National Museum of Nature and Science Tokyo 2007). The perception of high diversity could be partly an artefact of high sampling effort, as collecting has been conducted in the bay for roughly 125 years: by Döderlein, Doflein and Haberer, Mitsukuri and Owston, Emperor Showa, and the National Museum of Nature and Science, Tokyo (NSMT) (National Museum of Nature and Science 2007; Fujita 2008). These various collections have been well preserved in several museums in Japan and Europe (Mawatari 2009). Despite high sampling effort, Sagami Bay does actually seem to be exceptionally rich in benthic fauna. Examining all available collections, Hirose (2010) detected 232 species of cheilostomes occurring in the bay; if cyclostomes and ctenosomes are taken into account, the true bryozoan diversity could be close to 300 species, which seems remarkable for a near-shore area of only 70 by 90 km in extent.

In order to address factors responsible for the high biodiversity in Sagami Bay, we focused on species with large, erect bryozoan colonies, for two reasons. One is that due to their morphology; erect colonies are more consistently collected by trawl than flat, encrusting colonies, making comparisons of occurrence through time and among localities more reliable. In addition, it is possible that assemblages of erect bryozoans exist in Sagami Bay that are comparable to so-called bryozoan thickets reported from other parts of the world (Probert and Batham 1979; Carter et al. 1985; Batson and Probert 2000). We have no evidence that the erect species form thickets in Sagami Bay, but the high number of erect species compared to other localities sampled in Japan suggests that it is likely to be the case.

Döderlein (1883) was the first one referring to bryozoan assemblages in Japan. The author mentioned a higher diversity and abundance of large, erect bryozoans attributed to *Lepralia* and *Retepora*, from west of Miura Peninsula, and noted that he suspected a community of glass sponges, corals, and bryozoans in the area.

More than 120 years later, Grischenko et al. (2007) found a dense aggregation of roughly 20 large colonies of the erect phidoloporid *Phidolopora elongata* in a horizontal crevice low in the intertidal zone at Akkeshi Bay, Hokkaido, Japan; this may support our assumption that similarly dense aggregations of phidoloprids and other erect species actually construct bryozoan thickets in Sagami Bay.

## *Component Species Forming Bryozoan Thickets*

Bryozoan colonies are often classified on the basis of colony morphology alone. Classifications of erect colony morphology in bryozoans have varied according to authors and the intended purpose of the classification (e.g., Cuffey 1974, 1977; McKinney and Jackson 1989; Hageman et al. 1998; Scholz 2000; Scholz et al. 2005). Bryozoan thickets comprise various colony growth types, however, the classification of colony morphology varied among researchers until Cuffey (1974, 1977), who proposed a utilitarian classification for reef-associated bryozoan studies. Cuffey (1977) classified bryozoan colonies into four groups based on colony contributions to reefs: principal frame-builders, accessory frame encrusters, sediment-movement inhabitors, and sediment formers. Each category of Cuffey's classification is suitable for several different types of studies: e.g., frame-builders are easily estimated from historical collections and thus are suitable for comparisons of representative species in bryozoan assemblages through time and among localities.

Since erect bryozoan colonies enhance biodiversity by providing a habitat for other organisms (Stebbing 1971), analysis of potential bryozoan assemblages in Sagami Bay will provide critical information relevant to the relationship between bryozoan assemblages and associated marine organisms in Japan. In our study, we examine the species composition of erect bryozoans in five categories of colony morphology present in Sagami Bay; compare both the diversity of colony morphologies and species composition of erect bryozoans among Sagami Bay and more northern (Otsuchi Bay) and southern (Nansei Islands) areas in Japan; compare the taxonomic composition of erect bryozoans in Sagami Bay with bryozoan thickets observed off the Otago peninsula, New Zealand; and outline some of the factors controlling the considerable high diversity in Sagami Bay.

# Material and Methods

## *Study Areas and Collecting Effort*

### Sagami Bay

Sagami Bay, next bay to the west from Tokyo Bay on the Pacific coast of central Honshu Island, Japan, lies at 35°N latitude (Fig. 9.1a, d). Sagami Bay proper is

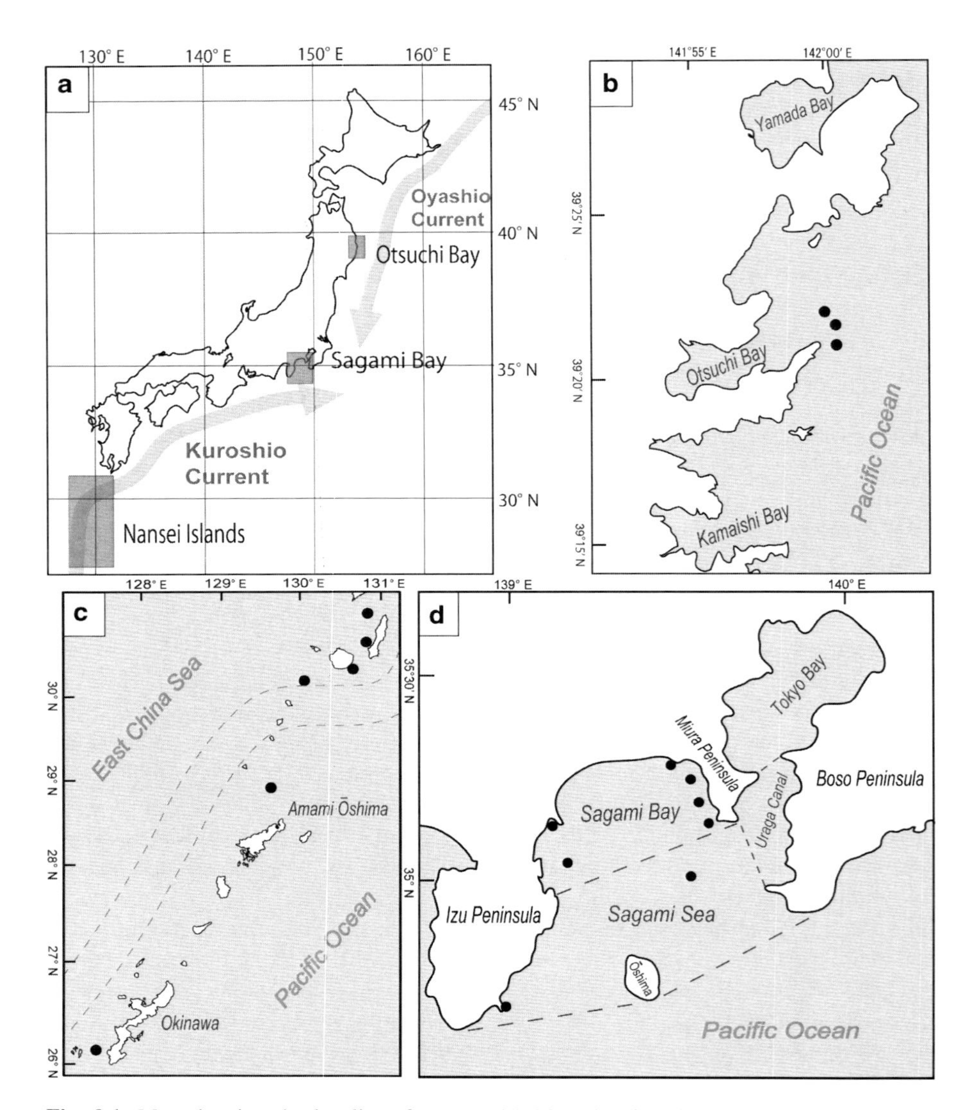


**Fig. 9.1** Map showing the locality of survey. (**a**) Map showing the locality of Sagami Bay, Otsuchi Bay, and Nansei Islands in Japan. *Arrows* indicate the major water currents. (**b**) Localities of collecting stations in Otsuchi Bay. (**c**) Localities of collecting stations in Nansei Islands. The *broken lines* showing the flowing way of Kuroshio Current. (**d**) Map showing the main localities where the historical collections collected in Sagami Bay

bounded by the Miura Peninsula to the east and the base of the Izu Peninsula to the west (Fig. 9.1a, d); it constitutes the northern part of the Sagami Sea, bounded by a line running from the tip of the Boso Peninsula to Oshima and thence to the tip of the Izu Peninsula. In this study, we consider both Sagami Bay proper and the Sagami Sea as constituting 'Sagami Bay.' The bottom along the east side of the

Izu Peninsula comprises a narrow shelf; that along the west side of the Miura Peninsula is a broad shelf with many peripheral submarine canyons and a number of knolls. The maximum depth in Sagami Bay is about 1,500 m. Subarctic intermediate water originating from the Oyashio Current is found in the 250–1,000 m depth interval in the bay, whereas the warm Kuroshio Current flowing from the southwest has a warming effect on water less than 250 m depth; around the Miura Peninsula, surface water less than 10 m in depth mixes with water from Tokyo Bay (Iwata 1979; Iwata and Matsuyama 1989; Kawabe and Yoneno 1987; Hinata et al. 2003). The surface water temperature in Sagami Bay is coldest from January to March (minimum ~15°C) and warmest from July to August (maximum ~25°C).

In this study, we identified erect bryozoans among more than 700 cheilostome specimens in bryozoan collections made in Sagami Bay by Döderlein from 1880 to 1881, Doflein and Haberer from 1900 to 1905, and Emperor Showa from 1928 to 1988, and 500 cheilostome specimens collected by NSMT from 2001 to 2005.

## Otsuchi Bay

Otsuchi Bay (Fig. 9.1b) lies on the coast of Iwate Prefecture, also known as the Sanriku Coast, on the Pacific side of northeastern Honshu (39°20′30″N, 141°56′40″E). It is a typical rias coastline formed from a submerged river valley; the maximum depth is 77 m, and has a surface seawater temperature range from 5°C in winter to 20°C in summer. The Sanriku Coast is influenced by the cold Oyashio Current (Fig. 9.1a) flowing from the Kurile Islands; it is coldest in the central part because neither the warm Kuroshio Current nor the warm Tsugaru Current (a branch of the warm Tsushima current flowing northward in the Sea of Japan and emerging to the Pacific from Tsugaru Strait) reaches this area to a significant extent (Horikoshi 1985).

The senior author examined approximately 90 specimens collected off the mouth of Otsuchi Bay in April or May from 2008 to 2010. Specimens were taken by a fisher's net using F/V *Taku-maru*, and by dredge by the senior author aboard R/V *Yayoi* of the Otsuchi Marine Research Center, University of Tokyo; we found erect colonies at four stations.

## Nansei Islands

The Nansei Islands (Fig. 9.1c), extending southwest from Kyushu Island between latitudes 30°51′N and 24°02′N, form a boundary between the East China Sea and the Pacific Ocean. They are situated west of the Ryukyu Trench, between the Philippine Sea and Eurasian Plates. The sea around the Nansei Islands is strongly affected by the warm Kuroshio Current (Fig. 9.1a), which originates east of the Philippines, flows northward in the East China Sea along the Ryukyu Islands, passes from the East China Sea to the Pacific through Tokara Strait just south of Kyushu, and flows northeastward along the coast of southern Japan.

The senior author collected approximately 260 specimens by dredge, slide net, and beam trawl from 12 localities in the Nansei Islands in May 2009 and 2010, aboard TR/V *Toyoshio-maru* of Hiroshima University; samples from six of these localities contained erect bryozoan colonies.

## *Erect Colony Morphologies and Relative Abundance*

In this study, we selected erect bryozoan colonies (species) based on three features; (1) colonies more than 5 cm high, (2) colonies attached to a rocky substratum, and (3) species of which the colonies were collected more than twice during the past collection. We discussed on the erect bryozoan species that satisfied these requirements.

One of the authors (Scholz 2000; Scholz et al. 2000, 2005) has classified colonies on the basis of their secondary structure or growth pattern into five groups: bryolith, bryostromatolite, fenestrate and/or tree-like, foliaceous and/or honeycomb-like, and robust branching. We revised this classification and defined five erect morphologies (Fig. 9.2) on the basis of overall appearance and presumed durability in the environment: (1) *robust branching* (thick, mostly cylindrical branches, with branch diameter large relative to internode distance), (2) *nodular* (thick, mound-like growth), (3) *reticulate* (zooids arranged in perforate sheets), (4) *foliaceous* (zooids arranged in non-perforate, often anastomosing sheets, unilaminar or bilaminar), and (5) *dendroid* (branching, with cylindrical branches narrow in diameter compared to internode distance).

In the results, we do not report the actual number of specimens detected as a measure of abundance, since it was often not clear whether fragments came from one broken colony or several colonies. Instead, we report only P for present (a single large colony or two large colonies with few fragments) and A for abundant (several large colonies or many fragments) (Table 9.1).

# Results

## *Erect Bryozoans in Sagami Bay*

We detected 17 species forming erect colonies in Sagami Bay (Fig. 9.2). Table 9.1 shows the distribution of these species among the four collections examined, made at different time intervals over ~125 years. Döderlein collected entirely, and Doflein & Haberer mostly, in the eastern Sagami Bay (Doflein 1906; Fujita 2008; Hirose 2010), which is reflected in the lack of occurrences in the western part in Table 9.1. Emperor Showa devoted collecting effort almost equally to the eastern and western parts, and detected 10 and 12 species, respectively, in the two areas. The NSMT survey also made collections in both parts; like Emperor Showa, NMST detected ten species in

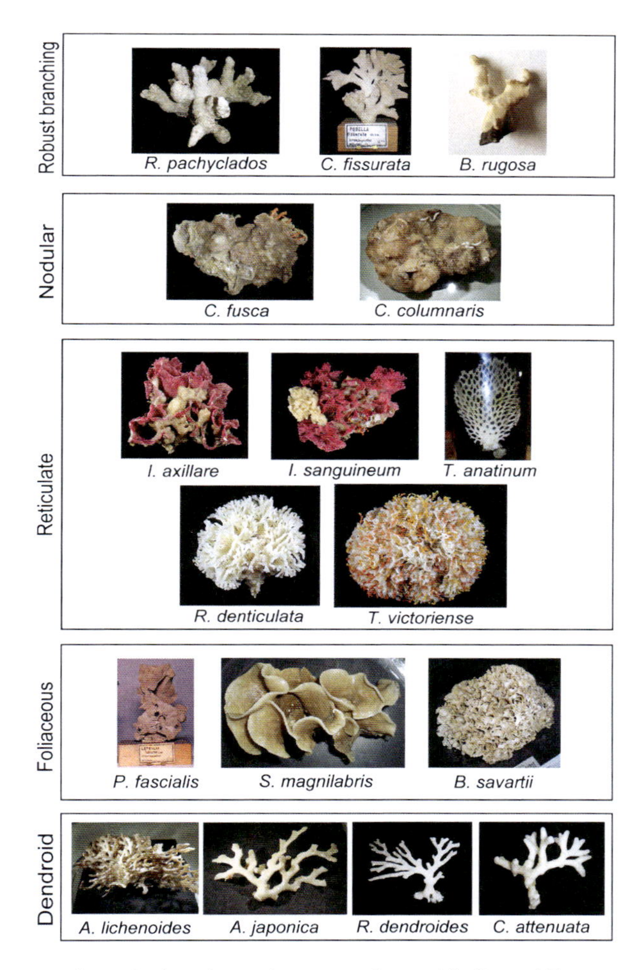


**Fig. 9.2** Representative colonies of erect bryozoans detected in Sagami Bay

the eastern part, though in six cases, one collection contained species not found in the other. NSMT detected markedly less diversity in western than in eastern Sagami Bay, and less than Emperor Showa collected on either side (Table 9.1).

Reticulate bryozoans is the morphological category, which comprised the most diverse species, with five species, all of them belonging to the Phidoloporidae. They were also the most consistently represented through time, with two of the five species detected in all time intervals, and the other three detected in two or three of the four time intervals (Table 9.1). Only two species (*Steginoporella magnilabris* and *Reteporella dendroides*) belonging to other form categories were detected in all four time intervals, and three species (*Cigclisula fissurata*, *Adeonellopsis japonica*, and *Celleporina attenuata*) in three of the four time intervals. Three species

**Table 9.1** Erect bryozoan species detected in the eastern (E) and western (W) parts of Sagami Bay. Each of the four collections examined representing different time intervals

| | Döderlein 1880–1881 | | Doflein–Haberer 1904–1905 | | Emperor Showa 1918–1971 | | NSMT 2001–2005 | |
|---|---|---|---|---|---|---|---|---|
| Category/species | E | W | E | W | E | W | E | W |
| Robust branching | | | | | | | | |
| *Rhynchozoon pachyclados* | P | – | – | – | – | A | – | – |
| *Cigclisula fissurata* | P | – | – | – | P | P | P | – |
| *Borgiola rugosa* | P | – | – | – | – | – | – | – |
| Nodular | | | | | | | | |
| *Celleporaria fusca* | – | – | P | – | – | A | P | – |
| *Celleporaria columnaris* | P | – | – | – | P | A | – | – |
| Reticulate | | | | | | | | |
| *Triphyllozoon victoriense* | P | – | P | – | P | A | – | – |
| *Iodictyum axillare* | P | – | P | – | A | P | A | P |
| *Iodictyum sanguineum* | P | – | P | – | – | A | P | – |
| *Reteporellina denticulata* | – | – | P | – | P | A | P | – |
| *Triphyllozoon anatinum* | P | – | P | – | – | – | – | – |
| Foliaceous | | | | | | | | |
| *Steginoporella magnilabris* | P | – | P | – | A | – | P | P |
| *Biflustra savartii* | – | – | – | – | P | P | – | – |
| *Pentapora fascialis* | P | – | – | – | – | – | – | – |
| Dendroid | | | | | | | | |
| *Adeonella lichenoides* | – | – | – | – | – | A | P | – |
| *Adeonellopsis japonica* | P | – | – | – | P | – | P | – |
| *Reteporella dendroides* | P | – | P | – | P | P | P | P |
| *Celleporina attenuata* | P | – | – | – | P | P | P | – |

*P* present, *A* abundant

(*Borgiola rugosa*, *Biflustra savartii*, and *Pentapora fascialis*) were restricted to a single collection, with two of the three species only in the Döderlein collection; these have not been seen in Sagami Bay for over a century.

Nine of the erect bryozoan species detected in Sagami Bay were collected only at depths less than 100 m; six species extended into the 100–200 m depth interval (Fig. 9.3). Only *Reteporella dendroides* and *Triphyllozoon anatinum* were detected at depths exceeding 200 m.

## Comparison Among Sampling Areas

Erect bryozoans were considerably more diverse in Sagami Bay than in the other areas. In the Nansei Islands, they were less than one-third as diverse as in Sagami Bay, but representatives of all except the foliaceous growth form occurred (Table 9.2). Among the species detected, *Adeonellopsis* sp. 1 was

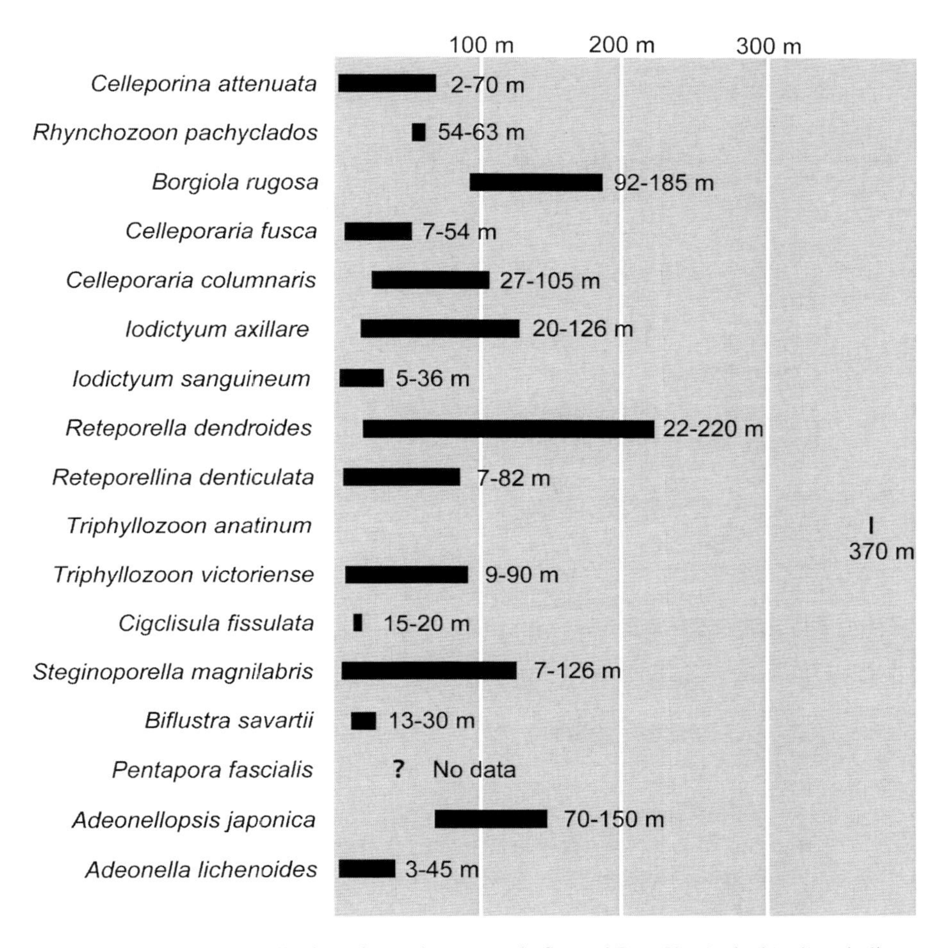


**Fig. 9.3** Bathymetric distribution of erect bryozoans in Sagami Bay. *Vertical white lines* indicate 100 m depth intervals

abundant in the area. Erect bryozoans were markedly less diverse near Otsuchi Bay than in Sagami Bay, and half as diverse as in the Nansei Islands (Table 9.2). Only three species were detected, representing only two of the five morphological categories.

The species composition was entirely different among the three localities, except for *Celleporaria fusca* occurring in both Sagami Bay and the Nansei Islands. Some genera, however, were common to form categories between localities, both Sagami Bay and the Nansei Islands, for example, showed a species of *Celleporaria* representing the nodular growth form, and a species of *Triphyllozoon* representing the reticulate form. All three areas had a species of *Adeonellopsis* representing the dendroid form.

**Table 9.2** Comparison of erect bryozoans among Sagami Bay (eastern and western areas combined), the Nansei Islands, Otsuchi Bay, and Otago Shelf

| Sagami Bay | Nansei Islands | Otsuchi Bay | Otago Shelf |
|---|---|---|---|
| Robust branching | | | |
| *Rhynchozoon pachyclados* | *Celleporaria fusca* | *Celleporina attenuata* | *Celleporina grandis* |
| *Cigclisula fissurata* | | *Heteropora* sp. 1 | |
| *Borgiola rugosa*[a] | | | |
| Nodular | | | |
| *Celleporaria columnaris* | *Celleporaria fusca* | – | *Celleporaria agglutinans* |
| *Celleporaria fusca* | | | |
| Reticulate | | | |
| *Iodictyum axillare* | *Dictyochasma* sp. 1 | – | *Hornera foliacea* |
| *Iodictyum sanguineum* | *Schizoretepora tumescens* | | |
| *Reteporellina denticulata* | *Triphyllozoon* sp. 1 | | |
| *Triphyllozoon anatinum*[a] | | | |
| *Triphyllozoon victoriense* | | | |
| Foliaceous | | | |
| *Steginoporella magnilabris* | – | – | *Hippomenella vellicata* |
| *Biflustra savartii* | | | |
| *Pentapora fascialis*[a] | | | |
| Dendroid | | | |
| *Adeonellopsis japonica* | *Adeonellopsis* sp. 1 | *Adeonellopsis japonica* | *Cinctipora elegans* |
| *Adeonella lichenoides* | | | |
| *Reteporella dendroides* | | | *Hornera robusta* |
| *Celleporina attenuata* | | | |

[a] Species not collected in the past 100 years

# Discussion

## *Diversity of Erect Bryozoans in Time and Space in Sagami Bay*

We detected few differences in the faunal composition between the eastern and western parts of the bay, perhaps in part due to differences in primary sampling localities (eastern or western) and total sampling effort among the collections representing the four time intervals in Sagami Bay. Among the 17 species of erect bryozoans, all but three were detected in both the eastern and western parts among the various collections, in one time interval or another. The exceptions were *Borgiola rugosa* and *Pentapora fascialis*, which were detected only in the Döderlein collection and only in the eastern part (the only area where Döderlein collected). Taking into account the considerable subsequent sampling in eastern Sagami Bay by Emperor Showa and NSMT, it appears that these species disappeared from the bay sometime between the late nineteenth and early to mid-twentieth centuries. The other exception is *Adeonellopsis japonica*, detected in three of the collections but only in eastern Sagami Bay, suggesting restriction to this side of the bay.

The most distinct pattern emerged from collections made across the middle twentieth century (Emperor Showa) and within the last decade (NSMT) (Table 9.1). In all, 14 species were detected in these two time intervals combined. Ten species were common to both intervals, but four were detected in the mid-twentieth century but not in the past decade, suggesting that there has been some loss of diversity between intervals. A more striking pattern emerged for western Sagami Bay, where Emperor Showa found 12 species but the NSMT survey detected only three remaining species, suggesting a considerable decrease in diversity. Although sampling effort by NSMT was less in western than in eastern Sagami Bay, several pieces of evidence suggest that these results reflect real changes. The reduction in diversity in the western part (from 12 down to 3 species) paralleled the overall reduction in diversity (14 to 10 species) recognized from a comparison between the Emperor Showa and NMST collections. In contrast, except for the three species found only in the early collections (*Borgiola rugosa* and *Pentapora fascialis*, in the Döderlein collection; *Triphyllozoon anatinum*, in the Döderlein collection and in the Doflein & Haberer collection), there was no reduction in the diversity of erect bryozoans between the late nineteenth century and mid-twentieth century: all the other frame-building species Döderlein (1883) detected also appeared in the Emperor Showa collection, as well as two additional species (*Celleporaria fusca* and *Reteporellina denticulata*). Total sampling effort by NSMT was roughly comparable to that by Emperor Showa, suggesting the apparent reduction in diversity (including some species from the better-sampled eastern part of the bay) by 2001–2005 is not an artefact.

## *Latitudinal Patterns in the Diversity of Erect Bryozoans in Japan*

We focused on the 14 present-day species of erect bryozoans in Sagami Bay for our comparison with the results of Otsuchi Bay and Nansei Islands. Although one would expect the diversity of shelf-dwelling bryozoan to decrease with increasing latitude in Japan, the erect species did not follow this pattern. The most northern locality (Otsuchi Bay, ~39°N) was least diverse both in species richness and growth forms, as expected, but Sagami Bay (~35°N) was markedly richer in species and somewhat richer in growth forms than the Nansei Islands (~24–30°N) farther south (Fig. 9.4, Table 9.2). Evidently, factors other than a latitudinal effect are responsible for the high diversity in Sagami Bay.

One possibility is the much higher sampling effort over a longer period; total numbers of specimens examined were 260, 1,200, and 90 for the Nansei Islands,

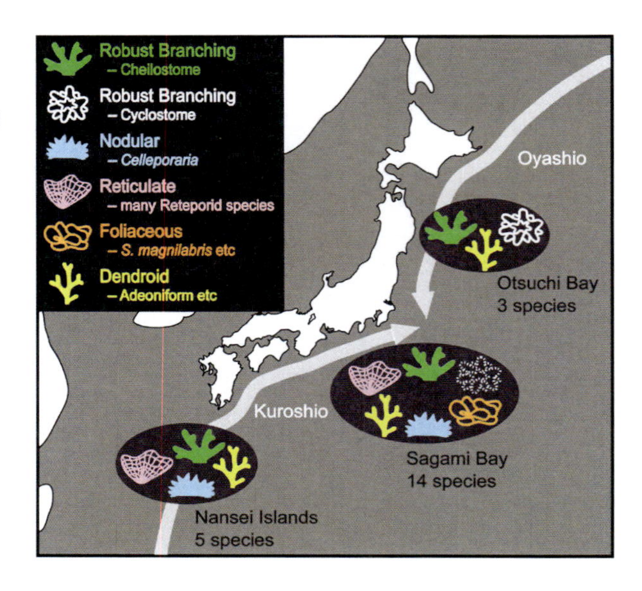


**Fig. 9.4** Distribution of colony forms detected in Sagami Bay, Otsuchi Bay, and the Nansei Islands. *Broad arrows* indicate the Kuroshio and Oyashio Currents

Sagami Bay, and Otsuchi Bay, respectively, a pattern of values that roughly parallels the number of species detected (5, 17, and 3, respectively). We are unable to assess the effect of this differential sampling effort on the diversities detected.

Another possible factor is a greater diversity of habitats and greater depth range in Sagami Bay compared to the other localities. Stations at which bryozoans were obtained in the vicinity of Otsuchi Bay ranged from 77 to 90 m, and in the Nansei Islands from 37 to 178 m, whereas erect bryozoans were collected in Sagami Bay at depths ranging from 2 to 370 m. Although we do not have exact data yet on the complexity of bottom habitats in Otsuchi Bay and the Nansei Islands, the bottom habitat in Sagami Bay is known to be differentiated, with an extensive shelf zone at depths less than 200 m, submarine canyons and knolls, environmental differences between the eastern and western parts of the bay (National Museum of Nature and Science Tokyo 2007).

This recalls that Cuffey et al. (1977) reported a unique distribution pattern of bryozoan reefs within Joulters Cays, Bahamas, which related to the fluctuation of water temperature, salinity, and tidal current. They mentioned that the habitat may have been too severe for corals but not for bryozoans, and especially cheilostomes tend to be more eurytopic or tolerant than corals and cyclostomes. Thus, the high diversity of large erect cheilostome colonies in Sagami Bay seems to reflect the variable environment of this area.

At the current stage in the development of our studies, we cannot make a direct link for the high diversity in Sagami Bay with the transport of faunal elements from the north and south by the Oyashio and Kurashio currents, respectively, because few of the erect species present in Sagami Bay were found in the other localities; *Celleporaria fusca* occurs both in Sagami Bay and off the Nansei Islands.

The species has previously been recorded from the upper bathyal of the Gulf of Aqaba, showing nodular growth, age correlated growth bands, and a very slow growth rate (Hillmer et al. 1996; Sobich 1996). *Celleporina attenuata* and *Adeonellopsis japonica* occurs in Sagami Bay and in Otsuchi Bay. Some genera, however, including *Rhynchozoon*, *Celleporaria*, *Steginoporella*, and most of the phidoloporid genera other than *Rhynchozoon*, are more diverse in the subtropics and at temperate latitudes than farther north; some of these genera were well represented in Sagami Bay and the Nansei Islands, but not at Otsuchi Bay.

At Otsuchi Bay, instead, representatives of more northern genera occurred: *Celleporina attenuata* and *Heteropora* sp. 1 replaced *Rhynchozoon pachyclados* and *Borgiola rugosa* seen in Sagami Bay in the robust branching growth morphology. In other words, *Celleporina attenuata* represents a northern faunal element in Sagami Bay. Horikoshi and Tsuchida (1981) reported *Borgiola rugosa* from inside Otsuchi Bay, though only in a species list; this species might also represent a northern element. The influences of currents from the north and south in Sagami Bay may thus be better viewed as a component of the environmental complexity there, permitting the occurrence of both southern and northern faunal elements. Thus, *Borgiola rugosa* is abundant in regions of Japan affected by the cold Oyashio Current, and the loss of erect colonies such as reported from Sagami Bay may have partly been caused by warming of the water temperature or a reduction in the influence of the cold current.

Due to the comparison of the diversity of erect growth forms, their distribution around Japan can be categorized as follows; (1) nodular forms represented by *Celleporaria* species and abundant reticulate forms are characteristic of southern localities in Japan, (2) robust branching forms represented by *Celleporina* and *Heteropora* species are characteristic of northern localities, and (3) colonies with dendroid branching growth represented by *Adeonellopsis* are widely distributed on the Pacific coast of Japan. Study of the bryozoan assemblage community provides useful information on the environment.

## *Comparison of the Erect Bryozoan Assemblage in Sagami Bay with Bryozoan Thickets at Otago, New Zealand*

At Otago Shelf (Batson and Probert 2000), one or two species represent each of the five categories of erect bryozoans detected in Sagami Bay (Table 9.2), and in one case the same genus is involved: species of *Celleporaria* contribute to the nodular category. Although, the overall taxonomic composition and diversity of rigidly erect bryozoans is generally showing more difference between the two localities, we found 14 species in the recent Sagami Bay compared to six species contributing to thickets at Otago Shelf. Also, at Sagami Bay erect bryozoans were dominated by cheilostomes, whereas at Otago Shelf cyclostomes (*Hornera* and *Cinctipora*) were proportionally much more prominent. According to Cuffey et al.

(1977), cyclostomes are stenotopic and cheilostomes are more eurytopic; thus, the erect bryozoan assemblages indicate that Sagami Bay includes more variable and severe habitats than Otago. The data show that in bryozoan assemblages, the morphology and size of the morphological groups of frame-builders are not taxonomically constrained. Ultimately, it will become necessary to employ underwater cameras or submersibles in Sagami Bay to check whether thickets occur or not. If they occur, they can be compared in detail with thickets in other parts of the world, and factors leading to bryozoan frame building or even bryozoans reefs can be investigated.

Bryozoan thickets are especially interesting because of the complex community associated with them. Canu and Bassler (1929: 361) noted for reticulate phidoloporids, "Never is a complete living colony encrusted by another organism. No larva has been able to develop here and disturb the harmony, no spores of the nullipores have been able to propagate themselves." Harmer (1934: 531) suggested that this generalization was incomplete and revised it to, "the living parts are generally free from encrusting organisms; a fact which I attribute to the efficiency of their avicularia in preventing the fixation of the larvae of encrusting and other organisms" with the exception "the dead and inactive parts of a reteporid colony form a good substratum for the attachment of organisms." We found many and various epizoans on three reticulate colonies (about 3–12 cm in diameter) in the collection of Emperor Showa, including poriferans, cnidarians (hydrozoans, anthozoans), crustaceans (amphipods, decapods, barnacles), polychaetes, bivalves, echinoderms (ophiuroids, echinoids), and colonies of other bryozoan species. Grischenko et al. (2007) found a similar community associated with the *Phidolopora elongata* aggregation at Akkeshi that included 11 species of cheilostomes, two of which were found only on *P. elongata*. This kind of bryozoan assemblage based on reticulate phidoloporid colonies as bioconstructors may be very abundant in the shelf and slope zones of the Sagami Bay, and this habitat may in turn play a crucial role in providing a habitat for many other organisms in the bottom environment.

**Acknowledgements** We thank Dr. Bernhard Ruthensteiner and Mrs. Eva Lodde (Zoologische Staatsammlung München) for their assistance in the study of Doflein's and Haberer's collections; Madame Marie-Dominique Wandhammer and Madame Marie Meister (Musée Zoologique Strasbourg) for assistance in observations and the loan of Döderlein's collection of bryozoans; Dr. Hiroshi Namikawa (Showa Memorial Institute of the National Museum of Nature and Science) for much advice and assistance, and the loan of the bryozoan collection of Emperor Showa; Prof. Susumu Ohtsuka (Hiroshima University) and the crew of T/V *Toyoshio-maru* for support during collecting around the Nansei Islands; Dr. Asako Matsumoto (Takurei University) and the stuff of the Otsuchi Marine Research Center, University of Tokyo for crucial support in collecting by the R/V *Yayoi* in Otsuchi Bay; Professor Matthew H. Dick (Hokkaido University) for suggestions on and detailed revisions of the manuscript; and especially Mrs. Brigitte Lotz (Forschungsinstitut Senckenberg) for great assistance during the stay of M. Hirose in Germany. This study was supported in part by a Fellowship from the Japan Society for Promotion of Science Fellows (No. 20-3856 to MH).

## References

Batson PB, Probert PK (2000) Bryozoan thickets off Otago Peninsula. New Zealand fisheries assessment report 2000/46:1–31

Canu F, Bassler RS (1929) Bryozoa of the Philippine region. US Natl Mus Bull 100:1–685

Carter RM, Carter L, Williams JJ, Landis CA (1985) Modern and relict sedimentation on the South Otago Continental Shelf, New Zealand. – N.Z. Oceanogr Inst Mem 93:1–43

Cuffey RJ (1974) Delineation of bryozoan roles in reefs from comparison of fossil bioherms and living reefs. In: Proceedings of the 2nd international coral reef symposium, vol 1. pp 357–364

Cuffey RJ (1977) Bryozoan contributions to reefs and bioherms through time. Stud Geol 4:181–194

Cuffey RJ, Fonda SS, Kosich DF, Gebelein CD, Bliefnick DM, Soroka LG (1977) Modern tidal-channel bryozoan reefs at Joulters Cays (Bahamas). In: Proceedings of 3rd international coral reef symposium, vol 2. pp 339–345

Döderlein LHP (1883) Faunistische Studien in Japan: Enoshima und die Sagami-Bai. Arch Naturgesch 49:102–123

Doflein F (1906) Ostasienfahrt. Erlebnisse und Beobachtungen eines Naturforschers in China, Japan und Ceylon. Verlag von B.G. Teubner, Leipzig/Berlin, p 511

Fujita T (2008) Echinoderms in Sagami Bay: past and present studies. In: Okada H, Mawatari SF, Suzuki N, Gautam P (eds) Origin and evolution of natural diversity, Proceedings of international symposium. The origin and evolution of natural diversity, Sapporo, 1–5 Oct 2007, pp 117–123

Grischenko AV, Dick MH, Mawatari SF (2007) Diversity and taxonomy of intertidal Bryozoa (Cheilostomata) at Akkeshi Bay, Hokkaido, Japan. J Nat Hist 41:1047–1161

Hageman SJ, Bock PE, Bone Y, McGowran B (1998) Bryozoan growth habits: classification and analysis. J Paleontol 72(3):418–436

Harmer SF (1934) The Polyzoa of the Siboga expedition. Part 3. Cheilostomata Ascophora, I. Family Reteporidae. Siboga Exped Rep 28c:502–640

Hillmer G, Scholz J, Dullo WC (1996) Two types of bryozoan nodules from the Gulf of Aqaba (Red Sea). In: Gordon DP, Smith AM, Grant-Mackie JA (eds) Bryozoans in space and time. Proceedings of the International Bryozoology Association conference, pp 125–132

Hinata H, Miyano M, Yanagi T, Ishimaru T, Kasuya T, Kawamura H (2003) Short-period fluctuations of surface circulation in Sagami Bay induced by the Kuroshio warm water intrusion through Ooshima West channel. Oceanogr Jpn 12:167–184

Hirose M (2010) Cheilostomatous Bryozoa (Gymnolaemata) from Sagami Bay, with notes on bryozoan diversity and faunal changes over the past 130 years. Ph.D. Thesis, Graduate School of Science, Hokkaido University, pp 270 with 261 Plates

Horikoshi M, Coastal Oceanography Research Committee, The Oceanographical Society of Japan (1985) Coastal oceanography of Japanese Islands. Tokai University Press, Tokyo, pp 242–252

Horikoshi M, Tsuchida E (1981) Benthic invertebrates recorded from Otsuchi Bay and the adjacent Sanriku Coast – primary catalogue of fauna – additions and corrections (1). Otsuchi Mar Res Cent Rep 7:47–70

Isono N (1988) Personnels related to Misaki Marine Biological Laboratory – birth of zoology in Japan. Gakkai Shuppan Center, Tokyo, p 230, in Japanese

Iwata S (1979) On the sea conditions in Sagami Bay – seasonal fluctuation of vertical structure and water type. Bull Jpn Soc Fish Oceanogr 34(1979):134–137 (in Japanese)

Iwata S, Matsuyama M (1989) Surface circulation in Sagami Bay: the response to variations of the Kuroshio axis. J Oceanogr Soc Jpn 45:310–320

Kawabe M, Yoneno M (1987) Water and flow variations in Sagami Bay under the influence of the Kuroshio path. J Oceanogr Soc Jpn 43:283–294

Mawatari SF (2009) Forschungsarbeiten an der Döderlein Sammlung. In: Scholz J (ed) Seesterne aus Enoshima. Ludwig Döderleins Forschung in Japan. Khoshid, pp 18–31

McKinney FK, Jackson JBC (1989) Bryozoan evolution. Studies in paleobiology. Unwin Hyman, Boston, p 238

National Museum of Nature and Science (ed) (2007) Fauna Sagamiana. Tokai University Press, Hatano, p 212

Probert PK, Batham EJ (1979) Epibenthic macrofauna off southeastern New Zealand and mid-shelf bryozoan dominance. NZ J Mar Fresh 13:379–392

Scholz J (2000) Eine Feldtheorie der Bryozoen, Mikrobenmatten und Sedimentoberflächen. Abh Senckenb Naturforsch Ges 552:1–193

Scholz J, Sterflinger K, Junge C, Hillmer G (2000) A preliminary report on bryostromatolites. In: Herrera Cubilla A, Jackson JBC (eds). In: Proceedings of the 11th International Bryozoology Association conference, Smithsonian Tropical Research Institute, Panama, pp 376–384

Scholz J, Ernst A, Batson P, Königshof P (2005) Bryozoenriffe. Denisia 16:247–262

Sobich A (1996) Analyse von stabilen Isotopen und Wachstumsstrukturen der Bryozoe *Celleporaria fusca* zur Rekonstruktion der Umweltbedingungen im nördlichen Golf von Aqaba. Diplomarbeit, Teil 2, im Fachbereich Geowissenschaften University, Bremen, pp 55

Stebbing ARD (1971) The Epizoic Fauna of *Flustra foliacea* [Bryozoa]. J Mar Biol Assoc 51:283–300

# Chapter 10
# Epizoic Bryozoans on Predatory Pycnogonids from the South Orkney Islands, Antarctica: "If You Can't Beat Them, Join Them"

## Fouled Pycnogonids from Antarctica

**Marcus M. Key, Jr., Joel B. Knauff, and David K.A. Barnes**

**Abstract** Antarctic bryozoans are poor spatial competitors compared to many sessile invertebrates. Antarctic bryozoans are frequently destroyed by ice scouring of the substratum during open water periods, and Antarctic bryozoans are specifically preyed upon by pycnogonids. Based on this, it was hypothesized that Antarctic bryozoans should foul pycnogonids more than other motile hosts and other sessile biotic and abiotic substrata. To test these hypotheses, 115 live pycnogonids were collected in the South Orkney Islands, Antarctica. Their carapaces were examined for epizoic bryozoans, and each colony's size was measured and its location mapped. Nine species of pycnogonids were identified containing 156 bryozoan colonies belonging to seven cheilostome species. Of the 115 pycnogonids, 26% were fouled by bryozoans. The bryozoan species richness on pycnogonids is similar to that on the adjacent boulders. Compared to other motile host animals, the number of bryozoan species per unit host surface area is an order of magnitude higher on pycnogonids. This may be attributed to carapaces of pycnogonids acting as refugia for the bryozoans from competition for space on hard substrata, ice scour, and predation by their host.

**Keywords** Ecology • Epibiosis • Bryozoans • Pycnogonids • Antarctica

---

M.M. Key (✉)
Department of Earth Sciences, Dickinson College, Carlisle, PA, USA
e-mail: key@dickinson.edu

J.B. Knauff
Controlled Hazardous Substance Enforcement Division, Maryland Department of the Environment, Baltimore, MD, USA

D.K.A. Barnes
British Antarctic Survey, National Environmental Research Council, Cambridge, UK

A. Ernst, P. Schäfer, and J. Scholz (eds.), *Bryozoan Studies 2010*,
Lecture Notes in Earth System Sciences 143, DOI 10.1007/978-3-642-16411-8_10,

## Introduction

Epibiosis, or fouling, is a common and well-documented occurrence on sessile hosts. It has also been documented amongst bryozoans on motile marine animals such as crabs (e.g., Key et al. 1997), isopods (Key and Barnes 1999), sea snakes (Key et al. 1995), pycnogonids (Wyer and King 1973), and even on floating gastropods (Taylor and Monks 1997).

Bryozoans are typically poor space competitors (Soule and Soule 1977), but some bryozoans escape from competition for substratum space by erect growth (McKinney and Jackson 1989). However, many bryozoans are extremely effective at rapid colonization of young surfaces including the external surfaces of both sessile and motile hosts (Moore 1973; Soule and Soule 1977; Key et al. 1996a). Such epibiosis, whether it is opportunism through to host specific commensalism, has the potential to be an advantage in the evasion of substratum competition. In most environments the 'lifespan' of abiotic substrata is likely to be greater than those of biotic substrata, but in the polar regions, animals' life spans are long and disturbance of abiotic substrata is frequent (Barnes et al. 1996). Epibiosis, therefore, can be of particular advantage in both the evasion of predators when the host is motile and in the evasion of substratum disturbance in the Antarctic, which is subject to massive and frequent ice scouring (Barnes and Clarke 1995; Key and Barnes 1999).

In the current study we aimed to: (1) quantitatively describe the prevalence of fouling of pycnogonids (sea spiders) by bryozoans in terms of the number, size, and percentage of hosts fouled, the number and size of colonies on fouled hosts, and the number of different bryozoan species on fouled hosts; (2) correlate host size with the prevalence and species richness of epizoic bryozoans (i.e., following the Theory of Island Biogeography); (3) test the hypothesis that bryozoans should foul pycnogonids more than other motile hosts and more than abiotic hard substrata as epibiosis on pycnogonids should remove the bryozoans from competition for space on hard substrata, substratum scouring, as well as predation.

### *The Hosts*

The host pycnogonids are basibionts in Wahl's (1989) terminology. The body of pycnogonids is usually thin, elongate, and segmented, with an elongate proboscis (Fig. 10.1). The trunk is made of four to six segments with the first acting as part of the head, whereas the remaining ones each attach to a pair of walking legs. Most pycnogonids have four pairs of walking legs, but may have as many as six in some species. Other appendages include chelifores, which are small clawed appendages used in feeding, and a pair of ovigerous legs occurring in both sexes in some species, which are used for cleaning appendages of epibionts and, on males, for egg-carrying (Arnaud and Bamber 1987). Ovigerous legs are absent from females of some species. The long and slender walking legs are composed of eight segments, each with a terminal claw (King 1973).

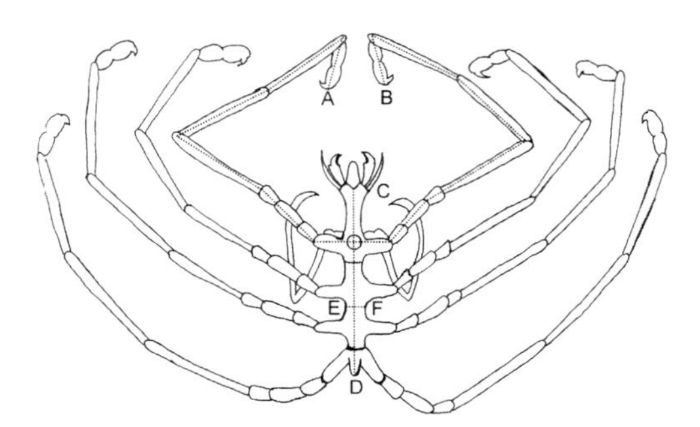


**Fig. 10.1** Template of a generic pycnogonid showing characters measured and segments onto which bryozoan colonies were mapped. AB = span of the first pair of walking legs (measured when the legs are both extended perpendicular to the body). CD = trunk length (measured from the base of the proboscis to the terminal end of the anus). EF = trunk width (measured directly behind the second walking legs)

Most pycnogonids are benthic, although some are able to swim. Pycnogonids are fairly slow-moving carnivorous grazers and thus feed on either sessile or slow-moving prey. Pycnogonids prey upon a variety of animals including sponges, cnidarians, molluscs (esp. nudibranchs), echinoderms, and polychaetes, but their main food sources are hydroids and bryozoans (King 1973; Wyer and King 1973; Ryland 1976; Arnaud and Bamber 1987; Bain 1991; Piel 1991; Mercier and Hamel 1994).

Molting frequency in pycnogonids decreases through ontogeny with increasing size, until sexual maturity at adulthood when it generally stops (Tomaschko et al. 1997). From the larval stage to sexually mature adult, male pycnogonids typically undergo 8 molts, whereas females undergo 9–11 molts, depending on the species. Males do not molt while carrying eggs (King 1973; Arnaud and Bamber 1987).

Known shallow water epibionts of pycnogonids include protozoans, algae, hydroids, postveliger bivalves, and bryozoans, whereas in deeper water they include foraminifera, sponges, hydroids, serpulids, cirripeds, brachiopods, and bryozoans (Wyer and King 1973; Arnaud and Bamber 1987; Sherwood et al. 1998).

## *The Epibionts*

Applying the nomenclature for marine organism-hard substratum relationships proposed by Taylor and Wilson (2002), the epibiotic bryozoans fouling the living pycnogonid hosts are termed sclerobionts in general and epizoozoans in particular. Bryozoans are preyed on by a variety of animals including platyhelminths, nematodes, arthropods, mollusks, echinoderms, chordates, and pycnogonids (Ryland 1970; McKinney and Jackson 1989; McKinney et al. 2003). Pycnogonids occurring on bryozoan colonies are well known from the literature (King 1973), but actual documented examples of pycnogonid predation of bryozoans are fewer (Table 10.1).

**Table 10.1** Documented examples from the literature of pycnogonid predation of bryozoans

| Pycnogonid | Bryozoan | Reference |
|---|---|---|
| *Ascorhyncus* sp. | *Serialaria* spp. | Döhrn (1881) |
| *Phoxichilidium femoratum* | *Crisia* sp. | Prell (1910) |
| *Anoplodactylus* sp. | *Bowerbankia* sp. | Lebour (1945) |
| *Austrodecus glaciale* | *Cellarinella foveolata* | Fry (1965) |
| *Pycnogonum littorale* | *Flustra foliacea* | Wyer and King (1973) |
| *Achelia echinata* | *Flustra foliacea* | Wyer and King (1973, 1974) |
| *Nymphon gracile* | *Bowerbankia imbricata* | King (1973); Wyer and King (1973, 1974) |
| *Tanystylum isabellae* | *Amathia distans* | Varoli (1994) |
| *Anoplodactylus stictus* | *Amathia distans* | Varoli (1994) |
| *Stylopallene longicauda* | *Amathia wilsoni* | Sherwood et al. (1998) |
| *Pseudopallene watsonae* | *Orthoscuticella* cf. *ventricosa* | Staples (2004) |
| *Pseudopallene reflexa* | *Orthoscuticella* cf. *ventricosa* | Staples (2004) |

Pycnogonids are zooid-level predators of bryozoans (McKinney et al. 2003), so the predation is generally not lethal at the colony level, as the entire colony is not consumed (Berning 2008). Four different methods of predation of bryozoans by pycnogonids have been documented. (1) Access through the frontal pores of the bryozoan. Fry (1965) reported how the pycnogonid *Austrodecus glaciale* uses its proboscis to penetrate through the frontal pores of *Cellarinella foveolata* zooids. (2) Access through the operculum of the bryozoan. Wyer and King (1973) reported how the pycnogonid *Achelia echinata* uses its proboscis to penetrate through the operculum of *Flustra foliacea* zooids. Staples (2004) reported that the digitiform process of the chela of the larval form of the pycnogonid *Pseudopallene watsonae* is used to manipulate the operculum covering the orifice of each *Orthoscuticella* cf. *ventricosa* zooid facilitating insertion of the proboscis. As a result of predation, many of the zooids are empty with the operculum displaced to a vertical position (Staples 2004). (3) Using acute spines at tips of the palps to bore into zooecia before inserting proboscis. Arnaud and Bamber (1987) reported the pycnogonid genera *Austrodecus* and *Rhynchothorax* using this method. (4) Access through the crushed zooecium of the bryozoan. The chela of the adult form of the pycnogonid *Pseudopallene watsonae* is used to crush the zooids before insertion of the proboscis (Staples 2004). As pycnogonids are suctorial, carnivorous, grazing predators (Arnaud and Bamber 1987), predation ends with the pycnogonid sucking out the internal organs of the bryozoan with its muscular pharynx located inside its long thin proboscis, shredding them with its chelae, and transferring them to the setal "oyster basket" mouth at the anterior end of the proboscis for ingestion (Wyer and King 1973, 1974). King (1973) reported the chelae of *Nymphon gracile* being used to macerate the bryozoan *Bowerbankia* and hold it against the mouth for ingestion. Those methods involving the proboscis depend on the predator having the right size and shape proboscis for the prey's morphology (e.g., operculum size and shape) (King 1973).

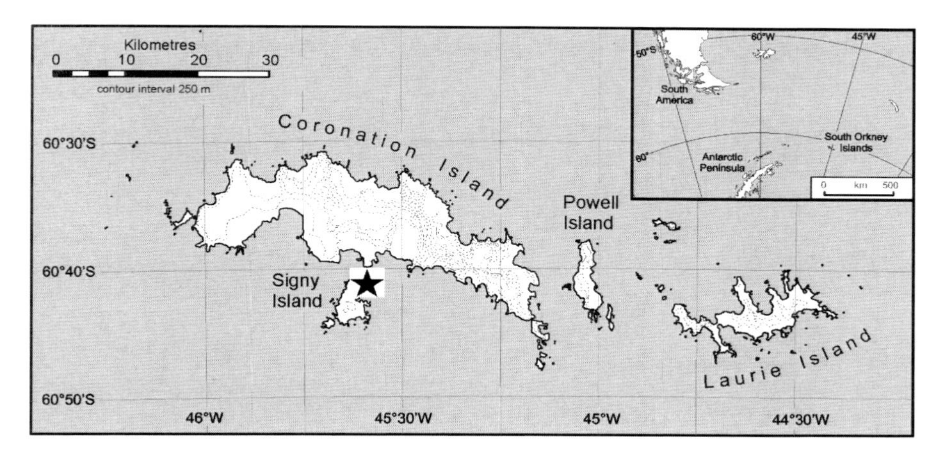


**Fig. 10.2** Map showing location of the study site (star) in the South Orkney Islands, Antarctica. Modified from http://www.antarctica.ac.uk/living_and_working/research_stations/signy/index.php

## Materials and Methods

A total of 115 living pycnogonids were collected for this study. They are housed in the British Antarctic Survey. They were collected in Outer Borge Bay, between Signy Island and Coronation Island in the South Orkney Islands, Scotia Arc, Southern Ocean, maritime Antarctic (Latitude 60° 42′ S, 45° 37′ W) (Fig. 10.2). Collection occurred during the austral summer from December 1991 to February 1992. Specimens were collected and sorted from Agassiz benthic trawls ranging from 50 to 150 m water depth towed for bottom durations of 15 min at each location. A combination of echo sounder and hydrographic charts was used to ensure that each sample came from a narrow range of depths (within 10 m of the nominal depth). The environment where these specimens were collected is classified as a South Atlantic coastal boulder community and is the site of frequent ice scouring of the substratum (Barnes 1999).

The size of each pycnogonid was determined using the following three standard morphometric proxies for age: span of the first pair of walking legs (measured when both legs are extended perpendicular to the long axis of the trunk, AB in Fig. 10.1), trunk length (measured from the base of the proboscis to the terminal end of the anus, CD in Fig. 10.1), and trunk width (measured directly behind the second walking legs, EF in Fig. 10.1). Missing limbs and any non-bryozoan epizoozoans were also noted.

The location of each bryozoan colony was mapped onto a generic template (Fig. 10.1), showing dorsal and ventral surfaces for each individual host. The number of individual colonies per host was counted, and the surface area of each individual colony was measured in dorsal and ventral view. As the host's body and leg segments are cylindrical, the entire colony could not always be measured digitally from just the dorsal and ventral views. In these cases, the area of individual colonies was determined by measuring the diameter and length of the colony. This was used to calculate the maximum surface area of the colony assuming a perfect cylinder, which pycnogonid legs approximate. Finally, any of the colony's surface area from this cylinder that was missing was subtracted. Total bryozoan colony surface area for a given host was determined by adding up individual surface area measurements of all colonies on a host. Repeatability experiments indicate the PC-based image analysis system using digitized video images of the pycnogonids and bryozoans has a maximum measurement error of 3.2%.

## Results and Discussion

The following nine species of pycnogonids were identified: *Nymphon australe* Hodgson (63% of individuals), *Decolopoda australis* Eights (9%), *Ammothea clausi* Pfeffer (7%), *Ammothea meridionalis* Hodgson (7%), *Pentanymphon antarcticum* Hodgson (7%), *Colossendeis megalonyx* Fry and Hedgpeth (3%), *Ammothea striata* (Mobius) (1%), *Pallenopsis pilosa* (Hoek) (1%), and *Pycnogonum gaini* Bouvier (1%). For *A. clausi*, *A. meridionalis*, *A. striata*, and *P. gaini* this is the first record for the South Orkney Islands (for a list of the known fauna of the South Orkney Islands, see Barnes et al. 2009, Appendix S1).

Of the 16 pycnogonids for which gender could be confidently determined, 56% were females, 44% were males, and 29% of the males were carrying eggs. As so few pycnogonids could be sexed, inter-gender comparisons were not tested.

Among the 115 pycnogonids, trunk length varied from 2.16 to 23.00 mm (mean = 7.52 mm, standard deviation = 4.19 mm). Trunk width ranged from 0.51 to 5.14 mm (mean = 1.58 mm, standard deviation = 1.12 mm). The span of the first walking legs ranged from 15.36 to 320 mm (mean = 65.89 mm, standard deviation = 59.15 mm).

Host surface area was not measured directly due to logistical constraints. Instead it was calculated assuming the trunk was a rectangular prism and each pair of legs was a cylinder as wide as the diameter of the encrusting bryozoan colonies (e.g., Fig. 10.3), which was measured, and as long as the span of the first pair of walking legs which was measured. Using Fig. 10.1, the host surface area was calculated as 2*CD*EF + 4*(2*$\pi$*diameter/2*(AB-EF)). This resulted in a mean surface area of the hosts of 727 $mm^2$ (range = 170–3,500 $mm^2$, standard deviation = 656 $mm^2$).

Though sponges and algae were also found living on the host pycnogonids, this study concerns itself only with bryozoans, as they were the most diverse and prevalent of the epizoozoans encountered. Seven cheilostome bryozoan species were identified: *Antarctothoa antarctica* Moyano and Gordon, *Antarctothoa dictyota* (Hayward) (Fig. 10.3), *Antarctothoa discreta* (Busk), *Crassimarginatella perlucida* (Kluge), *Osthimosia* sp. Jullien, *Thrypticocirrus phylactelloides* (Calvet), and *Xylochotridens rangifer* Hayward and Thorpe. The record of *C. discreta* is the first record for the South Orkney Islands. Two of the bryozoan species (*C. antarctica* and *C. dictyota*) found on the pycnogonids have also been reported from adjacent sites encrusting rocks (Barnes et al. 1996) and isopods (Key and

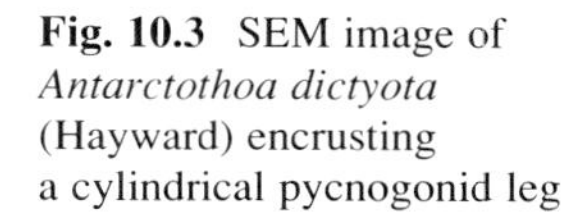

**Fig. 10.3** SEM image of *Antarctothoa dictyota* (Hayward) encrusting a cylindrical pycnogonid leg

Barnes 1999). *C. antarctica* and *Osthimosia* sp. have been reported as epizoozoans on Antarctic echinoids (Linse et al. 2008). *C. antarctica* and *X. rangifer* have been reported as epizoozoans on Antarctic brachiopods (Barnes and Clarke 1995). The most commonly occurring species (*C. antarctica*) has been reported on a total of 13 different types of biotic and abiotic substrata (Barnes and Clarke 1995; Key and Barnes 1999). *C. antarctica* and *X. rangifer* are considered low specificity epibiotic species as they occur widely on various host organisms, but rarely on rocks (Barnes and Clarke 1995).

Pycnogonids are fouled by a variety of organisms in addition to the bryozoans reported here. These include foraminiferans, sponges, hydroids, anemones, brachiopods, barnacles, tunicates, serpulids, and molluscs (Hedgpeth 1964; King 1973; Pipe 1982; Barnes et al. 2004). Epizoic bryozoans on pycnogonids are rarely reported in the literature. King (1973) reported the pycnogonid *Nymphon gracile* fouled by the bryozoan *Electra pilosa*. Key and Barnes (1999) reported two species of bryozoans (*Celleporella antarctica* and *C. bougainvillea*) as epibionts on pycnogonids. This study increases the reported number of bryozoan species fouling pycnogonids from three to nine.

Of the 115 host specimens, 30 (26%) were fouled by bryozoans. The total number of bryozoan colonies found was 156. The number of bryozoan colonies per pycnogonid ranged from zero (for unfouled hosts) to 25 (mean = 1.36, standard deviation = 4.09). For the 30 fouled hosts, the number of bryozoan colonies ranged from one to 25 (mean = 5.20, standard deviation = 6.71).

The 156 bryozoan colonies occurred on a total of 310 host trunk/leg segments. Of the 156 bryozoan colonies, 101 (65%) occurred on only one segment. For example, of the 25 bryozoan colonies fouling the pycnogonid in Fig. 10.4, 23 occur on single segments, and only two spanned two segments. The maximum number of segments covered by a single colony was eight, and the mean was 1.99 segments (standard deviation = 1.86 segments). The larger colonies covered significantly more segments (linear regression, $R^2 = 0.534$, $p < 0.001$). The restriction of most colonies to a single segment of their host may reflect the movement of

**Fig. 10.4** Pycnogonid *Decolopoda australis* Eights fouled by 25 bryozoan colonies indicated by lighter color on darker carapace of host. Specimen is 9 cm across

the host's exoskeleton at articulated segment joints (Manton 1978), which may restrict lateral growth of the bryozoan colony. This pattern has been documented on other arthropod hosts fouled by bryozoans (Key and Barnes 1999; Key et al. 1996b, 1999, 2000).

Most (95%) bryozoan colonies were found growing on the ovigerous and walking legs. Only 5% were on the central trunk region. This most likely is simply a function of the fact that the legs represent a much larger proportion (82%) of the surface area of the pycnogonids compared to the trunk (18%), based on the relative sizes of the segments in Fig. 10.1. As also reported by Barnes and Clarke (1995), the number of bryozoan colonies per leg segment was proportional to the length of the leg segment.

The exoskeleton of the pycnogonid is made of a hard cuticle, which in some species is covered by hair-like setae (King 1973). We noticed that those species with more setae, tended to have fewer and more fragmented colonies, so perhaps an additional function of the setae is to reduce fouling. In addition, the ovigerous legs are used for cleaning the walking legs, especially in the females, which don't carry eggs (King 1973). King (1973) reported the egg carrying males of the pycnogonid *Nymphon gracile* were more fouled by the bryozoan *Electra pilosa*. This was interpreted to be because the ovigerous legs were unavailable for cleaning, and the males not molting while carrying eggs. Thus, we could expect the number and size of bryozoan colonies to be greater on those hosts with eggs compared to those without, but no significant difference was found (t-tests, $p > 0.05$). However the sample size was small (only 11 of the 115 pycnogonids (10%) had eggs). Otherwise, our results support Arnaud and Bamber's (1987) suggestions that the ovigerous legs are not effective at removing encrusting bryozoans from appendages.

The surface area of individual colonies ranged from 0.19 to 171.53 mm$^2$, (mean = 18.84 mm$^2$, standard deviation = 27.92 mm$^2$). The combined total surface area of bryozoan colonies on individual pycnogonids ranged from zero (for unfouled hosts) to 1,039.32 mm$^2$ (mean = 25.45 mm$^2$, standard deviation = 119.88 mm$^2$). For the 30 fouled hosts, the combined total surface area of bryozoan colonies on individual pycnogonids ranged from 1.24 to 1,039.32 mm$^2$ (mean = 97.54 mm$^2$, standard deviation = 221.85 mm$^2$). Based on the fouled host areas calculated, the

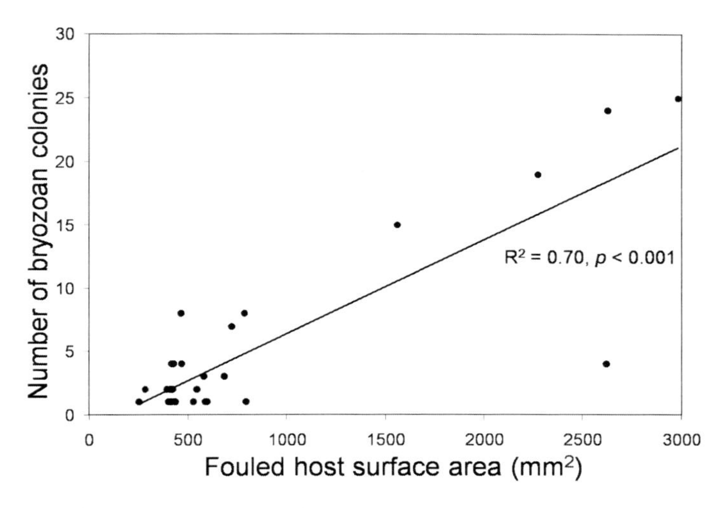


**Fig. 10.5** Plot of number of bryozoan colonies versus surface area of fouled host pycnogonids

percentage of host surface area covered by bryozoans ranged from 0.1% to 72% (mean = 10%, standard deviation = 17%).

The Target Area Effect Hypothesis of MacArthur and Wilson's (1967) Theory of Island Biogeography predicts that larger islands should have higher species richness (Lomolino 1990). This is because larger islands have (amongst other things) higher immigration rates simply because they are a bigger target for 'propagules' (Stracey and Pimm 2009). If one considers the host pycnogonids as hard substratum 'islands' in a 'sea' of soft subtratum muds, then it is predicted that there should be more bryozoans colonies on larger hosts as they provide a larger target for bryozoan larval settlement. As predicted by the Target Area Effect Hypothesis, the number of bryozoan colonies across species increases significantly with increasing host surface area when considering all hosts (linear regression, $R^2 = 0.202$, $p < 0.001$) or only fouled hosts (linear regression, $R^2 = 0.705$, $p < 0.001$; Fig. 10.5). As expected with more room to grow and more colonies, the larger hosts have a significantly larger total surface area of bryozoan colonies (linear regression, $R^2 = 0.062$, $p = 0.008$), but the correlation between host surface area and the percent of host surface area covered by bryozoans is not significant (linear regression, $R^2 = 0.021$, $p = 0.453$). The Target Area Effect has been documented on a variety of motile hosts fouled by bryozoans (Key and Barnes 1999; Key et al. 1996a, 2000).

## *Comparison with Other Substrata*

How does the bryozoan species richness on the pycnogonids (i.e., seven species) compare with that on abiotic hard substrata from adjacent sites? In trawl samples in

**Table 10.2** Comparison of fouling prevalence and epizoic bryozoan species richness among motile ephemeral hosts of varying size and latitude

| Host | Host surface area ($cm^2$) | # of hosts | % of hosts fouled | # of bryozoan species | % of hosts fouled per $cm^2$ | # of bryozoan species per $cm^2$ | Latitude | Source |
|---|---|---|---|---|---|---|---|---|
| Pycnogonids | 7 | 115 | 26 | 7 | 3.71 | 1.00 | 61°S | This study |
| Isopods | 73 | 60 | 42 | 10 | 0.58 | 0.14 | 61°S | Key and Barnes (1999) |
| Brachyuran crabs | 104 | 168 | 16 | 3 | 0.15 | 0.03 | 35°N | Key et al. (1999) |
| Horseshoe crabs | 504 | 56 | 77 | 2 | 0.15 | 0.00 | 1°N | Key et al. (2000) |
| Trilobites | 10 | 14,958 | 0.1 | 3 | 0.01 | 0.30 | 20°S[a] | Key et al. (2010) |
| Sea snakes | 398 | 1,364 | 1 | 1 | 0.00 | 0.00 | 2°N | Key et al. (1995) |

[a]Upper Ordovician paleolatitude

the same region and at similar depths, eight species of encrusting bryozoans were found on other substrata, with only one species (*Thrypticocirrus phylactelloides*) in common (Barnes 1995). Likewise, 20 encrusting cheilostome bryozoan species were reported on rocks from 42 m depth (Barnes et al. 1996) and seven species from five recent trawls at 150–250 m (Barnes et al. 2009). Thus, bryozoan species richness on pycnogonids is of similar magnitude to that on adjacent abiotic substrata on the shallow shelf. Strict comparisons of richness per unit area between pycnogonids and abiotic substrata would be complex to interpret, as they would require measurements of substratum stability, surface rugosity, biofilms, etc.

In the shallow subtidal (5 m water depth) in the same area as in this study, 84% of the pycnogonids were fouled by bryozoans (Barnes and Clarke 1995). This is higher than the 26% reported here from 50 to 150 m water depth (but by fewer species). In contrast, settlement panels at 25 m in the same area as this study revealed only 2% cover after 21 months (Barnes 1996), but 90% of rocks at 42 m at this site had bryozoans on them (Barnes et al. 1996). Smaller and shallower rocks are more frequently disturbed (i.e., turned over by scouring) (Barnes et al. 1996) which may be similar in some respects to the disturbance caused by molting of the pycnogonids.

How does this 26% fouling prevalence of pycnogonids by epizoic bryozoans compare to other motile ephemeral hosts from different environments? Table 10.2 shows the results from published studies using the same methodologies as this study on the prevalence and epizoic bryozoan species richness among host organisms of varying sizes and from varying latitudes. All are motile, and all are ephemeral due to molting of their carapace (i.e., horseshoe crabs, brachyuran crabs, isopods, trilobites, and pycnogonids) or shedding of their skin (i.e., sea snakes). They range in size from a mean surface area of 7 $cm^2$ (i.e., the pycnogonids of this study) to 504 $cm^2$ (i.e., horseshoe crabs). The general trend, as expected from the

Target Area Effect Hypothesis of the Theory of Island Biogeography, is that larger hosts in Table 10.2 should have a higher prevalence of fouling, but it was not significant (linear regression, $R^2 = 0.217$, $p = 0.352$). The one outlier is the one non-arthropod host (i.e., the sea snake). Sea snakes are air breathers and spend part of their life out of water, which is detrimental to the bryozoans resulting in a lower prevalence of fouling (Key et al. 1995). Removing this outlier, the correlation is more significant (linear regression, $R^2 = 0.759$, $p = 0.054$).

Due to the Target Area Effect, the data must be corrected for the different sizes of the host substrata. This was done by dividing the percentage of hosts fouled and the number of bryozoan species by the surface area of the host. For both the percentage of hosts fouled per $cm^2$ and the number of bryozoan species per $cm^2$, the pycnogonid hosts are an order of magnitude higher (Table 10.2). This may be attributed to one or more of the following factors. The carapaces of pycnogonids may be acting as refugia for the bryozoans from selective pressures of predation by their host, from ice scour, or from space competition on abiotic hard substrata.

The selective pressure for epibiosis on pycnogonid hosts may come from predation of bryozoans by the pycnogonids themselves. As discussed previously, some pycnogonids eat bryozoans (Table 10.1). As is true for most pycnogonids, it is not known if any of these species prey on the epizoic bryozoan species found in this study. Five of the nine pycnogonid species come from genera (i.e., *Ammothea*, *Colossendeis*, and *Pycnogonum*) known for having a well developed proboscis used to thrust into the tissue of their prey for feeding (King 1973). It is possible that by residing on an inaccessible part of their predator, such as on the trunk or the proximal segments of the legs, the bryozoans could avoid being consumed. Soule and Soule (1977) suggested that epibiotic bryozoans should in general experience reduced predation compared to those on abiotic hard substrata.

The selective pressure for epibiosis on pycnogonid hosts may come from the destruction of the benthic community by scouring from drifting, floating ice (Peck and Bullough 1993; Arntz et al. 1994; Conlan et al. 1998; Barnes 1999; Gutt 2000, 2001; Gutt and Piepenburg 2003; Barnes and Conlan 2007). This is especially true of bryozoans (Barnes 1999; Brown et al. 2004; Bader and Schäfer 2005). The sessile organisms cannot move out of the way to avoid the scouring (Gutt et al. 1996; Peck et al. 1999; Gutt 2001). In Antarctica, ice scour occurs down to 500 m (Barnes and Lien 1988; Lien et al. 1989; Gutt et al. 1996), well below the depths of this study (i.e., 50–150 m). Ice scour has been occurring over geological and evolutionary time scales, so the Antarctic shelf species are ice scour-adapted (Barnes and Conlan 2007). Motile animals, which are the first to recolonize scoured areas (Conlan et al. 1998; Peck et al. 1999; Lee et al. 2001), may also be the most adept at avoiding being killed by scours (Brown et al. 2004).

In polar environments, ice scour, not predation, is often the main structuring element of nearshore communities as almost all of the macrofauna is removed by the gouging and trampling effects of grounding of floating ice during the summer (Dayton 1990; Dayton et al. 1994; Gutt et al. 1996; Peck et al. 1999; Brown et al. 2004). This has been shown for bryozoans in particular (Barnes and Bullough 1996). Therefore epibiosis on motile hosts that eliminates competition for space on hard substrata as well as avoids ice scour should be selectively advantageous. Pycnogonids could avoid icebergs by walking or swimming, but as they do both so slowly (King 1973), they would most likely just be pushed out of the way due to their almost neutral buoyancy (Morgan 1977; Schram and Hedgpeth 1978).

This physically controlled (i.e., high frequency of substratum disturbance by ice) view of the Antarctic benthos (Barnes 1999; Barnes and Conlan 2007), especially of epibiosis on motile hosts, is in contrast to the biologically accommodated view of Gutt and Schickan (1998). The latter interpret Antarctic epibiosis as a result of evolution in a physically stable environment leading to a biologically accommodated community. Unfortunately, their study excluded encrusting forms, such as bryozoans, which are the most common epizoozoans.

Just as the actions of gouging, scraping, and crushing predators impacted the evolution of growth form in bryozoans (McKinney et al. 2003), the actions of gouging, scraping, and crushing icebergs may have impacted the evolution of epibiosis in polar bryozoans. If this is true, then the bryozoan diversity on motile hosts should be higher in polar regions. Table 10.2 shows this exact pattern. The number of epizoic bryozoans species per host is usually three or less, but there are two outliers (i.e., the isopods with ten species and pycnogonids with seven species, both of which are from high latitudes, in this case the Antarctic).

The selective pressure for epibiosis on pycnogonid hosts may come from space competition on abiotic hard substrata. In sessile communities, space is often a limiting resource (Dayton 1971; Paine 1974; Jackson 1977), and this is especially true for Antarctic bryozoans (Barnes and De Grave 2002; Barnes 2005). A more restricted but important substratum for sessile organisms in the Antarctic is motile animal substrata (Barnes et al. 2004) which may be a more stable substratum than easily overturned rocks (Gutt 2000), even despite the instability caused by molting. Epibiosis is a frequent solution to the problem of limited conventional (i.e., hard, abiotic) substratum space (Wahl 1989, 2009) as epibionts may benefit from improved survivorship (Lohse 1993) and/or feeding (Laihonen and Furman 1986) compared to those on abiotic hard substrata. Bryozoans are a common, and often the most abundant, faunal component of epibiota on a variety of host substrata, and most recently Wahl (2009: Fig. 4.2) reported that there are more epizoic bryozoan species than any other group of metazoans.

## Costs and Benefits of Epibiosis

Epibiosis creates a variety of costs and benefits to both the epizoozoan and the basibiont (Wahl 2009). Possible detrimental effects to the bryozoans include: (1) unstable substratum (e.g., due to movement of articulated joints and molting; molting is especially a problem if the host molts before the bryozoan reaches a size large enough for sexual reproduction), (2) risky habitat change (e.g., host movement to water of different pressure, temperature, and/or salinity not conducive to bryozoan growth), and (3) shared doom (e.g., predation of host resulting in accidental predation of its epizoic bryozoans, but see discussion below of minimal predation on pycnogonids).

Possible beneficial effects for the bryozoans include: (1) increased substratum area (e.g., reduced competition for space among bryozoans compared to abiotic hard substrata), (2) favorable hydrodynamics (e.g., host movement may improve filter feeding of bryozoans by lifting them out of the bottom boundary layer), and (3) free transport (e.g., avoiding ice scour by being on a motile host as compared to a abiotic hard substratum as well as improved gamete dispersal and increased geographic range).

Possible detrimental effects to the pycnogonids include: (1) increased weight; probably not a major problem due to the almost neutral buoyancy of the epizoozoans, (2) increased drag; probably not a major problem due to the encrusting nature of the epizoozoans and the slow walking movement of the hosts, (3) decreased flexibility; possibly a problem as 35% of the bryozoan colonies extended over at least two body segments, (4) shared doom (e.g., predation of bryozoans probably not a major problem as the epizoic bryozoan lifestyle probably reduces their susceptibility to predation via camouflage), and (5) increased susceptibility to predation (e.g., if the bryozoan colonies increase the host's weight and drag, decrease its flexibility, and attract predators of the bryozoans which may also eat the host). There were a few hosts where bryozoans had grown over the eyes of the pycnogonid, suggesting that the epibionts were impairing the host's sensory perception, which could increase susceptibility to predation.

The possible beneficial effect for the pycnogonids is camouflage. For example, Vance (1978) documented a decrease in predation by starfish on clams, which were covered by bryozoans. By camouflage we include Wahl's (2009) concept of insulation (i.e., hiding of the host from optical, tactile, and chemical detection by its prey). As only 10% of the host pycnogonids are covered on average by bryozoans, this benefit may not apply to many of the pycnogonids. The only other reported percent epibiont cover of pycnogonids was 1% from the North Sea (Pipe 1982). In addition, as pycnogonids have such little tissue and such a high proportion of cuticle to tissue, they are not a common prey item of predators which include anemones, isopods, crabs, shrimp, and fishes (King 1973; Arnaud and Bamber 1987).

## Conclusions

In this study of epizoic bryozoans fouling pycnogonids from Antarctica, nine species of pycnogonids were identified containing 156 bryozoan colonies belonging to seven cheilostome species. The percentage of host surface area covered by bryozoans averaged 10%. As predicted by the Target Area Effect Hypothesis of Island Biogeography Theory, the number of bryozoan colonies increased significantly with increasing host surface area. Of the 115 pycnogonids, 26% were fouled by bryozoans.

The bryozoan species richness on pycnogonids is similar to that on adjacent abiotic hard substrata. Compared to the other motile host animals, the number of bryozoan species per unit host surface area is an order of magnitude higher on pycnogonids. The carapaces of pycnogonids may be acting as refugia for the bryozoans from hard substratum competition, ice scour, and predation by their host. Living on a motile host, the bryozoans may also benefit from improved feeding and gamete dispersal. The pycnogonids may not accrue any benefits from the bryozoan fouling except possible reduced predation resulting from camouflage created by the patchy nature of the bryozoan colonies on the pycnogonids' carapaces.

**Acknowledgments** We would like to thank the following people who made this project possible. Specimens were collected by Stefan Hain (British Antarctic Survey). Kathy Nicholson (British Antarctic Survey) aided in shipping the specimens from the UK to the US. C. Allan Child (Smithsonian Institution's Natural History Museum) identified the pycnogonids. Scott Lidgard (Field Museum of Natural History) and David Staples (Museum Victoria) provided information on pycnogonid predation on bryozoans. Helpful reviews by Piotr Kuklinski (Institute of Oceanology, Polish Academy of Sciences) and Judy Winston (Virginia Museum of Natural History) greatly improved this manuscript.

## References

Arnaud F, Bamber RN (1987) The biology of Pycnogonida. Adv Mar Biol 24:1–96

Arntz WE, Brey T, Gallardo VA (1994) Antarctic zoobenthos. Oceanogr Mar Biol Ann Rev 32:241–304

Bader B, Schäfer P (2005) Bryozoans in polar latitudes: Arctic and Antarctic bryozoan communities and facies. Denisia 16:263–282

Bain BA (1991) Some observations on biology and feeding behavior in two southern California pycnogonids. Bijdr Dierkd 61:63–64

Barnes DKA (1995) Sublittoral epifaunal communities at Signy Island, Antarctica. II. Below the ice-foot zone. Mar Biol 121:565–572

Barnes DKA (1996) Low levels of colonisation in Antarctica: the role of bryozoans in early community development. In: Gordon DP, Smith AM, Grant-Mackie AJ (eds) Bryozoans in space and time. National Institute of Water and Atmospheric Research, Wellington, pp 19–28

Barnes DKA (1999) The influence of ice on polar nearshore benthos. J Mar Biol Assoc UK 79:401–407

Barnes DKA (2005) Life, death, and fighting at high latitude: a review. In: Moyano HI, Cancino JM, Wyse Jackson PN (eds) Bryozoan studies 2004. Balkema, Leiden, pp 1–14

Barnes DKA, Bullough LW (1996) The diet and distribution of nudibranchs at Signy Island, Antarctica. J Mollus Stud 62:281–287

Barnes DKA, Clarke A (1995) Epibiotic communities on sublittoral macroinvertebrates at Signy Island, Antarctica. J Mar Biol Assoc UK 75:689–703

Barnes DKA, Conlan KE (2007) Disturbance, colonization and development of Antarctic benthic communities. Philos Trans R Soc B 362:11–38

Barnes DKA, De Grave S (2002) Modelling multivariate determinants of growth in Antarctic bryozoans. In: Wyse Jackson PN, Buttler CJ, Spencer Jones ME (eds) Bryozoan studies 2001. Balkema, Lisse, 7–17

Barnes DKA, Rothery P, Clarke A (1996) Colonisation and development in encrusting communities from the Antarctic intertidal and sublittoral. J Exp Mar Biol Ecol 196:251–265

Barnes DKA, Warren NL, Webb K, Phalan B, Reid K (2004) Polar pedunculate barnacles piggy-back on pycnogona, penguins, pinniped seals and plastics. Mar Ecol Prog Ser 284:305–310

Barnes DKA, Kaiser S, Griffiths HJ, Linse K (2009) Marine, intertidal, freshwater and terrestrial biodiversity of an isolated polar archipelago. J Biogeogr 36:756–769

Barnes PW, Lien R (1988) Icebergs rework shelf sediments to 500 m off Antarctica. Geology 16:1130–1133

Berning B (2008) Evidence for sublethal predation and regeneration among living and fossil ascophoran bryozoans. In: Hageman SJ, Key MM Jr, Winston JE (eds) Bryozoan studies 2007, vol 15. Virginia Museum of Natural History Special Publication, Martinsville, pp. 1–8

Brown KM, Fraser KP, Barnes DKA, Peck LS (2004) Links between the structure of an Antarctic shallow-water community and ice-scour frequency. Oecologia 141:121–129

Conlan KE, Lenihan HS, Kvitek RG, Oliver JS (1998) Ice scour disturbance to nearshore benthic communities in the Canadian high Arctic. Mar Ecol Prog Ser 166:1–16

Dayton PK (1971) Competition, disturbance, and community organization: the provision and subsequent utilization of space in a rocky intertidal community. Ecol Monogr 41:351–389

Dayton PK (1990) Polar benthos. In: Smith WO (ed) Polar oceanography, part B: chemistry, biology, and geology. Academic, London

Dayton PK, Mordida BJ, Bacon F (1994) Polar marine communities. Am Zool 34:90–99

Döhrn A (1881) Die Pantopoden des Golfes von Naepel und der angrenzenden Meeresabschnitte. Monogr Fauna Flora Golfes Neapel 3:1–152

Fry WG (1965) The feeding mechanisms and preferred foods of three species of Pycnogonida. Bull Brit Mus Nat Hist 12:195–233

Gutt J (2000) Some "driving forces" structuring communities of the sublittoral Antarctic macrobenthos. Antarct Sci 12:297–313

Gutt J (2001) On the direct impact of ice on marine benthic communities, a review. Polar Biol 24:553–564

Gutt J, Piepenburg D (2003) Scale-dependent impact on diversity of Antarctic benthos caused by grounding of icebergs. Mar Ecol Prog Ser 253:77–83

Gutt J, Schickan T (1998) Epibiotic relationships in the Antarctic benthos. Antarct Sci 10:398–405

Gutt J, Starmans A, Dieckmann G (1996) Impact of iceberg scouring on polar benthic habitats. Mar Ecol Prog Ser 137:311–316

Hedgpeth JW (1964) Notes on the peculiar egg-laying habit of an Antarctic prosobranch (Mollusca: Gastropoda). Veliger 7:45–46

Jackson JBC (1977) Competition on marine hard substrate: the adaptive significance of solitary versus colonial strategies. Am Nat 111:743–767

Key MM Jr, Barnes DKA (1999) Bryozoan colonization of the marine isopod *Glyptonotus antarcticus* at Signy Island, Antarctica. Polar Biol 21:48–55

Key MM Jr, Jeffries WB, Voris HK (1995) Epizoic bryozoans, sea snakes, and other nektonic substrates. Bull Mar Sci 56:462–474

Key MM Jr, Jeffries WB, Voris HK, Yang CM (1996a) Epizoic bryozoans and mobile ephemeral host substrata. In: Gordon DP, Smith AM, Grant-Mackie JA (eds) Bryozoans in space and time. National Institute of Water and Atmospheric Research, Wellington

Key MM Jr, Jeffries WB, Voris HK, Yang CM (1996b) Epizoic bryozoans, horseshoe crabs, and other mobile benthic substrates. Bull Mar Sci 58:368–384

Key MM Jr, Jeffries WB, Voris HK, Yang CM (2000) Bryozoan fouling pattern on the horseshoe crab *Tachypleus gigas* (Müller) from Singapore. In: Herrera CA, Jackson JBC (eds) Proceedings of the 11th International Bryozoology Association conference. Smithsonian Tropical Research Institute, Balboa

Key MM Jr, Volpe JW, Jeffries WB, Voris HK (1997) Barnacle fouling of the blue crab *Callinectes sapidus* at Beaufort, North Carolina. J Crustacean Biol 17:424–439

Key MM Jr, Winston JE, Volpe JW, Jeffries WB, Voris HK (1999) Bryozoan fouling of the blue crab *Callinectes sapidus* at Beaufort, North Carolina. Bull Mar Sci 64:513–533

Key MM Jr, Schumacher GA, Babcock LE, Frey RC, Heimbrock WP, Felton SH, Cooper DL, Gibson WB, Scheid DG, Schumacher SA (2010) Paleoecology of commensal epizoans fouling *Flexicalymene* (Trilobita) from the Upper Ordovician, Cincinnati Arch region, USA. J Paleontol 84:1121–1134

King PE (1973) Pycnogonids. Hutchinson, London

Laihonen P, Furman ER (1986) The site of settlement indicates commensalism between blue mussel and its epibiont. Oecologia 71:38–40

Lebour MV (1945) Notes on the Pycnogonida of Plymouth. J Mar Biol Assoc UK 26:139–165

Lee HJ, Gerdes D, Vanhove S, Vincx M (2001) Meiofauna response to iceberg disturbance on the Antarctic continental shelf at Kapp Novegia (Weddell Sea). Polar Biol 24:926–933

Lien R, Solheim A, Elverhøi A, Rokoengen K (1989) Iceberg scouring and sea bed morphology on the eastern Weddell Sea shelf, Antarctica. Polar Res 7:43–57

Linse K, Walker LJ, Barnes DKA (2008) Biodiversity of echinoids and their epibionts around the Scotia Arc, Antarctica. Antarct Sci 20:227–244

Lohse DP (1993) The effects of substratum type on the population dynamics of three common intertidal animals. J Exp Mar Biol Ecol 173:133–154

Lomolino MV (1990) The target area hypothesis: the influence of island area on immigration rates of non-volant mammals. Oikos 57:297–300

MacArthur RH, Wilson EO (1967) The theory of island biogeography. Princeton University Press, Princeton

Manton SM (1978) Habits, functional morphology and the evolution of pycnogonids. Zool J Linn Soc 63:1–22

McKinney FK, Jackson JBC (1989) Bryozoan evolution. Unwin Hyman, Boston

McKinney FK, Taylor PD, Lidgard S (2003) Predation on bryozoans and its reflection in the fossil record. In: Kelley PH, Kowalewski M, Hansen TA (eds) Predator–prey interactions in the fossil record. Kluwer, New York, pp 239–261

Mercier A, Hamel J-F (1994) Deleterious effects of a pycnogonid on the sea anemone *Bartholomea annulata*. Can J Zool 72:1362–1364

Moore PG (1973) Bryozoa as a community component on the northeast coast of Britain. In: Larwood GP (ed) Living and fossil Bryozoa. Academic, London, pp 21–36

Morgan E (1977) The swimming of *Nymphon gracile* (Pycnogonida): the energetics of swimming at constant depth. J Exp Biol 71:205–211

Paine RT (1974) Intertidal community structure: experimental studies on the relationship between a dominant competitor and its principal predator. Oecologia 15:93–120

Peck LS, Bullough LW (1993) Growth and population structure in the infaunal bivalve *Yoldia eightsi* in relation to iceberg activity at Signy Island, Antarctica. Mar Biol 117:235–241

Peck LS, Brockington S, Vanhove S, Beghyn M (1999) Community recovery following catastrophic iceberg impacts in a soft-sediment shallow-water site at Signy Island, Antarctica. Mar Ecol Prog Ser 186:1–8

Piel WH (1991) Pycnogonid predation on nudibranchs and ceratal autotomy. Veliger 34:366–367

Pipe AR (1982) Epizoites on marine invertebrates: with particular reference to those associated with the pycnogonid *Phoxichilidium tubulariae* Lebour, the amphipod *Caprella linearis* (L.) and the decapod *Corystes cassivelaunus* (Penant). Chem Ecol 1:61–74

Prell H (1910) Beiträge zur Kenntnis der Lebensweise einiger Pantopoden. Bergens Mus Aarbok (NR) 1909(10):1–30

Ryland JS (1970) Bryozoans. Hutchinson University Library, London

Ryland JS (1976) Physiology and ecology of marine bryozoans. Adv Mar Biol 14:285–443

Schram FR, Hedgpeth JW (1978) Locomotary mechanisms in Antarctic pycnogonids. Zool J Linn Soc 63:145–169

Sherwood J, Walls JT, Ritz DA (1998) Amathamide alkaloids in the pycnogonid, *Stylopallene longicauda*, epizoic on the chemically defended bryozoan, *Amathia wilsoni*. Pap Proc R Soc Tasman 132:65–70

Soule JD, Soule DF (1977) Fouling and bioadhesion: life strategies of bryozoans. In: Woollacott RM, Zimmer RL (eds) Biology of bryozoans. Academic, New York, pp 437–457

Staples DA (2004) Pycnogonida from the Althorpe Islands, South Australia. Trans R Soc South Aust 129:158–169

Stracey CM, Pimm SL (2009) Testing island biogeography theory with visitation rates of birds to British islands. J Biogeogr 36:1532–1539

Taylor PD, Monks N (1997) A new cheilostome bryozoan genus pseudoplanktonic on molluscs and algae. Invertebr Biol 116:39–51

Taylor PD, Wilson MA (2002) A new terminology for marine organisms inhabiting hard substrates. Palaios 17:522–525

Tomaschko K-H, Wilhelm E, Bückmann D (1997) Growth and reproduction of *Pycnogonum litorale* (Pycnogonida) under laboratory conditions. Mar Biol 129:595–600

Vance RR (1978) A mutualistic interaction between a sessile marine clam and its epibionts. Ecology 59:679–685

Varoli FMF (1994) Feeding aspects of *Tanystylum isabella* and *Anoplodactylus strictus* (Pantopoda). Rev Bras Zool 11:623–627

Wahl M (1989) Marine epibiosis. 1. Fouling and antifouling: some basic aspects. Mar Ecol Prog Ser 58:175–189

Wahl M (2009) Epibiosis: ecology, effects and defenses. In: Wahl M (ed) Marine hard bottom communities: patterns, dynamics, diversity, and change. Springer series: ecological studies 206(1):61–72

Wyer DW, King PE (1973) Relationships between some British littoral and sublittoral bryozoans and pycnogonids. In: Larwood GP (ed) Living and fossil Bryozoa. Academic, London, pp 199–207

Wyer DW, King PE (1974) Feeding in British littoral pycnogonids. Estuar Coast Mar Stud 2:177–184

# Chapter 11
# Growth Rate of Selected Sheet-Encrusting Bryozoan Colonies Along a Latitudinal Transect: Preliminary Results

## Growth Rate of Bryozoans

**Piotr Kuklinski, Adam Sokolowski, Marcelina Ziolkowska, Piotr Balazy, Maja Novosel, and David K.A. Barnes**

**Abstract** Climate change driven alterations of sea-water temperature, salinity, acidity and primary production in many coastal regions will probably affect the ecophysiological performance of sedentary organisms. Despite bryozoan ubiquitous and often dominant occurrence in coastal zones there are few studies on their growth dynamics. Here we report growth rates of selected sheet-encrusting bryozoans from four contrasting (in mean annual water temperature) environments: Adriatic Sea (44° N), Baltic Sea (54° N), northern Norway (68° N), and Spitsbergen (78° N). Perspex panels were photographed underwater and colonies' growth rates analyzed backwards using digital images. We found a negative trend between growth rate and latitude. Congeneric bryozoan species from lower latitudes grew faster: the average growth rate of the cyclostome genus *Diplosolen* from the Adriatic Sea was 75 mm$^2$ after 5 months (~180 mm$^2$/year) while *Diplosolen arctica* from Spitsbergen grew only 4 mm$^2$/year. Similarly the average surface area of *Microporella arctica* individuals after 12 months in northern Norway was 12 mm$^2$

P. Kuklinski (✉)
Institute of Oceanology, Polish Academy of Sciences, ul. Powstanców Warszawy 55, Sopot 81-712, Poland

Department of Zoology, Natural History Museum, Cromwell Road, London SW7 5BD, UK
e-mail: kuki@iopan.gda.pl

A. Sokolowski • M. Ziolkowska
University of Gdansk, Institute of Oceanography, Al. Pilsudskiego 46, 81-378 Gdynia, Poland

P. Balazy
Institute of Oceanology, Polish Academy of Sciences, ul. Powstanców Warszawy 55, Sopot 81-712, Poland

M. Novosel
Laboratory of Marine Biology, Rooseveltov trg 6, 10000 Zagreb, Croatia

D.K.A. Barnes
British Antarctic Survey, High Cross, Madingley Road, Cambridge CB3 0ET, United Kingdom

A. Ernst, P. Schäfer, and J. Scholz (eds.), *Bryozoan Studies 2010*,
Lecture Notes in Earth System Sciences 143, DOI 10.1007/978-3-642-16411-8_11,

compared to 2 $mm^2$ in Spitsbergen. An exception from this general pattern was *Einhornia crustulenta* in the brackish environment of the Baltic Sea, which grew relatively rapid for this latitude and water temperature (surface area of up to 657 $mm^2$/month after settlement).

**Keywords** Bryozoa • Growth rates • Latitude

# Introduction

Understanding growth rates of bryozoans is important for several reasons. Among these is the fact that a number of species is recognized as producers of natural products of great biomedical potential. For example, bryostatins isolated from *Bugula neritina* show strong antitumour activity being in the final phases of clinical trials as cancer drugs (Madhusudan et al. 2003). Brominated alkaloids produced by another bryozoan (*Flustra foliacea*) may act as muscle relaxants (see Kahle et al. 2003). Understanding drivers of growth is necessary for an efficient and sustainable biomass production. Bryozoans are also important for the establishment and maintenance of local biodiversity and are thus a key group involved in conservation planning. Many arborescent species have a bioconstructional role providing shelter for other epibenthic taxa including juveniles of commercial fish (e.g. Bradstock and Gordon 1983). On the other hand, bryozoan larvae often recruit onto hard structures such as shellfish, boats and pier pilings, affecting underwater constructions and ships (e.g. Gordon and Mawatari 1992). In addition, species that form annual growth bands preserve an environmental record (e.g. water temperature or food availability with time) which can act as libraries of past conditions thus enabling measurement back across various temporal scales (Smith and Key 2004; Barnes et al. 2006; Schäfer and Bader 2008). In order to assess the role of bryozoans in present and past coastal ecosystems, it is necessary to understand the dynamics of their ecophysiological performance, including growth rate and expansion potential (e.g. Brey et al. 1998).

Over the last 50 years climate change has been shown to alter sea-water parameters such as near surface temperature, salinity, pH and primary production (e.g. Hoegh-Guldberg and Bruno 2010). At high latitudes, shallow waters form the ‘front line’ for such change as they are particularly vulnerable to climate-induced environmental alterations. Consequently, shallow-water organisms are likely to have more different energy requirements, reproductive cycles and growth rates than deep-water ones. Therefore, baseline studies are important to detect the impact of climate change on ecological performance.

Several studies have shown that linear growth increments among erect bryozoans range from 2.5 to 100 mm/year (e.g., Barnes et al. 2007; Smith 2007; Lombardi et al. 2006, 2008). However, despite their ubiquitous and often dominant (amongst bryozoans) occurrence in the coastal zone (see Bowden et al. 2006), there are few data on growth rates of sheet-encrusting species. Here we report growth rates of selected sheet-encrusting bryozoans from four contrasting environments

**Table 11.1** Characteristics of sampling locations (region, location, geographical position, range of water temperature and salinity) and taxonomic identification of the bryozoans collected at each site.

| Location | Geographical position | Water temperature (°C) | Salinity (PSU) | Bryozoans |
|---|---|---|---|---|
| Jurjevo (Adriatic Sea) | 44° 56′ N | 9.0–25.0 | 36–38 | *Diplosolen* cf *obelia* |
| | | | | *Callopora dumerilii* |
| | 14° 55′ E | | | *Puellina hincksi* |
| Gulf of Gdańsk (Baltic Sea) | 54° 29′ N | 0.0–22.0 | 5–8 | *Einhornia crustulenta* |
| | 18° 34′ E | | | |
| Fornes (Norwegian Sea) | 68° 52′ N | 3.5–8.0 | 33–35 | *Escharella immersa* |
| | | | | *Microporella arctica* |
| | 17° 28′ E | | | *Membraniporella nitida* |
| | | | | *Patinella* sp. |
| Isfjorden | 78° 12′ N | −1.8–4.0 | 25–35 | *Cribrilina annulata* |
| Spitsbergen | 15° 14′ E | | | *Diplosolen arctica* |
| | | | | *Microporella arctica* |
| | | | | *Tegella arctica* |

along a latitudinal gradient ranging from the warm temperate Adriatic Sea to the cold European Arctic (Gačić et al. 2001). We aim to establish baseline data on growth rates for particular species and highlight potential factors driving growth tempo of targeted bryozoans' species and assemblages.

## Material and Methods

### *Study Area*

Four study sites were located along a latitudinal profile from the Adriatic Sea in the South through the Baltic Sea and northern Norway up to Spitsbergen in the North (Table 11.1; Fig. 11.1). Selected sites have similar bottom topography, rugosity and profile. In each case substrata were characterised by rocks (cobbles to boulders) overlain by fine sediment (mud or sand). All locations had a pronounced seasonality in water temperature, light intensity, water dynamics, and primary production. For example, at the site in the Adriatic Sea (Jurjevo), surface temperature varies between 9°C in winter to more than 25°C in summer (Orlić et al. 2000). Salinity values for the Adriatic Sea range from approximately 36–38 PSU. In the Baltic Sea (Gulf of Gdansk), surface temperature at the study location varies between 0°C in winter to over 22°C in summer (Kruk-Dowgiallo and Szaniawska 2008) and salinity from 5 to 8 PSU. Water temperature at the choosen location in the northern Norway (Fornes) varies from approximately 3.5–8°C and rarely freezes (Loeng 1991) while salinity in the coastal zone ranges from 33 to 35 PSU. In the

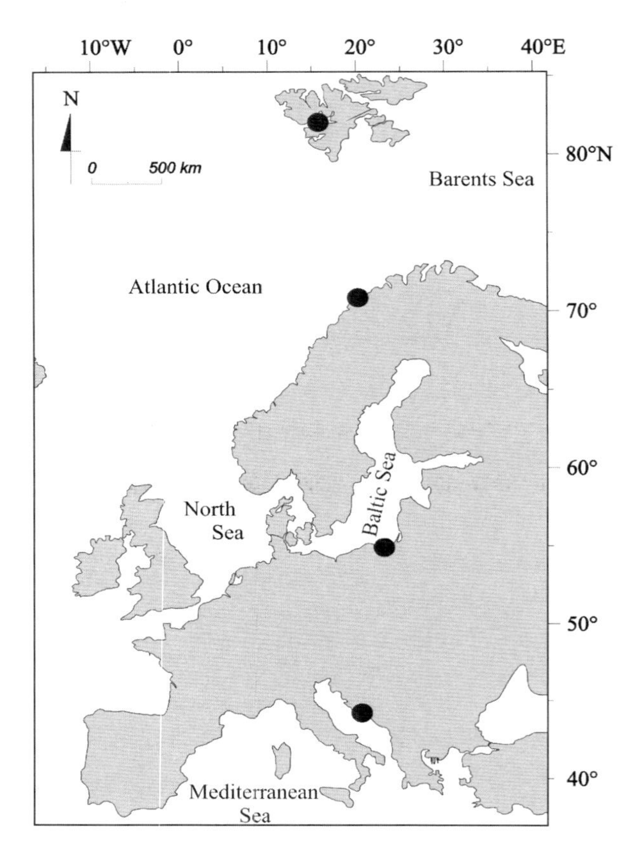


**Fig. 11.1** Locations (marked by *dots*) used in this study

Spitsbergen (Isfjorden) site, water temperature varies between −1.8°C and 4°C, but can periodically reach up to 7°C. The surface water along the west coast of the fjord freezes for a few months in winter and can be scoured by summer icebergs (Węsławski et al. 1988); salinity in the shallows ranges from 25 to 35 PSU (Table 11.1). During the study, mean water temperature was 15.4°C for the Adriatic Sea location, 10.5°C for the Baltic Sea, 6.7°C for northern Norway, and 0.5°C for Spitsbergen. Data was obtained with an Onset, HOBO pendant temp/light UA002-64 logger.

## *Protocol*

In order to minimise the influence of substratum heterogeneity and for ease of replication, three flat Perspex panels, of similar size and shape (15 × 15 cm), were submerged in water at 3.5 m depth in Balic Sea, 6 m in Northern Norway, 7 m on

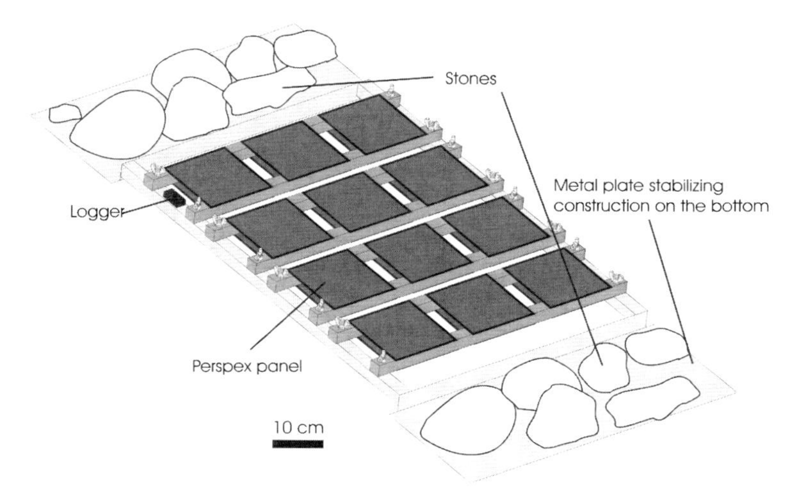


**Fig. 11.2** Experimental array of 12 Perspex panels (15 × 15 cm each) bolted into a frame and fixed to the seafloor with stones

Spitsbergen, and 8 m in Adriatic Sea. Panels were attached to a horizontal metal frame secured on the sea floor by stones (Fig. 11.2). In northern Norway, they were positioned directly on the gravelly sea bottom. Panels were photographed underwater on a monthly basis in the Baltic Sea (March 2008 to March 2009), 3-monthly intervals in the Adriatic Sea (March 2008 to October 2009), and annually in northern Norway and Spitsbergen (from 2005 to 2009). At the end of the study, panels were retrieved from water and species identified. For the purpose of this study, we selected only species occurring on panels across the whole study period. Phylogenetically related species were chosen whenever possible. Growth rates were analyzed backwards using underwater digital images and expressed as the increase in surface area of colonies over time. Colony surface measurements were done with Image J.

## Results

Three bryozoan species were measured in the Adriatic Sea (Fig. 11.3a, Appendix A). Panels were installed in March but all analysed species recruited between May and August. After 18 months of growth, average colony surface area was between 92 $mm^2$ (*Diplosolen* cf. *obelia*, n = 3) and 131 $mm^2$ (*Callopora dumerilii*, n = 2). All individuals exhibited a nearly constant growth across the entire study period.

Only one bryozoan species, *Einhornia crustulenta* (Fig. 11.3b, Appendix A), colonized the panels in the brackish environment of the Baltic Sea. Recruitment of *E. crustulenta* started in July, 4 months after the panels' installation. During the first

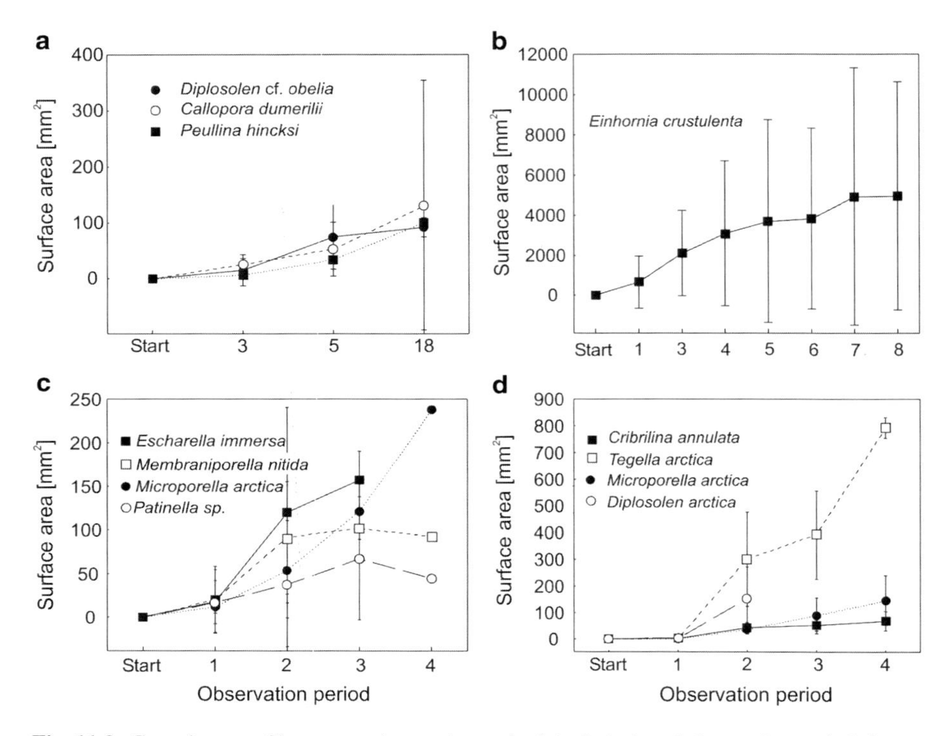


**Fig. 11.3** Growth rates of bryozoans (mean ± standard deviation) and observation periods in our locations along a latitudinal transect: (**a**) Adriatic Sea (observation period – months), (**b**) Baltic Sea (observation period – months), (**c**) Northern Norway (observation period – years) and (**d**) Spitsbergen (observation period – years). For number of samples at each location see Appendix A

month of growth, *E. crustulenta* colonies reached surface areas between 236 and 1,982 mm$^2$ (n = 6). None of the colonies exhibited constant growth during the study period (July 2008–March 2009). Colonies decreased periodically in size, presumably due to either mechanical damage or competitive interactions with conspecific colonies or the barnacle *Balanus improvisus*, which recruited in high numbers on the panels. Colony's size at the end of the observation period ranged from 2,376 to 8,290 mm$^2$.

In northern Norway we analysed the growth rate of four species over a period of 4 years (Fig. 11.3c, Appendix A). Due to the low observation frequency (only in summer), the exact time of species settlement is unknown. During the first year all species reached a similar size ranging on average from 12 (*Microporella arctica*, n = 4) to 20 mm$^2$ (*Membraniporella nitida*, n = 2). Over the next 3 years disproportion among species growth rates was observed. One individual of *Microporella arctica* reached 238 mm$^2$ in size after 4 years being the largest colony from this location whilst the cyclostome *Patinella* sp. had the smallest colony (45 mm$^2$). All species exhibited a near constant increase in colony size during the study period except for one colony of *Membraniporella nitida* and one of *Escharella immersa*, which decreased during the final year of observation due to mechanical damage.

At the Spitsbergen location, as in northern Norway, the growth rate of four species was analysed over a period of 4 years (Fig. 11.3d, Appendix A). Low observation frequency (only during summer) also precluded identification of settlement time. Colony's sizes after the first observation year was very similar among species ranging from 1 (*Microporella arctica*) to 5 $mm^2$ (*Tegella arctica, Diplosolen arctica*) in surface area. After the first year unevenness in tempo of growth was more apparent among species. Colonies of *T. arctica* reached up to 806 $mm^2$ after 4 years of study being the largest colonies present on all panels. *Cribrilina annulata* colonies reached a maximum size of 88 $mm^2$ after the same period. All colonies had near constant increases in growth across the study period.

If we exclude Baltic Sea data from the analysis a negative trend between growth and latitude is detected. In the Adriatic Sea after 5 months, colonies' mean surface area ($\pm$ standard deviation) was 61 $\pm$ 27.1 $mm^2$ (n = 6), in Northern Norway after 1 year 16 $\pm$ 12.4 $mm^2$ (n = 14) and in the Spitsbergen (also 1 year) 3 $\pm$ 2.7 $mm^2$ (n = 16). Within genera (as well as within other taxonomic units) bryozoans from lower latitudes also grew faster. Indeed, similarly to cheilostomes, cyclostomes' performance showed an order of magnitude difference in growth between the Adriatic and the Spitsbergen. The average growth rate during 5 months of the cyclostome *Diplosolen* cf. *obelia* from the Adriatic Sea was 75 $\pm$ 28.7 $mm^2$ (n = 3) while *Diplosolen arctica* from Spitsbergen grew 4 $\pm$ 1.4 $mm^2$ (n = 3) after 1 year. Likewise, Norwegian colonies grew faster than those in the Arctic waters. Indeed, the average surface area of *M. arctica* was 12 $\pm$ 3.9 $mm^2$ (n = 4) after the first year in the northern Norway compared to 2 $\pm$ 0.3 $mm^2$ (n = 4) in Spitsbergen.

## Discussion

We found a negative trend between initial growth rates and latitude. Studied species from the Adriatic Sea grew faster (in colony surface area) when compared with species from contrasting Northern Norway and Spitsbergen locations. In general, Arctic species grew an order of magnitude slower than temperate ones, which corresponds to the difference in general ecophysiological performance between polar and temperate bryozoans reported earlier (see Bowden et al. 2006 for encrusting and Barnes et al. 2007 for erect rigid morphologies). Direct comparisons are invariably complicated and confounded by taxonomic, temporal and environmental differences. Not surprisingly that different sets of species colonized the panels in most locations. Only one species (*Microporella arctica*) was found at two locations (northern Norway and West Spitsbergen); one genus (*Diplosolen*) also occurred at two locations (Spitsbergen and the Adriatic Sea). Colonies from northern Norway grew faster relative to those from West Spitsbergen during the same time period. The initial growth of *Diplosolen* in the Adriatic Sea was also faster than in West Spitsbergen. Growth rates of congeneric species may vary by an order of magnitude in the same region (e.g., see Barnes et al. 2007). However, growth form (morphology) may have a bigger impact on growth performance than spatial effects. Growth rate of *Einhornia crustulenta* (in the Baltic Sea) after 1 month was higher than that of other bryozoans in the whole study. In general, brackish waters have very few bryozoans. Indeed, only 11 bryozoan species have been recorded, mostly Ctenostomata (Winston 1977) in salinities between 3 and 8 PSU. *Einhornia crustulenta* is a brackish water specialist (Nikulina and Schäfer 2006) with an opportunistic life strategy that unables a rapid hard substrata colonization (pers. observations). In the study area, this species has few natural competitors for space except the barnacle *Balanus improvisus*. The rapid growth of *E. crustulenta* may be due more to the lack of competition for space than to growth form or latitude. For example, Barnes and DeGrave (2001) found that spatial competition (especially interspecific) in encrusting Antarctic species influenced growth more strongly than latitude, species identity or environmental conditions. In addition, it cannot be excluded that favourable eutrophic conditions of the southern Baltic (e.g. elevated primary production and sedimentation rate of suspended organic matter) also contributed to the exceptionally high growth rate noticed in bryozoans in the Gulf of Gdańsk.

Previous field and laboratory observations suggested that bryozoans from low latitude/warmer environments grow faster. Lombardi et al. (2006, 2008) showed that in *Pentapora fascialis* combined growth rates for winter and summer bands were larger in locations with higher mean annual water temperature. The Arctic species *Harmeria scutulata* also exhibits larger diameter of colonies in areas with warmer mean annual water temperatures (Kuklinski and Taylor 2006). Bowden et al. (2006) reported that growth rates of the Antarctic *Arachnopusia inchoata*, *Chaperiopsis protecta*, and *Fenestrulina rugula* were slower than those of (unrelated) temperate species. Laboratory studies seem to confirm field observations. Indeed, experiments with controlled conditions including temperature, salinity and food concentration show that colonies of encrusting bryozoans grow faster under higher temperatures (e.g. *Cryptosula pallasiana* reached nearly 300 zooids after 4 weeks at 18°C whereas it had only slightly more than 100 zooids during the same period at 14°C, Amui-Vedel et al. 2007). This higher growth tempo is likely to be the result of an increased overall metabolic rate and thus physiological processes (rates of food uptake and digestion) due to higher ambient temperature. Indeed, there was a clear gradient of water temperature along the latitudinal transect studied here supporting the hypothesis that temperature might be a key factor driving the observed patterns of faster growth towards warmer areas.

Although annual growth rates of polar bryozoan species are in general slower compared to temperate ones (Bowden et al. 2006; Barnes et al. 2007), they can be comparable during the actual period of growth (Barnes 1995; Smith 2007). This supports the idea of a reduced growth tempo in polar bryozoans driven by food limitation resulting from the restricted duration of the vegetative period. Laboratory studies also confirm that food availability strongly influences the tempo of growth (e.g. *Electra pilosa*, Bayer et al. 1994).

Another factor that may restrict bryozoan's development includes spatial competition with other encrusters (e.g., barnacles, polychaetes, other bryozoans). Higher variability in growth rates was especially noticeable in the later stages of assemblage development when all competitors were already established and space on the panels became limited. It is well recognized that several Arctic encrusting species are much better competitors for space than others (Barnes and Kuklinski 2003). Barnes and Arnold (2001) as well as Linse et al. (2006) showed a pattern of linear increase in growth rate of encrusting bryozoan colonies from low to high Antarctic latitudes (from 54° to 68° S). Decrease of spatial competition was therefore suggested as a factor for these increased growth rates.

Our study also indicates that growth rates are strongly species related (e.g. O'Dea and Jackson 2002; Barnes et al. 2007). This was especially marked in northern Norway and Spitsbergen where we had 4 years of data (see Fig. 11.3c–d). Indeed, species like *Diplosolen arctica* would grow rather deterministically reaching a certain size during the first 2 years and then would disappear as a result of competition or predation. While *Tegella arctica* would briefly increase exponentially in surface area, and later outcompete other species covering large proportions of the panels. Although encrusting assemblages develop faster in warmer areas in general (pers. observation), many individual species have faster development due to their specific life histories. Thus key determinants of growth tempo appear complex including many factors such as food quality, quantity and availability, temperature, salinity and competitive pressure as well as internal characteristics related to taxon evolutionary history.

**Acknowledgements** The authors wish to thank Dr. Suzanne Williams, Joachim Scholz and an anonymous reviewer for their comments and corrections leading to a more improved manuscript. This study has been completed thanks to a grant from the Polish Ministry of Science and Higher Education (No NN304 404038).

# Appendix A: All Measurements of Bryozoans Included in the Study

| Location | Bryozoans | Colony | | Observation period (in roman)/Surface area [mm$^2$] | | | | | | |
|---|---|---|---|---|---|---|---|---|---|---|
| Jurjevo (Adriatic Sea) | | | months | III | V | XVIII | | | | |
| | *Diplosolen* cf. *obelia* | 1 | | 13.0 | 81.1 | 95.4 | | | | |
| | | 2 | | 2.4 | 43.6 | 82.6 | | | | |
| | | 3 | | 30.4 | 100.2 | 98.7 | | | | |
| | *Callopora dumerilii* | 1 | | 21.4 | 70.0 | 210.3 | | | | |
| | | 2 | | 29.2 | 35.8 | 52.3 | | | | |
| | *Peullina hincksi* | 1 | | 6.0 | 34.2 | 100.5 | | | | |
| Gulf of Gdańsk (Baltic Sea) | | | months | I | III | IV | V | VI | VII | VIII |
| | *Einhornia crustulenta* | 1 | | 418.7 | 2309.8 | 1437.4 | 2115.1 | 2672.3 | 2501.4 | 2376.8 |
| | | 2 | | 236.2 | 2946.5 | 5538.2 | 7408.5 | 7005.6 | 9236.4 | 8290.1 |
| | | 3 | | 476.3 | 1610.8 | 2067.8 | 2079.7 | 1882.0 | 2465.7 | 2807.6 |
| | | 4 | | 1982.3 | 3682.3 | 3307.4 | 3167.5 | 3731.0 | 5486.3 | 6326.5 |
| | | 5 | | 485.0 | 1098.3 | | | | | |
| | | 6 | | 343.7 | 1008.2 | | | | | |
| Fornes (Norwegian Sea) | | | years | I | II | III | IV | | | |
| | *Escharella immersa* | 1 | | 6.3 | 37.6 | 151.7 | | | | |
| | | 2 | | 6.7 | 113.5 | 165.2 | | | | |
| | | 3 | | 45.0 | 175.8 | 175.1 | | | | |
| | | 4 | | 13.2 | 151.5 | 136.9 | | | | |
| | *Microporella arctica* | 1 | | 13.3 | 45.9 | 109.0 | | | | |
| | | 2 | | 9.0 | 33.1 | 114.5 | | | | |
| | | 3 | | 17.3 | 57.6 | 138.8 | 238.2 | | | |
| | | 4 | | 9.3 | 76.5 | | | | | |
| | *Membraniporella nitida* | 1 | | 6.3 | 68.1 | 101.4 | 92.2 | | | |
| | | 2 | | 33.4 | 113.7 | | | | | |

(continued)

| Location | Bryozoans | Colony | | Observation period (in roman)/Surface area [$mm^2$] | | | |
|---|---|---|---|---|---|---|---|
| | *Patinella* sp. | 1 | | 35.4 | 89.0 | 92.2 | |
| | | 2 | | 13.2 | 17.2 | | |
| | | 3 | | 7.5 | 7.5 | | |
| | | 4 | | 11.9 | 37.3 | 42.4 | 44.6 |
| Isfjorden (Spitsbergen) | | | years | I | II | III | IV |
| | *Cribrilina annulata* | 1 | | 2.9 | 57.2 | 60.5 | 53.7 |
| | | 2 | | 2.1 | 39.3 | 61.5 | 81.2 |
| | | 3 | | 2.6 | 41.6 | 60.3 | 88.2 |
| | | 4 | | 2.2 | 36.2 | 43.4 | 52.5 |
| | *Diplosolen arctica* | 1 | | 5.7 | 233.9 | | |
| | | 2 | | 3.0 | 150.7 | | |
| | | 3 | | 3.2 | 80.9 | | |
| | *Microporella arctica* | 1 | | 1.4 | 40.5 | 137.5 | 186.4 |
| | | 2 | | 2.2 | 43.0 | 78.1 | 108.6 |
| | | 3 | | 1.5 | 22.4 | 45.0 | 89.0 |
| | | 4 | | 1.6 | 38.8 | 82.2 | 192.3 |
| | *Tegella arctica* | 1 | | 4.6 | 276.0 | 376.1 | 779.1 |
| | | 2 | | 4.1 | 282.7 | 442.3 | 806.1 |
| | | 3 | | 3.9 | 349.4 | 495.5 | |
| | | 4 | | 4.9 | 179.5 | 276.3 | |
| | | 5 | | 5.1 | 414.9 | 369.2 | |

## References


Amui-Vedel A-M, Hayward PJ, Porter JS (2007) Zooid size and growth rate of the bryozoan *Cryptosula pallasiana* Moll in relation to temperature, in culture and in its natural environment. J Exp Mar Biol Ecol 353:1–12

Barnes DKA (1995) Seasonal and annual growth in erect species of Antarctic bryozoans. J Exp Mar Biol Ecol 188:181–198

Barnes DKA, Arnold R (2001) A growth cline in encrusting benthos along a latitudinal gradient within Antarctic waters. Mar Ecol Prog Ser 210:85–91

Barnes DKA, DeGrave S (2001) Modelling multivariate determinants of growth in Antarctic bryozoans. In: Wyse Jackson P, Buttler C, Spencer Jones M (eds) Bryozoan studies 2001. Balkema, Lisse, pp 19–27

Barnes DKA, Kuklinski P (2003) High polar spatial competition: extreme hierarchies at extreme latitude. Mar Ecol Prog Ser 259:17–28

Barnes DKA, Webb KE, Linse K (2006) Slow growth of Antarctic bryozoans increases over 20 years and is anomalously high in 2003. Mar Ecol Prog Ser 314:187–195

Barnes DKA, Webb KE, Linse K (2007) Growth rate and its variability in erect Antarctic bryozoans. Polar Biol 30:1069–1081

Bayer MM, Cormack RM, Todd CD (1994) Influence of food concentration on polipide regression in the marine bryozoan *Electra pilosa* (L.) (Bryozoa: Cheilostomata). J Exp Mar Biol Ecol 178:35–50

Bowden DA, Clarke A, Peck LS, Barnes DKA (2006) Antarctic sessile marine benthos: colonization and growth on artificial substrata over three years. Mar Ecol Prog Ser 316:1–16

Bradstock M, Gordon DP (1983) Coral-like bryozoan growths in Tasman Bay, and their protection to conserve commercial fish stocks. New Zeal J Mar Fresh 17:159–163

Brey T, Gutt J, Mackensen A, Starmans A (1998) Growth and productivity of the high Antarctic bryozoan *Melicerita obliqua*. Mar Biol 132:327–333

Gačić M, Poulain PM, Zore-Armanda M, Barale V (2001) Overview. In: Cushman-Roisin B, Gačić M, Poulain PM, Artegiani A (eds) Physical oceanography of the Adriatic Sea. Kluwer Academic, Dordrecht, 1–44

Gordon DP, Mawatari SF (1992) Atlas of the marine-fouling Bryozoa of New Zealand ports and harbours, vol 107. Miscellaneous Publications of the New Zealand Oceanographic Institute, Wellington, pp 1–52

Hoegh-Guldberg O, Bruno J (2010) The impact of climate change on the world's marine ecosystems. Science 328:1523–1528

Kahle J, Liebezeit G, Gerdes G (2003) Growth aspects of *Flustra foliacea* (Bryozoa, Cheilostomata) in laboratory culture. Hydrobiologia 503:237–244

Kruk-Dowgiallo L, Szaniawska A (2008) Gulf of Gdansk and Puck Bay. In: Schiewer U (ed) Ecology of Baltic Coastal Waters, vol 197, Ecological studies. Springer, Berlin/Heidelberg, pp 139–165

Kuklinski P, Taylor PD (2006) Unique life history strategy in a successful Arctic bryozoan, *Harmeria scutulata*. J Mar Biol Assoc UK 86:1035–1046

Linse K, Barnes DKA, Enderlein P (2006) Body size and growth of benthic invertebrates along an Antarctic latitudinal gradient. Deep-Sea Res Pt II 53:921–931

Loeng H (1991) Features of the physical oceanographic conditions of the Barents Sea. Polar Res 10:5–18

Lombardi C, Cocito S, Occhipinti-Abbrogi A, Hiscook K (2006) The influence of sea water temperature on zooid size and growth rate in *Pentapora fascialis* (Bryozoa: Cheilostomata). Mar Biol 149:1103–1109

Lombardi C, Cocito S, Hiscook K, Occhipinti-Abbrogi A, Setti M, Taylor PD (2008) Influence of seawater temperature on growth bands, mineralogy and carbonate production in a bioconstructional bryozoan. Facies 54:333–342

Madhusudan S, Protheroe A, Propper D, Han C, Corrie P, Earl H, Hancock B, Vasey P, Turner A, Balkwill F, Hoare S, Harris AL (2003) A multicentre phase II trial of bryostatin-1 in patients with advanced renal cancer. Brit J Cancer 89:1418–1422
Nikulina E, Schäfer P (2006) Bryozoans of the Baltic Sea. Meyniana 58:75–95
O'Dea A, Jackson JBC (2002) Bryozoan growth mirrors contrasting seasonal regimes across the Isthmus of Panama. Palaeogeogr Palaeocl 185:77–94
Orlić M, Leder N, Pasarić M, Smirčić A (2000) Physical properties and currents recorded during September and October 1998 in the Velebit Channel (East Adriatic). Period Biol 102:31–37
Schäfer P, Bader B (2008) Geochemical composition and variability in the skeleton of the bryozoans *Cellaria sinuosa* (Hassall): biological versus environmental control. In: Hageman SJ, Key MM, Winston JE (eds), Bryozoan studies 2007. Proceedings of the 14th International Bryozoology Association conference, Boone, North Carolina, 1–8 July 2007. Virginia Mus of Nat Hist Publ 15:269–279
Smith AM (2007) Age, growth and carbonate production by erect rigid bryozoans in Antarctica. Palaeogeogr Palaeclimatol Palaeoecol 256:86–98
Smith AM, Key MM (2004) Controls, variations and a record of climate change in detailed stable isotope record in a single bryozoan skeleton. Quaternary Res 61:123–133
Węsławski JM, Zajączkowski M, Kwaśniewski S, Jezierski J, Moskal W (1988) Seasonality in an Arctic fjord ecosystem: Horsunfjord, Spitsbergen. Polar Res 6:185–189
Winston JE (1977) Distribution and ecology of estuarine ectoprocts: a critical review. Chesapeake Sci 18:34–57

# Chapter 12
# Patterns of Magnesium-Calcite Distribution in the Skeleton of Some Polar Bryozoan Species

## Mineralogy of Polar Bryozoan Skeletons

**Jennifer Loxton, Piotr Kuklinski, James M. Mair, Mary Spencer Jones, and Joanne S. Porter**

**Abstract** Polar marine environments are already starting to exhibit the effects of climate change. The Arctic is the most rapidly warming place on Earth, and changes of the seawater chemistry of polar oceans have been recorded. Calcifying Bryozoa have diverse skeletal mineralogies making them an ideal model for investigating differences caused by environmental change. The aim of this study is to quantify the skeletal mineralogical diversity of polar bryozoans using X-ray diffraction (XRD). Six species of erect Bryozoa were analysed, three Arctic and three Antarctic species. Within each of the three species from each region, one has a cemented attachment point, one has flexible growth and the third is attached by chitinous rootlets. The analysis shows no significant difference in Mg-calcite distribution along the length of the six species but does show species-specific variation in both the consistency of Mg-calcite distribution along the length of a colony and the relationship between concentration of Mg-calcite in the root and growing tip. Analysis shows a statistically significant trend of increasing Mg-calcite concentration with increasing temperature. This adds further data to a growing body of published evidence for this mineralogy trend. The results of this study suggest that if bryozoan species are to be used as indicators of environmental change then it will be critical to have robust, replicated data of species-specific profiles for Mg-calcite distribution. This data, viewed alongside published mineralogy trends,

J. Loxton (✉) • J.M. Mair • J.S. Porter
Centre for Marine Biodiversity and Biotechnology, School of Life Sciences, Heriot-Watt University, Edinburgh EH14 4AS, UK
e-mail: jll13@hw.ac.uk

P. Kuklinski
Institute of Oceanology, Polish Academy of Sciences, Powstancow, Warszawy 55, 81-712 Sopot, Poland

M. Spencer Jones
Zoology Department, , Natural History Museum, Cromwell Road, London SW7 5BD, UK

A. Ernst, P. Schäfer, and J. Scholz (eds.), *Bryozoan Studies 2010*,
Lecture Notes in Earth System Sciences 143, DOI 10.1007/978-3-642-16411-8_12,

may allow the use of skeletal mineralogy as a register of environmental effects and may enable monitoring of future impacts of climate change in marine benthic ecosystems.

**Keywords** Polar • Mineralogy • Magnesium • Skeleton • Bryozoan

# Introduction

## *Ocean Chemistry and the Calcifying Process*

Since the industrial revolution the amount of $CO_2$ in the atmosphere has increased due to the burning of fossil fuels, leading to wide-ranging climatic impacts (Hoegh-Guldberg and Bruno 2010). The oceans act as a natural buffer for the uptake of $CO_2$, this in turn leads to changes in seawater chemistry. Climate change has led to sea temperature rises (Bindoff et al. 2007); a decrease in ocean pH (Hoegh-Guldberg and Bruno 2010) and a corresponding decrease in seawater carbonate saturation (Jacobson 2005).

Changes in ocean chemistry are predicted to have a range of effects on calcifying marine organisms, and increased acidity in the oceans may lead to shell (Feely et al. 2004) and skeletal dissolution (Morse 2002; Orr et al. 2005). Marine organisms such as echinoderms, molluscs and crustaceans may have aragonite, calcite or a mixture of these present in their skeletons and shells (Ries et al. 2010). Of these forms of calcium carbonate, aragonite (Turley et al. 2007) and high Mg-calcite (Wood et al. 2008) are more sensitive to dissolution than the more resilient low Mg-calcite.

Cold-water regions, with their greater solubility for $CO_2$, are likely to be affected first by the increase in ocean acidification (Feely et al. 2004). It is estimated that within 50 years, the acidity of Arctic waters will be sufficient to corrode aragonite (Feely et al. 2004) and the Southern Seas will reach this same position by the end of the century (Orr et al. 2005; Smith 2009). Calcifying fauna form a significant component of benthic ecosystems globally, in terms of their biodiversity and their physical structure. In order to understand how calcifying organisms will be impacted by the predicted changes in seawater chemistry, we have chosen to study the mineralogy of Arctic and Antarctic Bryozoa with different growth forms.

## *Bryozoan Mineralogy: An Indicator of Change*

The sessile nature of suspension-feeding organisms such as Bryozoa means that they are sensitive to changes in physical parameters and are, therefore, potentially good indicators of changes in environmental conditions (Kuklinski and Bader 2007). Mineralogical polymorphism and the large geochemical skeletal spectrum exhibited by bryozoans (Smith et al. 2006) makes them an ideal model organism for investigating differences caused by both spatial variation and environmental conditions.

A wide variety of Mg levels have been recorded in bryozoan calcite. This ranges from low (< 4 mol% $MgCO_3$) to high (>12 mol% $MgCO_3$) (Smith et al. 2006; Kuklinski and Taylor 2009) with most bryozoans featuring low to intermediate Mg-calcite (4–12 mol% $MgCO_3$) (Smith et al. 2006; Lombardi et al. 2008; Kuklinski and Taylor 2009).

Cheilostomes exhibit diverse mineralogy and some of this has been attributed to astogeny and differences within local structures of colonies (Smith and Girvan 2010). Mineralogical variation within colonies has been attributed to several different factors. There are documented trends of increasing Mg levels in bryozoan calcite at warmer sea temperatures (Davis et al. 2000; Kuklinski and Taylor 2009) and at lower latitudes (Borisenko and Gontar 1991; Taylor et al. 2009). Higher Mg levels within calcite have been attributed to greater zooid age (Poluzzi and Sartori 1975; Smith et al. 1998; Kuklinski and Taylor 2009). Aragonite, a harder crystalline material than calcite, has been observed as a reinforcing outer layer in bryozoan skeletons (Smith and Girvan 2010).

## *Current Status of Bryozoan Mineralogy Studies*

Although there has been a number of studies of bryozoan mineralogy (e.g. Borisenko and Gontar 1991; Smith et al. 1998; Lombardi et al. 2008; Schäfer and Bader 2008; Kuklinski and Taylor 2009; Smith and Girvan 2010) our knowledge is still far from complete, particularly for polar bryozoans.

Some areas, which have, to date, received little attention, are:

- The relationship of Mg-calcite levels to colony substrate attachment method.
- Mg-calcite distribution within the skeleton. There has been research into the distribution of Mg-calcite in the transverse cross section of colonies but only a few studies looking at variations in Mg-calcite levels at points along the longitudinal length of an erect colony (Saundray and Bouffandreau 1958; Poluzzi and Sartori 1975; Smith et al. 1998; Schäfer and Bader 2008; Kuklinski and Taylor 2009). Notable studies in this field include work by Smith et al. (1998) comparing the growing tip, centre and base of New Zealand bryozoans and by Kuklinski and Taylor (2009) comparing the growing tip and base of several polar specimens.

The aim of this study is to quantify the diversity of mineral distribution in Arctic and Antarctic bryozoans with different growth forms, in order to establish a baseline of evidence for further study of the effects of climate change in these marine calcifying organisms. This is the first study with a systematic analysis of mineralogy along the whole length of the colony branch.

## Material and Methods

### Description of Study Areas

Arctic and Antarctic regions share characteristics such as year round low temperature, extreme seasonality caused by fluctuation of light supply and growth and retreat of ice cover. There are, however, a number of significant differences between the two regions.

The Arctic Ocean is a relatively small, seasonally ice-covered water mass between temperate Atlantic waters and cold ice-covered polar seas. When compared with other oceans, the Arctic Ocean is isolated due to the separation of these water bodies by steep oceanic gradients and surrounding land (Comiso 2010). Although the Arctic is considered to be a geologically younger environment (Kaiser et al. 2005) with fewer endemic species than the Antarctic, there is a high level of adaptation of species as a result of the extreme environment (Kuklinski and Taylor 2008; Bader and Schäfer 2005).

Antarctica (>45°S) has a continental landmass, of approximately the same size as the Arctic Ocean (Comiso 2010). It is surrounded by the Southern Ocean, a well-mixed, deep-water system with upwelling from the Antarctic Coastal Current. The Southern Ocean is open to exchange with the Atlantic, Pacific and Indian Oceans. The Antarctic was effectively cut off from the rest of the world since 25 Ma by the formation of the Circum-Antarctic current (Kaiser et al. 2005). This isolation of the Southern Seas has resulted in a high degree of endemism and specialisation of benthic flora and fauna.

### Material

Bryozoans used in this study were alive when collected and preserved in 100% ethanol prior to preparation and subsequent analysis. Three species were chosen from the Arctic and three species from the Antarctic. From each region one species was chosen to represent each of the substrate attachment types from the following definitions:

- *Rootlet*: Calcified erect species attached to the substrate through chitinous rhizoids (rootlets)
- *Flexible:* Lightly calcified (damaged by extended (>10 min) exposure to 10% bleach) erect species attached to the substrate through chitinous rhizoids (rootlets)
- *Cemented*: Calcified erect species attached to the substrate through cemented zooids

Table 12.1 Profile of sampling localities. Temperature data taken from various sources as detailed

| Code | Locality | Latitude and longitude | Average depth (m) | Mean annual temp (°C) |
|---|---|---|---|---|
| 1 | Kongsfjorden (Arctic) | 78°59′N, 11°56′E | 10 | 2.5 (Cottier et al. 2005) |
| 2 | Isfjorden (Arctic) | 78°07,001′N, 14°22,601′E | 97 | 2 (Skogseth et al. 2005) |
| 2 | Heleysundet (Arctic) | 78°40,61′N, 21°24,44′E | 13 | 0.1 (Haarpaintner et al. 2001) |
| 4 | East Greenland (Arctic) | 79°21,01′N, 7°45,25′W | 213 | 0.4 (Berge et al. 2005) |
| 5 | Terre Adélie (Antarctic) | 66°34′S, 144°18′E | 187.5 | −1.85 (Koubbi et al. 2010) |
| 6 | Terre Adélie (Antarctic) | 66°34′S, 142°39′E | 261.5 | −1.85 (Koubbi et al. 2010) |
| 7 | Terre Adélie (Antarctic) | 66°53′S, 142°39′E | 346.5 | −1.85 (Koubbi et al. 2010) |

A minimum of three replicate individuals of each of the species was selected. Where possible, complete individuals comprising root and tip were sampled.

Arctic samples were collected by SCUBA diving or dredging from sites 1–4 between 2001 and 2009 from around Svalbard and East Greenland (Table 12.1). Some of these areas (e.g., West Spitsbergen sites 1 and 2) could be more accurately considered to have a sub-Arctic climate due to the warming effects of the current (>3°C) originating from the vicinity of the Gulf Stream (Kuklinski and Taylor 2009).

Arctic bryozoan species were identified using the monograph of Kluge (1975). Three Arctic erect cheilostome species, *Pseudoflustra solida*, *Cystisella saccata*, and *Carbasea carbasea* were analyzed (Table 12.2, Fig. 12.1).

Antarctic samples were collected by Beam trawl from sites 5 to 7 in 2008 by the Collaborative East Antarctic Marine Census. This monitoring project involved three ships from Australia, Japan, and France working as a sampling network on 132 sites in order to gather information on the East Antarctic continental shelf in relation to environmental parameters (Koubbi et al. 2010). All samples were collected from the vicinity of Terre Adélie (Table 12.1). Water temperature was measured at the time of sample collection.

Antarctic bryozoan species were identified using Hayward (1995), and by personal communication with Hayward. Three Antarctic erect cheilostome species, *Cellarinella margueritae, Cellarinelloides crassus* and *Isosecuriflustra angusta* were analyzed (Table 12.2, Fig. 12.2). The use of a Beam trawl as a collection technique caused damage to cemented bryozoan colonies and as a result it was not possible to analyze *C. crassus* individuals with intact basal attachments in this study.

Existing temperature data for the collection sites was compiled from published literature (Table 12.1) and, where possible, a mean annual temperature was calculated for the locality and depth. This was not possible for the Antarctic collection sites where only a single temperature data point, measured at the time of sample collection, was available.

**Table 12.2** Specimen details. See Table 12.1 for location codes. Samples column indicates the number of samples extracted along the colony branch for each specimen

| | Species | Growth form | Specimen code | Locality code | Samples |
|---|---|---|---|---|---|
| Arctic specimens | *Pseudoflustra solida* | Rootlet | ARI | 4 | 5 |
| | *Pseudoflustra solida* | Rootlet | ARII | 4 | 4 |
| | *Pseudoflustra solida* | Rootlet | ARIII | 4 | 4 |
| | *Cystisella saccata* | Cemented | ACI | 1 | 5 |
| | *Cystisella saccata* | Cemented | ACII | 2 | 5 |
| | *Cystisella saccata* | Cemented | ACIII | 2 | 5 |
| | *Carbasea carbasea* | Flexible | AFI | 3 | 5 |
| | *Carbasea carbasea* | Flexible | AFII | 3 | 5 |
| | *Carbasea carbasea* | Flexible | AFIII | 3 | 5 |
| Antarctic specimens | *Cellarinella margueritae* | Rootlet | AARI | 5 | 6 |
| | *Cellarinella margueritae* | Rootlet | AARII | 5 | 5 |
| | *Cellarinella margueritae* | Rootlet | AARIII | 5 | 5 |
| | *Cellarinella margueritae* | Rootlet | AARIV | 5 | 5 |
| | *Cellarinelloides crassus* | Cemented | AACI | 7 | 5 |
| | *Cellarinelloides crassus* | Cemented | AACII | 7 | 5 |
| | *Cellarinelloides crassus* | Cemented | AACIII | 7 | 5 |
| | *Isosecuriflustra angusta* | Flexible | AAFI | 6 | 5 |
| | *Isosecuriflustra angusta* | Flexible | AAFII | 6 | 5 |
| | *Isosecuriflustra angusta* | Flexible | AAFIII | 5 | 5 |

## *Sample Preparation and Mineralogical Analysis*

Samples were examined under a dissection microscope and care was taken to thoroughly remove any substrate (e.g. shell or algae) and epibionts that could contaminate results through their added mineralogies. Colonies were prepared and divided as detailed as possible (Figs. 12.1, 12.2, and Table 12.2), and imaged using a LEO 1455- VP SEM, capable of imaging large, uncoated colonies using back-scattered electrons.

Mineralogical analyses were conducted using a high-precision Nonius XRD with a position-sensitive-detector. Bryozoan samples were powdered using a quartz pestle and mortar, divided into three quotients and affixed using a drop of acetone to three single quartz crystal substrates. Quantitative XRD analysis was undertaken to determine the dominant biomineral, the proportions of calcite and aragonite in bimineralic species and the Mg content of the calcite.

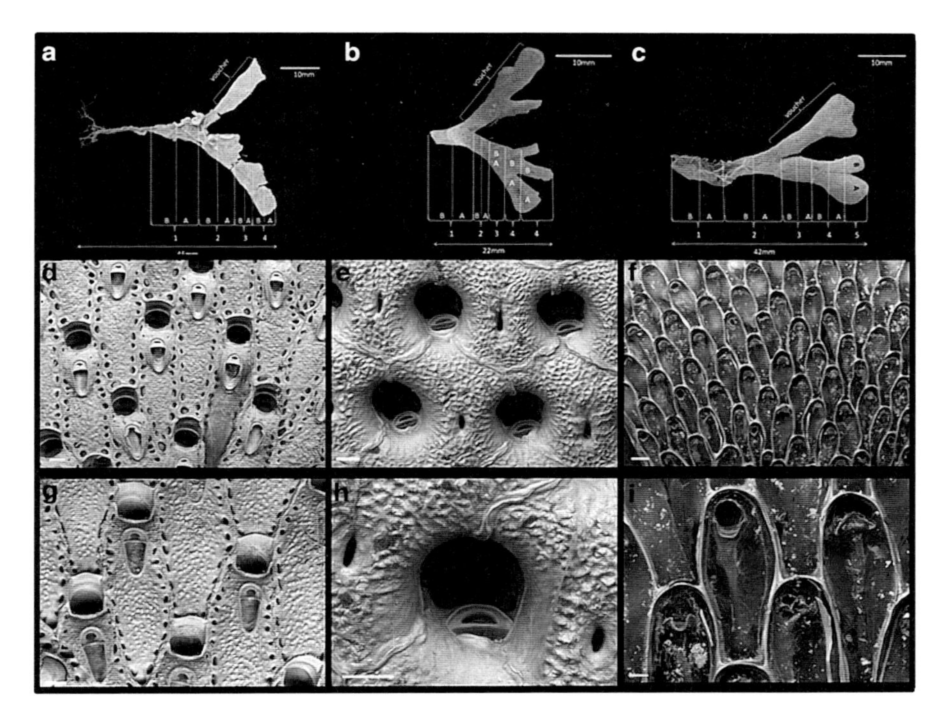


**Fig. 12.1** Arctic species profiles. (**a, d, g**). *Pseudoflustra solida*. (**a**) Colony division, *scale bar*: 10 mm. (**d**) Zooid cluster, *scale bar*: 100μm. (**g**) *Orifice* and *avicularia*, *scale bar*: 100μm. (**b, e, h**). *Cystisella saccata*. (**b**) Colony division, *scale bar*: 10mm. (**e**) Zooid cluster, *scale bar*: 100μm. (**h**) *Orifice* and *avicularia*, *scale bar*: 100μm. (**c, f, i**). *Carbasea carbasea*. (**c**) Colony division, *scale bar*: 10mm. (**f**) Zooid cluster, *scale bar*: 100μm. (**i**) *Orifice*, *scale bar*: 300μm

In order to determine the proportions of aragonite and calcite in bimineralic species, peak intensities were fitted to standard patterns generated from 100% aragonite and 100% calcite. The error associated with this method is estimated to be 2% based on repeatability studies of samples with a known aragonite proportion. In order to calculate mol% $MgCO_3$, the position of the d104 peak was measured, assuming a linear interpolation between $CaCO_3$ and $MgCO_3$. This composition information is accurate to within 2% on a well-calibrated instrument (Kuklinski and Taylor 2009). Pure silica (Si) and silver behenate ($AgC_{22}H_{43}O_2$) on quartz substrate were used as the instrument calibration standard (Blanton et al. 1995, 2000). Computations were undertaken using Nonius GUFI software (Dinnebier 1987).

## *Data Analysis*

Results were tested for normality and subjected to ANOVA and t-tests in order to analyse variance and linear regression to analyse linear trends.

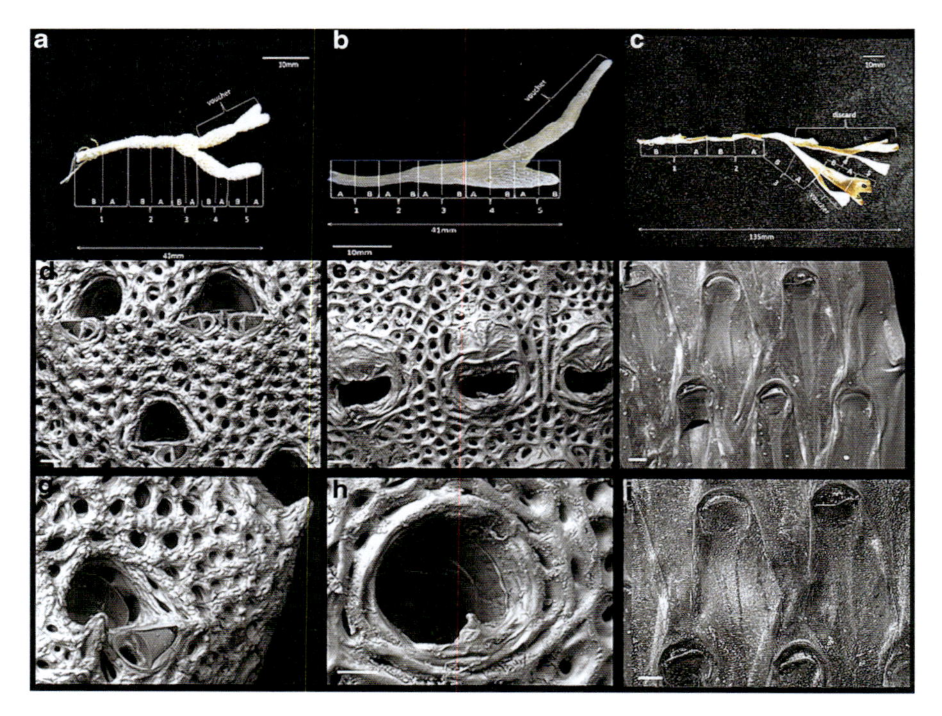

**Fig. 12.2** Antarctic species profiles. (**a, d, g**) *Cellarinella margeuritae*. (**a**) Colony division, *scale bar*: 10mm. (**d**) Zooid cluster, *scale bar*: 100μm. (**g**) *Orifice* and *avicularia*, *scale bar*: 100μm. (**b, e, h**) *Cellarinelloides crassus*. (**b**) Colony division, *scale bar*: 10mm. (**e**) Zooid cluster, *scale bar*: 100μm. (**h**) *Orifice*, *scale bar*: 100μm. (**c, f, i**) *Isosecuriflustra angusta*. (**c**) Colony division, *scale bar*: 10mm. (**f**) Zooid cluster, *scale bar*: 100μm. (**i**) *Orifice*, *scale bar*: 100μm

ANOVA, t-tests and linear regressions were conducted using SigmaStat software 2.03 (SPSS Inc. 1995).

# Results

## *Mineralogy*

The Arctic bryozoans tested were found to be intermediate-Mg calcite. The average calcite mol% $MgCO_3$ was found to be 8.09 (ranging from 7.16 to 10.23, stdev = 0.83) with all examined species found to be intermediate-Mg (4–11.99 mol% $MgCO_3$) calcitic species. None of the species was found to be bimineralic or aragonitic.

The Antarctic bryozoans tested were found to be intermediate-Mg calcite. The average mol% $MgCO_3$ was 5.45 (ranging from 3.69 to 6.81, stdev = 0.95) with all

**Table 12.3** Comparison of Mg-calcite between the colony base and tip. Statistical analysis conducted using *t*-test

| Species | | Average $MgCO_3$ | | Degrees of freedom (n − 1) | t-test | |
|---|---|---|---|---|---|---|
| | | Base | Tip | | P-value | t |
| *Cystisella saccata** | Arctic cemented | 8.76 | 10.23 | 16 | 0.03 | −2.36 |
| *Cellarinelloides crassus* | Antarctic cemented | 6.82 | 6.10 | 16 | 0.15 | −1.58 |
| *Pseudoflustra solida* | Arctic rootlet | 7.86 | 7.90 | 16 | 0.89 | 0.15 |
| *Cellarinella margueritae** | Antarctic rootlet | 4.17 | 5.37 | 18 | 0.00 | 4.16 |
| *Carbasea carbasea* | Arctic flexible | 7.48 | 7.45 | 16 | 0.92 | −0.11 |
| *Isoseculiflustra angusta* | Antarctic flexible | 6.13 | 5.69 | 16 | 0.21 | −1.35 |

*indicates statistical significance at $p < 0.05$

examined species found to be intermediate-Mg (4–11.99 mol% $MgCO_3$) calcitic species. One species, *Isosecuriflustra angusta*, was found to be bimineralic with levels of aragonite ranging between 11% and 33% (mean = 19.37, stdev = 8.79) of total calcium carbonate present.

Two species, the Antarctic rootlet species *Cellarinella margueritae* and the Arctic cemented species *Cystisella saccata*, were found to have a statistically significant difference (p = 0.002 and p = 0.031 consecutively) in mol% $MgCO_3$ between the base and the tip. Both species feature higher mol% $MgCO_3$ in the growing tip than in the base (Table 12.3).

## Mg Distribution in Erect Polar Bryozoans with Different Substrate Attachment Types

### Cemented Species

From replicate analysis, ANOVA results showed that there was no statistically significant difference in the mean $MgCO_3$ concentrations between individual sections sampled along the length of the colony branch (Fig. 12.3). All three *Cystisella saccata* specimens examined showed a consistency in Mg levels with a coefficient of variation (mean/STDEV*100) of 0.166 across the 45 tested samples. There is a statistically significant difference in mol% $MgCO_3$ between the base (mean mol% $MgCO_3$ = 8.76, stdev = 0.74) and the tip (mean mol% $MgCO_3$ = 10.23, stdev = 1.73) of the colony (P = 0.031, t = −2.359, degrees of freedom = 16).

All three *Cellarinelloides crassus* specimens examined showed a consistency in Mg levels with a coefficient of variation of 0.069 across the 45 tested samples, and no significant difference between the base and the tip.

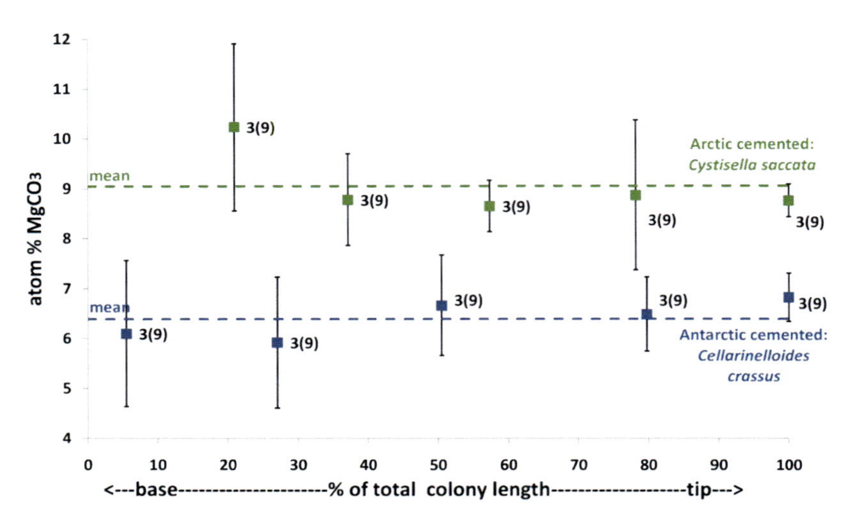


**Fig. 12.3** Mean values (± standard deviation) of mol% $MgCO_3$ along the length of cemented erect bryozoans in the Arctic (*green*) and Antarctic (*blue*). Specimen numbers are shown next to data points and sample sizes are shown in *brackets*

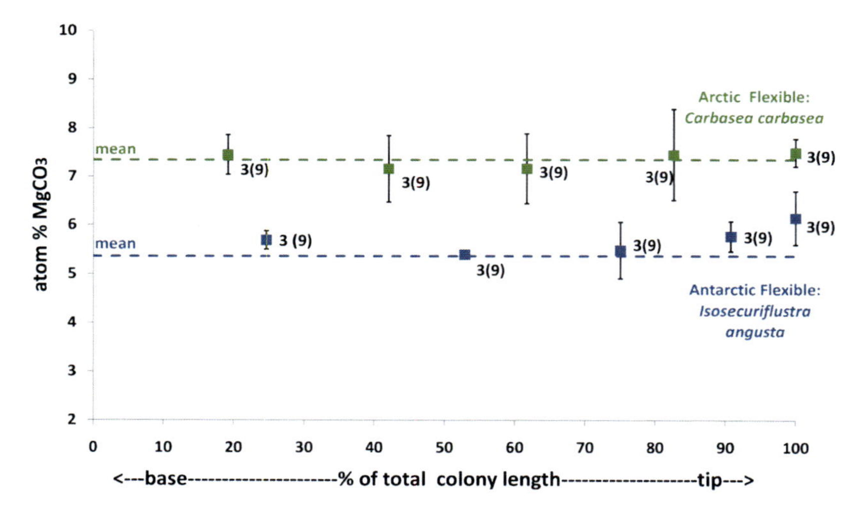


**Fig. 12.4** Mean values (± standard deviation) of mol% $MgCO_3$ along the length of flexible erect bryozoans in the Arctic (*green*) and Antarctic (*blue*). Specimen numbers are shown next to data points and sample sizes are shown in *brackets*

### Flexible Species

From replicate analysis, ANOVA results showed no statistically significant difference in the mean $MgCO_3$ concentrations along the length of the branch (Fig. 12.4). The Arctic flexible species, *Carbasea carbasea*, showed a high level of consistency in Mg levels with a coefficient of variation of 0.151 across the 45 tested samples.

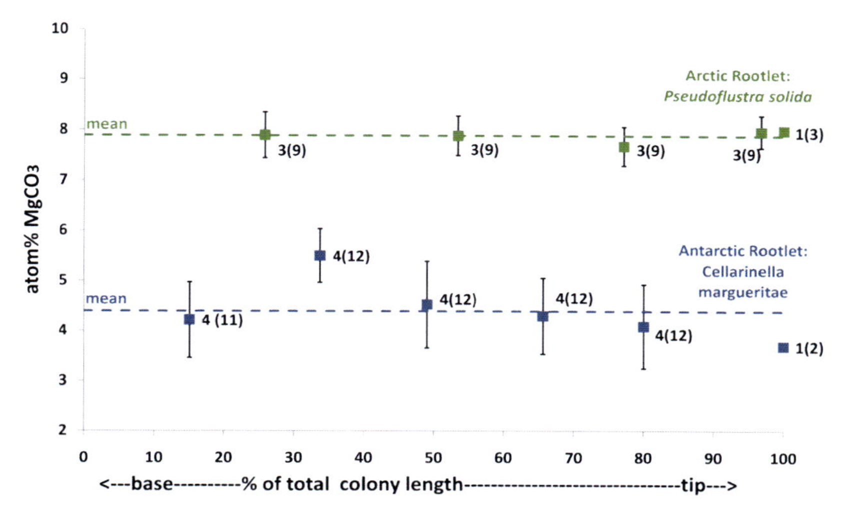


**Fig. 12.5** Mean values (± standard deviation) of mol% $MgCO_3$ along the length of rootlet erect bryozoans in the Arctic (*green*) and Antarctic (*blue*). Specimen numbers are shown next to data points and sample sizes are shown in *brackets*

The Antarctic flexible species, *Isosecuriflustra angusta*, showed an even higher level of consistency with a coefficient of variation of 0.25 across the 45 tested samples. *Isosecuriflustra angusta* is the only species examined featuring a bimineralic skeleton (aragonite content ranging from 11% to 33% of the total $CaCO_3$ present).

## Rootlet Species

From replicate analysis, ANOVA results showed no statistically significant difference in the mean $MgCO_3$ concentrations along the length of the branch (Fig. 12.5). In the Arctic rootlet species, *Pseudoflustra solida,* all four specimens examined showed a consistency in Mg levels with a coefficient of variation of 0.38 across the 60 tested samples.

In *Cellarinella margueritae*, all four specimens examined were consistent in Mg levels with a coefficient of variation of 0.097 across the 52 tested samples. *C. margueritae* features distinct annual growth check lines (Barnes et al. 2007) and as such it is possible to determine the year of calcite deposition by counting the bands. When the year of calcite deposition is plotted against mol% $MgCO_3$ a deviation from the mean can be seen in calcite deposited in 2003 with all eight measurements of mol% $MgCO_3$ deviating above the mean Mg-calcite of 4.39 mol% $MgCO_3$ by an average of 26.6% (Fig. 12.6). There is a statistically significant difference between mean Mg levels in the base (mean mol% $MgCO_3 = 4.16$, stdev $= 1.03$) and the tip (mean mol% $MgCO_3 = 5.37$, stdev $= 0.95$) of the *C. margueritae* ($p = 0.002$, $t = 4.16$, degrees of freedom $= 18$).

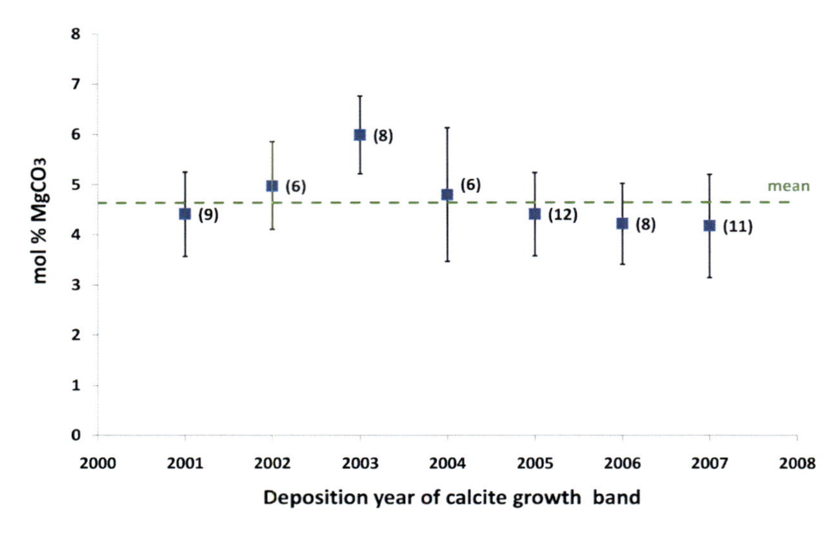


**Fig. 12.6** Mean values (± standard deviation) of mol% $MgCO_3$ at different positions/years of calcite deposition along the length of *Cellarinella margueritae*, an erect cemented bryozoan from the Antarctic. Sample sizes are shown in *brackets*

### Temperature Variation and Its Impact on Mg Content in Bryozoans

The relationship of mean mol% $MgCO_3$ in erect bryozoans to temperature (Fig. 12.7) was found to be statistically significant through both ANOVA ($F = 18.59$, P-value = $<0.001$), and linear regression ($R^2 = 0.779$, P-value = $<0.001$).

## Discussion

### *Magnesium Distribution in Erect Polar Bryozoans*

The results of XRD analysis showed that there was a general pattern of no significant difference in mean $MgCO_3$ distribution along the length of erect bryozoan colonies from the Arctic and the Antarctic.

In contrast to previous studies (Smith and Girvan 2010; Kuklinski and Taylor 2009), the six species studied here appear to show no increase in mean mol% $MgCO_3$ associated with astogeny. This may either indicate that this attribute is species dependent and does not apply to the species in this study or alternatively may be an indicator that the more susceptible high Mg-calcite has potentially either been dissolved by environmental conditions or removed through damage either in vivo or during collection, storage or processing.

In the case of the flexible species *Carbasea carbasea* and *Isosecuriflustra angusta*, it may be suggested that the typical faster growth rate of species with

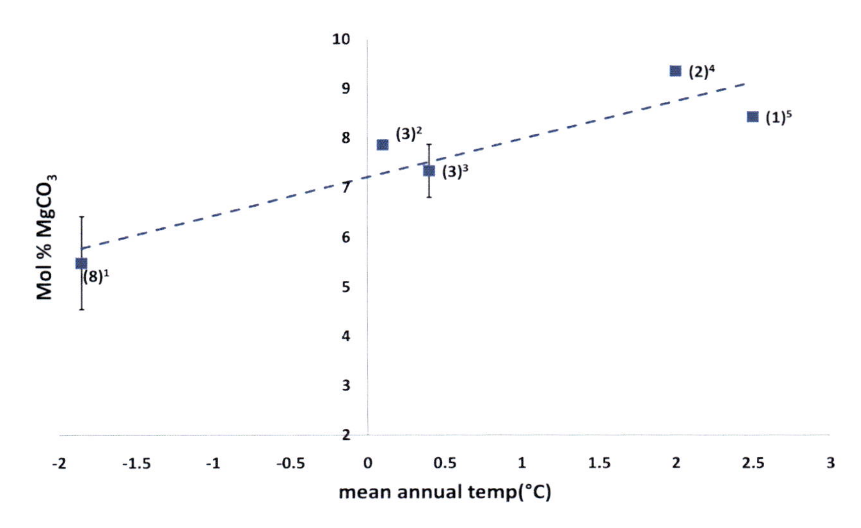


**Fig. 12.7** Mean values (± standard deviation) of mol% $MgCO_3$ and its relationship to mean annual temperature. Sample sizes are shown in *brackets*. Temperature data taken from the following sources: [1] "snapshot" temperature data collected at the time of specimen collection at each site (Koubbi et al. 2010). [2] mean annual temperature calculated from April 1998 and September 1998 data (Haarpaintner et al. 2001). [3] "typical" temperature as reported by Berge et al. (2005). [4] mean annual temperature calculated from April 2003 and September 2003 data (Skogseth et al. 2005). [5] mean annual temperature calculated from April 2002 and September 2002 data (Cottier et al. 2005)

these morphologies means that the zooids present are not old enough to show a measurable increase in mol% $MgCO_3$ between the older and younger zooids.

There are differences between species with respect to the consistency of Mg distribution along the length of the colony. Coefficients of variation range from 0.069 to 0.38. Future studies will help to further elucidate the consistency of magnesium distribution in polar and other erect bryozoans. If bryozoan species are to be used as indicators of environmental change then it will be critical to have robust, replicated data for the species-specific profiles of consistency of Mg distribution.

Two species in our study exhibited higher Mg-calcite in the tip than in the base (*Cysticella saccata* and *Cellarinella margueritae*). This observation is in agreement with the pattern reported by Kuklinski and Taylor (2009) for *C. saccata*. *C. saccata* is an Arctic species with a cemented base, whilst *C. margueritae* is an Antarctic species attached by chitinous rootlets. The presence of higher Mg-calcite in the tip compared to the base of these two species cannot be explained simply by attachment method and region of origin. This phenomenon may be more characteristic of individual species rather than attachment type or location.

The Antarctic rootlet specimens are the only species studied here that can be accurately aged due to the annual growth banding (Barnes et al. 2007). Specimens in this study range in age from 7 to 8 years (mean = 7.75, stdev = 0.5). Such

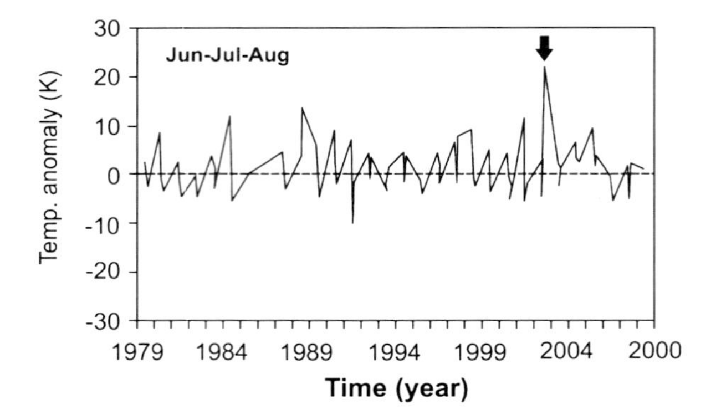


**Fig. 12.8** Mean temperature anomalies at 20 km at Dumont d'Urville between 1979 and 2008 for the three corresponding months in winter (one point for each month of a season, that is, three points per season). The "zero" line is *dashed* and the *grey line* is the estimated trend (Figure adapted from David et al. (2010) and shows that a temperature anomaly of approximately 22 K above the "zero" line occurred in July 2002)

specimens offer a valuable opportunity to investigate the effects of seasonal temperature on growth rates and mineralogy. XRD replicate samples were taken from location 2 (Fig. 12.2a) on the *C. margueritae* colonies. This point on the colonies is approximately 17.3 mm from the base and corresponds with growth occurring between 2002 and 2003. At this location for the four specimens tested, all eight measurements of mol% $MgCO_3$ from 2003 were above the mean Mg-calcite of 4.39 mol% $MgCO_3$ by an average of 26.6%.

It is recorded that the June-July-August (winter) period of 2002 was unusually warm with a mean July temperature of over 22 K above the 1979–2006 average (David et al. 2010). These observations, taken together, suggest a relationship between the mol% $MgCO_3$ levels measured in the colonies of *C. margueritae* and the exceptional environmental conditions during that time period (Fig. 12.8).

The trend of increasing mean mol% $MgCO_3$ with increasing temperature (Fig. 12.8) was found to be statistically significant. This study therefore concurs with and reinforces the patterns observed by previous authors (Lowenstam 1954; Davis et al. 2000; Ries 2005; Lombardi et al. 2008; Kuklinski and Taylor 2009).

## Conclusions

The result exhibited in this study relating mean mol% $MgCO_3$ levels to temperature adds more data to a growing body of evidence and reinforces previous work published by a number of authors (e.g. Lowenstam 1954; Davis et al. 2000; Ries 2005; Lombardi et al. 2008; Kuklinski and Taylor 2009; Taylor et al. 2009).

With polar oceans facing the potential double impact of increasing temperature (Hoegh-Guldberg and Bruno 2010) driving higher levels of more soluble high

Mg-calcite in combination with rising acidity (Hoegh-Guldberg and Bruno 2010) of the oceans increasing dissolution rates (Orr et al. 2005), it has never been more important to understand the potential consequences of climate change for calcareous marine organisms such as bryozoans.

Further studies in this area focussing on a larger number of species including a wider range of attachment types, in parallel with information on growth rates and full environmental data could potentially allow us to develop the interpretation of skeletal mineralogy into a useful gauge of past environmental effects on calcareous organisms (Taylor et al. 2009). This in turn may facilitate the use of mineralogy studies as an important tool for predicting the future impacts of climate change on bryozoans, an important component of calcareous benthic communities in polar regions.

**Acknowledgements** We would like to thank Jens Najorka and Gordon Cressey (Natural History Museum, London) for their assistance with XRD and mineralogical analysis. The authors would also like to thank Dave Barnes (British Antarctic Survey) and Pete Hayward (Swansea University) for their expertise on Antarctic species and regions. We would like also to acknowledge the voyage leader, Martin Riddle, the crew and the captain of the RV Aurora Australis. The CAML-CEAMARC cruise of RV Aurora Australis (IPY project n° 53) were supported by the Australian Antarctic Division, the Japanese Science Foundation, the French polar institute IPEV and the Muséum National d'Histoire Naturelle). This study has been completed thanks to grants from the Heriot-Watt Alumni Fund, the John Ray Trust and grant provided to PK by the Polish Ministry of Science and Higher Education (N N304 404038).

## References

Anisimov OA, Vaughan DG, Callaghan TV, Furgal C, Marchant H, Prowse TD, Vilhjálmsson H, Walsh JE (2007) Polar regions (Arctic and Antarctic). In: Parry ML, Canziani OF, Palutikof JP, van Der Linden PJ, Hanson CE (eds) Climate change 2007: impacts, adaption and vulnerability. Contribution of working group II to the fourth assessment report of the intergovernmental panel on climate change. Cambridge University Press, Cambridge, pp 653–685

Bader B, Schäfer P (2005) Impact of environmental seasonality on stable isotope composition of skeletons of the temperate bryozoan *Cellaria sinuosa*. Palaeogeogr Palaeclimatol Palaeoecol 226:58–71

Barnes DKA, Webb KE, Linse K (2007) Growth rate and its variability in erect Antarctic bryozoans. Polar Biol 30:1069–1081

Berge J, Johnsen G, Nilsen F, Gulliksen B, Slagstad D (2005) Ocean temperature oscillations enable reappearance of blue mussels *Mytilus edulis* in Svalbard after a 1000 year absence. Mar Ecol Prog Ser 303:167–175

Bindoff NL, Willebrand J, Artale VA, Cazenave A, Gregory J, Gulev S, Hanawa K, Le Quéré C, Levitus S, Nojiri Y, Shum CK, Talley LD, Unnikrishnan A (2007) Observations: oceanic climate change and sea level. In: Solomon S, Qin D, Manning M, Chen Z, Marquis M, Avery KB, Tignor M, Miller HL (eds) Climate change 2007: the physical science basis. Contribution of working group I to the fourth assessment report of the intergovernmental panel on climate change. Cambridge University Press, Cambridge, UK/New York, pp 385–432

Blanton TN, Huang TC, Toraya H, Hubbard CR, Robie SB, Louër D, Göbel HE, Will G, Gilles R, Raftery T (1995) JCPDS – International Centre for Diffraction Data round robin study of silver

behenate. A possible low-angle X-ray diffraction calibration standard. Powder Diffr 10 (2):91–95
Blanton TN, Barnes CL, Lelental M (2000) Preparation of silver behenate coatings to provide low- to mid-angle diffraction calibration. J Appl Crystallogr 33:172–173
Borisenko YA, Gontar VT (1991) Skeletal composition of cold-water bryozoans. Biol Morya 1:80–90
Comiso J (2010) Fundamental characteristics of the Polar Oceans and their sea ice cover. In: Polar Oceans from space, atmospheric and oceanographic sciences library 41, doi:10.1007/978-0-387-68300-3_2
Cottier F, Tverberg V, Inall M, Svendsen H, Nilsen F, Griffiths C (2005) Water mass modification in an Arctic fjord through cross-shelf exchange: the seasonal hydrography of Kongsfjorden, Svalbard. J Geophys Res 110:1–18
David C, Keckhut P, Armetta A, Jumelet J, Snels M, Marchand M, Bekki S (2010) Radiosonde stratospheric temperatures at Dumont d'Urville (Antarctica): trends and link with polar stratospheric clouds. Atmos Chem Phys 10:3813–3825
Davis KJ, Dove PM, Yoreo D (2000) The role of $Mg^{2+}$ as an impurity in calcite growth. Science 290:1134–1137
Dinnebier RE (1987) Nonius gufi software. http://www2.fkf.mpg.de/xray/html/gufi_software.html
Feely RA, Sabine CL, Lee K, Berelson W, Kleypas J, Fabry VJ, Millero FJ (2004) Impact of anthropogenic $CO_2$ on the $CaCO_3$ system in the oceans. Science 305:362–366
Haarpaintner J, Gascard J-C, Haugan PM (2001) Ice production and brine formation in Storfjorden, Svalbard. J Geophys Res 106:14001–14013
Hayward PJ (1995) Antarctic cheilostomatous Bryozoa. Oxford Science, New York
Hoegh-Guldberg O, Bruno JF (2010) The impact of climate change on the world's marine ecosystems. Science 328:1523–1530
Jacobson MZ (2005) Studying ocean acidification with conservative, stable numerical schemes for nonequilibrium air-ocean exchange and ocean equilibrium chemistry. J Res Atmos 110:17
Kaiser MJ, Attrill MJ, Jennings S (2005) Polar regions. In: Kaiser MJ, Attrill MJ, Jennings S, Thomas DN, Barnes DKA, Brierley AS, Polunin NVC, Raffaelli DG, Williams PJ le B (eds) Marine ecology: processes, systems and impacts. Oxford University Press, Oxford
Kluge GA (1975) Bryozoa of the Northern Seas of the USSR. Amerind, New Delhi
Koubbi P, Ozouf-Costaz C, Goarant A, Moteki M, Hulley P-A, Causse R, Dettai A, Duhamel G, Pruvost P, Tavernier E, Post AL, Beaman RJ, Rintoul SR, Hirawake T, Hirano D, Ishimaru T, Riddle M, Hosie G (2010) Estimating the biodiversity of the East Antarctic shelf and oceanic zone for ecoregionalisation: example of the ichthyofauna of the CEAMARC (Collaborative East Antarctic Marine Census) CAML surveys. Polar Science 4(2): 115–133
Kuklinski P, Bader B (2007) Comparison of bryozoan assemblages from two contrasting Arctic shelf regions. Estuar Coast Shelf Sci 73:835–843
Kuklinski P, Taylor PD (2008) Are bryozoans adapted for living in the Arctic? In: Hageman SJ, Key MM Jr, Winston JE (eds) Bryozoan studies 2007. Proceedings of the 14th International Bryozoology Association conference, Boone, North Carolina, 1–8 July 2007. Virginia Mus of Nat Hist Publ 15:101–110
Kuklinski P, Taylor PD (2009) Mineralogy of Arctic bryozoan skeletons in a global context. Facies 55:489–500
Lombardi C, Cocito S, Hiscock K, Occhipinti-Ambrogi A, Setti M, Taylor PD (2008) Influence of seawater temperature on growth bands, mineralogy and carbonate production in a bioconstructional bryozoan. Facies 54:333–342
Lowenstam HA (1954) Environmental relations of modification compositions of certain carbonate secreting marine invertebrates. Proc Natl Acad Sci U S A 40:39–48
Morse JW (2002) The dissolution kinetics of major sedimentary carbonate minerals. Earth-Sci Rev 58:51–84
Orr JC, Fabry VJ, Aumont O (2005) Anthropogenic ocean acidification over the twenty-first century and its impact on calcifying organisms. Nature 437:681–686

Poluzzi A, Sartori R (1975) Report on the carbonate mineralogy of Bryozoa. Doc Lab Geol Fac Sci Lyon 3:193–210
Ries JB (2005) Aragonite production in calcite seas: effect of seawater Mg/Ca ratio on the calcification and growth of the calcareous alga *Penicillus capitatus*. Paleobiology 31:445–458
Ries JB, Cohen AL, McCorkle DC (2010) A nonlinear calcification response to $CO_2$-induced ocean acidification by the coral *Oculina arbuscula*. Coral Reefs. doi:10.1007/S00338-010-0632-3
Saundray Y, Bouffandreau M (1958) Sur la composition chimique du systeme tégumentaire du quelques Bryozaires. Bull Inst Océogr Monaco 1119:1–13
Schäfer P, Bader B (2008) Geochemical composition and variability in the skeleton of the bryozoan *Cellaria sinuosa* (Hassall): biological versus environmental control. In: Hageman SJ, Key MM Jr, Winston JE (eds) Bryozoan studies 2007. Proceedings of the 14th International Bryozoology Association conference, Boone, North Carolina, 1–8 July 2007. Virginia Mus of Nat Hist Publ 15:269–279
Skogseth R, Haugan PM, Jakobsson M (2005) Watermass transformations in Storfjorden. Cont Shelf Res 25:667–695
Smith AM (2009) Bryozoans as southern sentinels of ocean acidification. Mar Freshw Res 60:475–482
Smith AM, Girvan E (2010) Understanding a bimineralic bryozoan: skeletal structure and carbonate mineralogy of *Odontionella cyclops* (Foveolariidae: Cheilostomata: Bryozoa) in New Zealand. Palaeogeogr Palaeclimatol Palaeoecol 289:113–122
Smith AM, Nelson CS, Spencer HG (1998) Skeletal carbonate mineralogy of New Zealand bryozoans. Mar Geol 151:26–27
Smith AM, Key M, Gordon DP (2006) Skeletal mineralogy of bryozoans: taxonomic and temporal patterns. Earth-Sci Rev 78:287–306
SPSS Inc (1995) SigmaStat for windows
Taylor PD, James NP, Bone Y, Kuklinski P, Kyser TK (2009) Evolving mineralogy of cheilostome bryozoans. Palaios 24:440–452
Turley C, Roberts JM, Guinotte JM (2007) Corals in deep-water: will the unseen hand of ocean acidification destroy cold-water ecosystems? Coral Reefs 26:445–448
Wood HL, Spicer JI, Widdicombe S (2008) Ocean acidification may increase calcification rates, but at a cost. Proc R Soc B 275:1767–1773

# Chapter 13
# Seagrass-Associated Bryozoan Communities from the Late Pliocene of the Island of Rhodes (Greece)

## Pliocene Seagrass Bryozoans

**Pierre Moissette**

**Abstract** On the island of Rhodes (Greece) the late Pliocene shallow-water siliciclastic deposits of the Kritika Member (Rhodes Formation) contain well-preserved leaves and rhizomes of the Mediterranean endemic marine phanerogam *Posidonia oceanica*. The leaf moulds are found on the bedding planes of fine-grained sands, whereas the rhizomes are preserved in life position within coarse-grained sediments.

The bryozoan community occurring either on the leaf moulds or as zoarial fragments in the sediment containing them consists of 25 species. Most of them belong to the membraniporiforms, while the cellariiforms are represented by numerous segments. Rare fragments of adeoniform and vinculariiform colonies also occur.

The community of bryozoans associated with the rhizomes comprises 49 species. The most common fragments belong to an adeoniform species, whereas membraniporiforms and cellariiforms are only moderately abundant. The vinculariiforms and celleporiforms are relatively uncommon, whereas the catenicelliforms, lunulitiforms and reteporiforms are scarce.

In the studied beds, bryozoans are the most diverse group of invertebrates and are represented by a total of 58 species, most of them extant. A comparison with present-day *Posidonia oceanica* meadows shows that many species are common to both seagrass communities. Three extant species are even obligative epiphytes of *P. oceanica* leaves. Several other species, although not associated exclusively with this marine phanerogam, generally thrive on its leaves and rhizomes.

**Keywords** Seagrass • Bryozoans • Pliocene • Rhodes • Greece • Mediterranean

---

P. Moissette (✉)
UMR 5276 CNRS, Laboratoire de Géologie de Lyon, Université Lyon 1, Géode, 69622 Villeurbanne Cedex, France
e-mail: Pierre.Moissette@univ-lyon1.fr

A. Ernst, P. Schäfer, and J. Scholz (eds.), *Bryozoan Studies 2010*,
Lecture Notes in Earth System Sciences 143, DOI 10.1007/978-3-642-16411-8_13,

## Introduction

Marine phanerogams appeared during the Late Cretaceous. Since then, together with their associated communities, they have been important components of temperate to tropical shallow-water marine environments. Due to their poor preservation potential, seagrasses are, however, rarely fossilized (Fritel 1909, 1913; Laurent and Laurent 1926; Voigt and Domke 1955; Radócz 1972; Voigt 1981; Di Geronimo 1984; Lumbert et al. 1984; Ivany et al. 1990; Froede 2002; Moissette et al. 2007).

Nevertheless, the late Pliocene shallow-water siliciclastic deposits of the Greek Island of Rhodes contain exceptionally well-preserved moulds of the leaves and casts of the rhizomes of *Posidonia oceanica*, an endemic Mediterranean seagrass. Associated with these fossil remains are relatively abundant and diverse skeletal organisms, among them bryozoans.

The three main purposes of this paper are to: (1) document the palaeontological characteristics of the bryozoan communities associated with the Pliocene seagrass leaves and rhizomes from Rhodes; (2) compare these results with those reported from the present-day Mediterranean *Posidonia oceanica* meadows; (3) reconstruct the palaeoenvironment in which these fossil communities developed.

## Geological Setting

The island of Rhodes belongs to the eastern margin of the Hellenic arc subduction zone marking the boundary between the African and the Eurasian plates (Fig. 13.1a). The basement rocks are mostly metamorphosed Mesozoic limestones and Tertiary carbonate and siliciclastic deposits. Pliocene-Pleistocene sediments (Fig. 13.1b) were deposited with complex facies patterns within a series of palaeovalleys along the coast of the island (Hanken et al. 1996; Cornée et al. 2006a).

The tectono-sedimentary organization of these deposits was studied by a number of authors (e.g., Mutti et al. 1970; Meulenkamp et al. 1972; Hanken et al. 1996; Nelson et al. 2001; Cornée et al. 2006a, b; Joannin et al. 2007; Titschack et al. 2008). Four formations, each comprising several members, are now recognised (Cornée et al. 2006a; Titschack et al. 2008): the Rhodes Formation, the Ladiko-Tsampika Formation, the Lindos Acropolis Formation, and the Plimiri Aeolianite Formation.

At the base of the Rhodes Formation, the Kritika Member (late Pliocene) comprises mostly siliciclastic deposits: clays and silts, sands and sandstones, gravels and pebbles. Fossil seagrass remains were found in several levels of two sections (Fig. 13.2) situated near the coastal village of Kritika (Moissette et al. 2007). Close to the base of one of the sections some of the clayey and fine-grained sandy beds yielded numerous *Posidonia oceanica* remains in the form of leaf moulds (Fig. 13.3). Seagrass rhizomes were found in both sections in coarse-grained layers, in the form of reddish vertical branches (Figs. 13.4, 13.5).

The skeletal organisms associated with these seagrass fossil remains are relatively abundant and diverse: crustose and geniculate coralline algae, benthic

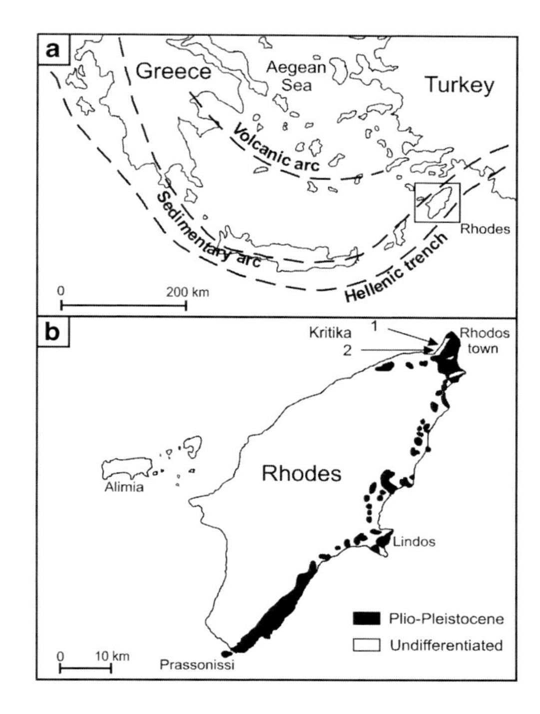


**Fig. 13.1** Location map of study area. (**a**) Location map of the island of Rhodes at the eastern margin of the Hellenic arc. (**b**) Simplified geological map of the island of Rhodes and location of the studied sections

foraminifers, serpulid worms, bivalves, gastropods, chitons, bryozoans, ostracods, crabs, and echinoids (Moissette et al. 2007).

## Material and Methods

The stratigraphic sections under study were first logged and measured (Fig. 13.2) and photographs of the outcrops were made during fieldwork.

One sample of sediment was collected from each of the three coarse-grained sand beds containing seagrass rhizomes, whereas two separate samples (loose sediment and blocks of laminated fine-grained sand or silty clay) were taken from the six levels containing leaf moulds.

The loose sediment samples were washed and sieved on a column of five sieves with diminishing mesh size (from 2 mm to 0.125 mm). After drying, fossils were picked from each fraction and an estimation of the number of specimens of each group was made. For the bryozoans, each colony fragment was counted as one specimen, although some morphotypes may break into numerous pieces (erect growth forms), whereas others (encrusting, nodular and discoidal) are mostly preserved as whole colonies.

The epiphytic organisms are found with their upper surface buried into the slightly indurated matrix containing the leaf moulds, and only their under surface is thus generally visible (Fig. 13.3). This material was prepared in the laboratory by first

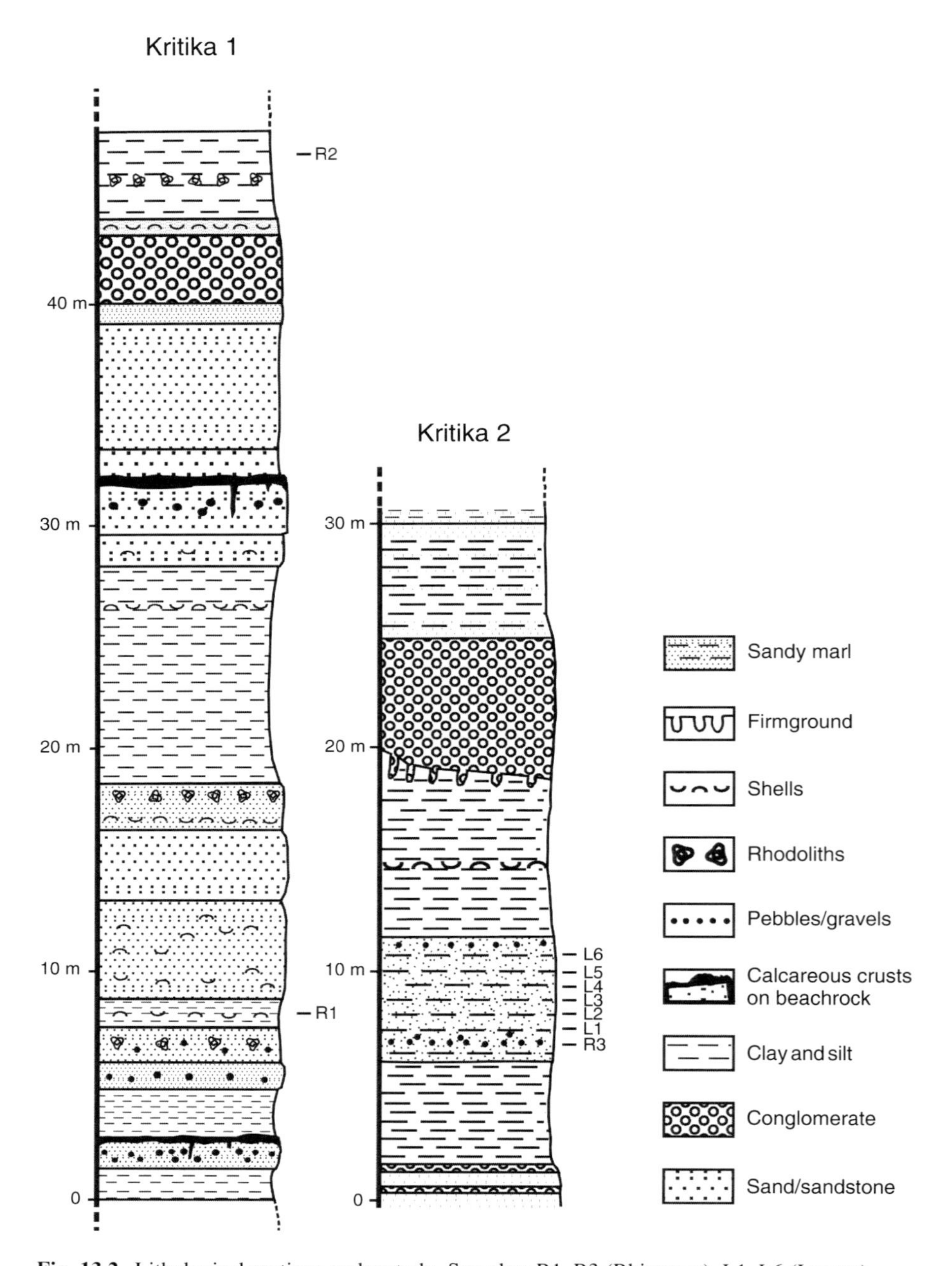


**Fig. 13.2** Lithological sections under study. Samples: R1–R3 (Rhizomes), L1–L6 (Leaves)

**Fig. 13.3** Photographs of *Posidonia oceanica* leaf moulds. (**a**) Photograph showing chaotically arranged moulds of *Posidonia oceanica* leaves in a fine-grained sandy sediment. (**b**) Detail view of a *P. oceanica* leaf mould showing parallel carbonaceous rib marks and crustose coralline algae (*light-coloured* patches). *Scale bars*: 5 mm

splitting the blocks with a knife and then peeling off the specimens from the moulds either with double-sided adhesive tape, or methyl acetate gel glue. This procedure worked relatively well in most cases, even if some of the specimens were incomplete.

Macrophotographs of a number of seagrass leaves and rhizomes were also taken in the laboratory and well-preserved specimens of bryozoans were selected and prepared for later observation and photography with a scanning electron microscope.

Bryozoan colonial morphotypes, or zoarial forms, have been correlated with various factors of the environments and their relative abundance has been shown as useful in palaeoenvironmental interpretations (e.g., Harmelin 1988; Nelson et al.

1988; Bone and James 1993; Smith 1995; Hageman et al. 1997; Moissette 2000). The bryozoans associated with the seagrass remains under study occur under eight morphotypes: membraniporiform (encrusting), celleporiform (nodular), adeoniform (flat robust branching), vinculariiform (delicate branching), reteporiform (fenestrate), cellariiform (articulated branching), catenicelliform (articulated zooidal), and lunulitiform (vagrant).

## Bryozoans Associated with the Fossil Leaves

A total of 25 species were found either encrusting the leaf moulds (Fig. 13.3) or in the fine-grained and laminated sands and silts containing them. Each of the 6 sampled levels yielded from 6 to 17 bryozoan species (Table 13.1).

With 17 species, the membraniporiforms are the most diverse, although the number of colonies is never abundant. On the contrary, the cellariiforms are represented by only 6 species, but numerous segments or fragments of segments occur (mostly belonging to the genus *Crisia*). Rare fragments of adeoniform (*Metrarabdotos moniliferum*) and vinculariiform (*Entalophoroecia deflexa*) colonies also occur.

Among the encrusting forms, four species are very fragile and were recorded as fossils for the first time by Moissette et al. (2007). Noteworthy is the fact that three of these extant species are obligative epiphytes of *Posidonia oceanica* leaves: *Electra posidoniae* (Fig. 13.6a), *Collarina balzaci* (Fig. 13.6b), and *Fenestrulina joannae* (Gautier 1954, 1962; Hayward 1975; Casola et al. 1987). The fourth species, *Haplopoma impressum* (Fig. 13.6c), lives either at depths of about 10–30 m on *P. oceanica* leaves or around 80–90 m on deep-water kelps (Gautier 1962; Hayward 1975). Several other membraniporiform species, although not associated exclusively with present-day seagrass communities, generally thrive on the leaves of this marine phanerogam (Kerneïs 1961; Gautier 1962; Harmelin 1973; Hayward 1975; Eugène 1978; Geraci and Cattaneo 1980; Casola et al. 1987). Among them, the most characteristic are probably *Aetea sica*, *Annectocyma major*, *Calpensia nobilis*, *Disporella hispida*, *Patinella radiata*, *Schizomavella auriculata*, *Tubulipora* spp., and *Watersipora subovoidea*. The extreme abundance of segments belonging to *Crisia* spp. is also remarkable.

The leave moulds are generally unordered and densely packed upon one another (Fig. 13.3a). They are preserved on the bedding planes of sands and silts intercalated with argillaceous horizons. These features indicate that sedimentation was predominantly controlled by storms and occurred after transport of the leaves into shallow-water environments, at depths less than 10 m, where they were rapidly buried.

## Bryozoans Associated with the Fossil Rhizomes

A total of 49 species (Table 13.2) were found in the coarse-grained sediment containing the *Posidonia oceanica* rhizome casts (Figs. 13.4, 13.5). Each of the 3 sampled levels yielded from 15 to 41 bryozoan species.

**Fig. 13.4** Field photograph of *Posidonia oceanica* rhizomes (*brownish* vertical stems) within a coarse-grained siliciclastic bed. *Scale bar*: 5 cm

**Fig. 13.5** *Posidonia oceanica* rhizome casts extracted from the sediment (**a**) and comparison with a present-day rhizome (**b**). *Scale bars*: 2 mm

The membraniporiform morphotype predominates in number of species, but none of the 22 species is abundant. The cellariiform segments (8 species) are relatively frequent, but the vinculariiforms (8 species) and especially the adeoniforms (4 species) are represented by numerous fragments. Among the adeoniforms, one species (*Metrarabdotos moniliferum*) largely predominates. Scarce small

**Table 13.1** Bryozoan zoarial forms and species associated with fossil *Posidonia oceanica* leaves (samples L1–L6). Faunal abundance: ● rare (1–10 specimens); ●●: present (11–20 specimens); ●●●: common (21–50 specimens); ●●●●: abundant (51–100 specimens); ●●●●●: very abundant (>100 specimens)

| Species/growth forms | L1 | L2 | L3 | L4 | L5 | L6 |
|---|---|---|---|---|---|---|
| Membraniporiform | | | | | | |
| *Aetea sica* | – | – | – | – | ● | ● |
| *Annectocyma major* | ● | – | ● | ●● | ● | ●●●● |
| *Calpensia nobilis* | ● | ● | – | ● | ● | ● |
| *Collarina balzaci* | ●● | ● | ● | ● | ● | ● |
| *Cryptosula pallasiana* | – | – | – | – | – | ● |
| *Disporella hispida* | ● | – | – | ● | ● | ● |
| *Electra posidoniae* | – | – | ● | ● | ● | ● |
| *Escharina vulgaris* | – | – | ● | – | – | – |
| *Fenestrulina joannae* | – | – | ● | – | – | – |
| *Haplopoma impressum* | ● | – | – | ● | ● | ● |
| *Patinella radiata* | ● | ● | – | ● | ● | ●●● |
| *Rhynchozoon pseudodigitatum* | – | – | – | – | ● | – |
| *Schizomavella auriculata* | – | – | – | – | ● | ● |
| *Schizoporella longirostris* | – | – | – | – | ● | – |
| *Tubulipora liliacea* | ● | – | – | ● | – | – |
| *Tubulipora plumosa* | ● | – | – | ● | ●●● | ●●●● |
| *Watersipora subovoidea* | ● | ● | – | – | – | – |
| Adeoniform | | | | | | |
| *Metrarabdotos moniliferum* | – | – | – | ● | ● | – |
| Vinculariiform | | | | | | |
| *Entalophoroecia deflexa* | – | – | – | – | ● | – |
| Cellariiform | | | | | | |
| *Crisia fistulosa* | ●●●● | ●●● | ●●● | ●●●● | ● | ●●●●● |
| *Crisia* cf. *occidentalis* | ●●●● | – | ●● | ●●●●● | ● | ●●●●● |
| *Crisia sigmoidea* | ●●● | – | ●●● | ● | – | ●●●●● |
| *Crisia* sp. 1 | – | – | – | ●●●●● | ● | ●●●●● |
| *Crisia* sp. 2 | ●●● | – | ● | ●● | – | ●● |
| *Margaretta cereoides* | – | ● | – | – | – | – |

celleporiform colonies (2 species) and catenicelliform segments also occur (2 species). One of the levels (Fig. 13.2, #R2) containing the rhizome remains also yielded a few colonies or colony fragments belonging to the lunulitiforms (2 species) and the reteporiforms (1 species).

Several species may be considered as more or less characteristic of the rhizomes of *Posidonia oceanica* (Kerneïs 1961; Gautier 1962; Harmelin 1973; Hayward 1975; Eugène 1978; Geraci and Cattaneo 1980; Casola et al. 1987). Among the membraniporiforms, this is the case of *Annectocyma major*, *Calpensia nobilis*, *Patinella radiata*, and *Tubulipora* spp. and among the vinculariiforms of *Entalophoroecia deflexa* and *Idmidronea atlantica*. The cellariiforms are mostly represented by species of the genus *Crisia*, but segments of *Cellaria salicornioides* and *Margaretta cereoides* are also relatively abundant.

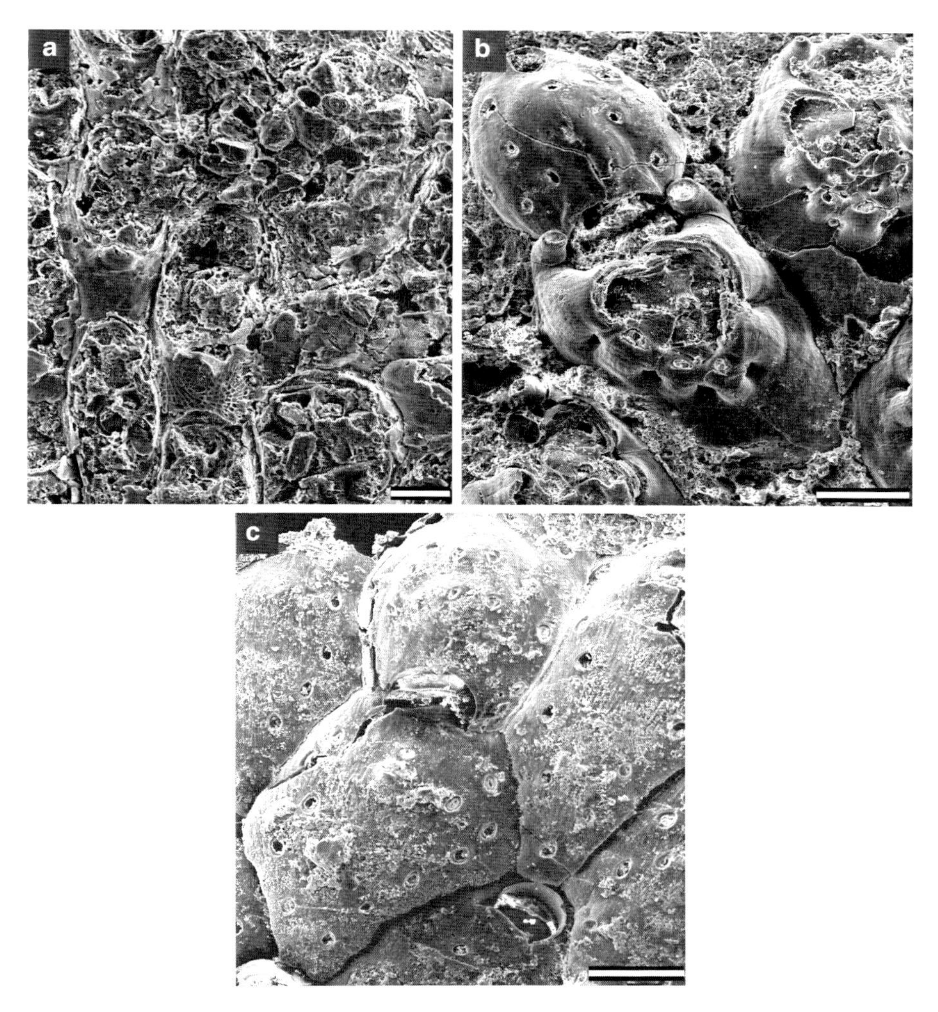


**Fig. 13.6** SEM photographs of characteristic membraniporiform bryozoans species. (**a**) *Electra posidoniae*. (**b**) *Collarina balzaci*. (**c**) *Haplopoma impressum*. *Scale bars*: 500 μm

Sample # R2 (see Fig. 13.2) is slightly different; it contains 41 bryozoan species, among them 2 lunulitiforms and 1 reteporiform, together with several deeper-water vinculariiforms (*Frondipora verrucosa*, *Idmidronea atlantica*, and especially *Tervia irregularis*).

Contrary to the fossil seagrass leaves, which were transported before accumulation and burial, the *Posidonia oceanica* rhizomes were found in life position within the sediment and the associated bryozoans may thus be considered as having lived in the same environment. Rapid burial in coarse-grained siliciclastic material was probably also controlled by storms and occurred in situ at water depths less than 20–40 m.

**Table 13.2** Bryozoan zoarial forms and species associated with fossil *Posidonia oceanica* rhizomes (samples R1–R3). Faunal abundance: ● rare (1–10 specimens); ●●: present (11–20 specimens); ●●●: common (21–50 specimens); ●●●●: abundant (51–100 specimens); ●●●●●: very abundant (>100 specimens)

| Species/growth forms | R1 | R2 | R3 |
|---|---|---|---|
| Membraniporiform | | | |
| *Annectocyma major* | ● | ●●●● | ● |
| *Buffonellaria divergens* | – | ● | – |
| *Calpensia nobilis* | ● | ●●●● | ●● |
| *Cosciniopsis ambita* | – | ●● | – |
| *Disporella hispida* | – | ●●● | – |
| *Ellisina gautieri* | – | ● | – |
| *Hagiosynodos latus* | – | ● | – |
| *Hippopleurifera pulchra* | – | ● | – |
| *Micropora coriacea* | – | ● | – |
| *Monoporella nodulifera* | – | ● | – |
| *Patinella radiata* | ● | – | ● |
| *Platonea stoechas* | – | ● | – |
| *Reptadeonella violacea* | – | ● | – |
| *Rhynchozoon neapolitanum* | – | ● | – |
| *Rhynchozoon pseudodigitatum* | ● | – | – |
| *Schizoporella longirostris* | – | – | ● |
| *Trypostega rugulosa* | – | – | ● |
| *Trypostega venusta* | – | ● | – |
| *Tubulipora liliacea* | – | ●● | ● |
| *Tubulipora plumosa* | – | ●● | ● |
| *Umbonula ovicellata* | – | – | ● |
| *Watersipora complanata* | – | – | ● |
| Celleporiform | | | |
| *Celleporina* cf. *canariensis* | – | ●●● | ● |
| *Turbicellepora coronopus* | – | ●●● | – |
| Adeoniform | | | |
| *Adeonella pallasii* | – | ●● | – |
| *Metrarabdotos moniliferum* | ●●●● | ●●●●● | ●●●●● |
| *Phoceana tubulifera* | – | ●● | – |
| *Smittina cervicornis* | – | ●●●● | – |
| Vinculariiform | | | |
| *Annectocyma arcuata* | – | ●●● | – |
| *Entalophoroecia deflexa* | ● | ●●●●● | ● |
| *Entalophoroecia gracilis* | – | ●● | – |
| *Frondipora verrucosa* | ● | ●●● | – |
| *Hornera frondiculata* | – | ●●●● | – |
| *Idmidronea atlantica* | ● | ●●●●● | – |
| *Tervia irregularis* | – | ●●●● | – |
| *Ybselosoecia typica* | ● | ●●●●● | – |
| Reteporiform | | | |
| *Reteporella* sp. | – | ●●●●● | – |

(continued)

**Table 13.2** (continued)

| Species/growth forms | R1 | R2 | R3 |
|---|---|---|---|
| Cellariiform | | | |
| *Cellaria salicornioides* | ● | ●●●●● | – |
| *Crisia fistulosa* | ●●● | ●●●●● | ●● |
| *Crisia* cf. *occidentalis* | | ●●●●● | |
| *Crisia sigmoidea* | ●●●● | ●●●● | ●● |
| *Crisia* sp. 1 | – | ●●●●● | – |
| *Crisia* sp. 2 | ● | – | ● |
| *Margaretta cereoides* | ● | ●●●●● | ●● |
| *Scrupocellaria scruposa* | – | ●●●●● | – |
| Catenicelliform | | | |
| *Halysisis diaphana* | ● | – | – |
| *Savignyella lafontii* | – | ●●●●● | – |
| Lunulitiform | | | |
| *Cupuladria canariensis* | – | ●●● | – |
| *Discoporella umbellata* | – | ●●●● | – |

## Bryozoan Communities Associated with Present-Day *Posidonia oceanica* Meadows

The marine phanerogam *Posidonia oceanica* lives at depths between 0 and 40 m throughout the Mediterranean where it often constitutes well-developed meadows. Living on the seagrass leaves and rhizomes are rich communities of algae, benthic foraminifers and invertebrates (e.g., Kerneïs 1961; Pérès and Picard 1964; Eugène 1978; Cinelli et al. 1984; Casola et al. 1987; Ribes and Gracia 1991; Balata et al. 2007).

The bryozoan faunas associated with the leaves and rhizomes of *P. oceanica* have been the subject of numerous studies (Kerneïs 1961; Gautier 1962; Pérès and Picard 1964; Harmelin 1973, 1976; Hayward 1974, 1975; Eugène 1978; Geraci and Cattaneo 1980; Zabala 1986; Casola et al. 1987). With an estimated total number of at least 140 species, the bryozoans are probably the most diverse group of organisms associated with the present-day Mediterranean *P. oceanica* meadows.

The membraniporiforms generally predominate, especially on *P. oceanica* leaves, but also often encrust the rhizomes. Apart from the four more or less obligative epiphytic species of *P. oceanica* leaves (*Collarina balzaci*, *Electra posidoniae*, *Fenestrulina joannae*, and *Haplopoma impressum*), the encrusting *Calpensia nobilis* often develops on the rhizomes of *P. oceanica* (Romero Colmenero and Sánchez Lizaso 1999; Cigliano et al. 2007). Among the erect taxa *Entalophoroecia deflexa*, *Cellaria salicornioides*, *Margaretta cereoides*, *Reteporella* spp., and *Scrupocellaria* spp. occur sufficiently frequently to be considered as specific to *P. oceanica* (Harmelin 1973; Hayward 1975; Geraci and Cattaneo 1980; Casola et al. 1987).

## Discussion

In the studied seagrass beds, bryozoans are the most diverse group of invertebrates and are represented by a total of 58 species, most of them (55) extant. A comparison with present-day *Posidonia oceanica* meadows shows that many species are common to both seagrass communities. Three extant species are even obligative epiphytes of *P. oceanica* leaves and a fourth species lives either on the leaves of this seagrass or on deep-water kelps (Gautier 1962; Hayward 1975). Several other species, although not associated exclusively with this marine phanerogam, generally thrive on its leaves and rhizomes.

However, a number of bryozoan taxa that are not strictly epiphytic, but which commonly occur in Recent *P. oceanica* meadows (and also in many Plio-Pleistocene deposits in Rhodes) are missing. This is the case of membraniporiform species such as *Chorizopora brongniartii*, *Cribrilaria innominata*, *Copidozoum tenuirostre*, *Fenestrulina malusii*, *Microporella ciliata*, *Puellina gattyae*, *Schizomavella discoidea*, *Schizotheca fissa* (Kerneïs 1961; Harmelin 1973; Hayward 1975; Eugène 1978; Geraci and Cattaneo 1980). Other encrusting or erect species are rare (or found in only one sample) in the studied material: *Escharina vulgaris*, *Cellaria* spp., *Reteporella* spp., *Scrupocellaria* spp., *Schizoporella* spp., *Schizomavella* spp.

Excepted for sample #R2, the absence or scarcity of these taxa suggests less than optimal environmental conditions, probably in shallow and turbid waters at depths of about 10–20 m. This is confirmed by the siliciclastic nature and the sedimentary structures of the deposits containing the seagrass remains (Moissette et al. 2007).

## Conclusions

This study documents the spectacular preservation of Pliocene *Posidonia oceanica* leaves and rhizomes within siliciclastic sediments. Associated with these seagrass remains are abundant and diverse epiphytic skeletal organisms, among them a few species without previous fossil record (Moissette et al. 2007). The exceptional preservation of these communities suggests rapid burial after minimal transport (for the leaves) or in situ growth and fossilization (for the rhizomes).

Bryozoans dominate the fauna and exhibit a relatively high biodiversity (58 species) with a marked increase in number of species/colonies relative to neighbouring (over- or underlying) soft bottom facies.

Although slightly less diverse than present-day *Posidonia oceanica* communities, the fossil bryozoan assemblages show distinctive characteristics with a number of species being considered specific or even obligative epiphytes of this marine phanerogam.

**Acknowledgments** I wish to thank Paula Desvignes for preparing the material, Arlette Armand for her help with scanning electron microscopy, Noël Podevigne for taking the macrophotographs, and Annie Dureu for logistical support during fieldwork. Yvonne Bone and Maja Novosel provided helpful reviews of an earlier version of the manuscript.

# References

Balata D, Nesti U, Piazzi L, Cinelli F (2007) Patterns of spatial variability of seagrass epiphytes in the north-west Mediterranean Sea. Mar Biol 151:2025–2035

Bone Y, James NP (1993) Bryozoans as carbonate sediment producers on the cool-water Lacepede Shelf, Southern Australia. Sediment Geol 86(3–4):247–271

Casola E, Scardi M, Mazzella L, Fresi E (1987) Structure of the epiphytic community of *Posidonia oceanica* leaves in a shallow meadow. PSZNI Mar Ecol 8(4):285–296

Cigliano M, Cocito S, Gambi MC (2007) Epibiosis of *Calpensia nobilis* (Esper) (Bryozoa: Cheilostomida) on *Posidonia oceanica* (L.) Delile rhizomes: effects on borer colonization and morpho-chronological features of the plant. Aquat Bot 86:30–36

Cinelli F, Cormaci M, Furnari G, Mazzella L (1984) Epiphytic macroflora of *Posidonia oceanica* (L.) Delile leaves around the island of Ischia (Gulf of Naples). In: Boudouresque CF, Jeudy de Grissac A, Olivier J (eds) International workshop *Posidonia oceanica* beds. GIS Posidonie, Marseille, pp 91–99

Cornée JJ, Moissette P, Joannin S, Suc JP, Quillévéré F, Krijgsman W, Hilgen F, Koskeridou E, Münch P, Lécuyer C, Desvignes P (2006a) Tectonic and climatic controls on coastal sedimentation: the Late Pliocene-Middle Pleistocene of northeastern Rhodes, Greece. Sediment Geol 187:159–181

Cornée JJ, Münch P, Quillévéré F, Moissette P, Vasiliev I, Krijgsman W, Verati C, Lécuyer C (2006b) Timing of Late Pliocene to Middle Pleistocene tectonic events in Rhodes (Greece) inferred from magneto-biostratigraphy and $^{40}Ar/^{39}Ar$ dating of a volcaniclastic layer. Earth Planet Sci Lett 250(1–2):281–291

Di Geronimo I (1984) Livelli a *Posidonia* nel Pleistocene inferiore della Sicilia. In: Olivier J, Boudouresque CF, Jeudy de Grissac A (eds) International workshop *Posidonia oceanica* beds. GIS Posidonie, Marseille, pp 15–21

Eugène C (1978) Etude de l'épifaune des herbiers de *Posidonia oceanica* (L.) Delile du littoral provençal. Unpublished Ph.D. Thesis, University of Marseille, Marseille, pp 141

Fritel PH (1909) Sur l'attribution au genre *Posidonia* de quelques *Caulinites* de l'Eocène du bassin de Paris. Bull Soc Géol Fr 9:380–385

Fritel PH (1913) Sur les Zostères du Calcaire grossier et sur l'assimilation au genre *Cymodoceites* Bureau des prétendues algues du même gisement. Bull Soc Géol Fr 13:354–358

Froede CR Jr (2002) Rhizolith evidence in support of a late Holocene sea-level highstand at least 0.5 m higher than present at Key Biscayne, Florida. Geol 30(3):203–206

Gautier YV (1954) Sur l'*Electra pilosa* des feuilles de Posidonies. Vie et Milieu 1:66–70

Gautier YV (1962) Recherches écologiques sur les bryozoaires chilostomes en Méditerranée occidentale. Recl Trav Station Marine d'Endoume 24(38):1–434

Geraci S, Cattaneo R (1980) Il popolamento a briozoi (Cheilostomata) della prateria a *Posidonia* di Procchio (Isola d'Elba). Ann Mus Civ Storia Nat Giacomo Doria di Genova 83:107–125

Hageman SJ, Bone Y, McGowran B, James NP (1997) Bryozoan colonial growth-forms as paleoenvironmental indicators: evaluation of methodology. Palaios 12:405–419

Hanken NM, Bromley RG, Miller J (1996) Plio-Pleistocene sedimentation in coastal grabens, north-east Rhodes, Greece. Geol J 31(3):271–296

Harmelin JG (1973) Bryozoaires de l'herbier de Posidonies de l'île de Port-Cros. Rap Comm Int Mer Méditerr 21(9):675–677

Harmelin JG (1976) Le sous-ordre des Tubuliporina (Bryozoaires Cyclostomes) en Méditerranée. Écologie et systématique. Mém de l'Inst Océanogr 10:1–326
Harmelin JG (1988) Les Bryozoaires, de bons indicateurs bathymétriques en paléoécologie ? Géol Méditerr 15(1):49–63
Hayward PJ (1974) Studies on the cheilostome bryozoan fauna of the Aegean island of Chios. J Nat Hist 8:369–402
Hayward PJ (1975) Observations on the bryozoan epiphytes of *Posidonia oceanica* from the island of Chios (Aegean Sea). In: Pouyet S (ed) Bryozoa 1974. Documents des laboratoires de géologie Lyon: H. S 3(2), pp 347–356
Ivany LC, Portell RW, Jones DS (1990) Animal-plant relationships and paleobiogeography of an Eocene seagrass community from Florida. Palaios 5:244–258
Joannin S, Cornée JJ, Moissette P, Suc JP, Koskeridou E, Lécuyer C, Buisine C, Kouli E, Ferry S (2007) Changes in vegetation and marine environments in the eastern Mediterranean (Rhodes Island, Greece) during the Early and Middle Pleistocene. J Geol Soc London 164:1119–1131
Kerneïs A (1961) Contribution à l'étude faunistique et écologique des herbiers de Posidonies de la région de Banyuls. Vie et Milieu 11(2):145–187
Laurent L, Laurent J (1926) Étude sur une plante fossile des dépôts du Tertiaire marin du sud des Célèbes, *Cymodocea micheloti* (Wat.) nob. Jaarbuch Mijnwezen Nederland und Indie 54:167–190
Lumbert SH, Den Hartog C, Phillips RC, Olsen FS (1984) The occurrence of fossil seagrasses in the Avon Park Formation (late Middle Eocene), Levy County, Florida (U.S.A.). Aquat Bot 20 (1-2):121–129
Meulenkamp JE, De Mulder EFJ, Van De Weerd A (1972) Sedimentary history and paleogeography of the Late Cenozoic of the Island of Rhodos. Z Dtsch Geol Ges 123:541–553
Moissette P (2000) Changes in bryozoan assemblages and bathymetric variations. Examples from the Messinian of northwest Algeria. Palaeogeogr Palaeclimatol Palaeoecol 155:305–326
Moissette P, Koskeridou E, Cornée JJ, Guillocheau F, Lécuyer C (2007) Spectacular preservation of seagrasses and seagrass-associated communities from the Pliocene of Rhodes, Greece. Palaios 22:200–211
Mutti E, Orombelli G, Pozzi R (1970) Geological studies on the Dodecanese Islands (Aegean Sea). IX. Geological map of the island of Rhodes (Greece); explanatory notes. Ann Géol Pays Héll 22:79–226
Nelson CS, Hyden FM, Keane SL, Leask WL, Gordon DP (1988) Application of bryozoan zoarial growth-form studies in facies analysis of non-tropical carbonate deposits in New Zealand. Sediment Geol 60(1–4):301–322
Nelson CS, Freiwald A, Titschack J, List S (2001) Lithostratigraphy and sequence architecture of temperate mixed siliciclastic-carbonate facies in a new Plio-Pleistocene section at Plimiri, Rhodes Island (Greece). Occas Report 25. Department of Earth Science, University of Waikato, Hamilton, pp 1–55
Pérès JM, Picard J (1964) Nouveau manuel de bionomie benthique de la Mer Méditerranée. Rec Trav Station marine d'Endoume 31(47):1–137
Radócz G (1972) *Zostera*-bryozoa-*Spirorbis* biocönózis a borsodi miocénből (A *Zostera*-bryozoa-*Spirorbis* biocoenosis from the Miocene of the Borsod Basin). Magyar Állami Földtani Intézet Évi Jelentése, pp 55–63 (in Hungarian, English abstract)
Ribes T, Gracia MP (1991) Foraminifères des herbiers de posidonies de la Méditerranée occidentale. Vie et Milieu 41(2/3):117–126
Romero Colmenero L, Sánchez Lizaso JL (1999) Effects of *Calpensia nobilis* (Esper 1796) (Bryozoa: Cheilostomida) on the seagrass *Posidonia oceanica* (L.) Delile. Aquat Bot 62 (4):217–223
Smith AM (1995) Palaeoenvironmental interpretation using bryozoans: a review. Geol Soc London Spec Publ 83: 231–243

Titschack J, Nelson CS, Beck T, Freiwald A, Radtke U (2008) Sedimentary evolution of a Late Pleistocene temperate red algal reef (Coralligène) on Rhodes, Greece: correlation with global sea-level fluctuations. Sediment 55:1747–1776

Voigt E (1981) Upper Cretaceous bryozoan-seagrass association in the Maastrichtian of The Netherlands. In: Larwood GP, Nielsen C (eds) Recent and fossil Bryozoa, Olsen & Olsen, Fredensborg, Denmark

Voigt E, Domke W (1955) *Thalassocharis bosqueti* Debey ex Miquell, ein strukturell erhaltenes Seagrass aus der holländischen Kreide. Mitt Geol Staatsinst Hambg 24:87–102

Zabala M (1986) Fauna dels briozous dels països catalans, vol 84, Arx Sec Ciénc. Inst d'Estud Catalans, Barcelona, pp 1–836

# Chapter 14
# A New Species of the Genus *Electra* (Bryozoa, Cheilostomata) from Southern Oman, Arabian Sea

## A New Species of Bryozoa from Southern Oman

**Elena A. Nikulina, Andrew N. Ostrovsky, and Michel Claereboudt**

**Abstract** A new cheilostome bryozoan *Electra omanensis* n. sp. is described in Southern Oman (Dhofar Region, Arabian Sea) and is also the first description of any bryozoan from this area. The new species morphologically most resembles *Electra indica* (Menon and Nair (Mar Biol Ass India 17:553–579, 1975), but differs in (1) colony shape – colonies consist of long narrow branching strips, and (2) in the shape of the gymnocystal spines. The combination of these features is sufficiently distinctive and allows discriminating the new species from all other known species in this genus. Genetic analyses of nuclear and mitochondrial ribosomal genes also support separate species status for this species of *Electra*.

**Keywords** Biodiversity • Molecular systematics • Taxonomy

## Introduction

Among malacostegan cheilostome bryozoans, the genus *Electra* is the most speciose, and includes currently more than 20 predominantly shallow-water taxa (Nikulina et al. 2007). Although the genus is known from the Maastrichtian (Taylor

---

E.A. Nikulina (✉)
Zentrum für Baltische und Skandinavische Archäologie, Stiftung Schleswig-Holsteinische Landesmuseen, Schloss Gottorf, D-24837 Schleswig, Germany
e-mail: elena.nikulina@schloss-gottorf.de

A.N. Ostrovsky
Department of Invertebrate Zoology, St. Petersburg State University, Universitetskaja nab. 7/9, St. Petersburg 199034, Russia

Department of Palaeontology, University of Vienna, Althanstrasse 14, A-1090 Vienna, Austria

M. Claereboudt
Department of Marine Science and Fisheries, College of Agricultural and Marine Sciences, Sultan Qaboos University, PO Box 34, Al-Khod 123, Sultanate of Oman

A. Ernst, P. Schäfer, and J. Scholz (eds.), *Bryozoan Studies 2010*,
Lecture Notes in Earth System Sciences 143, DOI 10.1007/978-3-642-16411-8_14,

and McKinney 2006), light calcification would have diminished fossilisation success (Nikulina and Taylor 2010); the vast majority of the known species are extant.

Recent studies show that the recognised number of *Electra* species is underestimated (Nikulina 2006, 2008a), but the high morphological similarity between sibling species, together with their ecophenotypic variability, often makes their differentiation difficult. Species identification was especially problematic in pre-SEM times when most of the species in this genus were described, and a revision of the genus using modern techniques is clearly necessary. For instance, combined morphological and molecular approaches revealed that *Electra pilosa* (Linnaeus, 1767), formerly considered as cosmopolitan is actually a complex, consisting of at least three distinct species (Nikulina et al. 2007; Nikulina 2008b).

Here we report on the finding of the first species from this group – *Electra omanensis* sp. nov. – recently discovered in the coastal waters of Southern Oman. This is the first description of any bryozoan from this area.

During field work in Southern Oman in January 2009, more than 120 bryozoan species were collected (Ostrovsky, Cáceres-Chamizo, Claereboudt, unpublished). Identification and descriptive work is currently in progress. (see http://palse2.pal.univie.ac.at/Bryozoa/Oman/Oman.html)

## Materials and Methods

### Material

This new species of *Electra* was found growing on fronds of the large seaweed *Nizamuddinia zanardinii* (Schiffner, 1934) and on non-identified red algae in Salalah, near Mirbat, Dhofar Region, Southern Oman (Arabian Sea). Samples were collected by SCUBA diving at 3–5 m depth by A.N. Ostrovsky and M. R. Claereboudt on 16.01.2090, 19.01.2010 and 23.01.2010.

For scanning electron microscopy (SEM), specimens were cleaned in a 7.5% solution of sodium hypochlorite, rinsed, air-dried, and coated with either gold or palladium-platinum. Some non-cleaned specimens were also coated with gold and scanned. Images were made using CAMSCAN-Serie-2-CS-44 (Kiel) and Jeol JSM-6400 (Vienna).

For molecular genetics, material was preserved in 98% ethanol and stored at 4°C. Beside the *Electra* from Oman, nine other electrids were involved in the study: *Electra pilosa* (Linnaeus, 1767), *Electra posidoniae* Gautier, 1954, *Electra scuticifera* Nikulina, 2008b, *Einhornia crustulenta* (Pallas, 1766), *Einhornia arctica* (Borg, 1931), *Einhornia moskvikvendi* (Nikulina, 2008a), *Einhornia korobokkura* (Nikulina, 2006), and *Aspidelectra melolontha* (Landsborough, 1852). This material was partially genetically studied previously (Nikulina 2006; Nikulina and Schäfer 2006; Nikulina et al. 2007; Nikulina 2008a, b). Details of the studied material are reported in Table 14.1.

**Table 14.1** Samples studied: species names, gene fragments with European Molecular Biology Laboratory (EMBL) accession numbers, and geographic information for each sample

| Species | 12S | 16S | 18S | 28S | Location |
|---|---|---|---|---|---|
| *E. omanensis* | FR754510 | FR754522 | FR754525 | FR754526 | Indian Ocean, Arabian Sea, Oman coast |
| *E. pilosa (1)* | FR754511 | AJ971066[a] | FR754527 | FR754528 | NW Atlantic, North Sea, Helgoland |
| *E. pilosa (2)* | FR754512 | AJ971065[a] | AM075768[a] | FR754529 | NW Atlantic, North Sea, Helgoland |
| *E. pilosa (3)* | FR754513 | AJ971067[a] | FR754530 | – | NW Atlantic, North Sea, Helgoland |
| *E. posidoniae (1)* | – | AJ971084[a] | AM75770[a] | – | Mediterranean, Adriatic Sea, Pab |
| *E. posidoniae (2)* | FR754514 | AJ971085[a] | AM75771[a] | – | Mediterranean, Mallorca |
| *E. scuticifera (1)* | – | AJ971068[a] | AM886854[a] | FR754531 | IW Pacific, Tasman sea, Maori Bay |
| *E. scuticifera (2)* | FR754515 | AJ971086[a] | FR754533 | FR754532 | IW Pacific, Tasman sea, West Coast |
| *E. korobokkura* | FR754516 | AJ853947[a] | AM158087[a] | – | NW Pacific coast of Hokkaido, Akkeshi Bay |
| *E. indica* | – | FR754523 | – | – | Indian Ocean, Cochin Bay |
| *E. crustulenta (1)* | – | AJ853844[a] | AM92413[a] | – | Western Baltic Sea, Lolland |
| *E. crustulenta (2)* | FR754517 | – | FR754535 | FR754536 | Arctic, White Sea, Onega Bay |
| *E. crustulenta (3)* | FR754518 | – | – | FR754537 | NW Atlantic, North Sea, Lawrentsmeer |
| *E. arctica* | – | – | AM773516[a] | – | Arctic, White Sea, Kandalaksha Bay |
| *E. moskvikvendi* | FR754520 | – | – | – | Baltic Sea, Lolland |
| *A. melolontha* | FR754519 | – | – | FR754538 | NW Atlantic, North Sea, Sylt |

[a]Depicts numbers, published in our previous studies

## Molecular Methods

Total DNA was extracted from the ethanol-preserved colonies (about 10–20 zooids) using the Qiagen DNEasy Tissue extraction kit. Fragments of mitochondrial 16S rDNA were amplified using the universal metazoan primers 16Sar and 16Sbr (Palumbi and Metz, 1991) or newly designed primers 16SF4 (CTCGGCAAAGAAGGGCTCCGCCTGTTTATCA AAAACAT-3) and 16SLr (TTCTCTTTTTCTGTTCCTTTCGTAAT-3). In addition, a fragment of the mitochondrial 12S rDNA of Cheilostomata was amplified. New primers were designed based on the sequences of Protostomia collected in the EMBL Nucleotide Sequence Database: 12SF-I (GGAAAAAATTGTGCCAGCADCCGCGGTTA-3) and 12SRLen (CACTTTCAAGTACGCCTACTGTGTTACGAC-3). Additionally, fragments of nuclear 18S and 28S rDNA were amplified using designed primers U-18-4F (AGGAGTGGAGCCTGCGGCTTAATTTGACTC-3), U-18-4R (AGGTTCACCTA-CGGAAACCTTGTT ACGAC-3); and the primer pair 6(28) (CTCTCTTAAGGTAGCC-AAATGCCTCGTC-3) and 10(28) (GTGAATTCTGCTTCATCAATGTAGGAAGAG CC-3) (Hillis et al. 1996).

PCR amplifications were performed in a total volume of 25 μl using Taq polymerase (New England Biolabs). Cycling parameters were as follows: 94°C for 2 min, followed by 35 cycles of 94°C for 30 s, 45°C (54°C for the 18S fragments) for 30 s, 72°C for 40 s, finishing with 7 min of elongation at 72°C. PCR products were purified using QIAquick PCR Purification Kits (Qiagen) and sequenced directly in both directions using the BigDye® Terminator v3.1 Cycle Sequencing Kit (Applied Biosystems) and an ABI3100 automatic sequencer.

*Sequence analyses.* Sequences were assembled and edited using SeqMan and EditSeq (DNASTAR Lasergene software). ClustalX with default settings was applied to perform automatic sequences alignment (Thompson et al. 1997). Phylogenetic analyses were undertaken using PAUP* v4.0b10 (Swofford 2000) and MrBayes 3.2.1 (Ronquist and Huelsenbeck 2003). Phylogenetic trees were obtained by the maximum parsimony, maximum likelihood, neighbour-joining, and Bayesian methods. Nodal support was estimated by a bootstrap procedure (Felsenstein 1985), with 1,000 replicates and a full heuristic search. MP analyses were conducted using a heuristic search with the tree-bisection-reconnection branch-swapping algorithm (Swofford and Olsen 1990). A first tree was obtained by random addition of sequences, 200 replicates were then generated, and ten trees were kept for a search of the most parsimonious trees. Gaps were treated as missing data.

In addition, the programme ModelTest v.3.6 (Posada and Crandall 1998) was used to find the model of DNA substitution that best fit the data (Posada and Buckley 2004). The parameters of this best-fit model selected with the Akaike Information criterion (Posada and Buckley 2004) were subsequently used for maximum likelihood and neighbour-joining analyses in PAUP* and Bayesian analyses in MrBayes. For ML analyses, we used a full heuristic search with 100 random-addition replicates and search parameters as described for the MP. We also applied Bayesian inference of phylogeny with MrBayes running four Metropolis coupled Markov Chain Monte Carlo chains (MCMC) simultaneously for 10,000,000 generations, sampling every 100th generation. Each Markov chain was started from a random tree. The MCMC output was analysed with TRACER v1.3 (Rambaut and Drummond 2004); 10% of the samples were discarded as a burn-in; 90,000 samples were used for estimating parameters, parameter variance, and posterior probabilities of particular nodes to construct a majority rule consensus tree.

Prior to combination of the datasets, a partition-homogeneity test (ILD test of Farris et al. 1994) implemented in PAUP* was conducted to evaluate phylogenetic congruence of gene fragments.

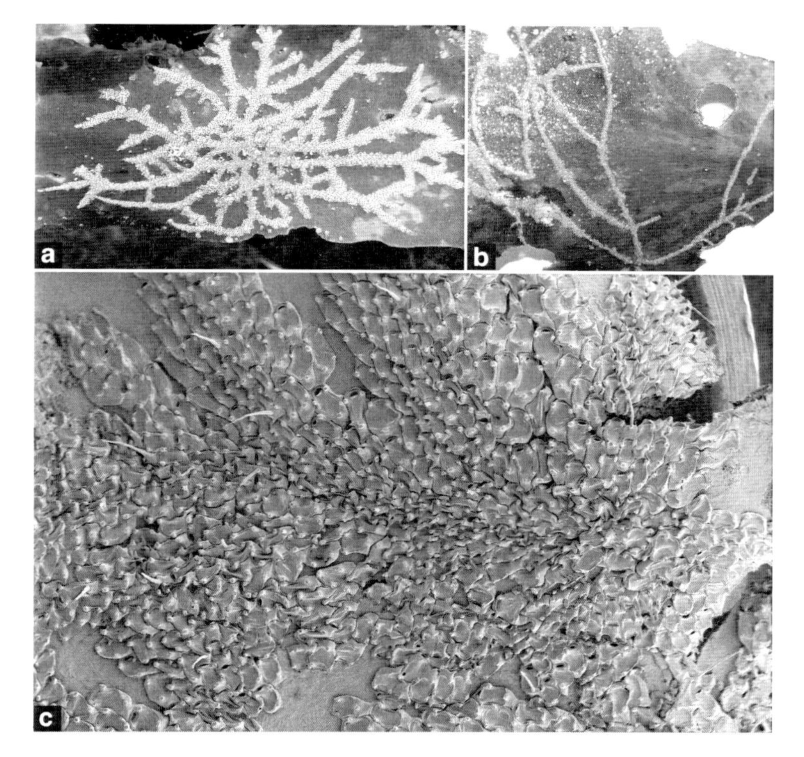


**Fig. 14.1** Colonies of *Electra omanensis* sp. nov.: (**a**, **b**) light photographs, (**c**) SEM picture

# Results

## *Systematic Account*

Suborder Malacostegina Levinsen, 1902

Family Electridae Stach, 1937

Genus *Electra* Lamouroux, 1816

*Electra omanensis* sp. nov.

Figures 14.1, 14.2, 14.3, and 14.4

Etymology: From the Sultanate of Oman, where the species was discovered.

Material: Holotype. Type SMF 1732 (Senckenberg Natural History Museum), alcohol-fixed, one part dried and coated with Pd-Pt; on red alga, 3 m, Salalah, near Mirbat, Southern Oman, Indian Ocean, 2009. Genetic data, Accession numbers in EMBL Data Base: FR754522 for 16S rDNA, FR754510 for 12S rDNA, FR754525 for 18S rDNA, and FR754526 for 28S nDNA.

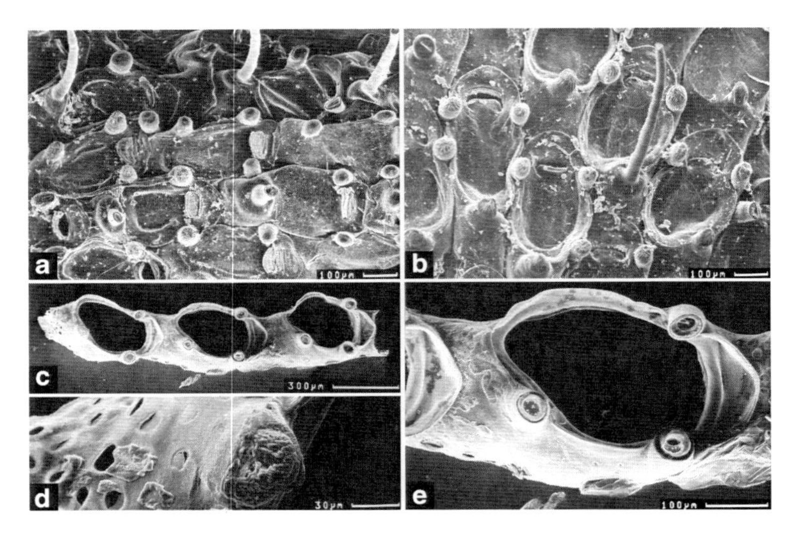


**Fig. 14.2** SEM figures of the type material: (**a**, **b**) holotype SMF 1732; (**c–e**) paratype SMF 1733, bleached zooids

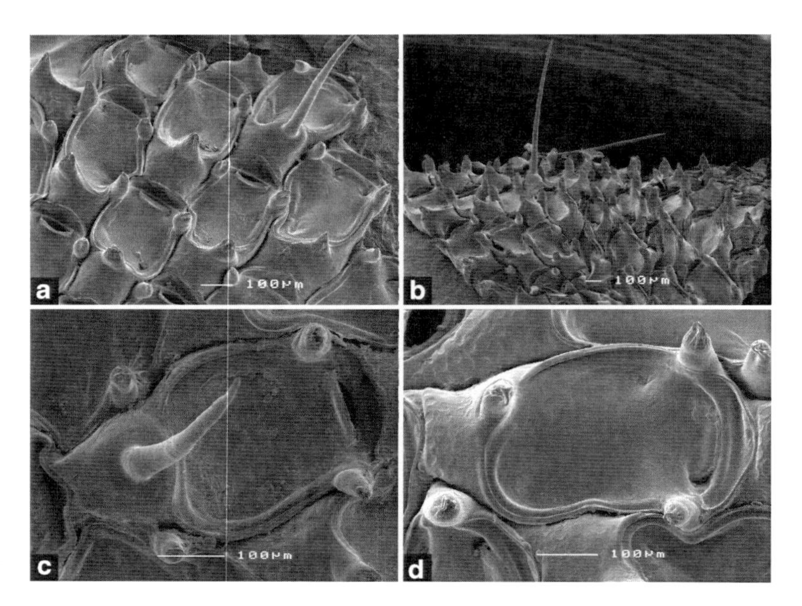


**Fig. 14.3** Zooid morphology of *Electra omanensis* sp. nov., SMF 1732: (**a**, **b**) morphology of the spines, (**c**) proximomedian and (**d**) proximolateral position of the proximal spine

Paratype: SMF 1733, bleached zooids coated with Pd-Pt; SMF 1735, fragment fixed in alcohol.

Other material: SMF 1734, colony fragment on red alga coated with Pd-Pt. Collecting data identical to that of the holotype.

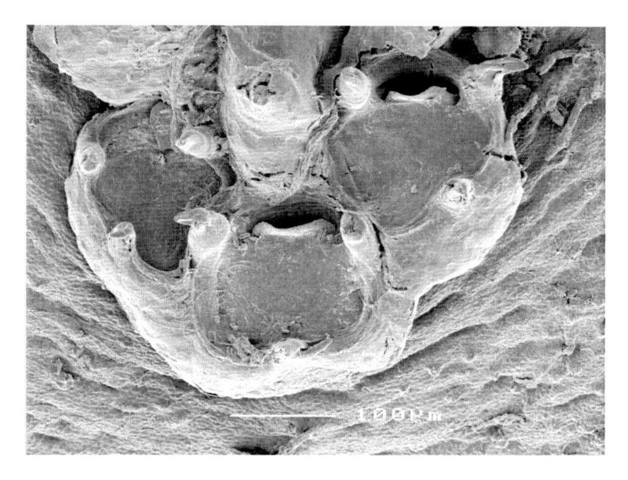

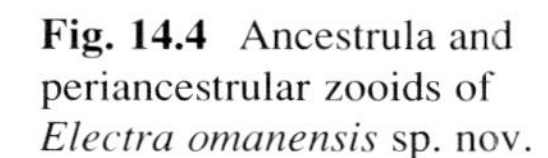

**Fig. 14.4** Ancestrula and periancestrular zooids of *Electra omanensis* sp. nov.

Diagnosis: Colony form of long narrow branching strips, moderately developed gymnocyst, which is about 1/3–1/4 zooidal length, and three gymnocystal spines. The two distolateral spines are short, thick and cylindrical, the proximal spine generally thick, short and conical, but sometimes long and setiform.

Description: Colonies encrusting macroalgae, white in colour, consisting of long strips of 3–8 zooids in width (Fig. 14.1). Strips are either parallel-sided or gradually widening distally, radiating from the colony centre and producing lateral and terminal branches of varying width and length with blunt, rounded or pointed tips. On wide algal fronds, the colonies often acquire a snow-flake shape with a dense network of anastomosing branches in its central part (Fig. 14.1a). On narrow fronds or in close proximity to a neighbouring colony, they are irregular in shape (Fig. 14.1b, c).

Zooids arranged in alternating, quincuncial series (Fig. 14.2a, b), pyriform, ovoid, rounded, rhomboid or subrectangular, typically narrower proximally. Gymnocyst occupying about 1/3–1/4 of the frontal surface, lightly calcified, translucent, with numerous round and oval pseudopores (Fig. 14.2c–e). Opesia oval, spacious, with a distinct mural rim (Fig. 14.1e). Distally this rim often extends in a shallow arch (hood-like structure when non-cleaned) recumbent on proximal gymnocyst of succeeding autozooid. Cryptocyst downward deflected, narrow and smooth. Operculum simple, with a thin chitinous sclerite. Basal wall weakly calcified, dissolving during cleaning in sodium hypochlorite (Fig. 14.2c, e).

Three spines occur around the opesia. Proximal spine normally thick, short and conical (Fig. 14.3a), but in many zooids it is slender and long (about 30 μm in diameter), setiform, attaining from half to two times the length of the zooid (Fig. 14.3b). Position of the proximal spine varies from proximomedian (Fig. 14.3c) to proximolateral (Fig. 14.3d). Distolateral spines short and thick (about 54 μm in diameter), cylindrical or slightly compressed laterally. Spines calcified, except their pointed tips, which are chitinous and brown. In dry non-cleaned specimens these tips normally shrink in radiating wrinkles. The setiform part of proximal spines is chitinous.

Ancestrula smaller than other zooids, with a reduced proximal gymnocyst. It buds three zooids, one distal and two distolateral (Fig. 14.4). Zooids are smaller in zone of astogenetic change.

*Measurements* (Sample size = 20, zooids measured in a zone of astogenetic repetition).
*Zooid length*: Mean = 488; Median = 491; Range = 400–580; SE = 12;
*Zooid width*: Mean = 244; Median = 244; Range = 216–280; SE = 5;
*Opesia length*: Mean = 442; Median = 441; Range = 391–491; SE = 9;
*Opesia width*: Mean = 253; Median = 250; Range = 209–297; SE = 8.

## Molecular Data

**16S.** The aligned ten 16S sequences had a length of 443 bp with 177 variable and 139 parsimony-informative sites. All sequences were unique, corresponding to six species: *E. omanensis* sp. nov., *E. indica*, *E. scuticifera*, *E. posidoniae*, *E. pilosa*, and *Einhornia crustulenta* as the outgroup. Equally weighted parsimony analysis yielded a single tree with a length of 283 steps; consistency index (CI) was 0.8693 and a retention index (RI) of was 0.8822. The Hasegawa-Kishino-Yano model of nucleotide substitution (HKY85) with transition/transversion ratio = 1.9467 and assumed nucleotide frequencies (A = 0.35590, C = 0.14740, G = 0.17650) was applied to conduct maximum likelihood, neighbour-joining and Bayesian analyses. All clades of the neighbour-joining, maximum likelihood, and maximum parsimony trees had identical topology. A Bayesian majority rule consensus tree was created. A Bayesian phylogram with the topology supported by all four methods is shown in Fig. 14.5a.

**12S.** The aligned eleven 12S sequences had a length of 580 bp with 383 variable and 290 parsimony-informative sites. The data set consisted of ten unique sequences, corresponding to eight species: *Electra omanensis* sp. nov., *E. scuticifera*, *E. posidoniae*, *E. pilosa*, *Einhornia crustulenta*, *E. korobokkura*, *E. moskvikvendi*, and *Aspidelectra melolontha* as the outgroup. Equally weighted parsimony analysis yielded a single tree with a length of 809 steps; CI = 0.7676 and RI = 0.7227. A general time-reversible model of nucleotide substitution (Rodriguez et al. 1990) with estimated base frequencies (A = 0.23840, C = 0.23780, G = 0.28970, and T = 0.23410) and a proportion of invariant sites of 0.9067 (GTR + I) was applied to conduct maximum likelihood analyses. The similar model was applied to construct neighbour-joining and Bayesian trees. All methods provided identical tree topology. A Bayesian phylogram with the topology supported by all four methods is shown in Fig. 14.5b.

**18S.** The aligned fourteen 18S sequences had a length of 510 bp with 58 variable and 36 parsimony-informative sites. The data set consisted of nine unique sequences, corresponding to nine species: *Electra omanensis* sp. nov., *E. scuticifera*, *E. posidoniae*, *E. pilosa*, *Einhornia crustulenta*, *E. korobokkura*,

16S rDNA

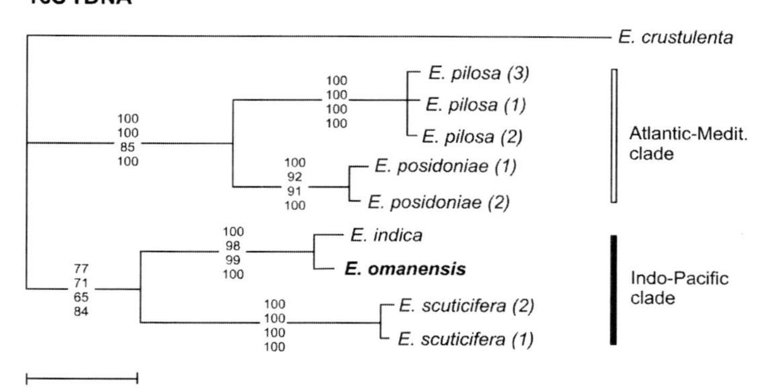


b

12S rDNA

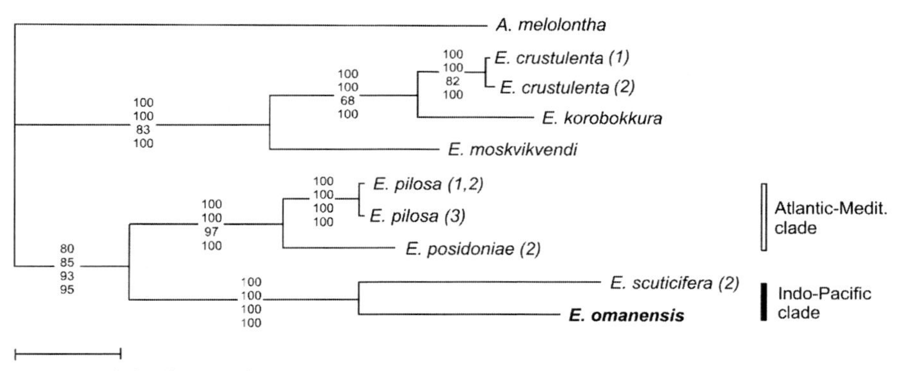


**Fig. 14.5** Phylograms constructed for the 16S and 12S rDNA fragments of the studied electrids. The topology depicts the strict consensus between MP, NJ, ML and Bayesian trees. Bootstrap values and clade posterior probabilities are given in the sequence above and provide statistical support for the particular node. Branch lengths calculated with Bayesian methods

*E. arctica*, *E. moskvikvendi*, and *Aspidelectra melolontha* as the outgroup. Equally weighted parsimony analysis yielded a single tree with a length of 81 steps; CI = 0.8889 and RI = 0.8816. A maximum likelihood and neighbour-joining analysis was completed using a HKY85 model of nucleotide substitution with gamma distribution of rates at variable sites (alpha = 0.0470), transition/transversion ratio = 1.6319 and assumed nucleotide frequencies (A = 0.1837, C = 0.2939, G = 0.2928) was applied to conduct maximum likelihood, neighbour-joining and Bayesian analyses. The maximum likelihood analysis

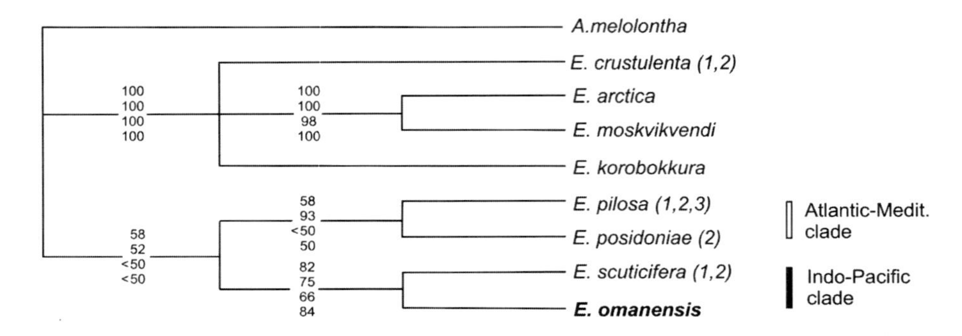


b
28S rDNA

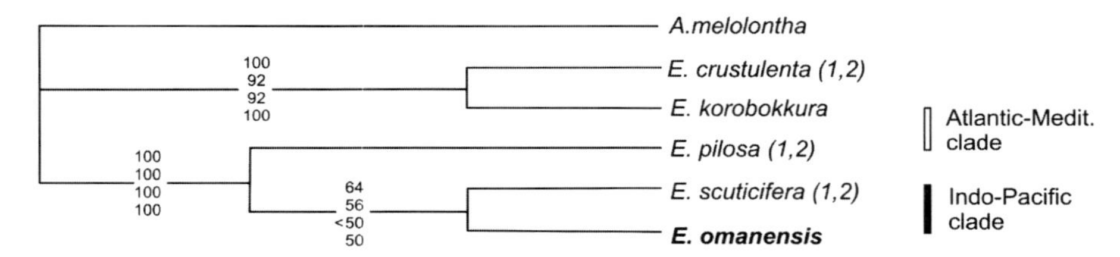


**Fig. 14.6** Cladograms constructed for the 18S and 28S rDNA fragments of the studied electrids. The topology depicts the majority-rule (**a**) or strict consensus (**b**) between MP, NJ, ML and Bayesian trees. Bootstrap values and clade posterior probabilities are given in the sequence above and provide statistical support for the particular node

resulted in a single tree, where *E. omanensis*-*E. scuticifera*-*E. pilosa*-*E. posidonia* clade was bad resolved (<50%). The same result provided Bayesian method (Bayesian support = 50%). In the rest, all applied methods generated similar trees, which not contradicted to those constructed from mitochondrial DNA. A cladogram with the topology supported by majority methods is shown in Fig. 14.6a.

**28S.** Nine 28S sequences were 670 bp long, with 78 variable and 41 parsimony-informative sites. The data set consisted of six unique sequences, corresponding to six species: *Electra omanensis* sp. nov. *E. scuticifera*, *E. pilosa*, *Einhornia crustulenta*, *E. korobokkura*, and *Aspidelectra melolontha* as the outgroup. A single maximum parsimony tree with a length of 28 steps (CI = 0.9655 and RI = 0.9524) was obtained. A maximum likelihood and neighbour-joining analysis was completed using a HKY85 model of nucleotide substitution with gamma distribution of rates at variable sites (alpha = 0.1296), transition/transversion ratio = 1.8361. Assumed nucleotide frequencies (A = 0.23210, C = 0.24510, G = 0.29080) were applied to conduct maximum likelihood, neighbour-joining and Bayesian analyses.

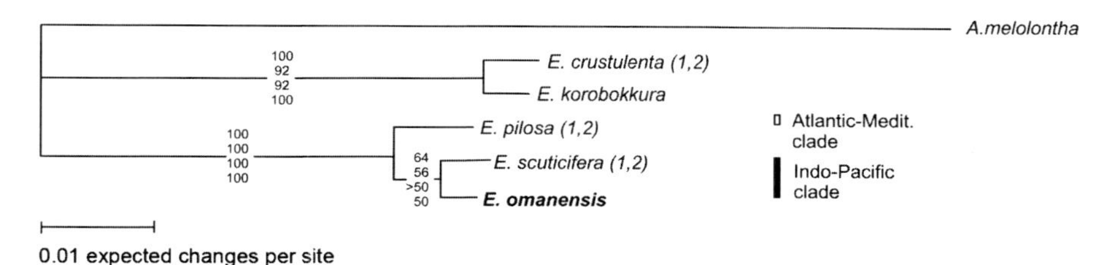


**Fig. 14.7** Phylogram constructed for the combined 18S and 28S rDNA data set, the strict consensus between MP, NJ, ML and Bayesian trees. Bootstrap values and clade posterior probabilities are given in the sequence above and provide statistical support for the particular node. Branch lengths calculated with Bayesian methods

The maximum likelihood analysis resulted in a single tree, where *E. omanensis-E. scuticifera* clade had bootstrap support <50%. The same result provided Bayesian method (Bayesian support = 50%). All applied methods generated similar trees, which did not contradict those constructed from mitochondrial DNA (Fig. 14.6b).

**18S + 28S.** Phylogenetic trees constructed for 18S and 28S provided very weak support for *E. omanensis-E. scuticifera* clade, especially in ML and Bayesian trees, therefore we combined both data sets to obtain more phylogenetic informative characters in the highly conserved DNA fragments.

Twelve sequences (six sequences of each of the two gene fragments) were combined and a data matrix of 1,180 positions with 136 variable and 77 parsimony-informative sites was obtained. The ILD test failed to reject the homogeneity assumption within the 95% probability interval (P = 1.00). The topology of the MP tree was stable to Goloboff's weighting (Goloboff 1993), providing trees with constant parameters: the length of 842 steps, CI = 0.621 and RI = 0.818. We used GK = 2 to calculate bootstrap presented in Fig. 14.7. A maximum likelihood and neighbour-joining analysis was completed using a HKY85 model of nucleotide substitution with gamma distribution of rates at variable sites (alpha = 0.1115), transition/transversion ratio = 1.5640 and assumed nucleotide frequencies (A = 0.2107, C = 0.2641, G = 0.2934) was applied to conduct maximum likelihood, neighbour-joining and Bayesian analyses. All the methods provided congruent results in respect to monophyly of the *E. omanensis-E. scuticifera-E. pilosa* and *E. omanensis-E. scuticifera* clades although with low statistical support for the latter (Fig. 14.7).

## Discussion

Our knowledge of the diversity of tropical Bryozoa is far from satisfactory. The number of newly described species is constantly growing, but even approximate estimates of the diversity of local faunas is hampered by the necessity of checking old descriptions using type specimens distributed in various museums (see, for instance, Tilbrook 2006). The case of the Indian Ocean is exemplary in this regard. Despite a rather large number of published studies, most of them were made in pre-SEM times and often cannot be considered as reliable sources of information without additional examination of the type material. In addition, most of the East African, South Arabian, Indonesian, and Western Australian coasts were never sampled.

## Morphological Comparison

At least half of the known species of *Electra* are described from tropical waters, and some of them are recorded from the tropical zone of the Indian Ocean and/or Western Indo-Pacific. The discovered new species is similar to the Australian *Electra triacantha* (Lamouroux, 1816), *E. oligopora* Gordon, 2009 from New Zealand, and, especially, *E. indica* (Menon and Nair, 1975) from Indian waters. All four species resemble the Atlantic species *Electra pilosa* in zooidal morphology, and should be grouped in the *Electra pilosa*-species complex.

The descriptions of *E. triacantha* made by Lamouroux (1816) stressed the presence of three spines, reflected in the species name. Specimens in the Natural History Museum, London (Busk collection) attributed to this species by Busk, revealed the following distinctive characters. The number of spines is more than three in these specimens, normally reaching 6–7. Some spines are smaller than the two distolateral and proximal spines, and can be easily overlooked when using stereomicroscope. The larger spines are placed on a distinct oval base, and all spines are slender and acute than in *E. omanensis.*

*E. oligopora* Gordon, 2009 possesses three spines in elongate rectangular zooids and has a strongly reduced gymnocyst with few pseudopores. Also, Gordon (1986, Pl. 2 A-B) illustrated another species of *Electra* (as *E. pilosa*) with three slender conical spines and the distal rim extending in a hood-like structure in pyriform zooids. The combination of characters in these two *Electra species* differs from that in *E. omanensis* described here.

*E. omanensis* sp. nov. morphologically most resembles *E. indica* (Menon and Nair, 1975), but differs in colony and spine shape. Direct comparison showed that colonies of the latter species are more compact, being formed of wide lobes. The number of spines is three, but they are acute and flattened. Also zooids of *E. indica* are shorter and wider on average (see Menon and Nair, 1975; Menon and Menon 2006) than in *E. omanensis* sp. nov.

## Molecular Comparison

Genetic analyses demonstrated that *Electra omanensis* sp. nov. belongs to the Pacific clade of the *E. pilosa* species complex, and that *E. omanensis* is closest relative of *E. indica*. Unfortunately, the specimen of *E. indica* was poorly preserved, and only a single16S fragment was successfully amplified from its DNA. Independent of missing data from other fragments, separate species status is supported by morphological data. All sequences of *E. omanensis* sp. nov., including the highly conserved nuclear ribosomal gene, are unique for *E. omanensis* and allow unambiguous discrimination of it from the morphologically related species.

*Electra pilosa* var. *"verticillata"* is mentioned in a list of species from Socotra (Scholz et al. 2001) and potentially may be our species.

**Acknowledgements** We are deeply indebted to Dr P.D. Taylor, the Natural History Museum London, for help in work with BNHM collection, and Dr S. Louis, Cochin University of Science and Technology, India, for sending us *Electra indica* samples. J.P. Cáceres-Chamizo, Department of Palaeontology, University of Vienna, sorted samples from Oman. A. Ostrovsky thanks FWF (grants P19337-B17 and P22696-B17), Austria, and RFBR (grant 10-04-00085a), Russia, for financial support. The work supported with DFG (Scha355/27) to E.A. Nikulina.

## References

Borg F (1931) On some species of *Membranipora*. Ark Zool 22A(4):1–35

Farris JS, Källersjö M, Kluge AG, Bult C (1994) Testing significance of incongruence. Cladistics 10:315–320

Felsenstein J (1985) Confidence limits on phylogenies: an approach using the bootstrap. Evolution 39:783–791

Gautier Y (1954) Sur *l'Electra pilosa* des feuilles de Posidonies. Vie Milieu 5:65–70

Goloboff PA (1993) Estimating character weights during tree search. Cladistics 9:83–91

Gordon DP (1986) The marine fauna of New Zealand: Bryozoa: Gymnolaemata (Ctenostomata and Cheilostomata Anasca) from the western South Island continental shelf and slope. New Zealand Oceanographic Institute Memoir, 95:1–121

Gordon DP (2009) New bryozoan taxa from a new marine conservation area in New Zealand, with a checklist of Bryozoa from Greater Cook Strait. Zootaxa 1987:39–60

Hillis DM, Moritz C, Mable BK (1996) Molecular systematics. Sinauer Associates, Sunderland

Lamouroux JVF (1816) Histoire des polypiers coralligènes flexibles, vulgairement nommés zoophytes. F. Poisson, Caen

Landsborough D (1852) A popular history of British zoophytes, or corallines. Reeve and Co, London

Levinsen GMR (1902) Studies on Bryozoa. Videnskabelige Mcddclclscr fra den natur- historiske Forening i Kjobcnhavn 54:1–31

Linnaeus C (1767) Systema Naturae per regna tria nature, secundum classes, ordines, genera, species, cum characteribus, differentiis, synonymis, locis. Salvius, Stockholm

Menon NR, Menon NN (2006) Taxonomy of Bryozoa from the Indian EEZ. A monograph OSTC marine benthos 01, ocean science & technology cell. CUSAT, Kochi

Menon NR, Nair NB (1975) Indian species of the division Malacostega (Polyzoa, Ectoprocta). Bryozoans from Indian waters. Mar Biol Assoc India 17:553–579

Nikulina EA (2006) *Electra korobokkura* sp. nov., a new species of cheilostome bryozoan from the Pacific coast of Hokkaido. Invert Zool 3(1):23–31

Nikulina EA (2008a) Taxonomy and ribosomal DNA-based phylogeny of the *Electra crustulenta* species group (Bryozoa: Cheilostomata) with revision of Borg's varieties and description of *Electra moskvikvendi* sp. nov. from the Western Baltic Sea. Org Divers Evol 8:215–229

Nikulina EA (2008b) *Electra scuticifera* sp. nov.: redescription of *Electra pilosa* from New Zealand as a new species (Bryozoa, Cheilostomata). Schr Nat Wiss Ver SH 70:91–98

Nikulina E, Schäfer P (2006) Bryozoans of the Baltic Sea. Meyniana 58:75–95
Nikulina EA, Taylor PD (2010) Two new species of *Electra* (Bryozoa, Cheilostomata) from the Miocene of the Aquitaine Basin, France. Geobios 43:219–224
Nikulina EA, Hanel R, Schäfer P (2007) Cryptic speciation and paraphyly in the cosmopolitan bryozoan *Electra pilosa* – impact of the Tethys closing on species evolution. Mol Phylogenet Evol 45:765–776
Pallas PS (1766) Elenchus zoophytorum, sistens generum adumbrationes generaliores et specierum cognitarum succinctas descriptiones, cum selectis auctorum synonymis. van Cleef, Den Haag
Palumbi SR, Martin A, Romano S, McMillan, WO, Stice L, Grabowski G (1991) The Simple Fool's Guide to PCR. Spec Publ, Dept Zool and Kewalo Mar Lab Univ Hawaii, Honolulu
Posada D, Buckley TR (2004) Advantages of AIC and Bayesian approaches over likelihood ratio tests for model selection in phylogenetics. Syst Biol 53:793–808
Posada D, Crandall KA (1998) Modeltest: testing the model of DNA substitution. Bioinformatics 14:817–818
Rambaut A, Drummond AJ (2004) Tracer v1.3. http://beast.bio.ed.ac.uk/Tracer. Accessed 30 Nov 2009
Rodriguez F, Oliver JL, Marin A, Medina JR (1990) The general stochastic model of nucleotide substitution. J Theor Biol 142:485–501
Ronquist F, Huelsenbeck JP (2003) MRBAYES 3: Bayesian phylogenetic inference under mixed models. Bioinformatics 19:1572–1574
Schiffner V (1934) Acystis, eine neue gattung der Sargassaceen und über einige algen aus dem Roten Meere. Hedwigia 74:115–118
Scholz J, Kadagies N, Böggemann M (2001) Bryozoen vom Sokotra-Archipel (Jemen). Nat u Mus 131:218–224
Stach LW (1937) Bryozoa of Lady Julia Percy Island. Proc R Soc Victoria 49:374–384
Swofford DL (2000) PAUP*. Phylogenetic analysis using parsimony (*and other methods). Version 4. Sinauer Associates, Sunderland
Swofford DL, Olsen GJ (1990) Phylogenetic reconstruction. In: Hillis DM, Moritz C, Mable BK (eds) Molecular systematics, 2nd edn. Sinauer Associates, Sunderland
Taylor PD, McKinney FK (2006) Cretaceous Bryozoa from the Campanian and Maastrichtian of the Atlantic and Gulf coastal plains, United States, vol 132, Scripta Geologica. National Natuurhistorisch Museum Naturalis, Leiden, pp 1–346
Thompson JD, Gibson TJ, Plewniak F, Jeanmougin F, Higgins DG (1997) The CLUSTAL_X windows interface: flexible strategies for multiple sequence alignment aided by quality analysis tools. Nucleic Acids Res 25(24):4876–4882
Tilbrook KJ (2006) Cheilostomatous Bryozoa from the Solomon Islands. Santa Barbara Museum of Natural History monographs 4, Stud Biodiv 3:1–386

# Chapter 15
# Molecular Phylogenetic Analysis Confirms the Species Status of *Electra verticillata* (Ellis and Solander, 1786)

## Species Status of *Electra verticillata*

**Elena A. Nikulina, Hans De Blauwe, and Oscar Reverter-Gil**

**Abstract** *Electra verticillata* was original described by Ellis and Solander ((1786) The natural history of many curious and uncommon zoophytes collected from various parts of the globe by the late John Ellis, systematically arranged and described by the late D. Solander, London), and since then the species status of this bryozoan has been in dispute. Many bryozoologists considered *E. verticillata* as one variety of colony morphology of *Electra pilosa* (Linnaeus 1767). To test the species status of *E. verticillata,* we analysed DNA sequences from material from the Bay of Douarnenez (near Morgat, France), together with sequences from *E. pilosa, E. posidoniae, E. scuticifera, E. indica*, and *Electra omanensis*. Phylogenetic analyses based on fragments of the 18S, 16S and 12S ribosomal RNA genes confirmed the status of *E. verticillata* as a separate species. We also examined the morphology of specimens of *E. pilosa* and *E. verticillata* in various institutions as well as in own collections. This study revealed morphological and ecological differences between these two species and clarified the geographical distribution of *E. verticillata*.

**Keywords** Taxonomy • Biodiversity • DNA • *Electra pilosa* • Ecology • Distribution

E.A. Nikulina (✉)
Zentrum für Baltische und Skandinavische Archäologie, Stiftung Schleswig-Holsteinische Landesmuseen, Schloss Gottorf, D-24837 Schleswig, Germany
e-mail: elena.nikulina@schloss-gottorf.de

H. De Blauwe
Royal Belgian Institute of Natural Sciences. rue Vautier 29–1000 Brussels, Watergang 6, 8380 Dudzele, Belgium
e-mail: deblauwehans@hotmail.com

O. Reverter-Gil
Departamento de Zooloxía e Antropoloxía Física, Universidade de Santiago de Compostela, 15782 Santiago de Compostela, Spain
e-mail: oscar.reverter@usc.es

A. Ernst, P. Schäfer, and J. Scholz (eds.), *Bryozoan Studies 2010*,
Lecture Notes in Earth System Sciences 143, DOI 10.1007/978-3-642-16411-8_15,

## Introduction

*Flustra verticillata* was originally described by Ellis and Solander (1786) from the Mediterranean Sea. In the same publication, the authors also mentioned two other species: *Flustra pilosa* (Linnaeus, 1767) and the new species *Flustra dentata*, both from the Atlantic Ocean. Several authors subsequently cited these species (e.g., Gmelin 1789; Bosc 1802; Lamarck 1816). Lamouroux (1816, 1821) erected the genus *Electra* for *F. verticillata*, the history of which is thus significant for members of this genus. Many species of *Electra* were described to be globally distributed, and differences in morphology can reflect geographical variation, but can also represent cryptic species (Nikulina et al. 2007; Nikulina 2008a). Some species of *Electra* were synonymised with other species after their description or, conversely, divided into several varieties (e.g., Farre 1837; Smitt 1867; Norman 1894; Borg 1931), some of which were subsequently accorded specific rank (Powell 1968; Gautier 1954; Nikulina 2008a, b).

*Electra* and other electrids exhibit a high degree of morphological plasticity, as well as high ecological tolerance. This combination may partially explain the existence of numerous morphological types (Norman 1894; Borg 1931). Since the middle of the nineteenth century, bryozoologists have tended to synonymize *E. verticillata*, *E. pilosa*, and *E. dentata*, or to view them as varieties or subspecies of *E. pilosa* (e.g., Farre 1837; Smitt 1867; Fischer 1870; Norman 1894; Hayward and Ryland 1998). Nevertheless, some zoologists doubted the synonymy of *E. pilosa* and *E. verticillata* (e.g., Bobin and Prenant 1960; Gautier 1962; Cook 1968; d'Hondt and Goyffon 2002). Bobin and Prenant (1960) studied *E. verticillata* from the Bay of Douarnenez and showed that it is similar to yet distinct from *E. pilosa*. Cook (1968) stated that the characters used by Bobin and Prenant (1960) were not representative, but came to the same conclusion. In contrast, investigating enzymatic polymorphism in *E. pilosa* and *E. verticillata* from the Bay of Biscay, d'Hondt and Goyffon (2002) found that the zymograms of the two were identical for the main enzymatic systems. Hence, the taxonomic status of *E. verticillata* and of other varieties or species of *Electra* remained unclear. The status of *E. verticillata* as a valid species is especially important, as it is the type species for *Electra*.

Recent studies employing DNA sequences have resolved similar taxonomic questions (Ryland et al. 2009), including those concerning electrids (Nikulina et al. 2007; Nikulina 2008a). A study of the geographic population structure of the putatively cosmopolitan species *Electra pilosa* and *Einhornia crustulenta* (Pallas, 1766) (formerly *Electra crustulenta*) revealed several morphologically similar species with restricted distributions, including *Electra scuticifera* Nikulina 2008b; *Einhornia korobokkura* (Nikulina, 2006); *Einhornia moskvikvendi* (Nikulina 2008a); and *Electra oligopora* Gordon, 2009, and confirmed the specific status of some varieties, e.g., *Einhornia arctica* (Borg, 1931), but failed to support other varieties (*typica* and *baltica*) as distinct species (Nikulina 2008a).

To test the species status of *E. verticillata*, we sequenced ribosomal RNA genes. Fragments of the mitochondrial 16S and 12S RNA gene were aligned and analysed with homologous sequences from *E. pilosa*. As the mitochondrial genome represents only a maternal perspective of evolution (Degnan 1993; Palumbi and Baker 1994), we validated our phylogeny based on mitochondrial data by adding a nuclear marker, part of the 18S rRNA gene, which has been broadly used in phylogenetic inference (Hillis and Dixon 1991; Grechko 2002; Halanych and Janosik 2006). Our study included four other species similar and closely related to *E. pilosa*: *Electra posidoniae* Gautier, 1954; *Electra scuticifera* Nikulina 2008b; *Electra indica* Menon and Nair, 1975; and *Electra omanensis* Nikulina et al. (this volume). The more distantly related *E. crustulenta* and *E. korobokkura* were used as outgroup taxa. Although phylogenetic relationships within electrids remain unknown, the genus *Einhornia* Nikulina 2007 was assumed to be most likely the sister taxon to our ingroup (Nikulina et al. 2007). Morphology was studied by using scanning electron microscopy (SEM) to reveal differences between *E. verticillata* and *E. pilosa*. The distribution of *E. verticillata* was revised using data from the literature, museum specimens, and our own material.

## Material and Methods

### *Sampling*

Colonies of *E. verticillata* were collected from the Atlantic coast of France (Morgat and St-Jean-de-Luz), Spain (Ria de Coruña), and Portugal (Praia de Falésia). A fragment of a colony in the Bay of Douarnenez, near Morgat (48°14′N 4°29′W) (Fig. 15.1) was preserved in 98% ethanol for molecular genetic analyses. The colony occurred in a hole eroded in the rock beneath a steep cliff bordering a sandy beach. The hole was situated among rocks at the edge of the beach, was partly filled with sand, and was probably only exposed to the air during extreme low tides. The zooids covered a slender, ramifying seaweed, with the colony forming a nearly complete sphere about 30 cm in diameter. Water entered and left the hole through a gully. The base of the colony extended into the sand, forming a mat of stolons about 2 cm thick and filled with sand grains.

Table 15.1 lists the samples included in the study. For the morphological and morphometric analyses, colonies of *E. pilosa* were collected from the Atlantic coast of France (Bay of Arcachon), The Netherlands (Eastern Scheldt estuary), and Ireland (Galway Bay).

### *Other Material Examined*

For revision of the geographic distribution of *E. verticillata*, material in various institutions was also examined: Muséum National d'Histoire Naturelle, Paris

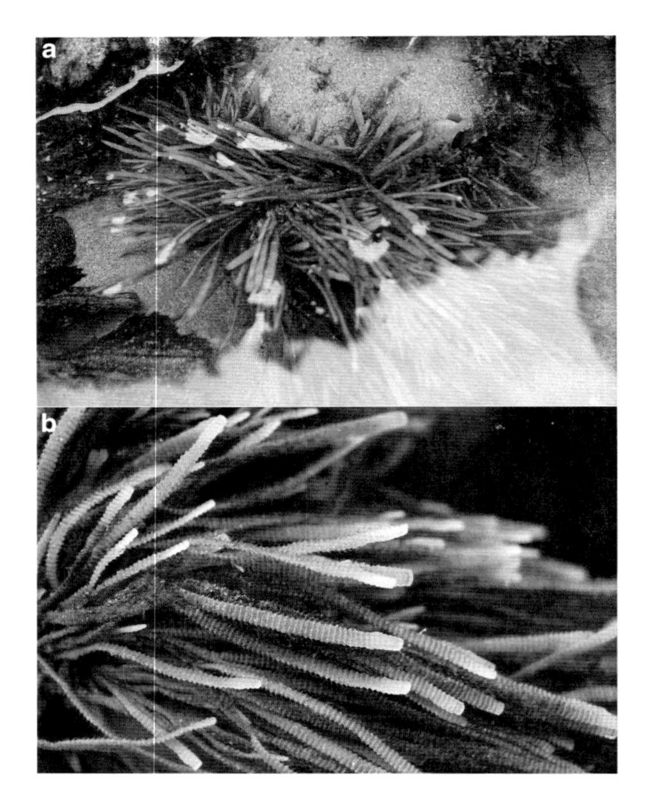


**Fig. 15.1** *Electra verticillata* in situ, Bay of Douarnenez, Morgat, France; we used this sample in the molecular phylogenetic study

(MNHN); Natural History Museum, London (NHMUK); Museo Nacional de Ciencias Naturales, Madrid (MNCN); Manchester Museum (MM).

## *Molecular Techniques*

Total DNA was extracted from ethanol-preserved colonies (about 10–20 zooids) using the Qiagen DNEasy Tissue Extraction Kit. Part of the 18S gene was amplified using primers we designed, U-18-4F (AGGAGTGGAGCCTGCGGCTTAA-TTTGACTC-3) and U-18-4R (AGGTTCACCTACGGAAACCTTGTTACGAC-3). Part of the 16S gene was amplified using the universal metazoan primers 16Sar and 16Sbr (Palumbi et al. 1991) or primers we designed, 16SF4 (CTCGGCAAAGAAGGGCTCCGCCTGTTTATCAAAAACAT-3) and 16SLr (TTCTCTTTTTCTGTTCCTTTCGTAAT-3). We designed 12S primers based on protostome sequences in the EMBL Nucleotide Sequence Database: 12SF-I

**Table 15.1** Specimens included in the molecular phylogenetic study, gene fragments with GenBank accession numbers, and sample locality information

| Species | 12S | 16S | 18S | Locality |
|---|---|---|---|---|
| *E. verticillata* | FR754521 | FR754524 | FR754534 | NW Atlantic, Bay of Douarnenez, Morgat |
| *E. pilosa (1)* | FR754511 | AJ971066* | FR754527 | NW Atlantic, North Sea, Helgoland |
| *E. pilosa (2)* | FR754512 | AJ971065* | AM075768* | NW Atlantic, North Sea, Helgoland |
| *E. pilosa (3)* | FR754513 | AJ971067* | FR754530 | NW Atlantic, North Sea, Helgoland |
| *E. posidoniae (1)* | – | AJ971084* | AM75770* | Mediterranean, Adriatic Sea, Pab |
| *E. posidoniae (2)* | FR754514 | AJ971085* | AM75771* | Mediterranean, Mallorca |
| *E. scuticifera (1)* | – | AM886854* | AM886854* | IW Pacific, Tasman Sea, Maori Bay |
| *E. scuticifera (2)* | FR754515 | AJ971086* | FR754533 | IW Pacific, Tasman Sea, West Coast |
| *E. omanensis* | FR754510 | FR754522 | FR754525 | Indian Ocean, Arabian Sea, Oman coast |
| *E. indica* | – | FR754523 | – | Indian ocean, Arabian Sea, Kerala Bay |
| *E. crustulenta* | – | AJ853844* | AM92413* | Western Baltic Sea, Lolland |
| *E. korobokkura* | FR754516 | AJ853947* | AM158087* | NW Pacific coast, Hokkaido, Akkeshi Bay |

An asterisk indicates a sequence from GenBank

(GGAAAAAATTGTGCCAGCADCCGCGGTTA-3) and 12SRLen (CACTTTCA-AGTACGCCTACTGTGTTACGAC-3).

Amplifications were performed in 25 μl of PCR mixture (20 mM Tris–HCl, 10 mM (NH4)2SO4, 10 mM KCl, 2 mM MgSO4, 0.1% Triton X-100, pH 8.8) containing 0.5 units Taq polymerase (New England Biolabs), 200 mM dNTPs, 0.5 mM primers, and 1 μl template DNA. Cycling parameters were as follows: 94°C for 2 min; 35 cycles of 94°C for 30 s, 45°C (54°C for 18S) for 30 s, and 72°C for 40 s; and 7 min at 72°C. PCR products were purified with the QIAquick PCR Purification Kit (Qiagen) and sequenced directly in both directions using the BigDye® Terminator v3.1 Cycle Sequencing Kit (Applied Biosystems) and an ABI3100 automatic sequencer.

## *Phylogenetic Analyses*

Sequences were assembled and edited using SeqMan and EditSeq software (DNASTAR Lasergene). ClustalX (Thompson et al. 1997) was used with default settings to perform automatic sequence alignments. Phylogenetic analyses were undertaken using PAUP* v4.0b10 (Swofford 2000) and MrBayes 3.2.1 (Ronquist and Huelsenbeck 2003). Phylogenetic trees were obtained using the maximum parsimony (MP), maximum likelihood (ML), neighbour-joining (NJ), and Bayesian (BA) methods. Nodal support was estimated by bootstrapping (Felsenstein 1985), with analyses of 1,000 pseudoreplicates by full heuristic searches. MP analyses were conducted by heuristic searches with the tree-bisection-reconnection branch-swapping algorithm (Swofford and Olsen 1990). A first tree was obtained by random addition of sequences, 200 replicates were then generated, and 10 trees were kept for a search for the most parsimonious trees. Gaps were treated as missing data. ModelTest v.3.6 (Posada and Crandall 1998) was used to find the model of DNA substitution that best fit the data (Posada and Buckley 2004). The parameters of this best-fit model selected with the Akaike information criterion (Posada and Buckley 2004) were subsequently used for ML and NJ analyses in PAUP* and BA analyses in MrBayes. For ML analyses, we used a full heuristic search, with 100 random-addition replicates and search parameters as described for the MP analysis. Bayesian inference of phylogeny was conducted using MrBayes running four Metropolis-coupled Markov-Chain Monte-Carlo chains (MCMC) simultaneously for 10,000,000 generations, sampling every 100th generation. Each Markov chain was started from a random tree. The MCMC output was analysed with TRACER v1.3 (Rambaut and Drummond 2004); 10% of the samples were discarded as burn-in; 90,000 samples were used to estimate parameters, parameter variance, and the posterior probabilities of particular nodes to construct a majority rule consensus tree.

### Morphological Study

Parts of colonies *of E. verticillata* and *E. pilosa* were dried, coated with Pd-Pt, and photographed with a SciScan scanning electron microscope. The following zooidal characters were measured from SEM images: zooid length and width, opesia length, and opesia inclination. Two specimens of *E. verticillata* (Bay of Douarnenez and Praia de Falésia) and three colonies of *E. pilosa* (Eastern Scheldt Estuary, Arcachon Bay, and Galway Bay) were measured. Zooid length and width, and opesia length, were measured on 20 zooids, and opesia inclination on 10 zooids, for each colony studied. All statistical calculations and tests to evaluate the significance of morphological differences were conducted with PAST software (Hammer et al. 2001). We prepared the permutation *t*-test with 10,000 permutations to test the equality of means, and the Mann–Whitney *U* test to test the equality of medians.

## Results

### Phylogenetic Analysis

*18S rRNA gene*. The alignment of ten 18S sequences was 510 bp long, with 18 variable and seven parsimony-informative sites. No variability was found within any of the species, and therefore the data set consisted of six unique sequences, corresponding to six species: *E. verticillata, E. omanensis, E. scuticifera* (1, 2), *E. posidoniae* (1, 2), *E. pilosa* (1–3), and the outgroup taxon *E. korobokkura*. An MP analysis using equally weighted characters yielded a single tree 32 steps long, with a consistency index (CI) of 1 and retention index (RI) of 1. ML, NJ, and BA analyses were conducted using the F81 model of nucleotide substitution (Felsenstein 1981) with a gamma shape parameter of 0.0062 and estimated nucleotide frequencies of A = 0.19080, C = 0.2771, and G = 0.2955. The ML and NJ analyses resulted in single trees. A BA majority rule consensus tree was created. All trees were completely resolved and had identical topologies, and included the highly supported (88–100%) clade (*E. verticillata* (*E. omanensis, E. scuticifera*)). *Electra posidoniae* and *E. pilosa* formed a sister clade with lower nodal support (76–95%). Figure 15.2a shows the Bayesian phylogram, with nodal support values from all four methods.

*16S rRNA gene*. The alignment of 11 sequences was 447 bp long, with 190 variable and 49 parsimony-informative sites. The data set consisted of 11 unique sequences obtained from seven species: *E. verticillata, E. omanensis, E. indica, E. scuticifera* (1, 2), *E. posidoniae* (1, 2), *E. pilosa* (1–3), and the outgroup taxon *E. korobokkura*. An equally weighted MP analysis yielded a single tree 331 steps long, with the CI = 0.8218 and RI = 0.8343. The tree topology was stable to Goloboff's weighting, and the best-fit tree was obtained with k = 0 (Goloboff fit = −112.8333). The HKY85 model of nucleotide substitution (Hasegawa-Kishino-Yano 1985) with a gamma shape parameter of 0.2648, a transition/transversion ratio of 2.0386, and assumed nucleotide frequencies of A = 0.3654, C = 0.1431, G = 0.1693 was used in the ML, NJ, and BA analyses. All trees from the NJ, ML, and MP analyses were identical in topology and consisted of two sister clades: the highly supported (98–100%) clade (*E. posidoniae, E. pilosa*), and the less well supported (77–94%) clade (*E. scuticifera* (*E. verticillata* (*E. omanensis, E. indica*))). A Bayesian majority rule consensus tree was created. Figure 15.2b shows the Bayesian phylogram, with nodal support values from all four methods.

*12S rRNA gene*. The alignment of eight sequences was 575 bp long, with 329 variable and 187 parsimony-informative sites. The data set consisted of six unique sequences obtained from *E. verticillata, E. omanensis, E. scuticifera* (2), *E. posidoniae* (2), *E. pilosa* (1–3), and the outgroup taxon *E. korobokkura*. An equally weighted MP analysis yielded a single tree 581 steps long, with the CI = 0.8468 and RI = 0.6888. The tree topology was stable to Goloboff's weighting, the best-fit tree was obtained with k = 0 (Goloboff fit = −143.5). The general time-reversible model of nucleotide substitution (Rodriguez et al. 1990) with estimated base frequencies of A = 0.3577, C = 0.1687, G = 0.1886, a specified substitution rate matrix, and a gamma shape parameter of 0.4938 was used in the ML analysis. A similar model was applied to construct NJ and BA trees. All methods yielded an identical tree topology, consisting of two sister clades with high nodal support (85–100%): (*E. posidoniae, E. pilosa*) and (*E. omanensis* (*E. scuticifera* (*E. verticillata*))). Figure 15.2c shows the Bayesian phylogram, with nodal support values from all four methods.

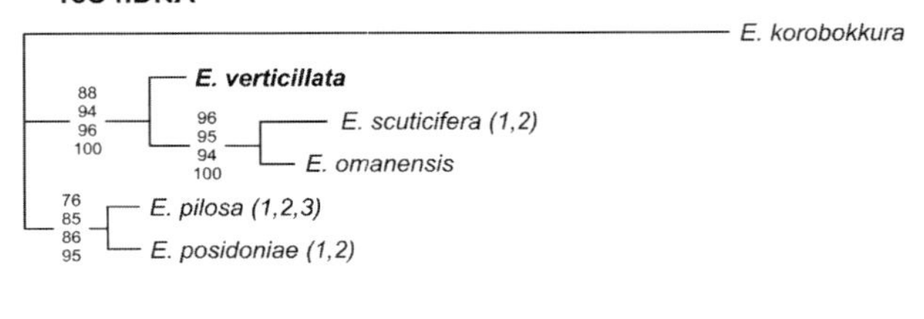


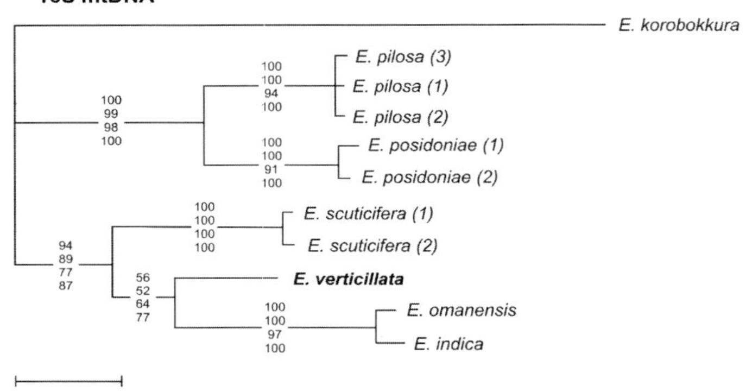


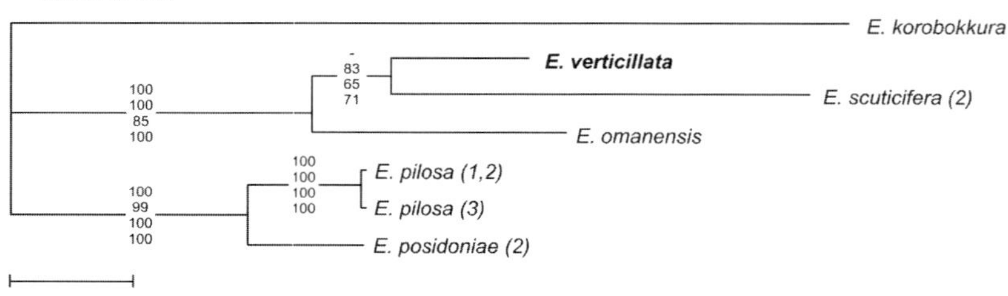


**Fig. 15.2** Phylograms constructed from 18S (**a**), 16S (**b**), and 12S (**c**) data. The topology depicts the strict (**a**, **b**) or majority rule (**c**; MP analysis provided 70% bootstrap support for the [*E. verticillata, E. omanensis*] clade) consensus between the MP, NJ, ML and Bayesian trees. The four-number columns at nodes show bootstrap support values and posterior probabilities from the MP, NJ, ML, and BA analyses, respectively (*top* to *bottom*). Branch lengths were calculated with the Bayesian approach

## *Morphology of* Electra verticillata

The colony of *E. verticillata* forms erect tufts (Fig. 15.1). The base of the colony consists of linear stolons, highly branched and matted, forming a more or less dense mat up to 2 cm thick. The stolons are made up primarily of kenozooids and more or less degraded autozooids. The kenozooids are sometimes very large, without visible differentiation, with simple ectocyst, non-calcified and devoid of pores. An underdeveloped opesia, operculum, pores, or spines may be present. The stolons attach to solid substrates (shells, seaweed, etc.) by means of young buds, each of which can develop into an autozooid. Numerous erect branches grow from the encrusting base; these are cylindrical to ribbon-like and consist of autozooids arranged in regular verticils, each composed of five autozooids in encrusting branches and up to 15 zooids in free tufts (Fig. 15.3a, b). The tufts are bilaminar (Fig. 15.3c). Branching may occur by simple separation of the autozooidal series into two branches. Most frequently, new branches arise from a lateral bud (Fig. 15.3d); this gives rise to two autozooids jointed back to back, which in turn give rise to four autozooids in the next generation.

Autozooids have the shape of an obliquely truncate cone, with the truncation corresponding to the opesia (Fig. 15.3e). The opesia is rounded rectangular, slightly elongate, and occupies about half the zooidal length. The edge of the opesia bears five (rarely six) spines, largely chitinous: one proximal, pointed, curved above the opesia, rarely exceeding the length of the autozooid, frequently more prominent in the zooids in the laterals of the verticil; a shorter pair at the proximolateral corners of the opesia; and a distolateral pair, poorly developed, on the lateral walls level with the hinges of the operculum. The rim around the opesia distal to this last pair of spines projects slightly distally. The marginal sclerite of the operculum is narrow, light brown, the bulge at the hinge bending sharply toward the midline and often decorated with various extensions. The cryptocyst is thin and transparent, well developed proximally, tapering laterally, sometimes surrounding the opesia distally. The gymnocyst is smooth, translucent, with numerous rounded pseudopores; pseudopores are absent in the most proximal area. Communication is via pore chambers or multiporous plates.

When the polypide is retracted, the long oesophagus forms a curve that prolongs the cardiac loop and represents the most proximal part of the polypide, because the cardia is relatively short. There are 9–13 (usually 11 or 12) short tentacles that do not exceed the proximal edge of the opesia when the polypide is retracted.

The ancestrula is smaller and flatter than autozooids, with a rounded base; its gymnocyst is uniformly perforated by 50–60 pores. It bears the five spines seen in astogenetically mature zooids, plus an additional two thin spines, one on each side lateral to the main spine (ancestrulae of *E. pilosa* invariably have five spines). The ancestrula buds two or three daughter autozooids, and subsequent growth tends to form linear branching series appressed to one another laterally and connecting by lateral septula; the autozooids are lined up in transverse rows starting from the second generation. This growth can lead to a very regular whorled arrangement of

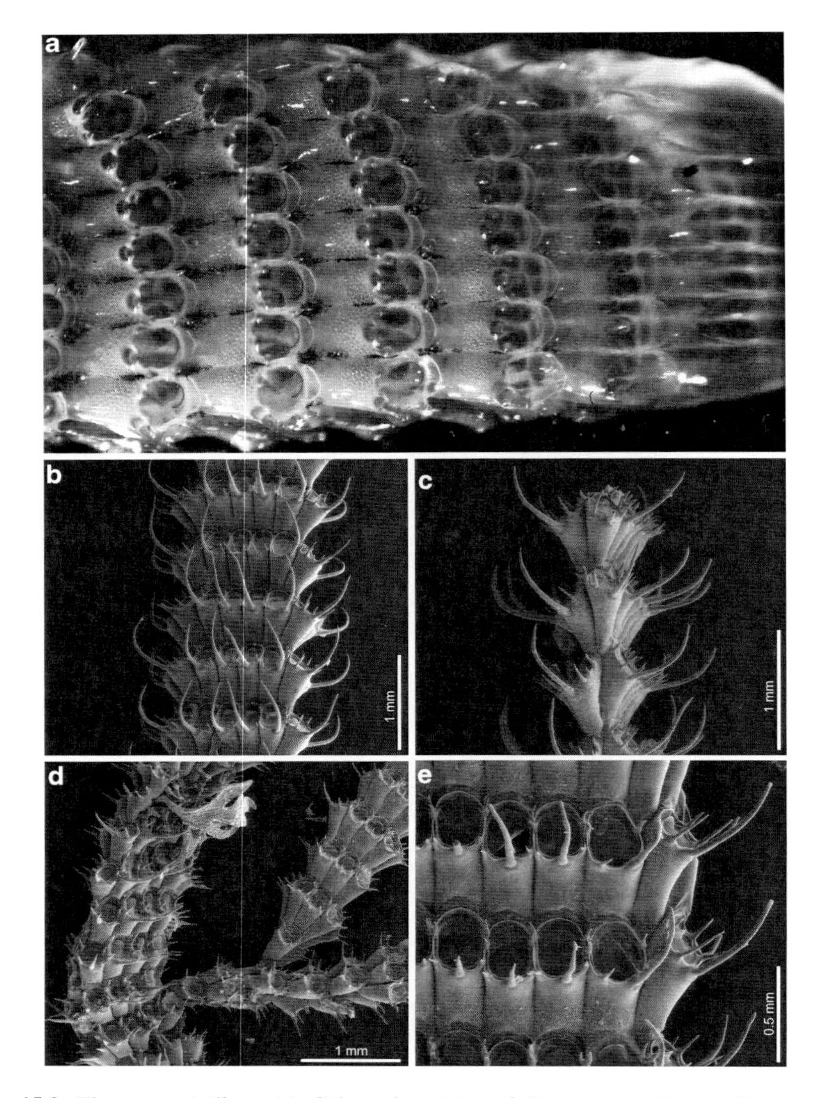


**Fig. 15.3** *Electra verticillata*. (**a**) Colony from Bay of Douarnenez, Morgat, France; light microscopy. (**b–e**) Details of the branch structure of *E. verticillata*; SEM images of bleached samples from Praia de Falésia, Portugal (**b**, **c**); Arcachon Bay, France (**d**); and Bay of Douarnenez, France (**e**) (From the collection of Hans De Blauwe)

tufts, parts of which encrust algae (Fig. 15.4b), with bilaminar branches arising from the crusts (Fig. 15.3). For comparison, the morphology of *E. pilosa* is shown in Fig. 15.4c–f.

Table 15.2 lists measurements of some zooidal characters for *E. verticillata* and *E. pilosa*. The statistical analyses indicated statistically significant differences

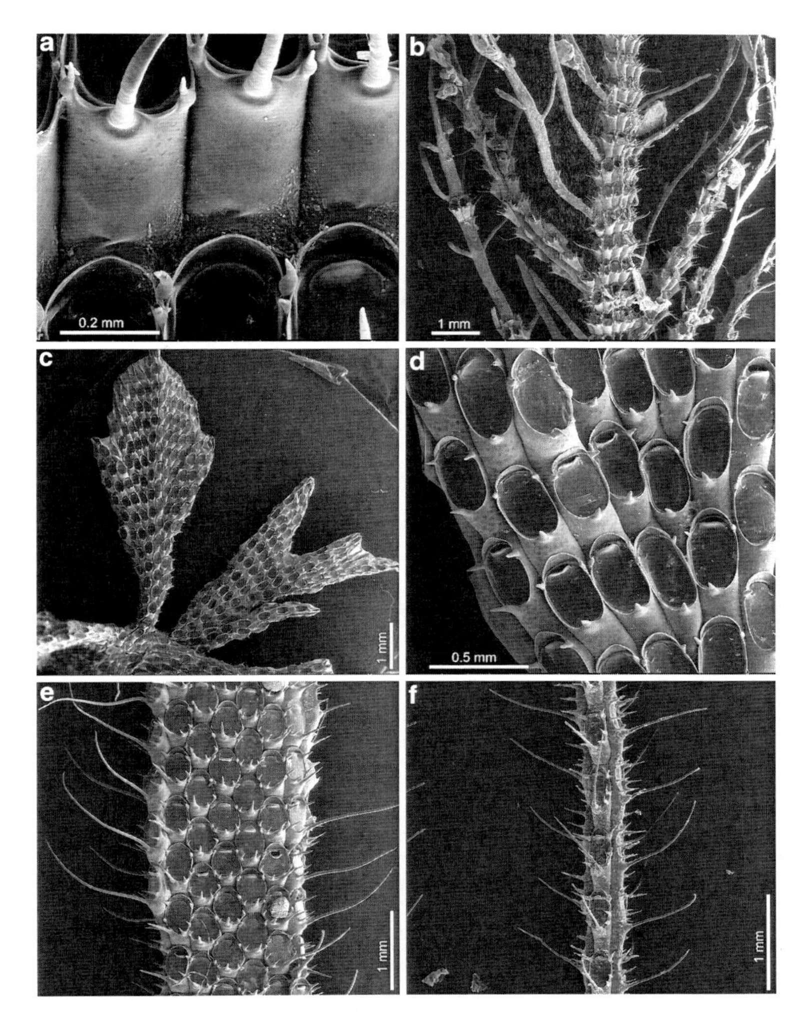


**Fig. 15.4** *Electra verticillata* (**a**, **b**) and *E. pilosa* (**c**–**f**). (**a**) Pseudopores on the gymnocyst of *E. verticillata* (Bay of Douarnenez, Morgat, France). (**b**) Colony of *E. verticillata* overgrowing an alga (Bay of Douarnenez, St-Jean-de-Luz, France). (**c**–**e**) Erect, bifoliate branches in *E. pilosa* incrusting an algal frond, demonstrating some variation in zooidal arrangement; specimens are from the Eastern Scheldt Estuary, The Netherlands (**c**, **d**) and Galway Bay, Ireland (**e**, donated by Marco Faasse). (**f**) *Electra pilosa* encrusting a hydroid stem (Arcachon Bay, France) (From the collection of Hans De Blauwe)

between *E. verticillata* and *E. pilosa* in three characters: zooid length (greater in *E. verticilata*), opesia length (lesser in *E. verticillata*), and opesia inclination (greater in *E. verticillata*) (Table 15.3), all at a significance level of $p < 0.001$. The assumption of similarity was not rejected for zooid width, as the significance level was $p > 0.05$.

**Table 15.2** Measurements for zooids of *E. verticillata* and *E. pilosa*. Two colonies of *E. verticillata* (Bay of Douarnenez, France; Praia de Falésia, Portugal) and three colonies of *E. pilosa* (Bay of Arcachon, France; Eastern Scheldt estuary, The Netherlands; Galway Bay, Ireland) were used for measurements

| | *E. verticillata* | *E. pilosa* |
|---|---|---|
| **Zooid length (mm)** | | |
| N | 40 | 60 |
| Range | 0.52–0.68 | 0.47–0.66 |
| Mean | 0.61 | 0.57 |
| Median | 0.63 | 0.57 |
| Standard error | 0.009 | 0.010 |
| Standard deviation | 0.041 | 0.045 |
| **Zooid width (mm)** | | |
| N | 40 | 60 |
| Range | 0.24–0.30 | 0.20–0.29 |
| Mean | 0.27 | 0.25 |
| Median | 0.27 | 0.25 |
| Standard error | 0.004 | 0.007 |
| Standard deviation | 0.017 | 0.029 |
| **Opesia length (mm)** | | |
| N | 40 | 60 |
| Range | 0.27–0.34 | 0.37–0.47 |
| Mean | 0.32 | 0.42 |
| Median | 0.32 | 0.42 |
| Standard error | 0.004 | 0.006 |
| Standard deviation | 0.017 | 0.028 |
| **Opesia inclination** (°) | | |
| N | 20 | 30 |
| Range | 42–70 | 30–45 |
| Mean | 57 | 39 |
| Median | 59 | 40 |
| Standard error | 3.0 | 1.6 |
| Standard deviation | 9.3 | 6.1 |

# Discussion

## *Phylogenetic Analyses*

Phylogenetic analyses of mitochondrial and nuclear ribosomal genes demonstrated not only that *E. verticillata* is a species distinct from *E. pilosa*, but also that these two species are not directly related. *Electra verticillata* belongs to a Pacific group of *pilosa*-like species (*E. scuticifera, E. indica,* and *E. omanensis*) with high nodal support (77–100%, mean 93%), although the relationship among these species remain unresolved, as the three genes provided different results regarding relationships (Fig. 15.2).

**Table 15.3** P-values in the permutation *t*-test and Mann–Whitney *U* test

| | Zooid length | Zooid width | Opesia length | Opesia inclination |
|---|---|---|---|---|
| Permutation *t*-test | <0.0001 | 0.08 | <0.0001 | <0.0001 |
| Mann–Whitney *U* test | $2.8 \cdot 10^{-5}$ | 0.06 | $1.8 \cdot 10^{-16}$ | $6.3 \cdot 10^{-9}$ |

## *Morphological Differences Between* E. verticillata *and* E. pilosa

*Electra verticillata* and *E. pilosa* are discriminated by zooidal characters and colony morphology. Zooids of *E. verticillata* have a more conical shape due to the steep incline of the opesia to the frontal plane; the zooids are significantly longer and opesiae are significantly shorter than in *E. pilosa* (Tables 15.2 and 15.3). The ratio of opesium to zooid length is about 0.5 in *E. verticillata* and 0.7 in *E. pilosa*. The most conspicuous differences in colony morphology are the very regular verticillate zooid arrangement in erect branches of *E. verticillata* and a well-developed stolonial system, compared to quincuncial zooid arrangement and absence of stolons in *E. pilosa*.

*Electra pilosa* encrusts various substrata, sometimes giving rise to two-layered branches (Fig. 15.4c–f) that are sporadically ribbon-like. De Blauwe (2009) illustrates an encrusting (his Fig. 159) and a folious colony part (his Fig. 158) of *E. pilosa*. The arrangement of autozooids is normally quincuncial, but may change with the nature of the substratum, and in contrast to *E. verticillata* only sporadically appears somewhat verticillate. The possibility for the colony to produce erect parts, where some zooids are arranged in more or less transversal rows, can lead to confusion with *E. verticillata*.

Bobin and Prenant (1960) and Prenant and Bobin (1966) first described stolons in *E. verticillata*. Although Marcus (1926) experimentally induced stolon production in *E. pilosa*, they are not normally present in this species. The structure of stolons and their function in propagation of the colony are similar between these two species and other electrids such as *Einhornia arctica* (Borg, 1931) and *E. korobokkura* (Nikulina 1999, 2006), in which stolons are, however, very rare.

## *Ecological Notes*

Both *E. verticillata* and *E. pilosa* produce cyphonautes larvae, though the larvae of *E. verticillata* are not well known (Bobin and Prenant 1960). In the southern North Sea, *E. pilosa* is the first species to colonise new substrates such as wrecks and foundations of offshore wind turbines (De Blauwe unpublished). Bobin and Prenant (1960) suggested that in *E. verticillata*, colony tufts are annual and are regenerated from the perennial stolonal system every year; colonies are therefore less likely to originate from ancestrulae. The long-lived larvae of *E. pilosa* and their potential to settle on almost any substrate makes this species a successful coloniser, whereas

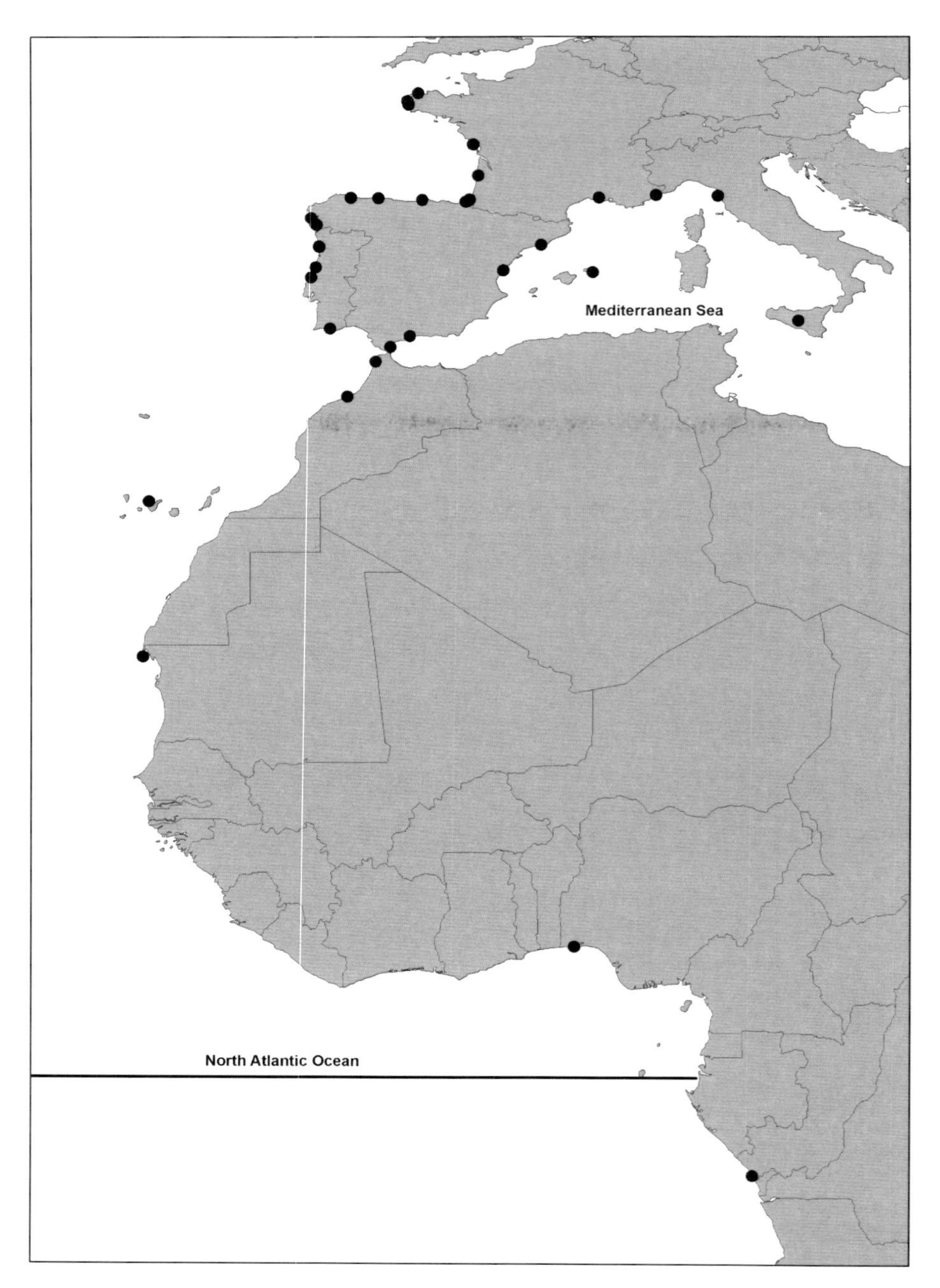


**Fig. 15.5** Distribution of *E. verticillata*, reconstructed from data in the literature data and material in collections (see Supplements 15.1 and 15.2)

*E. verticillata* seems to be more selective with regard to substrate choice (rocks, sand, and algae are needed), but the stolonial system is a good survival strategy once a colony is established in a suitable habitat.

*Electra verticillata* can adhere to stones, shell fragments, and especially algae, particularly *Gracilaria gracilis* (Stackhouse) M. Steentoft, L.M. Irvine and W.F. Farnham 1995, on which the most luxuriant colonies develop; we have never seen *E. verticillata* encrusting hydroids, as Hayward and Ryland (1998) described.

Bobin and Prenant (1960) provided detailed information on the ecology of *E. verticillata* at two stations in the Bay of Douarnenez, and described the habitat of the species and its commonly used substrates. The ecological conditions and morphology of the colony we found in the Bay of Douarnenez (see Material and Methods) near these previous two stations were identical to those Bobin and Prenant (1960) described and seem to be typical for this species.

## *Geographical Distribution of* E. verticillata

Analysis of material in museum collections, data from the literature, and some unpublished sources presented in Supplements 15.1 and 15.2 allowed us to map the geographic range of *E. verticillata* (Fig. 15.5). The species occurs along the Atlantic coast of the Iberian Peninsula, the Atlantic coast of North Africa, and in the Western Mediterranean. Cook (1968, 1985) reported *E. verticillata* at various stations along the West coast of Africa, reaching South Africa, but a detailed study of the original material will be necessary to confirm Cook's records. *Electra verticillata* has not been found north of Brittany, France (Bobin and Prenant 1960) and is thus absent from the North Sea. Populations in situ are probably absent between Hendaye near the French-Spanish border and Brittany, as suitable habitat is rare along this coast. Bobin and Prenant (1960) searched unsuccessfully for *E. verticillata* in the bay and channels of Arcachon, but found neither colonies nor ancestrulae on *G. gracilis*, the most common algal substrate for *E. verticillata*. Along this coast, colony fragments are beached regularly, probably originating from populations along the north coast of Spain.

**Acknowledgments** This project was funded by DFG research grant Scha355/27 (Germany). The work by Oscar Reverter-Gil was supported by the project "Fauna Ibérica: Briozoos I (Ctenostomados y Queilostomados Anascos)" (CGL2006-04167), co-financed by the Ministerio de Ciencia e Innovación, Spanish Government, and FEDER. ORG gratefully acknowledges a SYNTHESYS grant (GB-TAF-4454), which enabled a visit to the Natural History Museum London. We acknowledge Javier Souto for collecting specimens in Portugal and taking measurements, and Matthew Dick for reading the manuscript.

# Supplements

**Suppl. 15.1** Atlantic records of *E. verticillata* from North to South

| Station | Date | Source | Checked or identified by |
|---|---|---|---|
| France, Roscoff | No data | MNHN-12147, MNHN-12162 | Bobin, ORG |
| France, Brittany"Finistère" | 1826 | Bobin and Prenant (1960), pl I. Fig. 7 | P and B |
| France, Rade de Brest | No data | Guérin-Gavinet (1911) | No data |
| France, Bay of Douarnenez | No data | Bobin and Prenant (1960) | P and B |
| France, Brittany, Morgat | 29/03/2010 | This paper | HDB |
| France, Island Re | No data | de Beauchamps (1923) | P and B |
| France, Dept. Gironde, Arcachon and SW France | No data | Fischer (1870) | Fischer |
| France, Arcachon | 03/08/2001 | HDB (unpublished) | HDB |
| France Basque coast, Hendaye | No data | Bobin and Prenant (1960) | P and B |
| France Basque coast, Northeast of Hendaye | No data | d'Hondt and Goyffon (2002) | d'Hondt and Goyffon |
| France Basque coast, St-Jean-de-Luz | 12/08/2001 | HDB (unpublished) | HDB |
| France Basque coast, Bidart | 11/1935 | D'Hondt (1987) | D'Hondt |
| France Basque coast, Biarritz | 04/09/1909 | NHMUK-1882.7.7.10 | ORG |
| | | MNHN-5974 | ORG |
| France Basque coast, Biarritz | No data | Station Biologique d'Arcachon (in Bobin and Prenant 1960) | P and B |
| Spain, Santander | No data | Barroso (1912) | Álvarez |
| | | MMC 3/M/60 | |
| | | MMC 3/M/61 | |
| Spain, Asturias, Gijón | 1923 | MNCN-25.03/24 | Álvarez |
| Spain, Galicia: several localities in the North and South | No data | Reverter-Gil and Fernández-Pulpeiro (2001) | ORG |
| Portugal, not specified | No data | NHMUK-1897.5.1.486 | ORG |
| Portugal: Oporto | No data | NHMUK-1897.5.1.485 | ORG |
| Portugal: Matosinhos, Figueira da Foz, Nazaré | No data | ORG (unpublished) | Javier Souto |
| Portugal: Albufeira, Praia da Falésia | 03/2004 | HDB (unpublished) | HDB |
| Morocco: Mohammedia (Fedhala) | No data | Canu and Bassler (1925) | P and B |
| Morocco: Medina (Mogador) | No data | Canu and Bassler (1925) | P and B |
| Canary Islands | No data | Arístegui-Ruiz (1984) | Arístegui-Ruiz |
| Mauritania: Cabo Blanco | No data | O'Dea and Okamura (2000) | O'Dea and Okamura |
| Gabon, Pointe Noir | | Cook (1968, 1985) | Cook |
| Lagos | | Cook (1968, 1985) | Cook |

*MMC* Museo Marítimo del Cantábrico, *NHMUK* Natural History Museum, London, *MNHN* Muséum National d'Histoire Naturelle, Paris, *MNCN* Museo Nacional de Ciencias Naturales, Madrid, *P and B* Prenant and/or Bobin, *HDB* Hans De Blauwe, *ORG* Oscar Reverter-Gil

**Suppl. 15.2** Mediterranean records of *E. verticillata* from west to east

| Station | Date | Source | Checked or identified by |
|---|---|---|---|
| Spain, Andalusia: La Atunara | No data | López de la Cuadra and García-Gómez (1988) | ORG by photographs |
| | | López de la Cuadra (1991) | |
| Spain, Castellón | No data | Rioja lo Bianco (1920) | Rioja lo Bianco |
| | | Barroso (1921) | Barroso |
| | | Zabala (1986) | Zabala |
| Spain, Catalonia: Sitges | 05/1962 | NHMUK-1965.8.18.16 | ORG |
| | | NHMUK-1975.7.18.53 | ORG |
| Spain, Malaga | 1862 | Zoological museum of CAU, Kiel | No data |
| France, Camargue | 1966, 1980, 1983 | Harmelin | Harmelin |
| France, Nice (Nizza) | No data | Carus (1893) (Risso) | No data |
| Islas Baleares, Mahón | No data | Zabala (1986) | Zabala |
| Italy, Mare della Toscana | No data | Carus (1893) | No data |
| Sicily | No data | Gautier (1958) | Gautier |

*MMC* Museo Marítimo del Cantábrico, *NHMUK* Natural History Museum, London, *MNHN* Muséum National d'Histoire Naturelle, Paris, *MNCN* Museo Nacional de Ciencias Naturales, Madrid, *P and B* Prenant and/or Bobin, *HDB* Hans De Blauwe, *ORG* Oscar Reverter-Gil

# References


Arístegui Ruiz J (1984) Briozoos quilostomados (Ectoprocta, Cheilostomata) de Canarias: estudio sistemático, faunístico y biogeográfico. Tesis Doctoral, Facultad de Biologia, Universidad de La Laguna, Las Palmas

Barroso MG (1912) Briozoos de la Estación de Biología maritíma de santander. Trab Mus Cien Nat 5:1–63

Barroso MG (1921) Notas sobre algunas especies de Briozoos de España. Especies del golfo de Valencia. Boletín Real Soc Esp Hist Nat Sec Biol 50:68–78

Bobin G, Prenant M (1960) *Electra verticillata* (Ellis et Solander, 1786) Lamouroux 1816, (Bryozoaire Chilostome). Cah Biol Mar 1:121–156

Borg F (1931) On some species of *Membranipora*. Ark Zool 22A:1–35

Bosc L-A-G (1802) Histoire naturelle des Vers, III, 1st edn. Kessinger, Paris

Canu F, Bassler RS (1925) Les bryozoaires du Maroc et de Mauritanie. Mém Soc Sci Nat Maroc 10:1–79

Carus J-V (1893) Prodromus faunac mediterraneae. Vol. II. Part I. Bryozoa. E. Schweizerbarth'sche Verlagshandlung (E. Koch) Stuttgart pp 1–54

Cook PL (1968) Polyzoa from west Africa. The Malacostega, Part 1. Bull Br Mus (Nat Hist) Zool 16:113–160

Cook PL (1985) Bryozoa from Ghana. A preliminary survey. Ann Mus R Afr Cent Sci Zool Tervuren 238:1–315

de Beauchamps P (1923) Etude de bionomie intercotidale. Les îles de Ré et de'Yeu. Arch Zool Exp Gen 61:455–520

De Blauwe H (2009) Mosdiertjes van de Zuidelijke bocht van de Noordzee: determinatiewerk voor België en Nederland. Vlaams Instituut voor de Zee (VLIZ), Oostende

Degnan SM (1993) The perils of single gene trees: mitochondrial versus single-copy nuclear DNA variation in white-eye (Aves: Zosteropidae). Mol Ecol 2:219–226

Ellis J, Solander D (1786) The natural history of many curious and uncommon zoophytes collected from various parts of the globe by the late John Ellis, systematically arranged and described by the late D. Solander. Benjamin White and Son, London

Farre A (1837) Observations on the minute structure of some of the higher forms of Polypi, with views on a more natural arrangement of the class. Philos Trans R Soc Lond 127:387–426

Felsenstein J (1981) Evolutionary trees from DNA sequences: a maximum likelihood approach. J Mol Evol 17:368–376

Felsenstein J (1985) Confidence limits on phylogenies: an approach using the bootstrap. Evolution 39:783–791

Fischer P (1870) Bryozoaires, echinodermes et foraminifères de la Gironde. Act Soc Lin Bordeaux 27:1–15

Gautier Y-V (1954) Sur l'*Electra pilosa* des feuilles de Posidonies. Vie Milieu 5:65–70

Gautier Y-V (1958) Bryozoaires marins actuels de Sicile. Atti Societa Peloritana Sci Fis Mat Nat 4:46–68

Gautier Y-V (1962) Recherches écologiques sur les Bryozoaires Chilostomes en Méditerranée occidentale. Rec Trav Sta Mar Endoume 38(24):1–434

Gmelin J-F (1789) Systema naturae (Linné), 13 edition, cura Joh-Frid. Gmelin. Vol. I, Part. 6. Stockholm. Zoophyta, pp 3753–3871

Gordon DP (2009) New bryozoan taxa from a new marine conservation area in New Zealand, with a checklist of Bryozoa from Greater Cook Strait. Zootaxa 1987:39–60

Grechko VV (2002) Molecular DNA markers in phylogeny and systematics. Russ J Genet 38:851–868

Guérin-Gavinet G (1911) Contribution à l'étude des Bryozoaires des côtes armoricaines. I. Bryozoaires provenant de la rade de Brest et recueillis par le frères Crouan. Trav Sci Lab Zool Physiol Marit Concarneau 35:1–7

Halanych KM, Janosik AM (2006) A review of molecular markers used for annelid phylogenetics. Integr Comp Biol 46:533–543

Hammer O, Harper DAT, Ryan PD (2001) PAST: paleontological statistics software package for education and data analysis. http://palaeoelectronica.org/. Accessed 13 May 2001

Hasegawa M, Kishino H, Yano T (1985) Dating of the human-ape splitting by a molecular clock of mitochondrial DNA. J Mol Evol 22:160–174

Hayward PJ, Ryland JS (1998) Cheilostomatous Bryozoa: 1. Aeteoidea – Cribrilinoidea: notes for the identification of British species, vol 10, 2nd edn, Synopses of the British fauna (New Series). Field Studies Council, Shrewsbury

Hillis DM, Dixon MT (1991) Ribosomal DNA: molecular evolution and phylogenetic inference. Q Rev Biol 66:411–453

d'Hondt J-L (1987) Bryozoaires littoraux de la côte basque française. Bull Cent Etudes Rech Sci Biarritz 15:43–52

d'Hondt J-L, Goyffon M (2002) Observations complementaires sur le polymorphisme enzymatique d'*Electra pilosa* et d'*E. verticillata* (Bryozoaires Cheilostomes). Bull Soc Zool Fr 127:223–232

Lamarck J-B (1816) Histoire naturelle des animaux sans vertèbres, vol II. Verdière, Paris

Lamouroux J-V (1816) Histoire naturelle des polypiers coralligènes flexible, vulgairement nommès Zoophytes. F. Poisson, Caen

Lamouroux J-V (1821) Exposition méthodique des genres de l'ordre des polypiers avec leur description et celle des principales espèces, figures dans 84 planches, les 63 premières appartenant à l'histoire naturelle des zoophytes d'Ellis et Solander. Agasse, Paris

López de la Cuadra CM (1991) Estudio sistemático de los briozoos queilostomados (Bryozoa: Cheilostomida) del Estrecho de Gibraltar y áreas próximas. Ph.D. Thesis, Universidad de Sevilla

López de la Cuadra CM, García-Gómez JC (1988) Briozoos queilostomados del Estrecho de Gibraltar y áreas próximas. Cah Biol Mar 29:21–36

Marcus E (1926) Beobachtungen und Versuche an lebenden Meeresbryozoen. Zool Jahrb (Syst) 52:1–102

Menon NR, Nair NB (1975) Indian species of the division Malacostega (Polyzoa, Ectoprocta). Bryozoans from Indian waters. Mar Biol Ass India 17:553–579

Nikulina EA (1999) Vosvyshaushiesa tzepochki – neobychnaya forma rosta *Electra crustulenta* var. *arctica* Borg 1931 (Bryozoa, Cheolostomata) [Raising chains – unusual growth form of *Electra crustulenta* var. *arctica* Borg, 1931 (Bryozoa, Cheilostomata)]. Vestnik MGU 2:40–44 (In Russian)

Nikulina EA (2006) *Electra korobokkura* sp. n., a new species of cheilostome bryozoan from the Pacific coast of Hokkaido, Japan. Inver Zool 3:23–31

Nikulina EA (2007) *Einhornia*, a new genus for electrids formerly classified as the *Electra crustulenta* species group (Bryozoa, Cheilostomata). Schr Naturwiss Ver Schlesw-Holst 69:24–40

Nikulina EA (2008a) Taxonomy and ribosomal DNA-based phylogeny of the *Electra crustulenta* species group (Bryozoa: Cheilostomata) with revision of Borg's varieties and description of *Electra moskvikvendi* sp. nov. from the Western Baltic Sea. Org Divers Evol 8:215–229

Nikulina EA (2008b) *Electra scuticifera* sp. nov.: redescription of *Electra pilosa* from New Zealand as a new species (Bryozoa, Cheilostomata). Schr NaturwissVereins Schlesw-Holst 70:91–98

Nikulina EA, Hanel R, Schäfer P (2007) Cryptic speciation and paraphyly in the cosmopolitan bryozoan *Electra pilosa* – impact of the Tethys closing on species evolution. Mol Phylogenet Evol 45:765–776

Norman A-M (1894) A month on the Trondjhem Fiord. Ann Mag Nat Hist 8(6):112–133

O'Dea A, Okamura B (2000) Intracolony variation in zooid size in cheilostome bryozoans as a new technique for investigating palaeoseasonality. Palaeogeogr Palaeclimatol Palaeoecol 162:319–332

Pallas PS (1766) Elenchus Zoophytorum Sistens Generum Adumbrationes Generaliores et Specierum Cognitarum Succinctas Descriptiones cum Selectis Auctorum Synonymis. F. Varrentrapp, The Hague

Palumbi SR, Martin A, Romano S, McMillan,WO, Stice L, Grabowski G (1991) The Simple Fool's Guide to PCR. Spec Publ, Dept Zool and Kewalo Mar Lab Univ Hawaii, Honolulu

Palumbi SR, Baker CS (1994) Contrasting population structure from nuclear intron sequences and mtDNA of humpback whales. Mol Biol Evol 11:426–435

Posada D, Buckley TR (2004) Advantages of AIC and Bayesian approaches over likelihood ratio tests for model selection in phylogenetics. Syst Biol 53:793–808

Posada D, Crandall KA (1998) Modeltest: testing the model of DNA substitution. Bioinformatics 14:817–818

Powell NA (1968) Bryozoa (Polyzoa) of Arctic Canada. J Fish Res Board Can 25:2269–2320

Prenant M, Bobin G (1966) Bryozoaires, 2e partie. Chilostomes Anasca, vol 68, Faune de France. Fédération Française des Sociétés des Sciences Naturelles, Paris

Rambaut A, Drummond AJ (2004) Tracer v1.3. http://beast.bio.ed.ac.uk/Tracer. Accessed 30 Nov 2009

Reverter-Gil O, Fernández-Pulpeiro E (2001) Inventario y cartografía de los briozoos marinos de Galicia (N.O. de España). Nova Acta Cient Compostel Monogr 1:1–243

Rioja lo Bianco E (1920) Una campaña biológica en el Golfo de Valencia. Anales del Instituto General y Técnico de Valencia. Trab Lab Hidrobiol Esp 7:5–36

Rodríguez F, Oliver JF, Marín A, Medina JR (1990) The general stochastic model of nucleotide substitution. J Theor Biol 142:485–501

Ronquist F, Huelsenbeck JP (2003) MRBAYES 3: Bayesian phylogenetic inference under mixed models. Bioinformatics 19:1572–1574

Ryland JS, De Blauwe H, Lord R, Mackie JA (2009) Recent discoveries of alien *Watersipora* (Bryozoa) in Western Europe, with redescriptions of species. Zootaxa 2093:43–59

Smitt F-A (1867) Bryozoa marina in regionibus arcticis et borealibus viventia recensuit F.-A. Smitt. Öfvers Kongl Vetens Akad Förhand 24:443–487

Swofford DL (2000) PAUP*. Phylogenetic analysis using parsimony (*and other methods). Sinauer Associates, Sunderland

Swofford DL, Olsen GJ (1990) Phylogenetic reconstruction. In: Hillis DM, Moritz C, Mable BK (eds) Molecular systematics, 2nd edn. Sinauer Associates, Sunderland, pp 407–543

Steentoft M, Irvine LM, Farnham WF (1995) Two terete species of *Gracilaria* and *Gracilariopsis* (Gracilariales, Rhodophyta) in Britain. Phycologia 34:113–127

Thompson JD, Gibson TJ, Plewniak F, Jeanmougin F, Higgins DG (1997) The CLUSTAL_X windows interface: flexible strategies for multiple sequence alignment aided by quality analysis tools. Nucleic Acids Res 25:4876–4882

Zabala M (1986) Fauna dels briozous dels Països Catalans, vol 84, Arxius de la Secció de Ciències. Institut d'Estudis Catalans, Barcelona, pp 1–836

# Chapter 16
# Large Sediment Encrusting Trepostome Bryozoans from the Permian of Tasmania, Australia

## Sediment Encrusting Trepostomes

**Catherine M. Reid**

**Abstract** The Permian glaciomarine rocks of Tasmania contain unusually large trepostome bryozoan colonies that encrust soft sediments. These colonies have initially attached to hard substrates such as dropstones or brachiopod shells and have subsequently grown outwards across the sediment paleosurface. Specimen diameter is generally 150–350 mm, with one incomplete specimen with a radius of 400 mm. At least two species form this sediment encrusting morphology: *Stenopora crinita* Lonsdale 1845 and *S. ovata* Lonsdale 1844. Both of these species more commonly exhibit branching growth forms and the sediment encrusting forms are developed in offshore environments where slow sediment accumulation rates have allowed these low-profile colonies to flourish. This paper details the features of one well preserved 350 mm *S. crinita* specimen, along with a discussion of other material, and the significance of these forms with respect to depositional environment.

**Keywords** Free-lying • Cold-water • *Stenopora* • Growth form • Bryozoa • Permian

## Introduction

Free-lying, or sediment encrusting bryozoans are known from the Paleozoic, most from Ordovician to Carboniferous faunas. These Lower Paleozoic forms grow outwards, to about 50 mm diameter, as thin sheets (e.g., *Tabulipora*, *Lichenalia*), or by multilayering forming small mounds on sediments or hard substrates (e.g., *Prasopora, Fistulipora*) (Bigey 1981; McKinney and Jackson 1989). Some specimens of the Silurian *Lichenalia* are 150 mm in diameter encrusting soft

C.M. Reid (✉)
Department of Geological Sciences, University of Canterbury, Private Bag 4800, Christchurch 8140, New Zealand
e-mail: catherine.reid@canterbury.ac.nz

A. Ernst, P. Schäfer, and J. Scholz (eds.), *Bryozoan Studies 2010*,
Lecture Notes in Earth System Sciences 143, DOI 10.1007/978-3-642-16411-8_16,

carbonate muds (Watkins 1993). The demise of immobile free-lying soft sediment dwelling forms after the end-Paleozoic has been associated with an increase in bioturbation at the seafloor (Thayer 1979). Free-lying morphologies re-appeared in post-Paleozoic faunas, however, these colonies are usually very small and cap-shaped (Nelson et al. 1988; McKinney and Jackson 1989; Hageman et al. 1998), and unlike Paleozoic forms are able to adjust their position on soft substrates via avicularian mandibles (McKinney and Jackson 1989).

Large, Late Paleozoic, forms from the Permian of Tasmania are the focus of this paper. The Tasmanian specimens are immobile sediment encrusters that are proportionately much wider than high, and significantly larger than other Paleozoic free-lying forms. The Permian contains other examples of giant colonies (Madsen and Håkansson 1989; Key et al. 2001, 2005) and large colony size is also a distinctive feature of the Tasmania Basin fauna (Reid 2003, 2010). Previous Tasmanian sediment encrusting forms have been referred to *Stenopora ovata* Lonsdale, 1844 (Reid 2003), and this study describes new material assigned to *S. crinita* Lonsdale 1845 that shows a complete colony and the nature of initial colony attachment.

## Geological Setting and Bryozoan Fauna

Early Permian sediments in the Tasmania Basin were deposited in glaciomarine environments (Clarke et al. 1989; Rogala et al. 2007) and are richly fossiliferous from the Sakmarian. Faunas are dominated by brachiopods and bryozoans, along with mollusks and crinoids, and despite the cold-water environment bioclastic limestones are well developed in Sakmarian and Artinskian units (Clarke et al. 1989; Rogala et al. 2007). Onshore, inner shelf environments in the Sakmarian at Maria Island (Fig. 16.1) are recorded in fossiliferous siltstones, sandstones and dropstone conglomerates all containing abundant bryozoans and other invertebrates. Typical bryozoan growth forms are erect rigid fenestrate, erect rigid branching, hard substrate and erect rigid foliose. Contemporaneous fossiliferous silts and sands in offshore, basin central Hobart and Cygnet regions again contain abundant brachiopods and bryozoans. Bryozoan growth forms in these offshore deposits are typically erect rigid fenestrate and erect rigid branching, with some erect rigid foliose forms present. The later are formed by *Stenopora tasmaniensis* Lonsdale, 1844, and are abundant at Maria Island (Clarke and Baillie 1984; Reid 2003).

Following the widespread deposition of non-marine and marginal marine facies in the late Sakmarian, marine deposition resumed in the Artinskian with offshore fossiliferous siltstones of the lower Cascades Group (Fig. 16.1). These offshore siltstones are widespread from Maria Island to Cygnet and contain brachiopods and erect rigid fenestrate and branching bryozoans. In southern Tasmania, bioclastic limestones of the upper Cascades Group were developed in the early Artinskian offshore settings and contain well preserved brachiopods, crinoids, mollusks, and

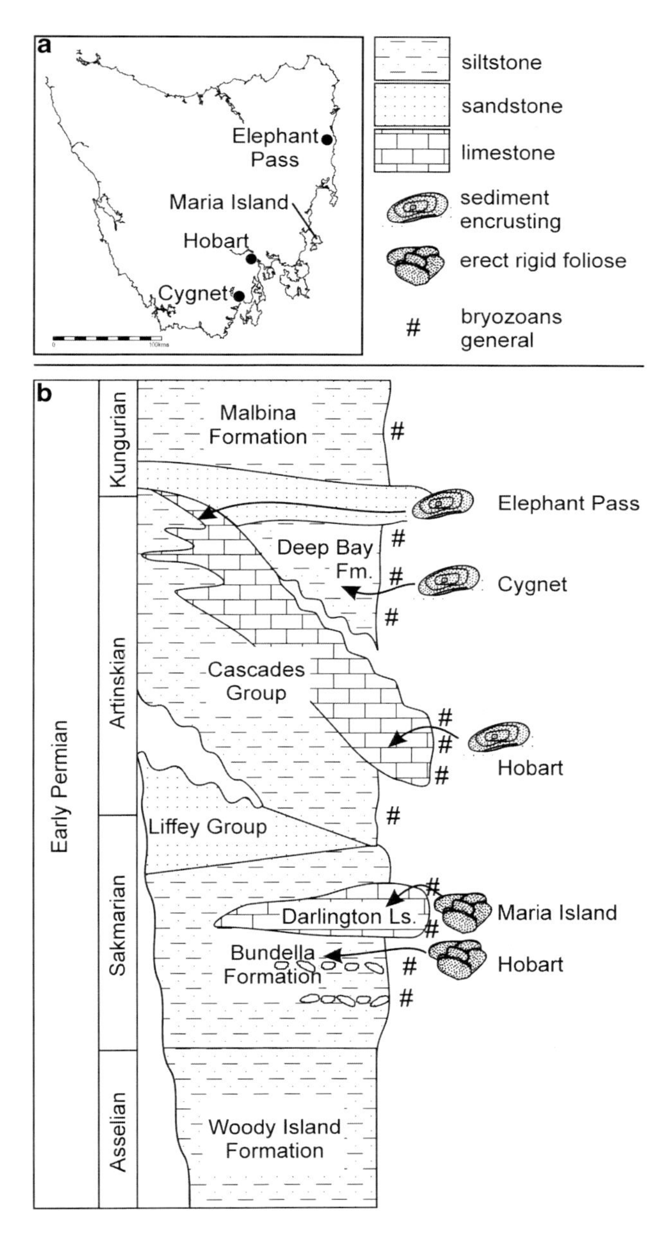


**Fig. 16.1** Site locations and generalized stratigraphy. (**a**) Location of significant bryozoan sites within Tasmania. (**b**) Early Permian stratigraphy of Tasmania, with stratigraphic occurrence of sediment encrusting and erect rigid foliose bryozoans marked (Modified from Reid (2010))

abundant bryozoans. Bryozoan growth forms are represented by erect rigid fenestrate and erect rigid branching along with hard substrate and sediment encrusting colonies (Fig. 16.1). These bioclastic limestone facies young, and thin, northward and in northern Tasmania are of late Artinskian age. The fossil content in the limestone facies remains similar, and bryozoan growth forms again include erect rigid fenestrate and branching forms and the sediment encrusting forms. At the same time, fossiliferous silts and sands were being deposited in offshore environments of the Deep Bay Formation in the Cygnet area of southern Tasmania (Farmer 1985) containing numerous brachiopods and bryozoans, all preserved as molds. Bryozoan growth forms include erect rigid fenestrate and branching forms, as well as sediment encrusting forms.

## Description of Material

Sediment encrusting bryozoans are found in several horizons and locations in Artinskian rocks of the Tasmania Basin. A few specimens with well preserved skeletons but incomplete colonies are known from Cascades Group limestones and have been assigned to *Stenopora ovata* Lonsdale, 1844 (Reid 2003). Additional large specimens, showing a colony radius up to 400 mm also occur in the Deep Bay Formation at Cygnet, described below, but skeletal material is not preserved and systematic identification is not possible. This study focuses on the description of a complete and well preserved specimen from the upper Cascades Group, Elephant Pass, and a summary of other material known from Hobart and Cygnet.

Specimens prefixed UC are housed in the University of Canterbury Geological Sciences collection. Specimens prefixed UTGD are housed in the University of Tasmania Geology Department collection.

### *Elephant Pass Material*

The specimen (UC19262) (Fig. 16.2a) was found within a loose boulder of fossiliferous silty limestone in the upper Cascades Group (late Artinskian) in a roadside outcrop in Elephant Pass (Fig. 16.1). The specimen is unique from other Tasmanian material in that the entire base and initial encrustation site are visible, and skeletal material is preserved. The colony is roughly circular, 240 by 350 mm in diameter, with a maximum thickness of 65 mm. The initial point of colony attachment is to a fragment of spiriferid brachiopod (Fig. 16.2b). The base of the colony shows concentric undulating growth bands 20–50 mm apart, that on closer examination are subdivided by finer bands 2–5 mm apart. In exposed cross sectional view the total thickness of the specimen is 65 mm, but this is divided into two sub-colonies,

**Fig. 16.2** Field photographs of sediment encrusting bryozoan UC19262 from Cascades Group limestone, Elephant Pass. (**a**) Overturned whole specimen showing initial encrustation point and successive growth rings, white arrows, total colony diameter (24 cm by 35 cm) and height, h (6.2 cm). Upper life position surface indicated by black arrows. Scale in centimeters. (**b**) Base of colony showing initial attachment to a spiriferid brachiopod, s, and subsequent unattached expansion of colony showing successive growth rings, arrows. Scale in centimeters

here designated A and B (Fig. 16.3a). The lower colony, of approximately 30 mm thickness, has been partially covered in sediment, and the surviving zoarium has overgrown itself in the same site (cf. Bigey 1981), as evidenced by growth relationships in one portion of the colony (Fig. 16.3b). Although the combined specimen height is up to 65 mm the height of an individual layer is 25–39 mm. The upper surface of the specimen is buried in well-cemented calcareous siltstone, and the presence or absence of monticules cannot be confirmed, but exposed cross sections show it has been bored (Fig. 16.3a) by unknown organisms.

## *Thin Section Description*

Acetate peel and thin sections were prepared of portions of the colony and internal features measured. In tangential section the autozooecia are oval, 0.39–0.63 mm

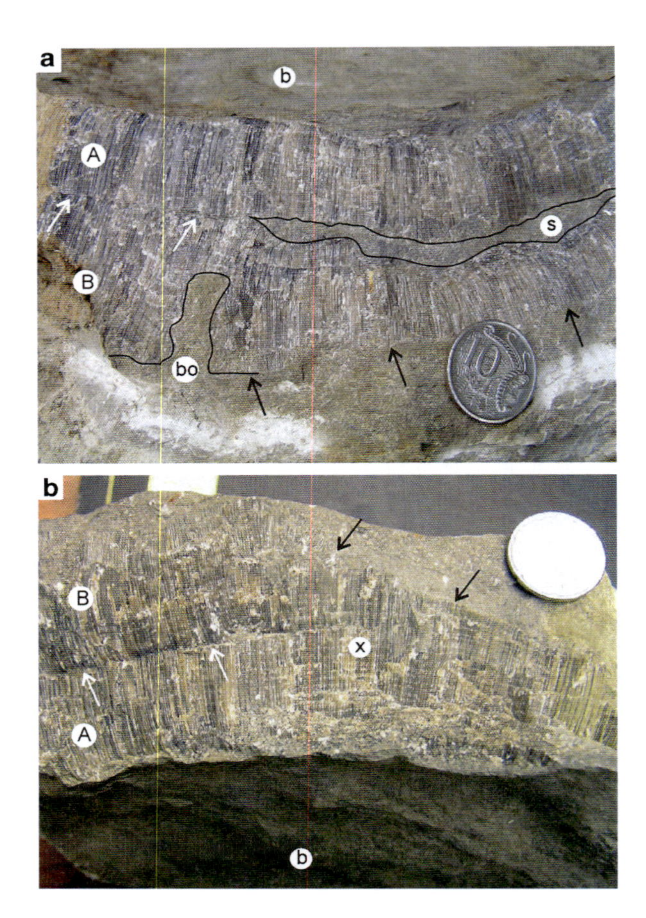


**Fig. 16.3** Photographs of sediment encrusting bryozoan UC19262 from Cascades Group limestone, Elephant Pass. (**a**) Detail showing successive growth of sub-colonies A and B, marked, note specimen is overturned and colony base is marked. (**b**) To the *left* of image sub-colony B grows directly on *top* of A, *white arrows*, but to the *right* of image the boundary is recorded by a 2–5 mm layer of sediment, s and outlined. The *top* surface of colony is undulose, *black arrows*, and contains a large boring, bo (*outlined*). Coin diameter 24 mm. (*b*) Detail of colony growth, specimen oriented in growth position and base marked, *b*. The boundary between sub-colony A and B is show to *left* of image, *white arrows*, and termination of this boundary occurs at x, where it can be seen the original colony, A, overgrows itself as sub-colony B to double the height of the specimen. The *top* of the colony is shown by *black arrows*, here 39 mm maximum thickness of single zoarium. Coin diameter 22 mm

diameter, surrounded by 5–7 small acanthostyles (Fig. 16.4a). Exilazooecia are 0.07–0.36 mm diameter and generally rare, however they do occur in small clusters in isolated areas of the zoarium. The zooecial tubes in the endozone are thin-walled and straight, with widely separated bands of thin monilae (Fig. 16.4b). Towards the

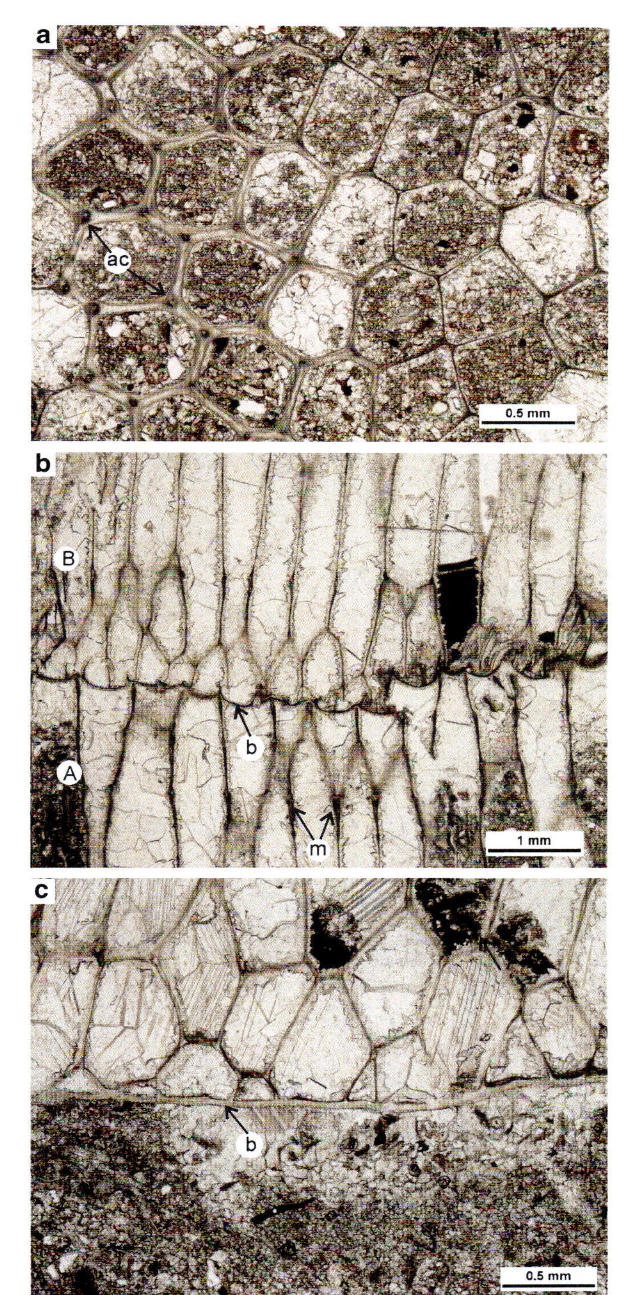


**Fig. 16.4** Thin section photomicrographs of *Stenopora crinita*, UC19262. *Scale bars* as shown. (**a**) Tangential thin section of UC19262 B showing thin endozonal wall on *right* of image, and exozonal wall with acanthostyles, ac, on *left* of image. (**b**) Transverse thin section through UC19262 A–B boundary, showing the widely spaced monilae in the base of B and the more closely spaced monilae, m, toward the *upper* part of the colony in A. Note the draping of the thin basal wall, b, over existing zooecia, *arrows*. Laterally this basal plate drapes over sediment covering UC19262 A (see Fig. 16.3). *Black* material is bitumen. (**c**) Transverse thin section of UC19262 A showing the base of colony, b. Note the thin basal skeleton of granular skeletal wall

upper part of the endozone, the monilae are slightly more closely spaced, but the exozone is thin and poorly defined, and zooecial walls here are 0.03–0.08 mm thick. Based on these internal features the specimen is identified as *Stenopora crinita* Lonsdale, 1845 (after Crockford 1945; Wass 1968; Reid 2003).

The base of the zoarium when encrusting sediment surfaces is marked by a straight single wall of granular skeletal material (Fig. 16.4c). The basal wall when occurring as an overgrowth of older zoarium, is draped over older zooecia that are an imperfectly flat surface (Fig. 16.4c). In thin section it is clear that diaphragms are typically absent and zooecial tubes that are not disturbed or repaired are 25–39 mm in height without the occurrence of diaphragms.

Portions of the colony that record disturbance by sediment cover, or are adjacent to sediment cover or other unknown causes of disturbance, show evidence of colony repair. Typically, the colony overgrows itself and basal wall is precipitated and new zooecia developed (Fig. 16.5a, b). The overgrowth of new basal wall is discontinuous in places and where some zooecia are replaced the adjacent zooecial tubes are continuous (Fig. 16.5b). The new colony base may overgrow the sediment cover with an undulating basal wall, but zooecial tubes remain more or less vertically oriented (Fig. 16.5a). Portions of the upper surface have been bored and the cavities are lined with skeletal material (Figs. 16.3a and 16.5c).

As discussed above, the bulk of the zoarium does not exhibit diaphragms, however, they do occur in some upper parts of the zoarium in areas where the colony has been disturbed. A clear set of simple diaphragms (Boardman 2001) occurs in the upper part of colony B, within 0.2–3.0 mm of the colony surface, adjacent to a portion of recent zoarium overgrowth, that has been bored (Fig. 16.5c).

## Other Material

Sediment encrusting bryozoans are also recorded from the early Artinskian Cascades Group, at Hobart, and the late Artinskian Deep Bay Formation, at Cygnet (Fig. 16.1). The Hobart material (UTGD20075), from the upper Cascades Group (Berriedale Limestone) preserves skeletal material and is tentatively identified as *Stenopora ovata* (Reid 2003). This material is of an incomplete zoarium, and the nature of the base of the colony indicates encrustation of a sediment surface rather than a hard substrate.

Additional specimens from the Deep Bay Formation at Cygnet are preserved as mold specimen only, and species identification is not possible. Colony sizes vary, but one incomplete colony shows a minimum radius of 400 mm (Fig. 16.6a), although most specimens are smaller at 150–200 mm diameter (Fig. 16.6b). The initial attachment point is not always evident, but one loose incomplete specimen (UC19263) shows encrustation to a cobble (Fig. 16.6c). Colony height of these specimens is about 10–15 mm, much thinner than the Elephant Pass material, but similar to the Hobart material assigned to *Stenopora* cf. *ovata*.

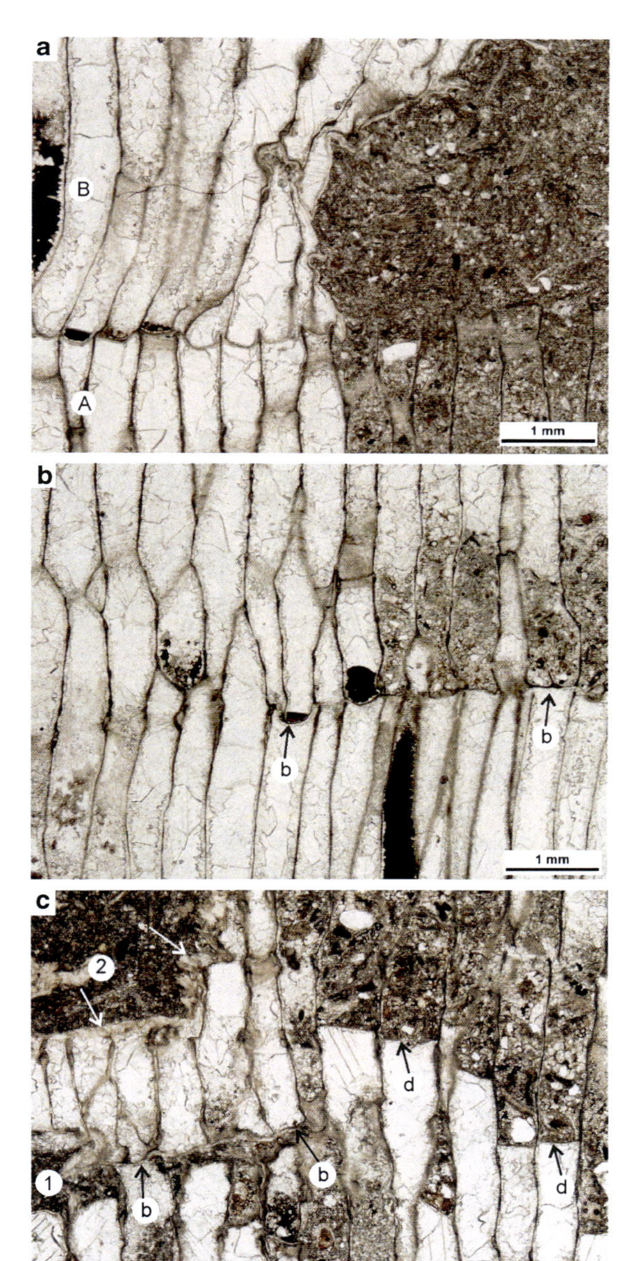


**Fig. 16.5** Thin section photomicrographs of colony repair features in *Stenopora crinita*, UC19262. *Scale bars* as shown. (**a**) Transverse thin section at boundary between colonies A and B, marked. *Left* side of image showing overgrowth of basal plate of colony B directly onto colony A, and *right* side of image showing colony B overgrowing sediment layer that has filled zooecial tubes of colony A. This sediment layer is laterally extensive (see Fig. 16.3a). (**b**) Transverse thin section of UC19262 B showing uninterrupted zooids on the *left* side of image, and the development of basal material, b, on the *right* side of image. (**c**) Transverse thin section of *upper* part of UC19262 B. *Lower-left* side of image shows development of new colony base, b, in response to a colony disturbance event, (1), in this case sediment cover. Clear diaphragms, d, are developed in the *upper* colony, at a similar colony level to borings, (2), into the recently developed new zooids. Zooecial tubes impacted by the boring have been closed by skeletal material, *arrows*. See text for further discussion

**Fig. 16.6** Field photographs of sediment encrusting bryozoans from the Deep Bay Formation, Cygnet. (**a**) Incomplete specimen (not collected) showing initial growth region, x, and scalloped growth rings, *black arrows*. Chisel 180 mm for scale. (**b**) In place large sediment encrusting colony (not collected) shown by growth rings, arrows, and colony centre, c. Hammer for scale. (**c**) Loose boulder specimen (UC19263) showing initial attachment of colony to a granite cobble, g, and growth of separate colony on *upper left, arrows*, shown by orientation of growth rings. Pocket knife is 9 cm

## Discussion

The bryozoan growth form of free-lying or sediment encrusting is well known from the Paleozoic, and the Tasmanian material does not display a new morphology. The Tasmanian material is, however, of much larger size than other examples. The colonies are both large in diameter at 200 to >400 mm, and thicker in height (25–39 mm for single layers of *Stenopora crinita,* 10–15 mm for *S. ovata*) than other Paleozoic forms. When exhibiting sediment-encrusting growth forms the colony height of both *S. crinita* and *S. ovata* are comparable to the radius of branching forms (15–32 mm for *S. crinita* and 5–15 mm for *S. ovata* (Crockford 1945; Wass 1968; Reid 2003)). The bryozoan colonies of the Tasmania Basin are commonly large, and this is not an unknown phenomenon in Permian faunas particularly within trepostomes such as *Tabulipora* (Madsen and Håkansson 1989; Nakrem 1995; Key et al. 2005) and *Stenopora* (Reid 2003). Within Tasmania *Stenopora tasmaniensis* occurs as large colonies of branching and foliose form, and other species of *Stenopora* form large branching colonies, and in the case of *S. ovata* and *S. crinita*, large sediment encrusting colonies. *Stenopora crinita* also forms large massive colonies of up to 250 mm diameter in the Sydney Basin with initial attachment to pebbles or shells (Crockford 1945).

Free lying, sediment-encrusting forms, by their low-profile morphology, will be prone to sediment coverage and burial, in comparison to higher-tiered erect forms (Reid 2010). As a result, this morphology will likely only be successful in environments with limited, or no, sediment accumulation. The sediment encrusting forms in the Tasmania Basin are found in offshore environments, where faunal composition and preservation indicates seafloor disturbance of fossils is rare, and sediment accumulation rates low (Reid 2010). In onshore environments such as Maria Island, erect branching, fenestrate, and foliose trepostomes predominate in siltstones over brachiopods and mollusks. Large foliose *S. tasmaniensis* are growing in situ and their erect form allows them to inhabit environments with moderate sediment accumulation rates but low water-energy (Reid 2010). The plasticity in growth form of *Stenopora* is allowing colonies to inhabit a wide variety of depositional environments, and the sediment encrusting form is well adapted to low energy and limited sediment accumulation. It is suggested here that if sediment accumulation is low or suspended over long periods encrusting colonies may attain large size.

While *S. tasmaniensis* exhibits erect branching, hard substrate encrusting and erect foliose forms it has not been recorded as sediment encrusting. In contrast, both *S. crinita* and *S. ovata* exhibit sediment encrusting, erect branching, and in the case of *S. crinita* at Black Head in the Sydney Basin, erect massive forms (Crockford 1945). Foliose forms observed by this author at Black Head may be *S. crinita*, but are not able to be collected. In Tasmania *S. tasmaniensis* is of Sakmarian age and foliose forms are developed in onshore and offshore environments where water energies are low, but sediment accumulation rates have precluded development of sediment-resting productid rich faunas (Reid 2010) and sediment encrusting forms.

*Stenopora ovata* occurs in Sakmarian and Artinskian rocks in Tasmania (Crockford 1945; Reid 2003), but only exhibits sediment encrusting morphologies in the Artinskian. *Stenopora crinita* occurs in Artinskian and younger rocks (Crockford 1945; Reid 2003) and along with *S. ovata* exhibits sediment encrusting form in offshore low-energy low sediment accumulation settings, co-occurring with productid brachiopods. While it remains unclear whether all species of the above species of *Stenopora* may be able to develop foliose, branching, and sediment encrusting forms, it appears that depositional environments are controlling the development of colony morphology rather than occurrence of species.

Diaphragms are absent from the diagnosis of *Stenopora* (see Wass 1968), and the Elephant Pass specimen does not exhibit diaphragms through most of the colony. However, clear simple diaphragms (cf. Boardman 2001) are seen in isolated parts of the upper zoarium of sub-colony B, lateral to borings into new zooids developed as overgrowths on sediment cover of the underlying colony surface (Fig. 16.5c). The internal features of the colony are otherwise typical of *Stenopora* and this specimen has been assigned to *S. crinita* on the basis of these internal features. Simple diaphragms are difficult to interpret (Boardman 2001), and in the specimen examined here occur within 0.5–3 mm of the colony surface. In life, the feeding polypide had to support itself near the colony surface, and in species such as *Stenopora crinita*, with long zooecial tubes unbroken by diaphragms, this must have been achieved by soft tissues. It is proposed that the occurrence of diaphragms in the Elephant Pass sediment encrusting specimen is a response to colony disturbance and the need to support regenerating feeding polypides in a colony under stress. The co-occurrence of diaphragms and new basal walls in this area of the colony (Fig. 16.5c) suggests that new feeding polypides developed in existing zooids, in response to disturbance of adjacent recently grown new zooids.

## Conclusions

1. Artinskian offshore sedimentary environments in the Tasmania Basin contain unusually large free-lying sediment encrusting colonies of *Stenopora crinita* and *Stenopora ovata*.
2. These sediment-encrusting colonies are best developed in offshore environments where both water energies and sedimentation rates are low.
3. Colonies establish themselves by initial attachment to shells or cobbles and are able to re-establish themselves in the same site of partially covered in sediment.
4. Diaphragms, usually absent in *Stenopora*, are locally developed in one colony, possibly as a strategy to overcome colony disturbance.

**Acknowledgements** Field activities were supported by a University of Canterbury Geological Sciences Internal Research Grant, and field work was carried out under a DPIWE Earth Materials collection permit (ES08091). Thin sections were prepared by Rob Spiers, Geological Sciences, UC.

## References


Bigey F (1981) Overgrowths in Palaeozoic Bryozoa: examples from Devonian forms. In: Larwood GP, Nielsen C (eds) Recent and fossil Bryozoa. Olsen and Olsen, Fredensborg, pp 7–17

Boardman RS (2001) The growth and function of skeletal diaphragms in the colony life of lower Paleozoic Trepostomata (Bryozoa). J Paleontol 75:225–240

Clarke MJ, Baillie PW (1984) Maria. Geological survey explanatory report, Geol Atlas 1:50,000 series, sheet 77 (8512N): Tasmania Department of Mines, Hobart, p 32

Clarke MJ, Forsyth SM, Bacon CA, Banks MR, Calver CR, Everard JL (1989) Late Carboniferous-Triassic. In: Burrett CF, Martin EL (eds) Geology and Mineral Resources of Tasmania. Geological Society of Australia Special Publication 15 Geological Society of Australia, Brisbane, pp 293–338

Crockford JM (1945) Stenoporids from the Permian of New South Wales and Tasmania. Proc Linn Soc NSW 70:9–24

Farmer N (1985) Kingborough. Geological survey explanatory report, Geological Atlas 1:50,000 series, sheet 88 (8311N). Tasmania Department of Mines, Hobart, p 97

Hageman SJ, Bock PE, Bone Y, McGowran B (1998) Bryozoan growth habits; classification and analysis. J Paleontol 72:418–436

Key MM Jr, Thrane L, Collins JA (2001) Space-filling problems in ramose trepostome bryozoans as exemplified in a giant colony from the Permian of Greenland. Lethaia 34:125–135

Key MM Jr, Wyse-Jackson PN, Håkansson E, Patterson WP, Moore MD (2005) Gigantism in Permian trepostomes from Greenland: testing the algal symbiosis hypothesis using $\delta^{13}C$ and $\delta^{18}O$ values. In: Moyano HIG, Cancino JM, Wyse-Jackson PN (eds) Bryozoan studies 2004. Taylor and Francis, London, pp 141–151

Lonsdale W (1844) Descriptions of six species of corals, from the Palaeozoic formations of Van Diemen's Land. In: Darwin C (ed) Geological observations on the volcanic islands visited during the voyage of the HMS Beagle. Smith, Elder, London, pp 161–169

Lonsdale W (1845) Palaeozoic fauna. Polyparia. In: de Strzelecki PE (ed) Physical description of New South Wales and Van Diemen's Land. Longman, Brown, Green and Longmans, London, pp 262–269, + 2 plates

Madsen L, Håkansson E (1989) Upper Palaeozoic bryozoans from the Wandel Sea Basin, North Greenland. Rapp Grønlands geol Unders 144:43–52

McKinney FK, Jackson JBC (1989) Bryozoan evolution. University of Chicago Press (reprint), Chicago, p 238

Nakrem HA (1995) Bryozoans from the Lower Permian Vøringen member (Kapp Starostin Formation), Spitsbergen (Svalbard), Skrifter vol 196, Norsk Polarinst, Oslo, p 92

Nelson CS, Hyden FM, Keane SL, Leask WL, Gordon DP (1988) Application of bryozoan zoarial growth-form studies in facies analysis of non-tropical carbonate deposits in New Zealand. Sediment Geol 60:301–322

Reid CM (2003) Permian Bryozoa of Tasmania and New South Wales: systematics and their use in Tasmanian biostratigraphy. Assoc Auatralas Palaeontol Mem 28:133

Reid CM (2010) Environmental controls on the distribution of Late Paleozoic bryozoan colony morphotypes: an example from the Permian of Tasmania, Australia. Palaios 25:692–702

Rogala B, James NP, Reid CM (2007) Deposition of polar carbonates during interglacial highstands on an Early Permian shelf, Tasmania. J Sediment Res 77:587–607

Thayer CW (1979) Biological bulldozers and the evolution of marine benthic communities. Science 203:458–61

Wass RE (1968) Permian Polyzoa from the Bowen Basin. Bur Min Res, Geol Geophys, Bull 90. Bureau Mineral Resources, Canberra, p 134

Watkins R (1993) The Silurian (Wenlockian) reef fauna of southeastern Wisconsin. Palaios 8:325–338

# Chapter 17
# Bryozoan Communities and Thanatocoenoses from Submarine Caves in the Plemmirio Marine Protected Area (SE Sicily)

## Bryozoans from Submarine Caves

**Antonietta Rosso, Emanuela Di Martino, Rossana Sanfilippo, and Vincenzo Di Martino**

**Abstract** Living and dead bryozoan communities from three caves in the "Plemmirio Marine Protected Area" (SE Sicily, Italy) were studied. Species richness from each cave and from the area as a whole (72 species) are comparable to those observed in other regions and caves within the Mediterranean. Communities consist largely of cave dwellers, sciaphilic and cryptic species, often related to coralligenous habitats, but include also some generalist species components. Bryozoans from hard surfaces (vaults, walls and floor) and bottom sediments were studied separately taking into account both living specimens and thanatocoenoses. According to previous data, communities of hard surfaces exhibit a trend of decreasing species richness towards the inner area and a clear patchiness, unlike those in sediments whose distribution appears strongly related to local sediment texture. Dead colonies and fragments from both hard surfaces and bottom sediments contribute valuable information concerning the pool of species potentially inhabiting caves. The usefulness and limits of different sampling methods for the study of cave bryozoans are discussed.

**Keywords** Bryozoans • Underwater caves • Hard-soft bottoms • Recent • Mediterranean

A. Rosso (✉) • R. Sanfilippo
Dipartimento di Scienze Biologiche, Geologiche e Ambientali, Università di Catania, Corso Italia, 55, Catania 95129, Italy

Consorzio Interuniversitario per le Scienze del Mare, Viale Isonzo, 32, Rome, Italy
e-mail: rosso@unict.it

E. Di Martino
Consorzio Interuniversitario per le Scienze del Mare, Viale Isonzo, 32, Rome, Italy

Dipartimento di Scienze Biologiche, Geologiche e Ambientali, Corso Italia, 55, Catania 95129, Italy
e-mail: manu.dimartino@hotmail.it

V. Di Martino
ISAFOM/Consiglio Nazionale Ricerche – OU Catania, Catania, Italy

A. Ernst, P. Schäfer, and J. Scholz (eds.), *Bryozoan Studies 2010*,
Lecture Notes in Earth System Sciences 143, DOI 10.1007/978-3-642-16411-8_17,

## Introduction

Bryozoans from Mediterranean cave environments have been intensively investigated in some geographical areas, especially near Marseille (Harmelin 1969, 1985a, b, 1986, 1997). Information is also available from Spain (Zabala and Gili 1985; Martí et al. 2004) and some Italian caves, mostly from the Tyrrhenian Sea, including the coasts of Campania (Balduzzi et al. 1980, 1985, 1989; Taddei-Ruggiero et al. 1996) and Ustica (Di Geronimo et al. 1993, 1997; Corriero et al. 1997). Several papers deal exclusively with settlement on panels (Harmelin 1980; Denitto et al. 2007). Sediments have been examined only in a few instances, mostly aiming to reconstruct the colonization history and/or the evolution of the caves relative to both sea level and morphological changes (Monteiro-Marquès 1981; Di Geronimo et al. 1993, 1997, 2000; Taddei-Ruggiero et al. 1996). Invariably, samples delivered only dead specimens, except for the semi-submerged Accademia Cave (Ustica), from where a few species encrusting pebbles from coarse-grained bottoms were reported (Di Geronimo et al. 1993).

This study was undertaken under the umbrella of a CoNISMa (National Inter-university Consortium for Marine Science) project aiming to analyse submarine cave environments within marine protected areas in Italy. This paper characterizes bryozoan faunas from three caves in the "Plemmirio Marine Protected Area" (PMPA), South of Syracuse (SE Sicily). Besides the living populations, thanatocoenoses from hard surfaces and bottom sediments were also examined in order to: (a) highlight similarities or differences between living and dead associations; (b) test the usefulness of different sampling methods and the information they give; (c) highlight compositional differences between caves and possible changes during the colonization history of each cave after its flooding, presumably at the beginning of the Holocene.

## Materials and Methods

The studied specimens come from three selected underwater caves (Granchi: GR, Gymnasium: GM, and Mazzere: MZ) located within the PMPA around the Maddalena peninsula, South of Syracuse, along the Ionian coasts of Sicily (Fig. 17.1). These karstic caves were formed presumably during the Quaternary lowstand phases in gently dipping Middle Oligocene-to-Tortonian limestones. Caves usually exhibit a nearly horizontal development and a flat floor with the openings situated within weaker layers, at about 20 m below sea level. The GR Cave is located on the more exposed NE side of the peninsula (15°19′40.2″N; 37°01′13.2″E), whereas the GM (15°18′48″N, 37°00′12″E) and the MZ (15°18′35.6″N, 37°00′18.3″E) caves lie on the opposite, more sheltered landward-facing SW side. Both GR and GM caves were mapped by Leonardi (1994). They are 53 and 65 m long respectively, do not exceeding 3 m in height and have blind ends. They consist of a suite of four chambers, more or less separated by locally

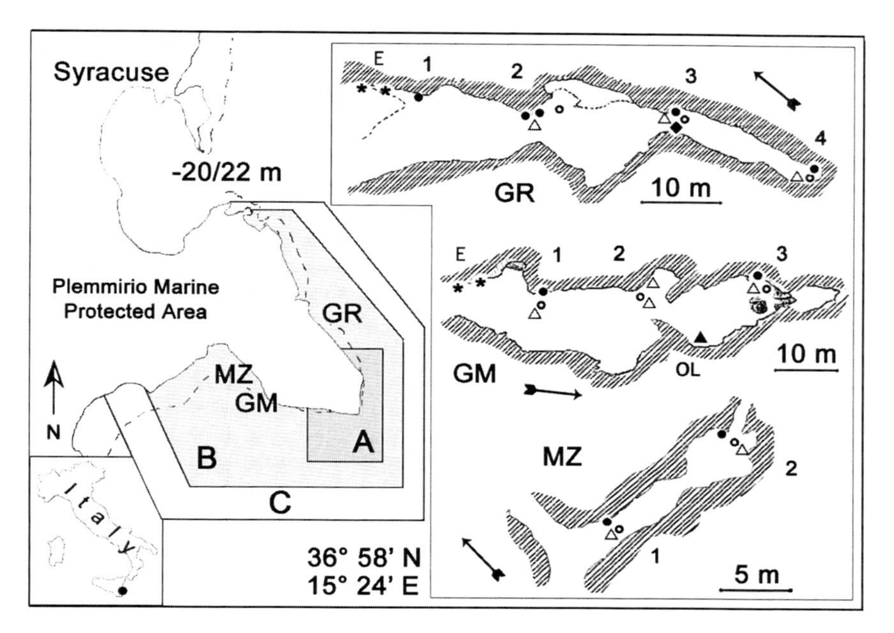


**Fig. 17.1** Map of the Plemmirio Marine Protected Area (PMPA) with location of the sampling sites within each investigated cave. *GM* and *GR* maps from Leonardi (1994); *MZ* map based on a preliminary relief by V. Di Martino. *A, B, C*: zonation of the PMPA. *Open circles*: vault samples; *solid circles*: wall samples; *solid diamonds*: hard bottom samples; *open triangles*: sediment samples; *solid triangles*: large sized sediment samples; *asterisks*: entrance algal samples; *E*: cave entrance stations; *1, 2, 3, and 4* stations at increasing distance from the entrance; *OL* previously collected large sediment sample

approaching lateral walls and concurrent lowering of the vaults and have wide (8 m wide × 3 m high) openings. In contrast, the MZ Cave (Di Martino V., pers. obs.) has a restricted entrance (3 m wide × 1.5 m high) and consists of one unexplored branch and a second, 35 m long and 3–4 m wide, progressively narrowing branch, that ends in a relatively wide chamber (ca. 1.8 m high) from which a very narrow tunnel (ca. 80 cm high) originates. Cave floors are largely covered with muddy deposits including bioclastic components and sparse large limestone clasts. After heavy rainy and stormy periods (as in the spring of 2009) it was noticed that the muddy bottom surface becomes extremely soupy. Sediments, including large amounts of *Posidonia* fibers, can accumulate inside the caves and mostly at their entrances where they form barriers. Small speleothemes are present as well as partly biogenic stalactites (Rosso et al., unpublished data 2010) hanging obliquely from the walls and perpendicularly from the vaults; rare stalagmites were observed in the deep part of the GM Cave. Additional information concerning the morphology, history and biology of GR and GM caves can be found in Leonardi (1994), Pitruzzello and Russo (2008), and Dutton et al. (2009).

Samples for this study were collected in September 2009 at the entrances and in some stations, at an increasing distance from the openings, from the walls, vaults

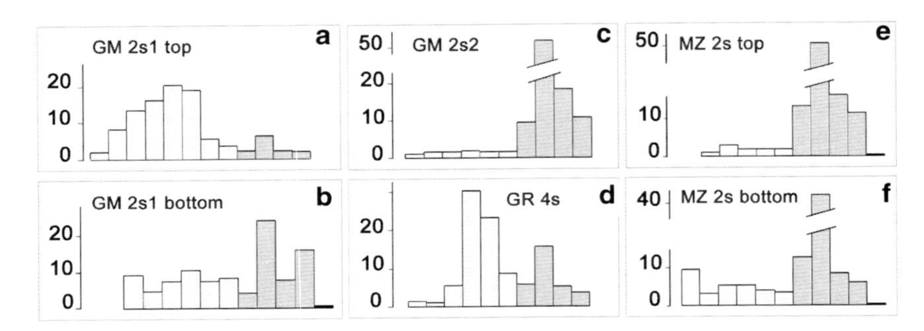


**Fig. 17.2** Frequency histograms for the texture of the most common sediments from cave floor surfaces (**c**, **e**) and for coarse sediments from selected sites and/or levels (**a**, **b**, **d**, **f**). *GM 2s1*: 15 m from the entrance; *GM 2s2*: 15 from the entrance; *GR 4s*: 35 m from the entrance; *MZ 2s*: 35 m from the entrance

and in the bottom sediments. On hard surfaces, limited (usually about 100 $cm^2$) but representative portions of the biogenic crust were scraped or broken off in order to minimize the possible sampling impact on the cave communities. Sediments were sampled using a hand-driven corer (diameter 8 cm; height 7–8 cm).

A total of 31 samples were collected (GR Cave: 14; GM Cave: 11; MZ Cave: 6): 8 from vaults, 9 from walls, 4 from the opening areas (irrespective of their location on the vaults or walls), 1 from a rocky floor, and 9 from bottom sediments before the scraping of walls and vaults. Another sample (ca. 2 $dm^3$), previously collected from the GM Cave, was also screened looking for additional species.

Samples collected on hard surfaces were kept in alcohol to avoid decomposition of organic matter. Sediments from minicores were usually treated as a single sample, except when different depositional layers (bottom and top) were identified based on textural and colour differences. Samples were washed and dried before sieving. Textural analyses of the sediments were performed on standard quantities, using sieving for fractions larger than 62.5 μm and an ELZONE granulometer for the mud component. Data is presented through frequency histograms (Fig. 17.2). Both living and dead bryozoans were identified under a stereomicroscope avoiding to extract fragments of biogenic concretions from hard surfaces, and picking colonies and their fragments from sediment fractions larger than 500 μm. Scanning electron micrographs were used to identify some species with very fine diagnostic morphological features.

Samples and specimens were deposited at the Palaeontological Museum, Catania University (PMC).

## Results

A total of 72 bryozoan species were found in the assemblages of the investigated caves, including 14 cyclostomes and 57 cheilostomes (18 anascans and 39 ascophorans). Diversity in each cave is lower (GR: 51 species, GM: 44 and MZ: 36) (Fig. 17.3a).

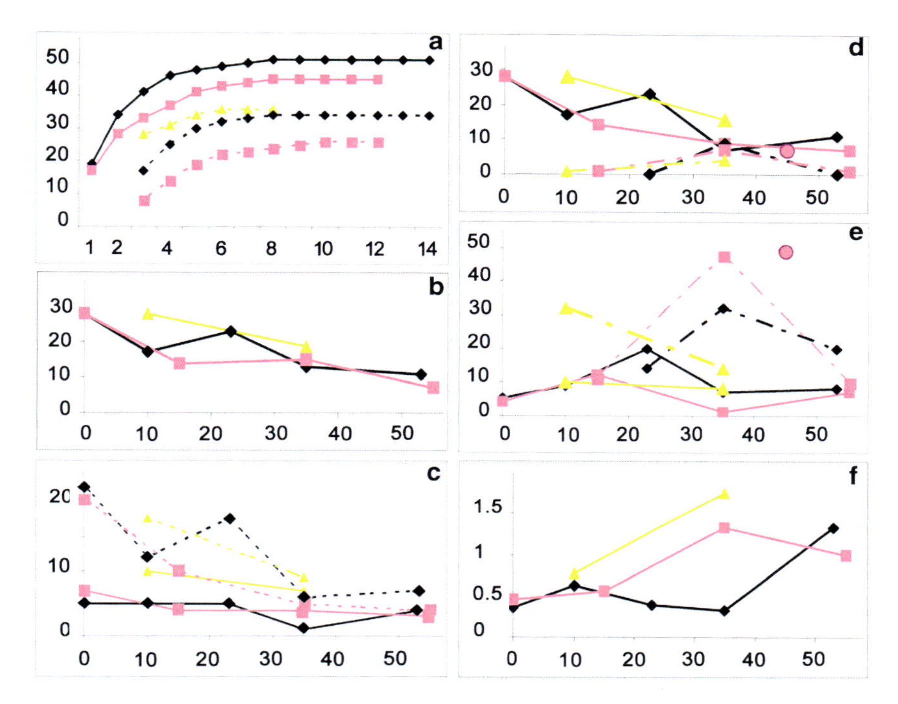


**Fig. 17.3** Plots of number of species in studied caves. *Black diamonds*: GR Cave; *dark gray squares*: GM Cave; *light gray triangles*: MZ Cave. (**a**) Cumulative curves of species richness of the bryozoan communities for each cave, including all samples (*continuous lines*) and omitting samples from the entrance, if present (*dashed lines*). (**b**) Total number of living species per site for individual caves (y-axis) found at increasing distance from the entrance (x-axis). (**c**) Total number of living species on hard surfaces per station site, separated for cheilostomes (*dotted lines*) and cyclostomes (*continuous lines*). (**d**) Total number of living species per site, separated for hard surface samples (*continuous lines*) and sediment samples (*broken lines*). (**e**) Total number of dead species per site, separated for hard surface samples (*continuous lines*) and sediment samples (*broken lines*). (**f**) Cyclostomes/ascophorans ratios per site in the hard surfaces communities

Taking in account only hard surfaces, total diversity is slightly lower (69 species) due to the absence of three ascophoran species. Moreover, the species richness in each cave is lower: 50, 39 and 33 species, in GR, GM and MZ caves, respectively. This implies that several species were present in only one or two out of the three caves. Among them, 24 species (4 cyclostomes and 20 cheilostomes) were recorded from a single sample. In contrast, only 17 species occurred in all caves. The most frequent taxa were runner *Aetea* species (frequency (F) = 68% of the examined samples), *Annectocyma major*, always represented by encrusting morphotypes (F = 63%), and *Puellina* (*Glabrilaria*) *pedunculata* (F ≈ 48%). Several other species had lower but still relatively high frequencies, ranging from 27% to 36%. These species were *Crisia pyrula*, with small erect flexible colonies, *Disporella hispida* (dome-shaped compound colonies), "*Microecia suborbicularis*" *sensu* Harmelin (1976) (fan-shaped branched colonies), *Onychocella marioni* (small-sized multilayered colonies), *Annectocyma indistincta*, *Escharina vulgaris*, and two *Puellina* species, *P.* (*Glabrilaria*) *corbula* and *P.* (*Cribrilaria*) *radiata*, both with minute colonies clustered in microcavities or encrusting cnidarian corallites. However, only a few of them (*E. vulgaris*, *O. marioni* and both *Puellina* species) are abundant, up to dominant in some samples.

Interestingly, several of the less frequent species were found only in samples from algae-dominated communities at the entrance of the caves. These are *Entalophoroecia deflexa*, *Crisia sigmoidea*, *Gregarinidra gregaria*, *Beania magellanica*, *Caberea boryi*, *Scrupocellaria* spp., *Mollia patellaria*, *Puellina* (*Puellina*) *setosa*, *Hippothoa flagellum*, *Chorizopora brongniartii*, *Reptadeonella violacea*, *Metroperiella lepralioides*, *Pentapora ottomuelleriana*, *Schizobrachiella sanguinea*, *Microporella* sp. B, and *Turbicellepora* cf. *magnicostata*. In contrast, few species were found in only one or few samples deeply inside the caves. They were *Puellina* (*C.*) *venusta*, *Fenestrulina malusii*, *Herentia hyndmanni*, and a *Bugula* species. These taxa are usually represented only by one or a few colonies.

Altogether, there was a trend of decreasing species richness, from the entrance to the end of each cave (Fig. 17.3b). This decreasing depends largely on the decreasing number of cheilostome species (Fig. 17.3c) and mostly ascophorans, except in the MZ Cave where the relatively rich cyclostome community has a decreasing trend similar to that of cheilostomes. Parallel decreases in coverage and often in the abundance of colonies have been observed and roughly estimated underwater (large areas of barren surface) and in the samples (coverage percentages for the innermost samples were always much less than 1% of the scraped surface).

Most bryozoans found formed small-sized encrusting colonies and a few erect flexible colonies, usually undetectable underwater (and often even in the laboratory with naked eye). Exceptions were some phidoloporid erect branching (vinculariform) and reticulate (reteporiform) species belonging to the genera *Reteporella* and *Schizoretepora*. Among them *Reteporella elegans* should be mentioned, as its up to 4–5 cm high, fan-shaped colonies form local populations near the entrance of the GM Cave.

Thanatocoenoses from the same samples account altogether for 41 species representing a decrease of over one/third relative to the living communities. Samples from near the cave entrances have the maximum decrease in species number. Indeed, it drops from 28 in the communities of the GR and the GM caves, to 4 and 5 in the corresponding thanatocoenoses, respectively (Fig. 17.3d). Species richness increases inwards (ranging from 10 to 20 species) but drops again to less than 8 in the deeper cave areas, although without a clear trend. Most dead specimens belong to species that are also present in the living associations, except for *Microecia occulta*, *Crassimarginatella crassimarginata*, *Calpensia nobilis*, *Escharoides coccinea*, *Schizoporella magnifica*, and *Reteporella feuerbornii*, which were found only in the thanatocoenoses.

Living bryozoans were also found in sediments largely consisting of silts (with a dominance of the medium component), and usually with a more or less marked tail due to the presence of gravelly-sandy fractions (Fig. 17.2c, e). Occasionally, sands and even coarser fractions prevail although the medium silt class remains evident (Fig. 17.2a, b, d, f).

Sediment communities include a total number of 17 species, but the number of species is much lower in each cave (9 in GR Cave, 7 in GM and 5 in MZ). Some species consist of entire colonies encrusting relatively large skeletons and fragments of invertebrates and other bryozoans. In addition, several species found consisted only of branches of erect colonies and variably sized fragments of encrusting colonies presumably detached from the rocky surfaces and able to survive for a certain time span. Most species found have a very low frequency, being present in only one out of the ten examined samples. Species such as *C. pyrula, O. marioni, Cosciniopsis ambita, E. vulgaris*, and *H. depressa* were present in two or three samples. Two species, *H. depressa* and *Hippomenella mucronelliformis*, were found exclusively in sediments. *Trematoecia ligulata* was present only in the large "old" sample from the GM cave.

Sediment samples show a peak in species richness in samples from 35 m (Fig. 17.3a), corresponding to the inner cave portions of the MZ Cave; the same is noted in samples from the intermediate sectors of the GR and MG caves. Interestingly, samples with living bryozoans always include coarse fractions.

Dead sediment assemblages include 68 species among which 17 are present neither in the living communities of the same cave nor in other caves in the area. These species are generally found in only one or two samples and mainly consist of small fragments of encrusting species. Thanatocoenoses along the caves show a trend comparable to that of the corresponding living communities, although there is a general increase in species richness with a pronounced peak at 35 m from the cave opening, with the exception of the MZ Cave where a decrease in species richness is obvious from 10 to 35 m (Fig. 17.3e).

## Discussion and Conclusions

The cumulative curves (Fig. 17.3) indicate that the number of investigated samples was enough to obtain all species presently living within the PMPA caves, despite their relatively small size. For hard surfaces, the sample size used was equivalent to about ¼ of that usually taken (see Bianchi et al. 2003) and comparable to less than 1/10 of those collected by Harmelin (1997).

After this study, the total number of cave-dwelling species in the area is set up to 72. This is rather low in comparison to the 116 species recorded by Harmelin (2000) from the French Mediterranean caves, including 77 and 78 species identified for the semi-dark (GSO) and dark (GO) cave communities, respectively. It should be noted however, that caves in that area have been intensively sampled (both the number of investigated caves and samples per cave) (Harmelin 1969, 1980, 1985a, b, 1986, 1997, 2000) in comparison to those of the PMPA (Di Geronimo et al. 2000, present study). The PMPA species number is slightly greater than that of Campania (65 species in two caves) (Balduzzi et al. 1980, 1989; Taddei-Ruggiero et al. 1996; Balduzzi and Rosso 2003), and Majorca areas (62 species in two caves) (Zabala and

Gili 1985); and it is twice as high as that found in two caves from Ustica Island, S Tyrrhenian Sea (34 species) (Di Geronimo et al. 1993; Corriero et al. 1997).

Otherwise, species richness recorded (50, 44 and 35 for the GR, GM and MZ caves, respectively) is comparable to that recorded for single caves from near Marseille (29–53 in the Moyade, Trémies, Triperie, and 3PP caves; Harmelin 1969, 1985a, b, 1986, 1997, 2000), Majorca (30 and 57 species recorded by Zabala and Gili (1985) from the Cathedral and J-1 caves), and Campania (35 and 46 from the Mitigliano and the Isca caves; Balduzzi et al. (1989) and Taddei-Ruggiero et al. (1996), respectively). Again, species richness values found in this study are higher than those from the shallower caves at Ustica (12 and 28 species from the Accademia and Grotta Verde caves; Di Geronimo et al. (1993) and Corriero et al. (1997), respectively).

The observed tendency of bryozoan cave communities to attain comparable species richness values could be tentatively explained through the MacArthur and Wilson's theory of island biogeography, considering each cave as an island and its species richness as the balance between immigration and local extinction (Harmelin 1997). On the other hand, the lower values found in the Ustica caves could be ascribed to the combination of their decidedly shallower location, involving a semi-submerged entrance, a higher light penetration, as well as possible differences in hydrodynamics caused by their near sea-level location, and sometimes the presence of a second submerged opening. Finally, for the Ustica Island caves, the isolation from the Sicily mainland leading to a more restricted pool of species as a possible near source of propagules could play an important role.

At first sight, it appears that the GR and the GM caves show a higher total species richness in relation to the MZ cave (Fig. 17.3a). Nevertheless, when samples collected at the cave entrances are omitted, species richness for single caves become more comparable (35, 26 and 34 species, for the GR, GM and MZ caves, respectively) despite differences in cave dimensions (the studied branch of the MZ Cave is narrow and only 35 m long contrasting with the 53 and 65 m of the wider GR and GM caves), exposition (the GM and MZ caves are more sheltered), opening size (larger for the GR and GM caves and small for the MZ cave), and presumed concomitant differences in the confinement degree. Finally, the sampling effort was also lower in the MZ Cave (6 samples, instead of 12 and 10 from the GR and GM caves, respectively, excluding the ones from near the entrances). The MZ Cave, with a small opening and a relatively sheltered exposition, is probably more confined than the other ones. Interestingly, this cave is not frequented by scuba divers. In contrast, the GM Cave, one of the favourite places for recreational diving, suffered a heavy research effort during the last years (Leonardi 1994; Taddei-Ruggiero et al. 1996; Pitruzzello and Russo 2008; Dutton et al. 2009, inter alias), exhibits relatively low species richness, presumably partly due to disturbance.

Looking at individual caves, a general impoverishment trend from the entrance towards the inner sectors is evident (Fig. 17.3b) and the drop at station 1 of GR Cave could be a sampling artefact resulting from only one sample. Also, the decreased species richness observed in the deeper areas could be caused by confinement and the concomitant food depletion as well as a decreased larval supply, as repeatedly suggested (Harmelin 1980, 1997; Balduzzi et al. 1989; Zabala et al. 1989). However, the observed pattern including a peak in species richness towards the middle, at the transition between the semi-dark and dark zones, doesn't match that recorded by Harmelin (1997, 2000) for bryozoan communities from underwater caves. Finally, the low number of species with obligatory small-sized colonies found in inward areas is somewhat consistent with the absence of bryozoan coverage recorded from about 40 m inward in the GR and GM caves (Pitruzzello and Russo 2008) and considered as one of the caves' general characteristic (Harmelin 1997).

## *Living and Dead Assemblages from Cave Walls and Vaults*

The trends delineated above depend largely on species richness, composition and distribution of the communities living on hard surfaces (Fig. 17.3c, d). Most species are known as typical cave dwellers and/or have been recorded as cryptic representatives in circalittoral and bathyal habitats. Some other taxa are more generalist already found in other Mediterranean caves. Among the species known to inhabit dark and semi-dark caves, *O. marioni*, *E. vulgaris*, *P. (G.) pedunculata* and several encrusting cyclostomes, particularly *A. indistincta* and *Diplosolen obelium* were found, the latter forming oligoserial lobate colonies typical of caves and cryptic habitats (Harmelin 1986). Some less sciaphilic species with affinity for coralligenous habitats, such as *C. pyrula*, *P. (C.) radiata*, *R. elegans* and *Schizoretepora solanderia,* are also present. The latter two species were earlier unknown from caves and could probably be added to the stock of diagnostic species in coralligenous and cave environments. A few other species of some interest were found for the first time. Among them, *C. ambita* and *Therenia rosei* (the latter recently described) are rare to common in the GM and MZ caves, but absent in the GR Cave. Their presence in caves in the Ionian Sea and other Italian regions (Rosso and Di Martino, unpublished data 2010) increases their geographical distribution and insight about their ecological requirements. Furthermore, *Bugula* spp., *Scrupocellaria* sp. A and *Microporella* spp. seemingly represent new species and/or alien species for the Mediterranean, whose status deserves further study (Rosso and Di Martino unpublished data 2010). Finally, *Synnotum aegyptiacum* was unexpectedly found, inside the GM Cave, at different distances from the entrance. This species seems to parallel the case of *Chlidonia pyriformis,* firstly found in the Mitigliano cave by Balduzzi et al. (1980) and subsequently recorded from other caves (Taddei-Ruggiero et al. 1996; Balduzzi and Rosso 2003; Corriero et al. 1997; Martí et al. 2004). Both species have small-sized, flexible, erect colonies, which are slender and lightly mineralised and have been recorded typically associated with littoral algae of the photophilic zone. In contrast, some species frequently recorded from caves, such as *Coronellina fagei*, *Callopora* spp., *Porella minuta*, were not found.

Excluding species present only in one sample, it appears that some species characterize single caves. For instance, *Escharoides mamillata* and *Parasmittina rouvillei* are restricted to the GR Cave, *Hippothoa flagellum* and *Reteporella* cf.

*mediterranea* to the GM Cave, and *Tubulipora hemiphragmata* and *Entalophoroecia gracilis* to the MZ Cave. Furthermore, species such as *D. obelium*, *P.* (*C.*) *radiata* and *Setosella cavernicola*, are extremely frequent (up to 100%) in one cave but absent in others, although, sometimes dead specimens were found. As a result, each cave has its own typical community sharing only a restricted pool of species with nearby ones, as already noticed for other regions (Harmelin 1997, 2000; Martí et al. 2004). Moreover, within each cave, and except for some taxa restricted to the cave entrances, most species exhibit a patchy distribution. These features, also noticed by Harmelin (1997, 2000) for caves and cryptic habitats could result from within-cave, small-scale, spatial heterogeneity, and external recruitment, possibly by subsequent events each favouring different species. This seems to be confirmed by the comparison between living and dead species found in the same sample, both in each cave and in the entire investigated area. The same surface can host different species at different times, but a single species may be present with living and dead colonies in different cave sectors. Finally, additional species found in the scant hard surface thanatocoenoses point to further compositional changes during time, caused by a continuous input of larvae supplied from the pool of species present on the neighbouring shelf.

As expected (Harmelin 2000) species with erect rigid morphotypes up to several centimetres high, are restricted to zones near the entrance and up to 10 m within, at least in the GM Cave. However, small-sized, flexible colonies mostly belonging to species with extremely slender, lightly mineralised colonies, such as *S. aegyptiacum*, *Scrupocellaria* spp., *Crisia* spp., *Caberea boryi* and *Savignyella lafonti* are also present in more recessed areas. The inferred adaptive strategy of these morphotypes is (a) to reduce drag and (b) to avoid siltation (Lagaaij and Gautier 1965). Consequently, their presence in cave environments has been related to both a relatively active water flow in tunnels (Harmelin 2000) and mud deposition (Zabala and Gili 1985), the latter being the most likely in the present instance.

Of particular interest is the presence of bryozoan nodules or "microbioherms" (see Jackson 1983 and Harmelin 2000) formed by several cyclostome and cheilostome species. In the PMPA caves, such an adaptation is exhibited by species growing usually as unilaminar encrusters, such as *Disporella hispida, O. marioni* and *Stephanollona armata*. The latter two in particular (usually growing as extensive colonies in neighbouring shelf settings) form small multilaminar colonies through self-overgrowth and/or frontal budding. This strategy could be an adaptation to elevate zooids from the cave surfaces subject to sediment smothering and/or for allowing the access to more effective water flow (and transported food particles) as suggested by Harmelin (2000).

## Living and Dead Assemblages from the Cave Sediments

The community thriving in the cave sediments is very scant, consisting of a total of 23 species: 9, 12 and 7 in the GR, GM and MZ caves, respectively. Each species is restricted to a few samples, typically from the middle part of the caves (Fig. 17.3d). Bryozoans colonize cave sediments when and where large (sand to gravel sized) clasts are present, consisting usually of authigenic bioclasts, such as solitary and colonial coral fragments, serpulids, molluscs, and disarticulated carapaces of crustacean decapods, which are extremely abundant in the GR Cave ("granchi" is the Italian word for "crabs"). From the sparse available data it appears that the presence of bioclasts may be extremely discontinuous in caves, also during time, as shown by the presence of superimposed sediment layers with different textures at certain sites (Fig. 17.2a, b, e, f). In particular, the poor sorting of sediments, often showing a negatively skewed bimodal, usually leptokurtic-to-highly-leptokurtic distribution, points to the mixing of selected fine allochthonous and *in situ* coarse biogenic components. The latter comes from the accumulation of skeletons and their fragments, such as carapaces and coxal segments of decapods in the GR Cave.

The living bryozoan communities consist almost entirely of encrusting species, some represented only by colony fragments, and by branches of phidoloporids, among which some *Reteporella* species and *S. solanderia* occur. They originate presumably, at least partly from colonies living on the walls and vaults which can be relatively easily detached and/or fragmented when they grow on exposed soft tissues of organisms such as sponges as observed in some samples, but also through mechanical actions by mobile organisms and by divers. Interestingly, sediment communities are more abundant and diversified in some samples from the GM cave, where dive impact caused by recreational and research activity is particularly intense (Leonardi 1994; Di Geronimo et al. 2000; Pitruzzello and Russo 2008; Dutton et al. 2009). Such colonies and fragments are probably able to survive for some time on the bottom constituting a sort of pseudo-population.

A few other species probably colonize the bottom clasts directly. This hypothesis is supported by the colonization pattern of some granules and small pebbles, paralleling the "small substrata" on shelf bottoms both in this area (Rosso 1996) and elsewhere (Harmelin 1976, 2000; McKinney 1996). Larvae may be supplied by fertile colonies dwelling on the cave walls or even from outside (at least for species not found on hard surfaces) together with silty sediments and *Posidonia* fibers transported by the incoming water flow and subsequently settling on the bottom. Of particular interest are species such as *H. depressa*, *H. mucronelliformis* and *T. ligulata* (the latter recently described) found exclusively in the sediments, but never on hard surfaces in any of the caves. This means that the analysis of sediments can add valuable information about cave communities. Moreover, the above mentioned species, as well as *S. cavernicola*, *P.* (*G.*) *corbula*, *T. rosei* and *C. ambita* increase the lists of living species in single caves due to their finding in sediment samples. The data set of species from each site is shown in Table 17.1.

Although poor in both species and specimens, the Plemmirio caves' communities are presently the only example of bryozoans recorded from soft bottom cave environments, except for the even poorer communities from Accademia cave, mostly colonizing volcanic pebbles (Di Geronimo et al. 1993).

Dead bryozoans form a relatively abundant part of the biogenics. Both the number of species and specimens is high compared to the living community, as

**Table 17.1** Distribution of bryozoans collected in the PMPA caves. *E* stations at the entrance and *1, 2, 3, 4*: stations at increasing distance from the entrance, *H* hard bottom sample, *S* sediment sample, *X* presence in both hard (vault/wall) and sediment samples, *C* living community, *T* thanatocoenosis, OL old sample

| Cave | Granchi (GR) | | | | | | | | | | Gymnasium (GM) | | | | | | | | | | Mazzere (MZ) | | | |
|---|---|---|---|---|---|---|---|---|---|---|---|---|---|---|---|---|---|---|---|---|---|---|---|---|
| Station | E | | 1 | | 2 | | 3 | | 4 | | E | | 1 | | 2 | | OL | | 3 | | 1 | | 2 | |
| Number and type of sample | 2 H | | 1 H | | 3 H,1 S | | 3 H,1 S | | 2 H,1 S | | 2 H | | 2 H,1 S | | 1 H,2 S | | 1 S | | 2 H,1 S | | 2 H,1 S | | 2 H,1 S | |
| Community/Thanatocoenosis | C | T | C | T | C | T | C | T | C | T | C | T | C | T | C | T | C | T | C | T | C | T | C | T |
| Cyclostomes | | | | | | | | | | | – | | | | | | | | | | | | | |
| *Annectocyma indistincta* (Canu-Bassler, 1929) | – | – | H | – | H | H | S | S | H | H | – | – | H | X | H | S | – | S | – | S | H | X | – | S |
| *Annectocyma major* (Johnston, 1847) | H | – | H | H | H | X | H | X | H | H | H | – | H | H | H | – | – | S | H | – | H | H | H | H |
| *Crisia fistulosa* Heller, 1867 | – | – | – | – | – | – | – | S | H | – | H | – | – | – | – | S | – | – | – | S | H | S | H | – |
| *Crisia pyrula* Harmelin, 1990 | H | – | – | – | H | H | S | – | – | – | H | – | – | – | – | S | – | S | X | – | H | – | H | – |
| *Crisia ramosa* Harmer, 1891 | – | – | H | – | – | S | – | – | – | S | H | – | – | – | – | – | – | S | – | – | – | – | – | – |
| *Crisia sigmoidea* Waters, 1916 | – | – | – | – | – | – | – | S | – | – | H | – | – | – | – | S | – | S | – | – | – | S | – | – |
| *Desmeplagioecia violacea* Harmelin, 1976 | – | – | – | – | – | – | – | – | – | – | – | – | – | – | – | – | – | – | – | – | H | – | – | – |
| *Diplosolen obelium* (Johnston, 1838) | H | – | – | – | H | H | – | – | H | H | – | – | – | – | – | S | – | S | – | – | H | H | H | H |
| *Disporella hispida* (Fleming, 1828) | H | – | – | – | H | S | S | – | – | – | H | H | H | – | H | S | – | S | – | H | H | S | H | S |
| *Disporella* sp. A | – | – | – | – | – | – | – | – | – | – | – | – | – | S | – | S | – | S | – | S | – | S | – | – |
| *Entalophoroecia deflexa* (Couch, 1844) | – | – | – | – | – | – | – | – | – | – | H | – | – | – | – | – | – | – | – | – | – | – | – | – |
| *Entalophoroecia gracilis* Harmelin, 1976 | – | – | – | – | – | – | – | – | – | – | – | – | – | – | – | – | – | – | – | – | H | – | H | – |
| *Idmidronea triforis* (Heller, 1867) | – | – | – | – | – | – | – | S | – | – | – | – | – | – | – | – | – | – | – | – | – | – | – | – |
| *Mecynoecia delicatula* (Busk, 1875) | – | – | – | – | – | – | – | – | – | – | – | – | – | – | – | – | – | S | – | – | – | – | – | – |
| *Microecia occulta* (Harmelin, 1976) | – | – | – | – | – | S | – | S | – | H | – | – | – | – | – | – | – | – | – | – | – | – | – | – |
| *Microecia "suborbicularis" sensu* Harmelin, 1976 | – | – | H | – | – | – | – | H | – | – | – | – | H | – | H | – | – | – | H | – | H | – | H | H |
| *Patinella mediterranea* de Blainville, 1834 | – | – | – | – | – | – | – | – | – | – | – | – | – | – | – | – | – | S | – | – | – | – | – | – |
| *Patinella radiata* (Audouin, 1826) | – | – | – | – | – | – | – | S | – | S | – | – | – | – | – | S | – | S | – | – | – | S | – | – |
| *Plagioecia patina* (Lamarck, 1816) | – | – | H | – | – | – | – | – | – | – | – | – | – | – | – | – | – | – | – | – | – | – | – | – |
| *Platonea stoechas* Harmelin, 1976 | – | – | – | – | – | S | – | S | – | – | – | – | – | – | – | S | – | S | – | – | – | S | – | – |
| *Tubulipora hemiphragmata* Harmelin, 1976 | – | – | – | – | – | – | – | – | – | – | – | – | – | – | – | – | – | S | – | – | H | – | – | – |

| | | | | | | | | | | | | | | | | | | | | | | | | |
|---|---|---|---|---|---|---|---|---|---|---|---|---|---|---|---|---|---|---|---|---|---|---|---|---|
| Cheilostomes | | | | | | | | | | | | | | | | | | | | | | | | |
| *Aetea* spp. | H | – | H | H | H | H | X | S | H | H | H | H | H | H | H | – | – | – | – | H | H | H | H | H |
| *Arthropoma cecilii* (Audouin, 1826) | – | – | – | – | – | – | – | S | – | – | – | – | – | – | – | – | – | – | – | – | – | – | – | – |
| *Beania magellanica* (Busk, 1852) | H | – | – | – | – | – | – | – | – | – | – | – | – | – | – | – | – | – | – | – | – | – | – | – |
| *Bugula* cf. *flabellata* (Thompson in Gray, 1848) | – | – | – | – | – | – | H | – | – | – | – | – | – | – | – | – | – | – | – | – | – | – | – | – |
| *Bugula* cf. *fulva* Ryland, 1960 | H | – | – | – | – | – | – | – | – | – | – | – | – | – | – | – | – | – | – | – | H | – | – | – |
| *Caberea boryi* (Audouin, 1826) | H | – | – | – | – | – | – | S | – | – | H | – | – | – | – | S | – | S | – | – | – | S | – | – |
| *Calpensia nobilis* (Esper, 1796) | – | – | – | – | – | – | – | X | – | S | – | – | – | – | – | S | – | S | – | – | – | S | – | – |
| *Cellaria salicornioides* Lamouroux, 1816 | – | – | – | – | – | – | – | S | – | S | – | – | – | – | – | S | – | S | – | – | – | S | – | – |
| *Celleporina caliciformis* (Lamouroux, 1816) | – | – | – | – | H | – | – | S | – | – | – | – | H | – | – | S | – | S | – | – | H | H | – | – |
| *Celleporina caminata* (Waters, 1879) | H | H | – | – | – | X | – | – | – | S | H | H | – | – | – | S | S | S | – | – | H | – | – | – |
| *Celleporina* cf. *caminata* (Waters, 1879) | – | – | – | – | – | – | – | – | – | – | – | – | – | – | – | – | – | – | – | – | – | S | – | – |
| *Celleporina canariensis* Aristegui, 1989 | – | – | H | – | H | – | – | S | – | – | H | – | – | – | – | – | – | – | – | – | H | – | – | – |
| *Celleporina siphuncula* Hayward-McKinney, 2002 | – | – | – | – | – | – | – | – | – | – | – | – | – | – | – | S | – | S | – | – | – | – | – | – |
| *Chaperiopsis annulus* (Manzoni, 1870) | – | – | – | – | – | – | – | – | – | S | – | – | – | – | – | – | – | – | – | – | – | – | – | – |
| *Chlidonia pyriformis* (Bertoloni, 1810) | – | – | – | – | H | – | – | – | – | – | – | – | – | – | – | – | – | – | – | – | – | – | – | – |
| *Chorizopora brongniartii* (Audouin, 1826) | H | – | – | – | – | – | – | – | – | – | – | – | – | – | – | – | – | – | – | – | H | – | – | – |
| *Copidozoum planum* (Hincks, 1880) | – | – | – | – | – | – | – | – | – | – | – | – | – | – | – | S | – | – | – | – | – | S | – | – |
| *Copidozoum tenuirostre* (Hincks, 1880) | – | – | – | – | – | – | – | – | – | – | – | – | – | – | – | – | – | S | – | – | – | – | – | – |
| *Cosciniopsis ambita* (Hayward, 1974) | – | – | – | – | – | – | – | – | – | – | – | – | H | X | H | X | S | S | – | S | – | – | S | X |
| *Crassimarginatella crassimarginata* (Hincks, 1880) | – | – | – | – | – | H | – | – | – | – | – | – | – | – | – | – | – | – | – | – | – | – | – | – |
| *Crassimarginatella maderensis* (Waters, 1898) | H | H | – | – | H | H | – | – | – | – | – | – | – | – | – | – | – | S | – | – | H | H | – | – |
| *Crassimarginatella solidula* (Hincks, 1860) | – | – | H | H | – | – | – | – | – | – | – | – | – | – | – | S | – | S | – | – | – | – | – | – |
| *Escharina vulgaris* (Moll, 1803) | H | – | – | – | H | X | X | S | – | – | H | – | – | – | S | S | S | S | – | – | H | S | – | – |
| *Escharoides coccinea* (Abildgaard, 1806) | – | – | – | – | – | H | – | S | – | S | – | – | – | – | – | – | – | – | – | – | – | – | – | – |
| *Escharoides mamillata* (Wood, 1844) | – | – | H | H | H | H | S | S | – | – | – | – | – | – | – | S | – | S | – | – | – | – | – | – |
| *Fenestrulina malusii* (Audouin, 1826) | – | – | – | – | H | – | – | – | – | – | – | – | – | H | – | S | – | – | – | – | – | H | – | – |
| *Gregarinidra gregaria* (Heller, 1867) | H | – | – | – | – | – | – | – | – | – | – | – | – | – | – | – | – | – | – | – | – | – | – | – |
| *Hagiosynodos latus* (Busk, 1856) | – | – | – | – | – | S | – | – | – | S | – | – | – | – | – | S | – | S | – | S | – | – | – | – |
| *Haplopoma sciaphilum* Silén-Harmelin, 1976 | – | – | H | – | – | H | – | – | – | – | – | – | – | – | – | – | – | – | – | – | – | – | – | – |

(continued)

**Table 17.1** (continued)

| Cave | Granchi (GR) | | | | | | | | | | Gymnasium (GM) | | | | | | | | | Mazzere (MZ) | | | | |
|---|---|---|---|---|---|---|---|---|---|---|---|---|---|---|---|---|---|---|---|---|---|---|---|---|
| *Herentia hyndmanni* (Johnston, 1847) | – | – | H | H | – | – | – | – | – | – | – | – | – | – | – | – | – | – | – | – | – | – | H | – |
| *Hippaliosina depressa* (Busk, 1854) | – | – | – | – | – | – | – | S | – | – | – | – | – | S | S | S | S | S | – | S | – | S | S | S |
| *Hippomenella mucronelliformis* (Waters, 1899) | – | – | – | – | – | – | S | S | – | – | – | – | – | – | – | S | – | – | – | – | – | – | – | S |
| *Hippopleurifera pulchra* (Manzoni, 1870) | – | – | – | – | – | – | – | – | – | – | – | – | – | – | – | – | – | S | – | – | – | – | – | – |
| *Hippothoa flagellum* Manzoni, 1870 | – | – | – | – | – | – | – | – | – | – | H | – | – | – | – | – | – | – | – | – | – | – | – | – |
| *Margaretta cereoides* (Ellis-Solander, 1786) | – | – | – | – | – | S | – | S | – | S | – | – | – | S | – | S | – | S | – | – | – | – | – | S |
| *Metroperiella lepralioides* (Calvet, 1903) | H | – | – | – | – | – | – | – | – | – | – | – | – | – | – | S | – | – | – | – | – | – | – | – |
| *Microporella* sp. A | – | – | – | – | – | – | – | – | – | S | – | – | – | – | – | – | – | – | – | – | – | – | – | – |
| *Microporella* sp. B | H | – | – | – | – | – | – | – | – | – | – | – | – | – | – | – | – | – | – | – | – | – | – | – |
| *Microporella* sp. C | – | – | – | – | – | – | – | – | – | – | – | – | – | – | – | – | – | – | – | – | – | S | – | – |
| *Mollia patellaria* (Moll, 1803) | H | – | – | – | – | – | – | – | – | – | H | – | – | – | – | – | – | – | – | – | – | – | – | – |
| *Myriapora truncata* (Pallas, 1766) | – | – | – | – | – | – | – | – | – | – | – | – | – | – | – | – | – | S | – | – | – | – | – | – |
| *Onychocella marioni* Jullien, 1881 | – | – | H | H | H | H | S | – | H | – | – | – | H | X | X | S | – | S | – | X | H | S | X | X |
| *Onychocella vibraculifera* Neviani, 1895 | – | – | – | – | – | – | – | – | – | – | – | – | – | – | – | S | – | – | – | – | – | S | – | – |
| *Parasmittina raigii* (Audouin, 1826) | – | – | – | – | – | – | – | – | – | – | – | – | – | – | – | – | – | – | – | – | – | S | – | – |
| *Parasmittina rouvillei* (Calvet, 1902) | H | – | – | – | H | – | – | – | – | – | – | – | – | – | – | S | – | – | – | – | – | S | – | – |
| *Pentapora ottomuelleriana* (Moll, 1803) | H | – | – | – | – | S | – | S | – | S | – | – | – | – | – | S | – | S | – | – | – | – | – | – |
| *Plesiocleidochasma mediterraneum* Chimenz-Soule, 2003 | – | – | – | – | – | – | – | – | – | – | – | – | – | – | – | – | – | – | – | – | – | S | – | S |
| *Puellina (Cribrilaria) innominata* (Couch, 1844) | – | – | H | – | – | – | – | – | – | – | – | – | – | – | – | – | – | – | – | – | – | – | – | – |
| *Puellina (C.) radiata* (Moll, 1803) | H | – | H | – | H | H | – | S | – | – | – | – | – | S | – | S | – | S | – | – | X | S | – | S |
| *Puellina (C.) setiformis romana* Harm-Arist., 1988 | H | – | – | – | H | - | – | – | – | – | H | – | – | – | – | S | – | S | – | – | – | – | – | – |
| *Puellina (C.) venusta* (Canu-Bassler, 1925) | – | – | – | – | H | – | – | S | – | – | – | – | – | – | – | S | – | – | – | – | – | – | – | – |
| *Puellina (Glabrilaria) corbula* Bishop-Hous., 1987 | – | – | – | H | H | – | – | S | H | X | – | – | – | S | S | S | – | – | – | – | H | H | H | S |
| *Puellina (G.) pedunculata* Gautier, 1956 | – | – | H | – | H | H | H | H | H | H | H- | – | H | H | – | – | – | – | H | H | H | H | H | H |
| *Puellina (Puellina) setosa* (Waters, 1899) | H | – | – | – | – | – | – | – | – | – | – | – | H | – | – | – | – | – | – | – | – | – | – | – |
| *Reptadeonella violacea* (Johnston, 1847) | H | H | – | – | – | – | – | S | – | S | – | – | – | – | – | S | – | S | – | – | – | S | – | S |
| *Reteporella* cf. *complanata* (Waters, 1894) | – | – | – | – | – | – | – | – | – | – | H | – | – | – | S | S | – | S | – | – | – | – | – | – |

| | | | | | | | | | | | | | | | | | | | | | | | | |
|---|---|---|---|---|---|---|---|---|---|---|---|---|---|---|---|---|---|---|---|---|---|---|---|---|
| *Reteporella elegans* Harmelin, 1976 | – | – | – | – | – | – | – | S | – | – | – | – | H | S | – | S | – | S | – | S | – | X | S | S |
| *Reteporella feuerbornii* (Hass, 1948) | – | – | – | – | – | – | – | H | – | – | – | – | – | – | – | – | – | – | – | – | – | S | – | – |
| *Reteporella grimaldii* (Jullien, 1903) | – | – | – | – | – | – | – | S | – | – | H | – | – | – | – | – | – | S | – | – | H | – | – | – |
| *Reteporella* cf. *mediterranea* (Smitt, 1867) | – | – | – | – | – | – | – | – | – | – | H | – | H | H | S | S | – | S | – | – | – | – | – | – |
| *Rhynchozoon revelatus* Hayward-McKinney, 2002 | – | – | – | – | – | – | – | – | – | – | – | – | – | – | – | – | – | S | – | – | – | – | – | – |
| *Schizomavella cornuta* (Heller, 1867) | H | H | – | – | H | H | S | – | – | S | H | – | – | – | – | S | – | – | – | – | H | X | – | – |
| *Schizomavella discoidea* (Busk, 1859) | H | – | H | – | – | – | – | – | – | – | H | – | – | – | – | – | – | – | – | – | – | – | – | – |
| *Schizomavella hastata* (Hincks, 1862) | – | – | – | – | – | – | – | – | – | S | – | – | – | – | – | S | – | – | – | – | – | – | – | – |
| *Schizomavella linearis linearis* (Hassall, 1841) | – | – | – | – | – | S | – | – | – | – | – | – | – | – | – | – | – | – | – | – | – | – | – | – |
| *Schizobrachiella sanguinea* (Norman, 1868) | H | – | – | – | – | S | – | – | – | – | H | – | – | – | – | – | – | S | – | – | – | – | – | – |
| *Schizoporella dunkeri* (Reuss, 1848) | – | – | – | – | – | – | – | – | – | – | – | – | – | – | – | – | – | S | – | – | – | – | – | – |
| *Schizoporella magnifica* (Hincks, 1886) | – | – | – | H | – | – | – | – | – | – | – | – | – | – | – | – | – | – | – | – | – | – | – | – |
| *Schizoretepora serratimargo* (Hincks, 1886) | – | – | – | – | – | – | – | – | – | – | H | H | – | – | – | – | – | S | – | – | H | – | – | – |
| *Schizoretepora solanderia* (Risso, 1826) | – | – | – | – | – | – | – | S | – | – | – | – | – | S | – | S | – | S | – | S | – | H | S | S |
| *Schizotheca fissa* (Busk, 1856) | – | – | – | – | – | – | – | H | – | – | – | – | – | – | H | – | – | – | – | – | – | – | – | – |
| *Scrupocellaria* cf. *aegeensis* Harmelin, 1969 | H | – | – | – | – | – | – | – | – | – | – | – | – | – | – | – | – | – | – | – | – | – | – | – |
| *Scrupocellaria delilii* (Audouin, 1826) | H | – | – | – | – | – | – | – | – | – | – | – | – | – | – | – | – | – | – | – | – | – | – | – |
| *Scrupocellaria scrupea* Busk, 1852 | – | – | – | – | – | – | – | – | – | – | H | – | – | – | – | – | – | – | – | – | – | – | – | – |
| *Scrupocellaria scruposa* (Linnaeus, 1758) | – | – | – | – | – | X | – | S | – | S | H | – | – | X | – | S | – | S | – | S | – | S | – | – |
| *Scrupocellaria* sp. A | – | – | – | – | – | – | – | S | – | S | H | – | – | – | – | – | – | S | – | – | H | – | – | – |
| *Setosella cavernicola* Harmelin, 1977 | – | – | H | H | H | H | – | – | H | X | – | – | – | – | S | – | – | – | – | S | – | – | H | S |
| *Smittina cervicornis* (Pallas, 1766) | – | – | – | – | – | S | – | – | – | – | – | – | – | – | – | – | – | – | – | – | – | – | – | – |
| *Stephanollona armata* (Hincks, 1861) | H | H | – | – | – | H | – | – | – | – | H | – | – | – | – | S | S | S | – | – | H | X | – | – |
| *Synnotum egyptiacum* (Audouin, 1826) | – | – | – | – | – | – | – | – | – | – | – | – | H | H | H | – | – | – | H | H | – | – | H | – |
| ? *Synnotum* sp. A | – | – | – | – | – | – | – | – | H | – | – | – | – | – | – | – | – | – | – | – | – | – | H | – |
| *Therenia rosei* Berning-Tilbrook-Rosso, 2008 | – | – | – | – | – | – | – | – | – | – | – | – | H | H | – | S | – | S | H | H | – | S | S | – |
| *Trematooecia ligulata* Ayari-Taylor, 2008 | – | – | – | – | – | – | – | – | – | – | – | – | – | – | – | S | S | S | – | – | – | – | – | – |
| *Turbicellepora* cf. *magnicostata* (Barroso, 1919) | – | – | – | – | – | – | – | – | – | – | H | – | – | – | – | – | – | – | – | – | – | – | – | – |
| *Umbonula ovicellata* Hastings, 1944 | – | – | – | – | – | – | – | – | – | S | – | – | – | – | – | S | – | – | – | – | – | S | – | – |
| *Watersipora subovoidea* (d'Orbigny, 1852) | – | – | – | – | – | – | – | S | – | S | – | – | – | – | – | S | – | S | – | – | – | S | – | – |

expected for soft bottom thanatocoenoses (Rosso 1996). Likewise the sediment lithic component, the bioclastic material suggests a bimodal provenance. A large part is clearly autochthonous and its species richness is derived presumably from the accumulation of fragments of rare and patchily distributed species and/or colonizing the caves at subsequent times. This is pointed out by the presence of taxa in the sediment thanatocoenoses typical of caves which were however absent from living communities, and some even from hard surfaces thanatocoenoses, in single caves or in the area as a whole. Nevertheless, thanatocoenoses from the few collected samples included only about 80% of all species presently living in the caves, probably due to the systematic absence of taxa with unpreservable non-mineralised, skeletons or low profiled, heavy mineralized colonies strongly adhering to the rocky surfaces, and hence hardly detachable and, consequently not supplying fragments to the bottom sediments. Furthermore, some species found in the thanatocoenoses, mostly in samples from near the cave entrance, are added to the sediments, probably through *post-mortem* transport from outside. These species (i.e. *Umbonula ovicellata*, *Chaperiopsis annulus*, *Watersipora subovoidea*, *Hagiosynodos latus*, and *Margaretta cereoides*) are known as common representatives of photophilic communities, mostly associated with shallow water soft algae and *Posidonia* meadows (Harmelin 1976; Zabala 1986; Rosso personal observations). Some specimens even show evidence of adhesion to these vegetal substrata on their under-surfaces (immuration casts).

Thanatocoenoses from sediments are particularly rich in comparison to those from hard surfaces. Like living communities from sediments, thanatocoenoses show a marked peak in species richness at about 35 m from the entrance (Fig. 17.2e). Richness values are high in comparison to caves from near Marseille located at about the same depth (Monteiro-Marques 1981), but are decidedly low when compared with the dead assemblages listed from the semi-submerged Accademia cave that include a wide range of dead specimens from the cave outside (Di Geronimo et al. 1997).

## Sampling Methods

The scraping of limited (10 × 10 cm) surface areas at certain intervals along underwater caves seems sufficient to evaluate species richness and total biodiversity in such environments, as suggested by results obtained for the Plemmirio caves. This sample size, only ¼ of the 20 × 20 cm usually adopted (Balduzzi et al. 1989; Bianchi et al. 2003), and from comparable to less than 1/10 of what was collected by Harmelin (1997), is thus suggested in order to reduce the possible sampling impact, above all in particularly interesting and/or protected areas.

Furthermore, it is clear from our data that both the analysis of bottom sediments and their thanatocoenoses can be an additional source of information about cave communities. Examination of both living and dead communities in small soft-bottom samples is suggested as the least destructive method to achieve a general idea about the species pool characterizing a given cave. However, further studies, possibly quantitative are also needed to better test the real potential of such samples. Treated with caution, sediment thanatocoenoses may be important, mediating the effect of patchiness in space distribution, and allowing to find species that only sporadically colonize caves or are rare (see Kidwell and Bosence 1991; Rosso 1996). For instance, fragments of *Myriapora truncata* found exclusively in the sediment thanatocoenoses from the GM Cave, suggest that this species is present in the cave community at about 20 m from the entrance, as actually confirmed by the examination of vault photographies (Pitruzzello and Russo 2008).

Nevertheless, both the photographic methods suggested by Balduzzi et al. (1982, 1985) and a mere visual census seem to be absolutely inefficient to study bryozoan community composition, and even general coverage, as most species (small-sized and cryptic) can be completely overlooked. This is confirmed in the earlier erroneously inferred complete absence of bryozoans in the inner parts of the GR and GM caves by Pitruzzello and Russo (2008) where, however, 24 species were found. Furthermore, most species cannot be identified (even from high resolution photos), whereas especially large-sized, mostly erect, and generally more obvious species (falling within the "conspicuous taxa" of Hiscock 1987) are selectively recognised and, hence, overestimated.

**Acknowledgements** Underwater sampling was made possible through E. Incontro (Director of the Plemmirio Marine Protected Area). Additional samples were available from the algological team lead by G. Furnari (Dipartimento di Scienze Biologiche, Geologiche e Ambientali, Univ. Catania). Thanks are due to J. Harmelin (Station Marine d'Endoume, Marseille), B. Berning (Oberösterreichische Landesmuseen, Linz), and P. Moissette (Univ. Claude Bernard, Lyon) for critical reviews. A. Viola and M. Vagliasindi (Dipartimento di Scienze Biologiche, Geologiche e Ambientali, Univ. Catania) are thanked for SEM assistance and textural analyses, respectively. This paper was financially supported by the CoNISMa project "Study of submarine cave environments – CODICE HABITAT 8330 – in the Marine Protected Areas of Pelagie, Plemmirio and Capo Caccia", led by S. Fraschetti (Univ. Lecce). Catania Palaeontological Research Group: contribution n. 372.

## References

Balduzzi A, Rosso A (2003) Briozoi. In: Cicogna F, Bianchi NC, Ferrari G, Forti P (eds) Grotte marine: cinquant'anni di ricerca in Italia. CLEM-ONLUS, Min Amb Tutela Territ, Rapallo, pp 195–202

Balduzzi A, Boero F, Cattaneo R et al (1980) Ricerche sull'insediamento dello zoobenthos in alcune grotte marine della penisola sorrentina. Mem Biol Mar Ocean suppl. 10:121–127

Balduzzi A, Boero F, Cattaneo R et al (1982) An approach to the study of the benthic fauna of some marine caves along the Penisola Sorrentina (Naples, Italy). In: Blanchard J, Mair J, Morrison I (eds) Proceeding of the sixth international scientific symposium of the underwater federation, Edinburgh, pp 176–182

Balduzzi A, Pansini M, Pronzato R (1985) Estimation par relèvements photographiques de la distribution de spongiaires et bryozoaires dans une grotte sous-marine du Golfe de Naples. Rapp Comm Int Mer Médit 29(5):131–134

Balduzzi A, Bianchi CN, Boero F et al (1989) The suspension-feeder communities of a Mediterranean Sea cave. In: Ros JD (ed) Topics Mar Biol Scient Mar 53 (2–3):387–395

Bianchi CN, Pronzato R, Cattaneo-Vietti R et al (2003) I fondi duri. In: Gambi MC, Dappiano M (eds) Manuale di metodologie di campionamento e studio del benthos marino mediterraneo. Biol Mar Mediterr 10(suppl):199–232

Corriero G, Scalera Liaci L, Gristina M et al (1997) Composizione tassonomica e distribuzione della fauna a Poriferi e Briozoi in una grotta semisommersa della riserva naturale marina "Isola di Ustica". Biol Mar Mediterr 4(1):34–43

Denitto F, Terlizzi A, Belmonte G (2007) Settlement and primary succession in a shallow submarine cave: spatial and temporal benthic assemblage distinctness. Mar Ecol 28(suppl 1):35–46

Di Geronimo I, La Perna R, Rosso A et al (1993) Popolamento e tanatocenosi bentonica della Grotta dell'Accademia (Ustica, Mar Tirreno meridionale). Nat Sicil ser.4, 17(1–2): 45–63

Di Geronimo I, Allegri L, Improta S et al (1997) Spatial and temporal aspects of benthic thanatocoenoses in a Mediterranean infralittoral cave. Riv Ital Paleontol Stratigr 103(1):15–28

Di Geronimo I, La Perna R, Rosso A et al (2000) Associazioni bentoniche da sedimenti di grotte carsiche in Sicilia. Speleol Iblea 8:97–102

Dutton A, Scicchitano G, Monaco C, Desmarchelier JM, Antonioli F, Lambeck K, Esat TM, Fifield LK, McCulloch MT, Mortimer G (2009) Uplift rates defined by U-series and $^{14}$C ages of serpulid-encrusted speleothemes from submerged caves near Siracusa, Sicily (Italy). Quat Geochronol 4:2–10. doi:10.1016.quageo.2008.06.003

Harmelin JG (1969) Bryozoaires des grottes sous-marines obscures de la région Marseillaise: faunistique et écologie. Téthys 1(3):793–806

Harmelin JG (1976) Le sous-ordre des Tubuliporina (Bryozoaires Cyclostomes) en Méditerranée: Ecologie et systématique. Mem Inst Océanogr Monaco 10:1–326

Harmelin JG (1980) Etablissement des communautés de substrats durs en milieu obscur. Résultats préliminaires d'une expérience à long terme en Méditerranée. Mem Biol Mar Oceanogr suppl. 10:29–52

Harmelin JG (1985a) Bryozoan dominated assemblages in Mediterranean cryptic environments. In: Nielsen C, Larwood GP (eds) Bryozoa: Ordovician to Recent. Olsen & Olsen Fredensborg, Denmark, pp 135–143

Harmelin JG (1985b) Organisation spatiale des communautés des grottes sous-marines de Méditerranée. Rapp Comm Int Mer Médit 29(5):149–153

Harmelin JG (1986) Patterns in the distribution of bryozoans in the Mediterranean marine caves. Stygologia 2(1/2):10–25

Harmelin JG (1997) Diversity of bryozoans in a Mediterranean sublittoral cave with bathyal-like conditions: role of dispersal processes and local factors. Mar Ecol Prog Ser 153:139–152

Harmelin JG (2000) Ecology of cave and cavity dwelling bryozoans. In: Herrera Cubilla A, Jackson JBC (eds) Proceedings of the 11th International Bryozoology Association conference, Smithsonian Tropical Research Institute, Balboa, pp 38–55

Hiscock K (1987) Subtidal rock and shallow sediments using diving. In: Baker JM, Wolff WJ (eds) Biological surveys of estuaries and coasts. Cambridge University Press, Cambridge, pp 198–237

Jackson JBC (1983) Biological determinants of present and past sessile animal distributions. In: Tevesz JSM, McCall PM (eds) Biotic interactions in Recent and fossil benthic communities. Plenum Press, New York/London, pp 39–120

Kidwell SM, Bosence DWJ (1991) Taphonomy and time-averaging of marine shelly faunas. In: Allison PA, Briggs DEG (eds) Taphonomy. Releasing the data locked in the fossil record. Plenum Press, New York/London, pp 115–209

Lagaaij R, Gautier YV (1965) Bryozoan assemblages from marine sediments of the Rhône delta, France. Micropaleontology 11(1):39–58

Leonardi R (1994) Contributo alla conoscenza delle grotte sommerse della Penisola della Maddalena (SR). Boll Accad Gioenia Sci Nat Catania 27(348):599–620

Martí R, Uriz MJ, Ballesteros E, Turon X (2004) Benthic assemblages in two Mediterranean caves: species diversity and coverage as a function of abiotic parameters and geographic distance. J Mar Biol Assoc UK 84:557–572

McKinney FK (1996) Encrusting organisms on co-occurring disarticulated valves of two marine bivalves: comparison of living assemblages and skeletal residues. Paleobiology 22(4):543–567

Monteiro-Marques V (1981) Peuplements des planchers envasés de trois grottes sous-marines obscures de la région de Marseille. Etude préliminaire. Téthys 10(1):89–96

Pitruzzello P, Russo GF (2008) Particolarità ambientali di due grotte sottomarine dell'area marina protetta del Plemmirio (Siracusa, Sicilia). Biol Ital 11(2008):41–47

Rosso A (1996) Popolamenti e tanatocenosi a Briozoi di fondi mobili circalitorali del Golfo di Noto (Sicilia SE). Nat Sicil 4 20(3–4):189–225

Taddei-Ruggiero E, Annunziata G, Rosso A et al (1996) Il benthos della Grotta sottomarina dell'Isca (Penisola Sorrentina): evidenze faunistiche della sua evoluzione recente. VII Congr Naz Soc Ital E Napoli 17:329–332

Zabala M (1986) Fauna dels bryozous dels Països Catalans. Inst Est Catalans Sec Cienc Barcelona 84:1–833

Zabala M, Gili JM (1985) Distribution des bryozoaires le long d'un gradient sédimentaire dans deux grottes sous-marines du littoral de Majorque. Rapp Comm Int Mer Médit 29(5):137–140

Zabala M, Rieira T, Gili JM, Barange M, Lobo A, Peñuelas J (1989) Water flow, trophic depletion, and benthic macrofauna impoverishment in a submarine cave from the western Mediterranean. PSZNI Mar Ecol 10(3):271–287

# Chapter 18
# The Genus *Sparsiporina* d'Orbigny, 1852 (Bryozoa, Cheilostomata): Late Eocene to Holocene

## The Genus *Sparsiporina* d'Orbigny

**Antonietta Rosso and Giampietro Braga**

**Abstract** *Sparsiporina elegans* (Reuss, 1848), type species of the Phydoloporidae genus *Sparsiporina,* typical of late Eocene to early Oligocene sediments from northern Italy and central Europe, is redescribed mostly based on topotype material. A second species is recognised as belonging to the same genus, *S. ramulosa* (Calvet 1906), first described from doubtfully living material from the Capo Verde Islands and now found to be relatively common in Gelasian sediments from Southern Italy. The two species show close similarities in zooid morphology, zooid arrangement along the branches and growth form, but differ mostly in the morphology of the avicularia, the morphology and location of the ovicells and the zooidal dimensions. The stratigraphical and geographical distributions of both species included in the genus are discussed, as well as their ecological requirements, which probably changed slightly during geological time.

**Keywords** Taxonomy • Cenozoic • Europe • Recent • Atlantic • Ecology

## Introduction

The genus *Sparsiporina* d'Orbigny, 1852 was originally introduced by d'Orbigny (1852) for *Retepora elegans* Reuss, 1848 and included within his family Porinidae (see Gordon 2002). Nevertheless, it was subsequently placed into the family

A. Rosso (✉)
Dipartimento di Scienze Biologiche, Geologiche e Ambientali, Sezione Scienze della Terra, Università di Catania, Corso Italia, 55, Catania I-95129, Italy
e-mail: rosso@unict.it

G.P. Braga
Formerly at the Università di Padova, Dipartimento di Geoscienze, Via Gradenigo 6, Padova I-35131, Italy
e-mail: giampietro.braga@gmail.com

A. Ernst, P. Schäfer, and J. Scholz (eds.), *Bryozoan Studies 2010*,
Lecture Notes in Earth System Sciences 143, DOI 10.1007/978-3-642-16411-8_18,

Phydoloporidae by Canu and Bassler (1929) where it has been traditionally accommodated as its features fit well within the family's characters. They include: the erect rigid but delicate branching colony with zooids opening on a single side, the presence of extrazooidal calcification on dorsal and basal colony portions, the slightly sculptured frontal wall with a single marginal pore, the primary orifice obscured by a protruding peristome bearing a pseudospiramen, the hyperstomial ovicell initially prominent but becoming immersed by secondary calcification, the presence of frontal avicularia. But, notwithstanding its long history, the genus remained monotypic (see Bock 2002) after more than one century and a half.

The finding of several specimens in sediments of Gelasian age from Sicily, strongly recalling both the fossil *Retepora elegans* from north Italy, and *Retepora ramulosa* Calvet 1906 described from North Atlantic deep-waters, fostered the comparison of these taxa and the placement of the latter species in *Sparsiporina*. This paper (1) re-describes and figures both *Sparsiporina elegans* (Reuss) and *S. ramulosa* (Calvet) through high quality SEM images, and (2) discusses the stratigraphic distribution of the genus as well as the ecological requirements of its constituent species.

## Materials and Methods

The material was mostly collected during field activities performed by Braga in the Veneto and south Trentino localities, NE Italy, and by Rosso in the Capo Milazzo Peninsula, NE Sicily (Fig. 18.1). Furthermore, type material of both analysed species was examined.

Specimens of *Sparsiporina elegans* (Reuss) derive from sedimentary formations particularly important both from stratigraphical and palaeogeographical perspective, which were deposited during the upper Eocene (Priabonian) – lower Oligocene (Rupelian) time span, largely corresponding to the C13/P 17-18/NP 21 zones. These Paleogene sediments rich in bryozoans were well known since the nineteenth century, when they were studied extensively by Reuss (1848, 1869) and Waters (1891, 1892) and designated as "Bryozoen Mergel" or "Bryozoen Schichten", mostly by the former author. Such bryozoan-bearing sediments, usually referred to as "Marne a Briozoi" or "Marne di Brendola", are isochronous and easily recognizable, thus constituting an important index-layer in the region. Nevertheless, marls locally grade into marly limestones or silty marls, and bodies they form are variable in thickness and laterally heteropic with coarser and shallower deposits, in relation to the environmental situations along the Veneto-Trentino palaeo-shelf margin, where deposition happened (Fig. 18.1b). Sedimentologic features, faunal content, bryozoan growth form and ecological requirements of still extant species point to an outer shelf environment in less than 100 m depth, in the photic zone (Braga 1963; Braga and Barbin 1988).

The examined *S. elegans* specimens originate from the classical Val di Lonte site and the Priabona locality, in the eastern side of the Lessini Mountains, and from near Brendola, Toara, and Monteccio di Costozza (Fig. 18.1b). The associated bryozoans belong to nearly 80 species, mostly forming delicate erect both rigid and articulate colonies, whereas fragments of encrusting species are rare.

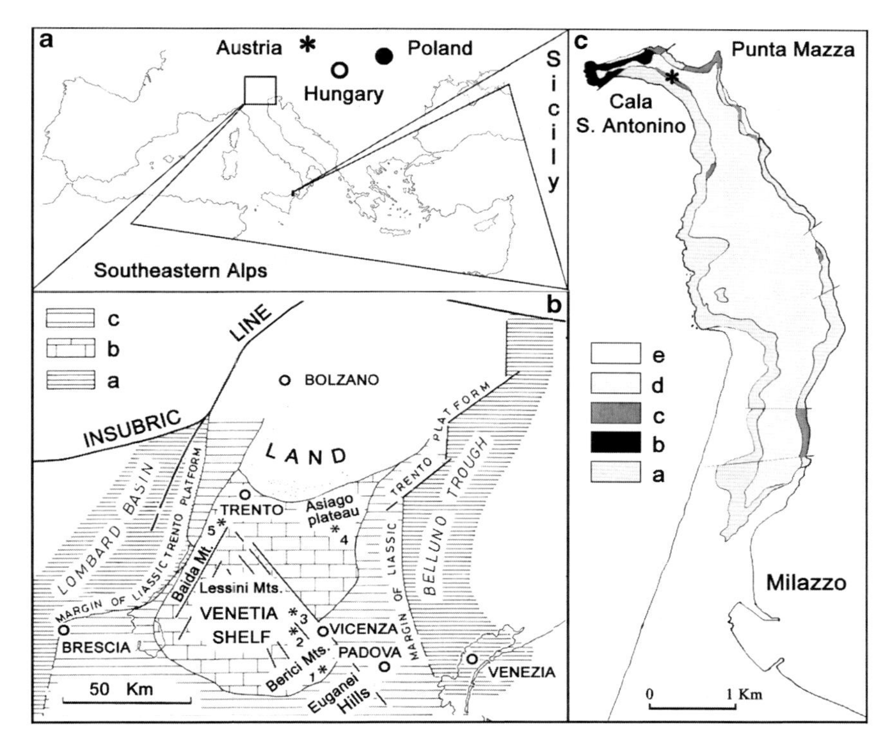


**Fig. 18.1** *Sparsiporina*-bearing localities. (**a**) European outcrops. *Asterisk*: Vienna area in the Austrian basin; *open circle*: Hungarian localities; *solid circle*: NW Transylvania (Romania). (**b**) upper Eocene – lower Oligocene successions of the Venetia shelf (Modified from Gordon and Braga (1994)). *a*: marly limestones of the Scaglia Cinerea Formation, westward, and marly, sandy and calcareous shales (flysch facies), eastward; *b*: shelf-facies deposits; *c*: hemipelagic deposits. *Asterisks* indicate bryozoan localities: *1*: Brendola, Toara, Monteccio di Costozza; *2*: Val di Lonte; *3*: Priabona; *4*: Crosara; *5*: Pannone. (**c**) Pleistocene sediments from the Capo Milazzo area, in Sicily (Modified from Fois (1990) and Lentini et al. (2000)). *a*: metamorphic basement; *b*: coralgal biolitites and calcareous breccias – late Tortonian-early Messinian; *c*: calcareous siltstones and marls – early Pliocene to early Pleistocene; *d*: loose biogenic cobblestone and volcanic ashes – middle to late Pleistocene; *e*: Holocene sediments. *Asterisk* indicates the Cala S. Antonino outcrop

Fossil material of *Sparsiporina ramulosa* (Calvet) originates from several samples collected in the so-called "yellow calcareous marls" (YCM) cropping out at Capo Milazzo Peninsula, in the north-eastern part of Sicily. YCM show an extremely high $CaCO_3$ content as they are mostly formed by planktonic (globigerinids and *Orbulina*) foraminifers. Sediments are actually coarser, consisting of silts, sandy silts, silty sands and muddy sands, and locally including thin lenses with abundant bioclasts (isidids, bryozoans, serpuloideans and echinoids) in the sandy and gravely fractions (see Rosso 2002a, b). Outcrops are usually a few metres thick, discontinuous and marked by erosive basal surfaces. Sediments fill small basins and local depressions of previous coastal areas locally lying on the crystalline basement, the Miocene reefal terrains or the basal conglomerates (Fois 1990). Palaeoenvironment was at relatively shallow bathyal depths probably less than 600 m as inferred from brachiopods (Gaetani and Saccà 1984; Gaetani 1986), foraminifers (Violanti 1988), ostracods (Sciuto 2003) and also from bryozoans (Rosso 2002a, b), corals and other faunas (Rosso pers. observations). Deposition occurred during and after the transgression following the Messinian salinity crisis (Fois 1990).

Particularly, specimens of *S. ramulosa* were found in layers cropping out at Cala S. Antonino, on the northwestern coast of the Capo Milazzo Peninsula (Fig. 18.1c), sedimentation having occurred during the MPl5 and MPl6 zones (Violanti 1988), largely overlapping with the Gelasian Stage of Rio et al. (1994). There, bryozoan palaeocommunities include nearly 60 species, most with erect rigid or flexible colonies.

Specimens were picked from residues coarser than 250 μm, due to the small size of the studied species. Selected fragments were treated with $H_2O_2$ and accurately washed to eliminate adhering clay and silty sediments. Specimens were observed uncoated under low vacuum conditions in a LMU Tescan Vega Scanning Electron Microscope. Some *S. elegans* and *S. ramulosa* images were taken by D. Gordon at NIWA, N. Vávra at Vienna University and P. Louzet at the Museum National d'Histoire Naturelle (MNHN) in Paris. Measurements were taken on SEM photos and reported as follows: min, max, mean, standard deviation, and, in brackets, the number of measurements. ZL: zooidal length; ZW: zooidal width; OvL: ovicell length; OvW: ovicell width; AvL: avicularial length; AvW: avicularial width; AnD: ancestrula diameter.

Type material from the Paris (MNHN) and from the "Naturhistorischen Museum Wien" has been examined. Fossil material of *S. ramulosa* is housed in the Catania Palaeontological Museum (CPM).

## Results

### *Systematic Part*

Phylum Bryozoa Ehrenberg, 1831

Class Gymnolaemata Allman, 1856

Superfamily Celleporoidea Johnston, 1838

Family Phidoloporidae Gabb and Horn, 1862

Genus *Sparsiporina* d'Orbigny, 1852

The original description of the genus by d'Orbigny (1852) concisely reports the main colonial and zooidal features: "Colony mineralized, not joined, fixed through a calcareous basal expansion from where compressed branches develop, consisting of four longitudinal series of zooids on a single side; zooids well distinct, swollen, with a distal circular opening and a pore placed proximally". This diagnosis needs to be amended to include further features, especially the arrangement of the zooids on the dorsal surface, which appears to be distinctive for the genotype species, as already remarked by Reuss (1848) and Waters (1891), and also for *R. ramulosa*.

Amended generic diagnosis: Colony erect, rigid and branched, small-sized, attached to the substratum through a minute encrusting base. Branches slender, with distinct frontal and dorsal surfaces, the former one exposing four longitudinal rows of zooids with the median two ones opening frontally, and the additional rows opening laterally but originating dorsally, on the side opposite to that of the orifice. This causes an apparent single series of zooids to be visible on the dorsal surface, resulting from strongly inclined, alternating zooids from the two lateral rows, separated by a zigzag suture. Primary orifice orbicular and sinuate, hidden at the base of a more or less projecting, often denticulate peristome, with a labial pore at the end of a long labial suture. Frontal surface finely sculptured with a single median pore. Ovicell more or less prominent, partly immersed in the distal zooid, with a calcified ectoecial rim leaving visible a calcified imperforate entoecium. Frontal avicularia sporadic, elongated, triangular to spathulate.

Type species: *Retepora elegans* Reuss, 1848

*Sparsiporina elegans* (Reuss, 1848)
Figures 18.2 and 18.3

*Retepora elegans* Reuss 1848: 48, pl. 6, Fig. 38; Waters 1891: 30, pl. 4, Figs. 9 and 10.

*Sparsiporina elegans*: Braga 1963: 37; Braga 1975: 147, pl. 2, Fig. 7; Braga 1980: 57, Fig. 42; Braga and Barbin 1988, 526, pl. 9, Figs. 7 and 8; Bizzarini and Braga 1997: 121, pl. 4, Figs. 18.1, 18.2, and 18.3; Braga and Crihan 2006, Table 1, pl. 2, Figs. 16 and 17; Braga 2008: Table 1, Fig. 89a–b.

Material examined: Type series: n. 68 Acq Post k.k. Mineralien Kabinet 1870, XIII at the Vienna Museum, labelled as *Retepora cellulosa* L/R 1860 VL (2) and containing two fragments.

Additional material: Late Eocene: Val di Lonte: 7 specimens; Crosara: 15 specimens; Brendola: 12 specimens; Possagno: 2 specimens; Priabona, including Boro and Granella: 8 specimens; Pannone and Val di Gresta: 19 specimens; Monteccio di Costozza: 2 specimens; Toara: 10 specimens. All this material is part of Braga's personal collections and housed at the Rovereto Museum.

Description: Colony erect, rigid and dichotomously branching, attached through a basal expansion. Branches slender, usually 400–600 μm in maximum diameter, formed by four longitudinal series of zooids all facing the same side, thus forming a frontal and a dorsal surface (Figs. 18.2 and 18.3a, f). Frontal surface including the two central rows and the frontal sides of the marginal ones, and dorsal side consisting of the hammered sculptured, swollen back-surfaces of the alternating zooids from the lateral rows, well marked by a furrowed zigzag suture, well evident even in late ontogeny (Figs. 18.2b and 18.3f).

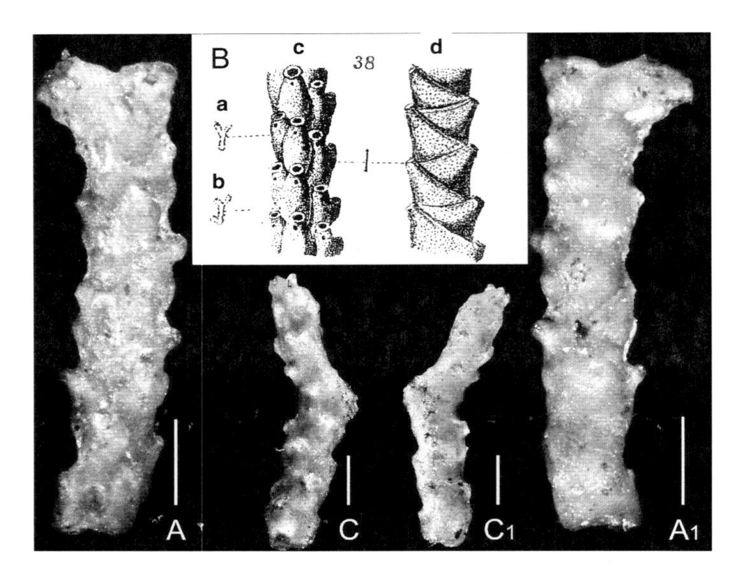


**Fig. 18.2** *Sparsiporina elegans* (Reuss, 1848). Types in Reuss' collection, "Naturhistorisches Museum Wien". *Scale bars*: 500 μm. (**A**, $\mathbf{A}_1$) Frontal and dorsal side of the lectotype. (**B**) Original images of the frontal and dorsal sides of *S. elegans* from Reuss (1848). (**C**, $\mathbf{C}_1$) Frontal and dorsal side of the only paralectotype

Zooids elongated, usually truncated at proximal and distal ends and slightly wider at mid length, or sinuous in areas with common avicularia, separated by slightly raised edges with a median deep groove (Fig. 18.3b). Frontal wall convex, fairly hammered, with a median pore located proximally at 1/4–1/3 of the zooidal length. Primary orifice immersed within the peristome, not visible on the examined specimens. Peristome usually level with the branch surface, but presumably longer and apparently often broken, with a drop-like spiramen at the base of a longitudinal spiraminal suture, up to 200 μm long, extending from 1/3 to 1/2 the frontal wall length (Figs. 18.2a and 18.3b, c, g).

Ovicell swollen but frontally depressed, recumbent on the frontal wall of the distal zooid; nearly as long as wide or slightly longer than wider in zooids from the central rows; elongated and more or less transversely inclined towards the dorsal surface in zooids from the lateral rows (Fig. 18.3e), often compressed if followed by an avicularium (Fig. 18.3h); deeply opening in the peristome, seemingly with a short labellum. Entoecium nearly flat showing possible remnants of a median suture, surrounded by a narrow rim of calcified smooth ectoecium (Fig. 18.3b).

Frontal avicularia sporadic, originating from the proximal pore, located on proximal mid part of zooids, on swollen avicularia, typically on zooids from the marginal rows and elsewhere on branches including fertile zooids; with a complete cross bar, a wide cordiform-to-acutely triangular rostral foramen (Fig. 18.3j), completely or largely occupying the bluntly triangular to slightly expanded, proximally directed rostrum, which is raised and upward pointing in zooids from the

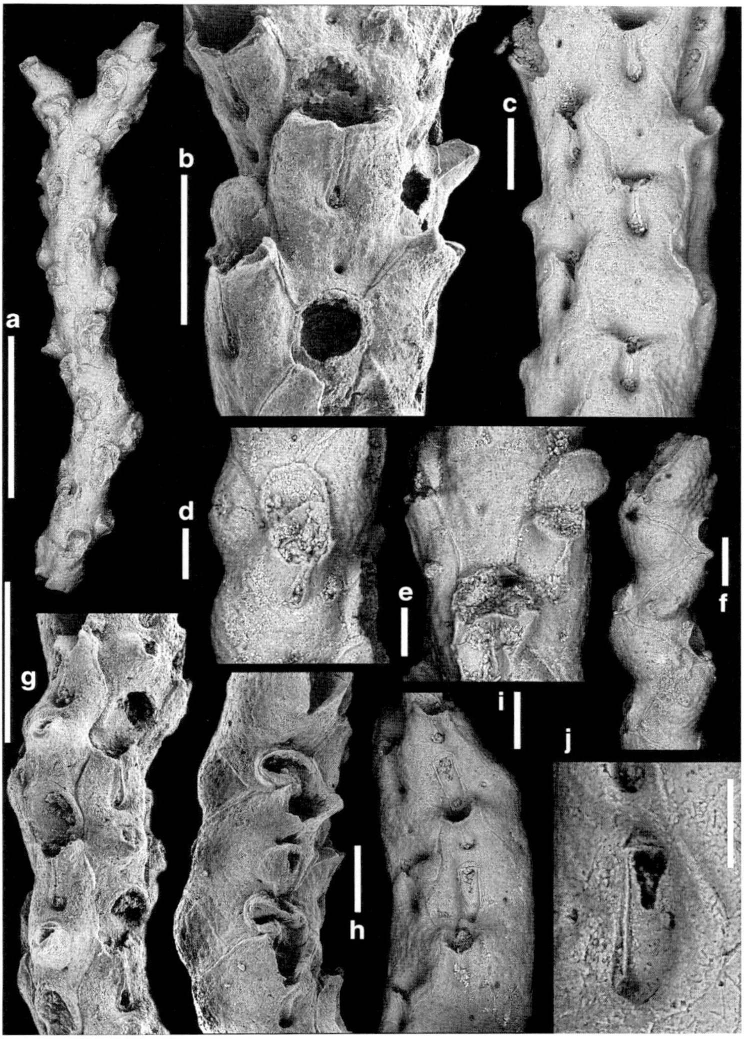

**Fig. 18.3** *Sparsiporina elegans* (Reuss, 1848) from the Veneto Region, late Eocene. (**a**): Costozza; (**b-c**) and (**f-j**): Val di Lonte; (**d-e**): Brendola. Photos b, g-h kindly supplied by DP Gordon. (**a**) Large fertile fragment. *Scale bar*: 1 mm. (**b**) Close-up of the frontal side with relatively young zooids. *Scale bar*: 200 μm. (**c**) Zooids exhibiting secondary calcification. *Scale bar*: 200 μm. (**d**) Broken ovicell of a lateral zooidal row. *Scale bar*: 100 μm. (**e**) Some ovicelled zooids. *Scale bar*: 100 μm. (**f**) Dorsal side showing the zigzag suture. *Scale bar*: 200 μm. (**g**) Frontal side of a branch with several ovicells and swollen avicularia with triangular raised rostra. *Scale bar*: 500 μm. (**h**) Lateral view of a branch showing avicularia and deeply inclined ovicells. *Scale bar*: 200 μm. (**i**) Avicularia from the lateral row, immersed in frontal calcification. *Scale bar*: 200 μm. (**j**) Close-up of an avicularium. *Scale bar*: 100 μm

central rows (Fig. 18.3g), and comparatively slender, raised but parallel to the zooidal surface on zooids from the lateral rows (Fig. 18.3h), becoming immersed in late ontogeny, after secondary calcification (Fig. 18.3i, j).

Measurements (in μm): ZL: 361–566, 437.47, 58.39 (17); ZW: 193–333, 267.08, 54.53 (12); OvL: 133–163, 151.83, 11.67 (6); OvW: 120–133, 130.40, 5.81 (5); AvL: 186–227, 209.50, 20.79 (4); AvW: 86–90, 88.00, 2.83 (2).

Remarks: *Sparsiporina elegans* was first described and figured by Reuss (1848) in his Fig. 38a–d (Fig. 18.2a–c), basing upon specimens from a not well-identified locality in the Vienna Basin, but subsequently specified by Reuss himself (1874) as actually originating from Casa Fortuna in the Val di Lonte, as also reported by Waters (1891) and Braga (1991). Material exhibiting the *S. elegans* features is presently located in Reuss' collections in the Vienna Museum (Braga 1991). It consists of two fragments in a single tube, labelled as *Retepora cellulosa* (L), although with an interrogation mark, and has probably been renamed, as suggested by the reported dates, later than 1848, when the species was first described.

The original description is given in Latin and in German as follows: "*? R. elegans m.,* stirpe parum ramose, tenuissima; facie anteriore porosa; cellulis oblique seriatis, quaternis, ovatis, supra attenuates, ostio terminali subrotundo, infraposito poro accessorio minimo; facie postica diagonaliter sulcata; tota superficie punctulata." The description was completed with some remarks in German, mostly concerning the zooid variability, and the peculiar branch morphology and zooid arrangement.

Owing to this last feature, Reuss (1848) placed his new species within the genus *Retepora* only provisionally. Agreeing with Reuss's opinions, d'Orbigny (1852) erected the new genus *Sparsiporina* with *R. elegans* Reuss as genotype, stressing differences in respect to species of *Retepora*. But oddly, Waters (1891) still ascribed the species to *Retepora*, seemingly disregarding or being unaware of the d'Orbigny's genus. The features of *S. elegans* appear to be highly distinctive, and can be easily detected even at low magnification, mostly the special pattern of the zooids along the branches and the oblique arrangement of zooids from the lateral rows, resulting in a zigzag suture on the dorsal side; the unique frontal pore and the long spiramen. SEM examination of topotypic specimens from Val di Lonte allowed frontal avicularia and ovicells to be described and figured. Both features were overlooked by Reuss (1848), but avicularia were described and ovicells also figured by Waters (1891: pl. 4, Fig. 10), as later repeatedly reported by Canu and Bassler (1929: 375, Fig. 143) and Bassler (1953: 213, Fig. 18.4a, b). Nevertheless, Braga (1975, 1980, 2008), Braga and Barbin (1988), Bizzarini and Braga (1997), and Braga and Crihan (2006) figured specimens completely lacking avicularia, which are actually rare in all examined material. Furthermore, after SEM examination both frontal and dorsal surfaces, when well preserved, are characterised by a honey-comb like sculpture, which could correspond to the "*punctulata*" appearance described by Reuss (1848). In contrast, true pores on the frontal surface are actually lacking, although a somewhat porous appearance could result from diagenetic processes on some specimens.

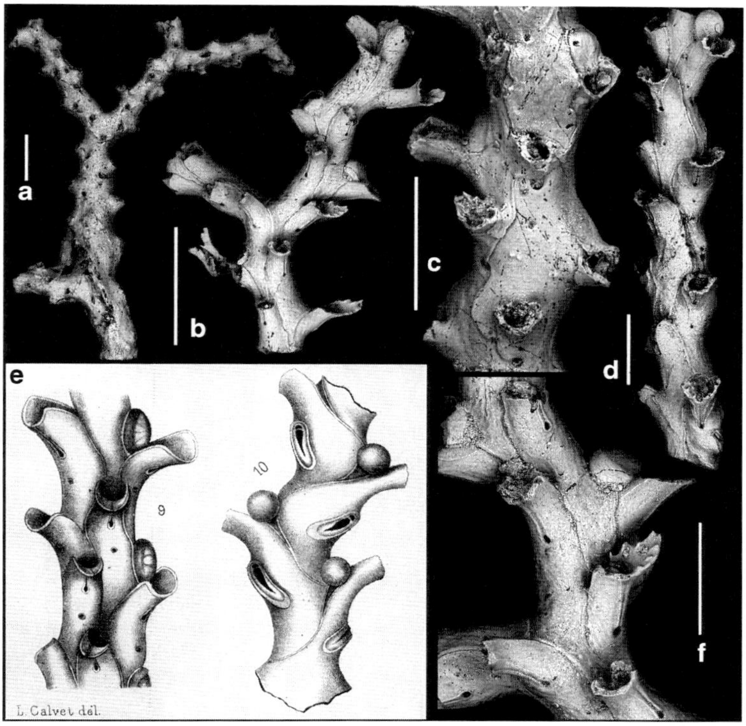

**Fig. 18.4** *Sparsiporina ramulosa* (Calvet 1906). Lectotype and paralectotypes in the Calvet's collection at the MNSN in Paris. Coll. Number: 2622. (**a**) Fragment from the basal part of a colony. *Scale bar*: 1 mm. (**b**) Branch, here chosen as lectotype (2622 C), with long protruding peristomes. *Scale bar*: 1 mm. (**c**) Close-up of an ovicell embedded in secondary calcification. *Scale bar*: 1 mm. (**d**) An ovicell and an avicularium, on the topmost and basalmost zooids of a young branch, respectively. *Scale bar*: 500 μm. (**e**) Original figure of the frontal and dorsal sides of *S. ramulosa* from Calvet (1906: pl. 28, Figs. 9 and 10). (**f**) Close-up of b with young prominent scalloped peristomes and an ovicell with its flat frontal tabula. *Scale bar*: 500 μm

The concept of *S. elegans*, as reported above, is not matched by descriptions in Zágoršek (2001, 2003), including (a) biserial to triserial branches, (b) up to 3–6 orificial spines, (c) an ascopore near the orifice and (d) porous walls of zooids and ovicells. Nevertheless, some figures given by Zágoršek (2001, pl. 33, Fig. 5; but probably also Fig. 4, as *Iodictyum rubeschi* (Reuss 1848)) and Zágoršek (2003, pl. 30, Fig. 4, as *Reteporella simplex* (Busk 1859)) could doubtfully represent worn specimens of *S. elegans*.

Distribution: *S. elegans* is known from the late Eocene to the early Oligocene of a widespread area, including the prealpine mountains and hills of north eastern Italy and areas of the Carpathian regions and central-eastern Europe such as Poland, Slovakia and Romania (Braga et al. 1996; Braga and Crihan 2006; Braga 2008) and, doubtfully (see above), also Austria and Hungary. The palaeogeographic

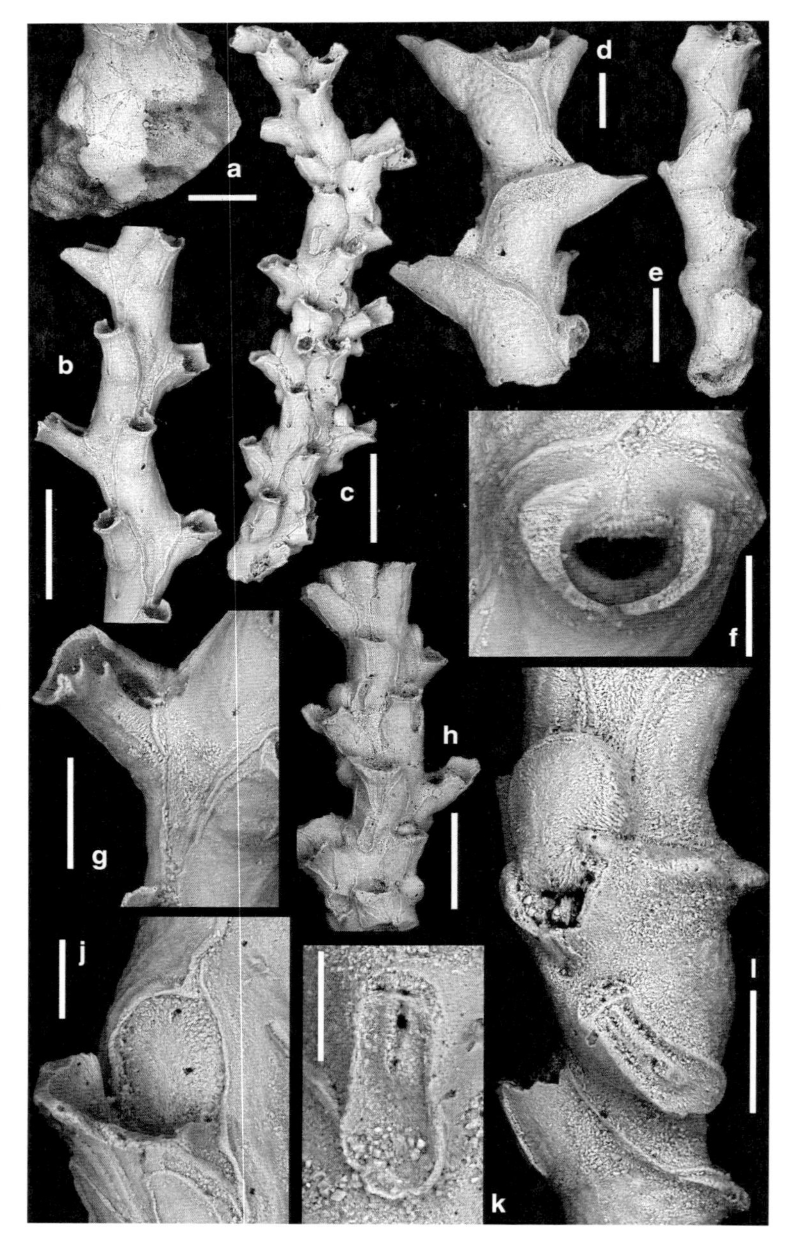


**Fig. 18.5** *Sparsiporina ramulosa* (Calvet, 1906). All from Cala S. Antonino samples, Gelasian. (**a**) Base of a colony adhering to a sandy element. *Scale bar*: 500 μm. (**b**) Young branch showing the long adnate peristomes of the central zooidal rows. *Scale bar*: 200 μm. (**c**) Frontal side of the largest fertile fragment. *Scale bar*: 1 mm. (**d**) Dorsal side of a well preserved young branch. *Scale bar*: 200 μm. (**e**) Dorsal side of an old branch. *Scale bar*: 500 μm. (**f**) Primary orifice. *Scale bar*: 100 μm.

distribution of the species, and of some of its associated taxa, is notably Tethyan. Interestingly, many of these western Tethyan species have more recent or even living counterparts at generic level, in bryozoan associations of Australasia sites, thus pointing to the migration along a corridor from western to eastern Tethys realm (Braga 1987; Gordon and Braga 1994).

*S. elegans* was a species that lived in relatively shallow-water quiet palaeoenvironments lacking coarse sedimentary input. Particularly, associations from upper Eocene outcrops of north eastern Italy point to depths of 50–90 m (Braga 1963; Braga and Barbin 1988), presumably comparable to those inferred for coeval sedimentary successions from Hungary and Austria (Zágoršek 2001, 2003). In contrast, bryozoans associated with *S. elegans* from lower Oligocene sediments of NW Transylvania indicate a depth of 100–200 m, corresponding to the shallow bathyal, or at least to the transition from continental shelf to the bathyal (Braga and Crihan 2006). This suggests a relatively wide bathymetrical distribution of the species or the deepening of its bathymetric range during time. Nevertheless, detailed analyses are needed for a better confidence.

*Sparsiporina ramulosa* (Calvet, 1906) comb. new
(Figures 18.4 and 18.5)

*Retepora ramulosa* Calvet 1906: 454, pl. 28, Figs. 9 and 10

*Retepora ramulosa* (Calvet, 1906): Hayward 1981: 58

*Reteporella ramulosa* (Calvet 1906): Tricart and d'Hondt 2009: pl. 2, Figs. 9–10

Material examined: Type series of specimens from NE Atlantic deep waters, deposited as syntypes at the MNHN n. 990, 991, 992, 993, 2622.

Additional fossil material: Milazzo Peninsula. Cala S. Antonino Ovest: samples 1 (1998): 141 specimens; sample 2 (1998): 28 specimens. Cala S. Antonino centre: sample 4 (1998): 38 specimens; sample 1 (1999): 9 specimens; sample 2 (1999): 6 specimens; sample 17 (2000): 6 specimens; sample 5 (2002): 302 fertile and sterile fragments, some including the expanded calcified base; sample 7 (2002): 20 specimens; sample 8 (2002): 2 specimens. Cala S. Antonino Est: sample 14 (2000): 9 specimens; Punta Mazza: sample 8: 4 specimens. Picking: about 50 fragments. All these specimens are housed as PMC. I.Pl. Rosso.Milazzo Coll. 1B.

Description: Colony erect, seemingly not exceeding some millimetres, rigid and dichotomously branching, as obvious from some fragments bifurcating at 1–3 mm intervals, attached to the substratum through a thin expansion of extrazooidal calcification, occasionally with radiciform expansions (Fig. 18.5a). Branches subcylindrical in cross section and with a serrated outline (Figs. 18.4b, e and

**Fig. 18.5** (continued) (**g**) Peristome of a zooid from a lateral row. *Scale bar*: 200 μm. (**h**) Fertile fragment with ovicells restricted to the lateral zooidal rows and a frontal avicularium on a zooid from the central rows. *Scale bar*: 500 μm. (**i**) Fertile zooid with its twisted avicularium. Note the prominent ovicell with its flat tabula. *Scale bar*: 200 μm. (**j**) Close-up of an ovicell immersed in the secondary calcification from neighbouring zooids. *Scale bar*: 100 μm. (**k**) Close-up of a frontal avicularium. *Scale bar*: 100 μm

18.5b–d), slender (usually about 350–400 μm in diameter, but up to 600 μm in some thick, basal fragments, not including protruding peristomes), formed by four longitudinal series of alternating zooids all facing the same side, individuating a frontal and a dorsal surface. Frontal surface entirely exposing two central rows and the frontal sides of the marginal ones and dorsal side exclusively consisting of an apparently unique row of superimposed zooidal back-surfaces (Figs. 18.4e and 18.5d), actually the result of alternating zooids from both marginal rows, well separating each other, due to a raised keel following the distal zooid terminations and the distal margin of the protruding peristomes, and a different sculpture, i.e. hammered below the keel and smooth upon it, in the axillary region. As zooids and their peristomial continuations are sharply inclined alternatively at the right and the left sides of the branch, the succession of keels forms a zigzag picture on the dorsal surfaces, which remains visible also in late ontogeny (Fig. 18.5e).

Zooids elongated, usually proximally pointed and distally truncated, separated by slightly raised edges and deep grooves. Frontal wall convex, smooth to fairly hammered (or with an orange skin-like sculpture), with a nearly centrally placed small pore. Primary orifice observed on fossil specimens (Fig. 18.5f) deeply immersed within the peristome, slightly inclined distally, larger than longer, with a straight, presumably beaded edge and an arcuate proximal lip, proximomedially knocked by a shallow not defined sinus. Peristome long (up to about 400 and 300 μm in Holocene and Gelasian zooids, respectively), tubular, flared at its end, with a minute round spiramen centrally located at the base of a longitudinal spiraminal suture, formed by two wings leaned against the frontal wall of the next zooids in zooids from the central rows (Figs. 18.4b, d–f and 18.5b); spout-like and protruding at an angle of 50–70° in zooids from the lateral rows, the wings scalloped and leaving a large distal longitudinal slit (Figs. 18.4b and 18.5g). Well preserved peristomes from the Atlantic show the edge surface internally ribbed. In late ontogeny peristomes, mostly from the central rows, become progressively coated by a layer of extending frontal calcification the distal edge of which forms a wide V-shaped sculpture on the zooid frontal surfaces.

Ovicells only observed on zooids from the marginal rows; swollen but frontally flat, longer than wider, slightly widening distally, recumbent on the frontal wall of distal zooids; opening deeply within the peristome, with a short, proximally straight labellum laterally marked by two large and shallow indentations (Figs. 18.4b–f and 18.5h–j). Entoecium smooth, entirely imperforate surrounded by a narrow rim of calcified ectoecium. In late ontogeny, ovicells become progressively embedded by frontal secondary calcification and partly covered by expansions of the distal zooids (Figs. 18.4c and 18.5j).

Frontal avicularia sporadic (Fig. 18.4d) but typically common on ovicellate zooids and on zooids from branches with fertile zooids (Fig. 18.5c), developed from the median frontal pore; the avicularium swollen but indistinct, longitudinally elongated and lined on the frontal surface, and proximally directed, so coaxial with the branch on zooids from the central rows and usually curving, distinctly inclined to nearly transversal, on the dorsal surface in zooids from the lateral rows (Fig. 18.5k); rostrum roughly parallel-sided for all or at least more than one half length, ending in a round to slightly spathulate tip; cross-bar complete and slightly arched; palate extensive with a slight linear foramen evolving into a median suture marked by raised rims.

Measurements (in μm): ZL: 714–858, 790, 46 (15); ZW: 210–362, 285.00, 47.11 (10); OvL: 160–265, 214.71, 36.7 (7); OvW: 145–200, 174.75, 22.66 (4); AvL: 208–286, 236.33, 43.15 (3); AvW: 92–115, 105.00, 11.79 (3); AnD: 151–153, 152, 1 (2).

Remarks: This species was first included in the presently unrecognizable genus *Retepora* Lamarck, 1801 by Calvet (1906) and Hayward (1981) and recently considered as belonging to *Reteporella* Busk, 1884 by Tricart and d'Hondt (2009). Nevertheless, it cannot be included in this latter genus, neither in *Reteporellina* Harmer, 1933, mostly due to the features of its labellate ovicell (lacking a median permanent fissure but possessing a marginal calcified ectoecium and an imperforate entoecium only marked by extremely faint lines radiating from the middle), and the peculiar dorsal side not exposing zooidal dorsal surfaces but resulting from the proximal frontal walls of the twisted lateral zooids. Additional differences include the extrazooidal calcification of the basal portions, the primary orifice morphology and the absence of peristomial avicularia and spines. In contrast, the organization of the dorsal side, as well as the features of zooids, ovicells and avicularia fit well the diagnosis of the genus *Sparsiporina*. *S. ramulosa* strongly recalls *S. elegans* in the general appearance but differs by the larger and comparably elongated zooids, the shape and size of the frontal avicularium and also by its frequency, the general morphology of the ovicell and its location, always restricted to the zooids of the lateral rows.

Distribution: The original material of *S. ramulosa* was collected during the Talisman 1883 cruise, in the deep-water station 96, at Capo Verde Isles (Lat. 19°19′–19°16′ N, Long. 20°22′–20°20′ W, −2330–2320 m) in grey-green muds. The material was later described by Calvet (1906) who argued that the collected branches were not surely alive and they could have been displaced by currents. The species, which is probably rare, has never been found since then, and simply recorded within stocks of deep-water bryozoans (Hayward 1981). Specimens from the Capo Milazzo area represent the second finding of *S. ramulosa* after more than 100 years, and also the only one known fossil record. Fragments come from sediments deposited during the MPl5 and MPl6 zones, nearly completely falling within the Gelasian stage and point to a Mediterranean geographic distribution of *S. ramulosa* at that time. Nevertheless, the species has not been found in post-Gelasian sediments in the same area, as well as in living and dead Mediterranean associations, after examination of a huge amount of samples (Rosso, personal observations). Information is very scant and hypotheses about the present and past distribution of this species need to be better substantiated by further findings. Nevertheless, it is reasonable that *S. ramulosa* appeared in the Atlantic-Mediterranean area, where it maintained a geographical range wide but probably changing and/or reducing during time (possible present-day absence from this relatively well investigated basin), also in relation to hydrological changes, mostly in the Mediterranean during and after the Gelasian, driven by the interplay of climate and tectonics (see Rosso and Di Geronimo 1998; Rosso 2005).

No special information is available for the unique Atlantic colony except that it was associated with other species such as *Myriozoum strangulatum* Calvet, *Haswellia alternata* Calvet, *Setosellina roulei* Calvet, and *Ichtyaria aviculata* Calvet, all exhibiting rigid erect slender colonies, except for the free living *S. roulei*. Analogously, palaeocommunities include mostly species exhibiting an erect rigid growth habitus, among which *Tessaradoma boreale* (Busk), *Metrarabdotos elegantissimus* Rosso and *Tervia irregularis* (Meneghini) are the more common together with some rooted species, such as *Batopora rosula* (Reuss). Interestingly, fossil proximal colony fragments are mostly detached from their original substrata. Bases are small, usually not larger than twice the basal branch diameter, with a more or less concave attachment surface and a nearly circular outline. Such general morphology and the bioimmuration casts often visible on the underside of basal expansions suggest that their commonly encrusted substrata were presumably organic. Only a few specimens still adhere to sandy clasts, which are nearly completely enveloped by root-like thin digitations causing an irregular outline (Fig. 18.5a).

## Discussion

The recognition of a second species, *S. ramulosa* besides *S. elegans*, within the genus *Sparsiporina* widens significantly both its geographic and stratigraphic distribution. Nevertheless, at present knowledge, the genus seems restricted to the western Tethysian region and its remnants from the late Eocene to the Gelasian and the present-day Atlantic. The two species appear closely related to each other and probably *S. ramulosa* is the Gelasian-(?)Recent counterpart of the late Eocene-early Oligocene *S. elegans*, although information is lacking for the intervening period, probably as a consequence of the scantiness in outcrops of deep-water sediments of these ages all over Europe. The only not surely living representative of the genus was collected in very deep waters (more than 2,000 m), although, again, some doubts remain about its autochthony. Mediterranean Gelasian specimens of *S. ramulosa* presumably lived at shallower depths seemingly not exceeding 600 m or a little more. Furthermore, palaeohabitat reconstructions for associations including *S. elegans* point to quiet conditions and limited sedimentary input and, particularly, palaeobathymetric inferences document a slight deepening during time from less than 100 m, during the late Eocene, to 100–200 m for the early Oligocene, from mid shelf to shelf break or very shallow slope environments. Consequently, the genus probably appeared in shelf habitats (stemming from *Reteporella-Reteporellina* species?) and subsequently adapted to colonize progressively deeper environments.

Interestingly, species of the genus *Sparsiporina* are morphologically reminiscent of those belonging to the genus *Bryorachis* Gordon and Arnold, 1998, included in the family Phidoloporidae. Affinities rely on the small colony size, the zooidal arrangement in four longitudinal rows, only two visible on the dorsal side, the peristomial features deeply marked by a pseudospiramen and the branch appearance like vertebral column with the peristomes of the lateral rows broadly protruding laterally. Nevertheless, differences are obvious, mostly in the frontal wall, the ovicell morphology, the number and distribution of avicularia, and the twinning of zooids on the dorsal side. *Bryorachis* is represented by only two species (*B. pichoni* Gordon and Arnold and *B. curiosa* Gordon and Arnold), known exclusively as living (the latter from a unique specimen) in the NE Australian and Norfolk Ridge waters, at decidedly bathyal depths, ranging from 970 to nearly 1,800 m (Gordon and Arnold 1998). Both species were considered as derived from *Reteporellina* and particularly adapted to colonize directly soft bottoms, probably through rootlets (Gordon and Arnold 1998). No hypothesis can be done about the time of divergence of these species from the Austral hemisphere, as they have no fossil representatives.

The genera *Sparsiporina* and *Bryorachis* seem to represent a case of nearly convergent morphologies and adaptations for the conquest of deep environments, with species of the former genus having a small attachment base and being able to encrust soft-bodied organisms, the exposed soft tissues of benthic organisms and even directly sediment sandy clasts.

**Acknowledgements** D. Gordon (NIWA) is thanked for discussions and for making available photos of some topotypes of *S. elegans*. Photos of the types of this species were kindly supplied by N. Vávra (Department of Paleontology, Vienna) whereas those of *S. ramulosa* were made available by P. Louzet (MNHN, Paris). J.-L. d'Hondt and J.-G. Harmelin are kindly acknowledged for information about *S. ramulosa*, A. Viola (Scienze Biologiche, Geologiche e Ambientali, Catania) for SEM analysis, S. Castelli and L. Franceschin (Dipartimento di Geoscienze, Padova University) for the composition of Fig. 18.1 and laboratory assistance. A special thank is due to N Vávra and P Moissette (Université Claude Bernard, Lyon) for reviewing the manuscript. Research was funded by Catania PRA grants to A. Rosso. Catania Palaeontological Research Group: contribution number 371.

## References

Allman GJ (1856) A monograph of the freshwater Polyzoa, including all the known species, both British and foreign. The Ray Society, London, pp 1–119

Bassler RS (1953) *Bryozoa* Vol. Part G. In: Moore RC (ed) Treatise on invertebrate paleontology. Geological Society of America/University of Kansas Press, Lawrence

Bizzarini F, Braga G (1997) I briozoi priaboniani dei dintorni di Crosara (Vicenza-Italia). Ann Mus Civ Rovereto, Sez Arch St Sci nat 13:91–126

Bock PE (2002) Recent and fossil Bryozoa: *Sparsiporina* d'Orbigny, 1852. http://www.bryozoa.net/cheilostomata/Phidoloporidae/*Sparsiporina*.html. Accessed 1 Sept 2009

Braga Gp (1963) I Briozoi del Terziario Veneto. 1° contributo. Boll Soc Paleontol Ital 2(1):16–55

Braga Gp (1975) I Briozoi dell'Eocene di Possagno. Schweiz Pälaont Abh 97:141–148

Braga Gp (1980) In: Antolini P, Braga G, Finotti F (ed) I Briozoi dei dintorni di Rovereto e Valle di Gresta. Museo Civico di Rovereto, pubblicazione, 82, Arti Grafiche Manfrini: Calliano, pp 103

Braga Gp (1987) Tethyan migration of some Tertiary Bryozoa. In: McKenzie KG (ed) Proceedings of international symposium of shallow Tethys, Wagga Wagga,15–17 Sept 1986. AA Balkema: Rotterdam/Boston, pp 379–385

Braga Gp (1991) Reuss' collection of cheilostome bryozoans from Venetia stored in Natur-Historisches Museum Wien: a proposed revision. In: Bigey FP (ed) Bryozoaires actuels et fossiles: Bryozoa living and fossil, Bull Soc scienc natur Ouest Fr, Nantes, Mém HS, pp 49–59

Braga Gp (2008) Atlas of Cenozoic Bryozoa of the north-eastern Italy (Venetia region). Lav Soc Veneta Sci Nat 33:71–92

Braga Gp, Barbin V (1988) Les bryozoaires du Priabonien stratotypique (Province Vicenza, Italie). Rev Paléobiol 7:495–556

Braga Gp, Crihan IM (2006) Up-dating of the taxonomy, stratigraphy and palaeoecology of Bryozoa rich sediments from Mera (N.W. Transylvania-Romania). Cour Forsch Inst Senckenberg 257:21–33

Braga Gp, Zágoršek K, Kazmer M (1996) Comparisons between Venetian and Western Carpathian Late Eocene bryozoan fauna. In: Braga Gp, Finotti F, Piccoli G (eds) Shallow Tethys 4 . Annali musei civici di Rovereto, Sezione archeologia-storia-scienze naturali, Suppl 11 (1995): 259–270

Busk G (1859) A monograph of the fossil Polyzoa of the Crag. Palaeontograph Soc, London, pp 1–136

Calvet L (1906) Bryozoaires. In: Expéditions scientifiques du "Travailleur" et du "Talisman" pendant les années 1880, 1881, 1882, 1883. Direction A Milne-Edwards et E Perrier, Masson and Cie (eds), Paris, pp 353–483

Canu F, Bassler RS (1929) Bryozoa of the Philippine region. Bull US Natl Mus 100(9):1–685, + 94 pls

Ehremberg CG (1831) Symbolae Physicae, seu Icones et descriptiones Corporum Naturalium novorum aut minus cognitorum, quae ex itineribus per Libyam, Aegyptum, Nubiam, Dongalam, Syriam, Arabiam et Habessiniam... studio annis 1820-25 redirerunt Berolini?, Berlin. (Pars Zoologica, v.4. Animalis Evertebrata exclusis Insectis.)

Fois E (1990) Stratigraphy and palaeogeography of the Capo Milazzo area (NE Sicily, Italy): clues to evolution of the southern margin of the Tyrrhenian Basin during the Neogene. Palaeogeogr Palaeoclimatol Palaeoecol 78:87–107

Gabb WM and Horn GH (1862) The fossil Polyzoa of the Secondary and Tertiary formations of North America. Jour Acad Nat Sci Philadelphia 5(2):111–179

Gaetani M (1986) Brachiopod palaeocommunities from the Plio/Pleistocene of Calabria and Sicilia (Italy). In: Racheboeuf PR, Emig CC (eds) Biostrat Paléoz 4, pp 281–288

Gaetani M, Saccà R (1984) Brachiopodi batiali nel Pliocene e Pleistocene di Sicilia e Calabria. Riv Ital Paleontol Strat 90(3):407–458

Gordon DP (2002) Late Cretaceous-Paleocene "porinids"-mixed frontal shields and evidence of polyphyly. In: Wyse Jackson PN, Battler CJ, Spencer Jones ME (eds) Bryozoan studies 2001. Swets and Zeitlinger, Lisse, pp 113–124

Gordon DP, Arnold PW (1998) *Bryorachis* (Phidoloporidae) and *Retelepralia* (Cheiloporinidae): two new genera of Indo-Pacific Bryozoa. Mem Qld Mus 42(2):495–503

Gordon DP, Braga GP (1994) Bryozoa: living and fossil species of the catenicellid subfamilies Ditaxiporinae Stach and Vasignyellinae nov. In: Crosnier A (ed) Résultats des campagnes MUSORSTROM, 12, vol 161, Mémoires du Muséum national d'Histoire naturelle. Ed. du Muséum, Paris, pp 55–85

Hayward PJ (1981) The Cheilostomata (Bryozoa) of the deep-sea. Galathea Rep 15:21–68

Johnston G (1838) A history of the British zoophytes. 2nd ed Lizars WH, Edimburgh, London & Dublin, pp 1–341

Lentini F, Catalano S, Carbone S (2000) Carta Geologica della Provincia di Messina, scala 1: 50,000. SELCA Firenze

Orbigny d' A (1852) Paléontologie Française. Terrains Crétacés, 5: Mollusques Bryozoaires. Masson V (ed), Paris

Reuss AE (1848) Die fossilen Polyparien des Wiener Tertiárbeckens. Haid Natur Abh 2:1–10

Reuss AE (1869) Paläontologische Studien ueber die aelteren Teriärschichten der Alpen. Die fossilien Anthozoen und Bryozoen der Schichtengruppe von Crosara. Denkschr K Akad Wiss 29(2):215–299

Reuss AE (1874) Die fossilen Bryozoen der Oestr. Ungarischen Miocäns. Denkschr K Akad Wiss 33(1):1–46

Rio D, Sprovieri R, Di Stefano E (1994) The Gelsasian stage: a proposal for a new chronostratigrafic unit of the Pliocene series. Riv Ital Paleontol Strat 100(1):103–124

Rosso A (2002a) *Terataulopocella borealis* gen. et sp. nov., a deep-water Pliocene lekythoporid (Bryozoa) from the Mediterranean area. Mem Sci Geol 54:65–72

Rosso A (2002b) *Bryobaculum carinatum* sp. n., gen. n., a new Mediterranean Pliocene deep-sea bryozoan. In: Wyse Jackson PN, Battler CJ, Spencer Jones ME (eds) Bryozoan studies 2001, Proceedings of the 12th International Bryozoology Association conference, AA Balkema Publishers Leiden, London/New York/Philadelphia/Singapore, pp 275–283

Rosso A (2005) Bryozoan facies in deep-sea Pleistocene environments of southern Italy. In: Cancino J, Moyano H, Wyse-Jackson PN (eds) Bryozoan studies 2004. Proceedings of the 13th International Bryozoology Association conference, AA Balkema Publishers Leiden, London/New York/Philadelphia/Singapore, pp 257–269

Rosso A, Di Geronimo I (1998) Deep-sea Pleistocene Bryozoa of southern Italy. Géobios 30 (3):303–317

Sciuto F (2003) Dati preliminari sulla ostracofauna pliocenica di Capo Milazzo (Sicilia NE). Boll Soc Paleontol Ital 42(1–2):179–184

Tricart S, d'Hondt JL (2009) Catalogue des spécimens-types de bryozoaires, brachipodes, ptérobranches et entéropneustes du departement "Milieu et Peuplements Aquatiques" (Muséum National d'Histoire Naturelle, Paris). Mém Soc Linn, Bordeaux, pp 1–110

Violanti D (1988) I foraminiferi plio-pleistocenici di Capo Milazzo. Boll Mus Reg Sci Nat Torino 6(2):359–392

Waters AW (1891) North-Italian Bryozoa. Q J Geol Soc Lond 47:1–34

Waters AW (1892) North-Italian Bryozoa. Pt. II. Cyclostomata. Q J Geol Soc Lond 48:152–162

Zágoršek K (2001) Eocene Bryozoa from Hungary (part II. Systematic palaeontology). Cour Forsch-Inst Senckenberg 231:19–159

Zágoršek K (2003) Upper Eocene Bryozoa from Waschberg Zone (Austria). Beitr Paläont 28:101–263

# Chapter 19
# Species of *Alcyonidium* (Ctenostomatida) from the Pacific Coast of North America: A Preliminary Account

## *Alcyonidium* from the Pacific Coast

**John S. Ryland and Joanne S. Porter**

**Abstract** We have collected at least eight species of *Alcyonidium* along the Pacific coast between southern California and British Columbia. One resembles, but cannot be confirmed as, *A. parasiticum*; the remainder are undescribed. Earlier records from the literature are examined and supposed identifications with European species discredited. Characters of taxonomic utility in a genus with no calcified skeletal parts are reviewed. The characteristic features of the collected species are summarised and problems discussed, to facilitate further study of this genus along the Pacific coast.

**Keywords** *Alcyonidium argyllaceum* • *A. mytili* • *A. polyoum* • Oviparity • Larviparity • Ancestrula

## Introduction

### *Historical Review*

The earliest reports of *Alcyonidium* on the Pacific coast of North America are by Robertson (1900, 1902). In 1900 she recorded *A. gelatinosum* (L.) (in the sense of *A. diaphanum* (Hudson), see Thorpe and Winston (1984)), *A. polyoum* Hassall (for identity see Ryland and Porter (2006)), and *A. cervicornis* [sic!]

J.S. Ryland (✉)
Department of Biosciences, Swansea University, Swansea SA2 8PP, UK
e-mail: j.s.ryland@swansea.ac.uk

J.S. Porter
School of Life Sciences, Centre for Marine Biodiversity and Biotechnology, Heriot-Watt University, James Muir Building, Gait 1, Edinburgh EH14 4AS, UK
e-mail: j.s.porter@hw.ac.uk

A. Ernst, P. Schäfer, and J. Scholz (eds.), *Bryozoan Studies 2010*,
Lecture Notes in Earth System Sciences 143, DOI 10.1007/978-3-642-16411-8_19,

(= *Flustrella* (now *Flustrellidra*) *cervicornis* (O'Donoghue and O'Donoghue, 1925)). Later (Robertson 1902), she provided an incomplete description of *A. pedunculatum* sp.n. More records and additional species were later added by Charles and Elsie O'Donoghue (1923, 1926). In 1923 they recorded *A. mamillatum* (= *A. mammillatum* Alder), *A. parasiticum* (Fleming), *A. mytili* (for identity see Cadman and Ryland (1996b)), and *A. spinifera* [sic!] (= *Flustrella* (now *Flustrellidra*) *spinifera*, see Osburn and Soule (1953)); and in 1926 they added *A. columbianum* sp.n. In making these identifications, both Robertson and the O'Donoghues relied solely on Hincks' *British marine Polyzoa* (Hincks 1880). For most bryozoans, Hincks provided remarkably reliable descriptions and illustrations (Ryland 1969), but *Alcyonidium* presented exceptional difficulties to him: quite apart from creating confusion over the names *diaphanum* and *gelatinosum*, the encrusting species were not properly described. Robertson and the O'Donoghues could not, therefore, have made accurate identifications from Hincks' (1880) accounts.

*Alcyonidium* is a large ctenostomate genus. Horowitz (1989) listed over 70 nominate species considered valid and subsequent research has increased this number to nearer 100. The taxonomy of the genus is difficult, given the lack of the skeletal characters that form the basis of cyclostomate and cheilostomate classification. In terms of morphology, this leaves colony form, zooid shape and size, absence/presence and length of the peristome, the number of tentacles, characteristics of the digestive tract, reproductive mode (oviparous or larviparous), the pattern of brooding (where applicable), and the form and initial budding pattern of the ancestrula as characters to distinguish species. Some aspects of ecology (e.g. habitat and the timing of reproduction) may also be helpful. The use of many of these characters still presents considerable difficulties. More recently, genetic methods for species' discrimination have been applied. Enzyme electrophoresis proved only partly successful (e.g. Thorpe and Ryland 1979), owing to the absence of sufficient polymorphic loci, while the electrophoresis of total proteins developed by J-L d'Hondt and collaborators (e.g., d'Hondt and Goyffon 1989), besides being methodologically flawed by the mixing of products from different colonies, failed to detect the co-occurrence of *A. gelatinosum* and *A. polyoum*.

Molecular methods based on DNA have more promise, given recent developments in marker development arising from mitochondrial genome sequencing projects (Jang and Hwang 2009; Waeschenbach et al. 2006, 2009) and success in taxon level phylogeographic studies (Dick et al. 2003; Mackie et al. 2006; Nikulina et al. 2007; Schwaninger 2008). We shall review the criteria useful for species discrimination in the hope that future descriptions of new species will be more nearly complete.

Where the colony form or gross morphology is distinctive (as in *A. pedunculatum* Robertson, 1902 and *A. enteromorpha* Soule, 1951), species may be more easily recognized (though we have no evidence either way regarding the validity of identifications of *A. diaphanum* (as *A. gelatinosum*) or *A. mammillatum* from the Pacific northwest). The greatest difficulties in Europe have arisen over species with thinly encrusting colonies, having no obviously distinctive characters (see Ryland and Porter 2006). Hincks (1880) included *A. mytili* and *A. polyoum* as separate but poorly differentiated species. Robertson (1900) listed both from Alaska but O'Donoghue and O'Donoghue (1923, 1926) recorded only *A. mytili* in the waters surrounding Vancouver Island. In California, S F Light (1941), in the first edition of his *Manual*, following O'Donoghue and O'Donoghue (1926), who had given a description and illustration, listed only *A. mytili* (apart from *A. cervicorne*). In the second edition (Rattenbury and Smith 1954) both *A. mytili* and *A. polyoum* were mentioned but not included in the key, while a quite different *Alcyonidium* sp. – possibly *A. pedunculatum* – is keyed and illustrated (as Fig. 121*i*). Meanwhile, elsewhere, several authors – following Harmer (1915) – decided that the two species could not be separated (Borg 1930; Marcus 1940; Silén 1942, 1943) and adopted the earlier name, *A. polyoum*, to cover both; wrongly, as it later transpired. The specific validity of *A. mytili* was clarified by Cadman and Ryland (1996b) and the differences between *A. polyoum* and *A. gelatinosum* (L.) non Hincks (1880) by Ryland and Porter (2006).

Osburn and Soule (1953) produced a comprehensive account of Pacific coast *Alcyonidium*. They included *A. polyoum* (to include both *A. mytili* sensu Robertson and the O'Donoghues), and *A. columbianum*; the former decision was discussed but not the latter, despite the fact that *A. columbianum* was described and illustrated with peristomes, and compared at the time to *A. albidum* Alder (O'Donoghue and O'Donoghue 1926). The other species were *A. disciforme* (Smitt) (from Alaska), *A. enteromorpha*, *A. mammillatum* (the variant spelling *mamillatum* appears to have been inadvertently introduced by Hincks (1880)), *A. parasiticum*, and *A. pedunculatum*. There was no mention of *A. gelatinosum* (sensu *A. diaphanum*). It then seems surprising to find that the only *Alcyonidium* discussed in the fourth edition of *Between Pacific Tides* (Ricketts and Calvin 1968) was *A. mytili*. The third edition of *Light's Manual* (Soule et al. 1975) reflects the later changes, including both *A. polyoum* and *A. parasiticum*, while Kozloff's *Marine invertebrates of the Pacific Northwest* (Bergey and Denning 1996) also refers to the species as *A. polyoum*. While it is possible that *A. mytili*, which reaches Iceland (Ryland and Porter 2006), and *A. gelatinosum*, which probably extends to the Arctic (if Kluge's (1962, 1975) embryo-containing '*A. mytili*' is *A. gelatinosum*), might occur in the northern Pacific, *A. polyoum* most certainly does not (Ryland and Porter 2006). From this early literature it could be concluded only that one or more species of encrusting *Alcyonidium* occurs on the coast from Point Barrow, Alaska (Osburn and Soule 1953), 71.35°N, through British Columbia to San Francisco Bay, though occurrences further south, through Monterey at least to the vicinity of San Diego, 32.7°N (Jeff Goddard and John Pearse, personal communications), are well-known to local biologists.

Our own collections from southern California to Vancouver Island have been serendipitous, facilitated by visits made for other purposes. They have established beyond doubt that numerous species are present – on open and sheltered coasts, on floats in marinas, and in lagoons (e.g. Elkhorn Slough) – several of which are obviously undescribed. It is difficult in this way to obtain enough comprehensive data for detailed descriptions, but some preliminary information is presented here.

This, together with the review of literature and of features useful for identification, will demonstrate the diversity of *Alcyonidium* and habitats on this coast, and will stimulate much needed further study.

## *Characters Useful for Species Discrimination in* Alcyonidium

### Colony

Colonies may be erect (*A. pedunculatum*), and occasionally unattached (*A. disciforme* Smitt), or form sessile spreading colonies (most). The range is well shown by the illustrations in d'Hondt (1983). Encrusting colonies may be unilaminar or become thicker by frontal budding (*A. mytili*). Extensive, flat colonies may display exhalant chimneys (Banta et al. 1974) as in *A. mytili*, or the surface may be raised in regular monticules (the South African *A. nodosum* O'Donoghue & de Watteville (Ryland 2001)).

### Zooid Shape and Size

Zooids are often initially more or less flat and hexagonal, for maximum close packing (Thorpe and Ryland 1987) but, especially if frontal budding occurs, tend with age to become deeper and columnar, while the orifice becomes central rather than distal in position. Size in plan view may therefore vary throughout the colony, especially along the age gradient, and in relation to season (Ryland and Porter 2005). Size difference may provide a useful specific character if very marked (e.g. *A. candidum* Ryland (1963) has zooids >1 mm in length). Size within a species or between closely related species may vary in relation to ambient temperature, decreasing as temperature rises (Ryland and Porter 2005). This is important when considering a long coastline: the sea surface temperature difference between the Queen Charlotte Islands and southern California is ~10°C (http://www.osdpd.noaa.gov/ml/ocean/sst/contour.html). A peristome may (e.g., *A. mammillatum*) or may not be present, and the zooid surface may be ornamented with kenozooids (*A. hirsutum* (Fleming)) or elongated papillae (*A. parasiticum*), which create a sediment trap.

### Tentacle Number

Tentacle number is an important specific character (Porter et al. 2000) and should be determined using counts from different parts of living colonies, preferably from different locations. Alternatively, the number can be determined from serial sections, as was the practice of J D Soule (Osburn and Soule 1953). The ancestrula and initial autozooids have fewer tentacles (Cadman and Ryland 1996a, b) but the

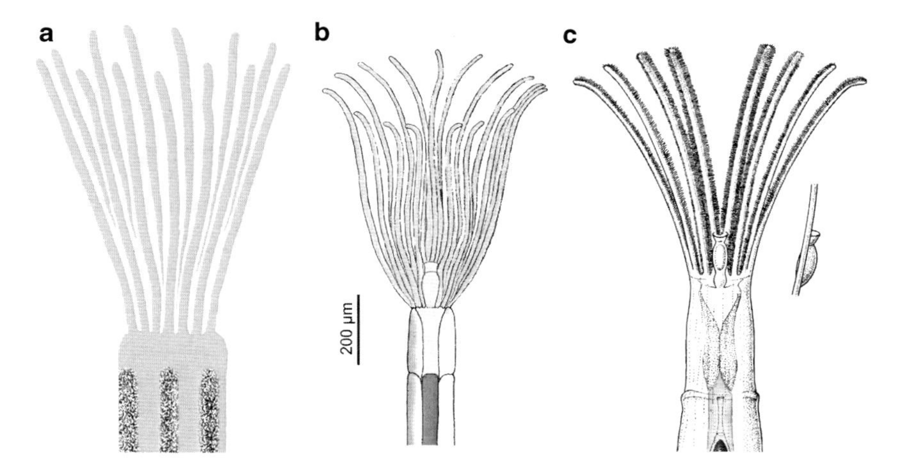


**Fig. 19.1** Freehand drawings of lophophores: (**a**) sp. #2 (Elkhorn Slough); (**b**) sp. #4 (Coos Bay); (**C**) sp. #8 (Bamfield), only dorsal-side tentacles drawn. Scale applies to (**b**) only, which was drawn using accurate measurements. Note inter-tentacular organs in (**b**) and (**c**)

adult number is soon reached. In live material the shape of the lophophore, the extent of the introvert, the occurrence of an intertentacular organ (ITO, see Fig. 19.1b, c), and any other unusual characters (such as the stripes on the introvert our *A.* sp. #2, Fig. 19.1a) should be noted.

## Morphology of the Digestive Tract

Le Brozec (1955) found differences in the proportions of gut parts in the zooids of *A. gelatinosum* and *A. polyoum*. Describing such characters has not generally been done (cf. Soule 1951), requiring the use of microscopic preparations, but remains desirable.

## Reproductive Mode

Whether zooids shed ova into the sea to develop into cyphonautes larvae (Reed 1991; Cadman and Ryland 1996a), or whether they brood embryos after degeneration of the polypide, is a fundamental character of *Alcyonidium* species. It is not yet established whether or not this is a phylogenetic attribute that would enable us to distinguish two clades. It is therefore premature, as well as impractical, to treat oviparous species as belonging to a separate genus, *Alcyonidioides* d'Hondt (2001). If zooids possess an intertentacular organ (normally only at particular times of the year) they are oviparous; if large yolky ova or brooded embryos are present they are larviparous (Fig. 19.3c). The manner in which the embryos are arranged (e.g. the circular disposition in *A. hirsutum*), and their number (Ryland and Porter 2006), may be diagnostic. Unfortunately, unidentified colonies often lack both intertentacular organs and embryos. Most species along the Pacific coast of North America appear oviparous.

### The Ancestrula and Early Astogeny

Ancestrular morphology and budding pattern may differ between congeners (e.g. in *Hippothoa* sensu lato) (Ryland and Gordon 1977) and is true in some larviparous species of *Alcyonidium* (Ryland and Porter 2006). These features remain unknown in oviparous species (even the well-studied *A. mytili*). They are extremely difficult to observe in established colonies, the ancestrula rapidly becoming unrecognizable, obscured by overgrowth of the surrounding autozooids. One of our species appears to have a distinctive initial budding pattern (Fig. 19.3d).

### Reproductive Phenology

In the British Isles, species toward the northern end of their geographical distribution tend to breed in summer while those toward the southern end of their distribution are winter breeders. It was this difference that led to the separation of the (?Arctic-) boreal *A. gelatinosum* and the temperate *A. polyoum* (de Putron and Ryland 1998; Ryland 2002). *A. mytili*, an oviparous boreal species, is also a winter breeder in the British Isles (Cadman and Ryland 1996a). Over the long latitudinal range of the Pacific coast similar differences are likely and, since most of the species appear oviparous, would best be established by recording presence/absence of intertentacular organs or an abundance of small ova (see below).

## Material and Methods

Our collections on the Pacific coast have been made during short visits over a number of years. JSR visited the Friday Harbor and Bamfield laboratories in 1986, making collections from submerged objects on harbor floats. Further material from Friday Harbor was sent to JSR by the late CG Reed in 1989. JSP visited Monterey, Friday Harbor and Bamfield in 1995, making intertidal collections.

JSR had a more extended visit to California and Oregon in June–July 2005, collecting from intertidal habitats over the low water period of spring tides, on every site accompanying a local, permit-carrying biologist. On rocky shores, *Alcyonidium* colonies were sought on stones and easily removable pieces of rock. Macro-algal thalli were inspected. Pier supports or other pilings, and the sessile fauna thereon, were examined when present. Colonies on wood were removed with a knife but those on stone surfaces were too thin and filmy to be collected in that way; splitting off a thin flake of rock was then necessary. Stones were packed in seaweed for transport and all samples returned to the laboratory aquarium (Hopkins Marine Station; Oregon Institute of Marine Biology) for maintenance and examination. Representative samples were eventually brought back to Swansea.

Stones and rock splits were photographed beside a calibrated scale, accompanied by close-up images where possible, using paired, angled flash heads fitted with diffusers. Microscopic examination was conducted using an Olympus dissecting'scope incorporating an objective diaphragm to increase depth of field and an Optronics Microfire camera. Particular *Alcyonidium* colonies were recorded as 'species #1' etc., since most clearly belonged to undescribed species.

Part of the material attached to thick plastic sheet (Fisherman's Bay, Lopez I.), which had been fixed in Bouin's Fluid and stored in 70% ethanol by CG Reed, was detached from the thick plastic sheet on which it had grown and was used for sectioning.

## Results

### *Collecting Sites*

Details of collection sites (approximate tidal heights, LW on the day, from NOAA: http://tidesandcurrents.noaa.gov/tides) are:

Friday Harbor Laboratory (FHL), University of Washington, San Juan Island, 48.55° N, 123.01° W; Jensen's Marina and FHL dock, 19 July 1986, JSR (no extant material), March 1989 and August 1989 (CG Reed); Fisherman's Bay, Lopez Island, 48.51° N, 122.91° W, colonies on black plastic sheet, 24 August 1989 (CG Reed).

British Columbia, Bamfield, 48.83° N, 125.14° W. The Marine Sciences Centre (formerly Marine Station) is in Bamfield Inlet, on the south side of the island-strewn Barkley Sound. Collection made from the line of floats below the Centre, 12–16 August 1986, JSR. Collection by JSP from a rocky area at LWST, Pachena Bay, 48.79° N, 125.12° W, August 1995.

California, Carmel, 36.54° N, 121.93° W, intertidal; rocky wall around Santa Cruz yacht marina, 36.97° N, 122.00° W; both August 1995, JSP. California, June 2005, intertidal, JSR: Cayucos, 35.46° N, 120.91° W, 22 June, −0.55 m; Old Salinas River–Elkhorn Slough river mouth, near Hwy 1 bridge, Moss Landing, 36.82° N, 121.74° W, 24 June, −0.46 m; Whistlestop Lagoon inlet/outlet culvert, Elkhorn Slough, 24 June; pilings of Santa Cruz wharf, 36.96° N, 122.02° W, 25 June, −0.43 m; Pescadero, 37.18° N, 122.39° W, just east of Pigeon Point lighthouse, mainly bedrock, 25 June, −0.49 m (7.47 AM, Ano Nuevo I.), 09.30–10.30 m.

Oregon, July 2005, JSR. Outside the rocky inner mole, North Spit, Coos Bay entrance (opposite Charleston), 43.36° N 124.33° W, 20 July, tide −0.6 m; shore just north of Fossil Point, Charleston, Coos Bay, 43.35° N, 124.315° W, 21 July, −0.7 m.

**Table 19.1** Characteristics of Pacific Coast *Alcyonidium* species (*ITO* intertentacular organ)

| | Localities | Colony form | Zooid and polypide | Tentacle number | Figure |
|---|---|---|---|---|---|
| 1 | Cayucos (Cay) Pigeon Point (PP) Coos Bay (NS) | Unilaminar | Zooids horizontal, distinct, when older often with a white (?lipid) lining | 18–19 Cay ~18 PP | Fig. 19.2a–c |
| 2 | Elkhorn Slough (ES) Coos Bay (CB) | Thick, muddy- or silty-looking | Zooids vertical; ITO and ova seen; introvert with earthy stripes | ~14 ES 14–15 CB | Fig. 19.1a |
| 3 | Santa Cruz wharf | Multilaminar with irregular, rounded prominences | Zooids jumbled, mixed orientation; ITO seen | 16 | Fig. 19.2d, e |
| 4 | Pigeon point; Coos Bay (CB) | Unilaminar. (Possibly >1 species) | Horizontal; in some colonies the zooids scarcely longer than broad, jumbled, with polypides upright in older region; ITO seen (CB) | 15–16 PP 15–16 CB | Fig. 19.1b |
| 5 | Coos Bay (CB) | Unilaminar; surface variously papillate | Zooids, horizontal but short at colony margin, more or less vertical in older parts, with raised peristomes occupying most of the frontal area | ~18 CB | Fig. 19.3a, b |
| 6 | Pigeon Point (PP); Carmel | Unilaminar | Larviparous; embryos pink | ~15 PP | Fig. 19.3c, d |
| 7 | Friday Harbor | Multilaminar | Oviparous (ITO and ova seen) | | Fig. 19.3e |
| 8 | Bamfield | Multilaminar, with mammillae | Oviparous (ITO seen) | ~16 | Figs. 19.1c and 19.2f |

## *The Species*

The collected samples belong to at least eight different species. All were encrusting and most were thick and multilaminar, with intercalary budding. To assist with future recognition, the characteristics are summarised in Table 19.1 and species are illustrated in Figs. 19.1, 19.2, and 19.3.

**Fig. 19.2** *Alcyonidium* species from the Pacific coast. (**a–c**) species #1; (**d, e**) species #3; (**f**) species #8. In (**e**) the *Alcyonidium* is overgrowing *Pollicipes polymerus*. No scale was used in (**f**) but the photographed section of rope extends 5–6 cm

## Discussion

We have shown that the Pacific coast of North America has a rich diversity of intertidal *Alcyonidium* species, our very limited collecting having obtained at least eight species. Only one of these may have been already described. The habitats range from bedrock platforms on the wave-exposed open coast, to protected rocky shores, sheltered embayments and fixed or floating structures associated with wharfs and marinas. All the species have been encrusting, as opposed to erect, but range from unilaminar to multilaminar, essentially smooth to raised in mammillae or protuberances. Some are readily recognizable (from information in Table 19.1 and Figs. 19.1, 19.2, and 19.3), others only with difficulty.

*Alcyonidium* sp. #1 (Fig. 19.2a–c) is unilaminar with clearly demarcated zooids made conspicuous by a whitish accumulation adjacent to the zooidal walls. This is likely to be lipid, as described by Cadman and Ryland (1996a) in *A. mytili*.

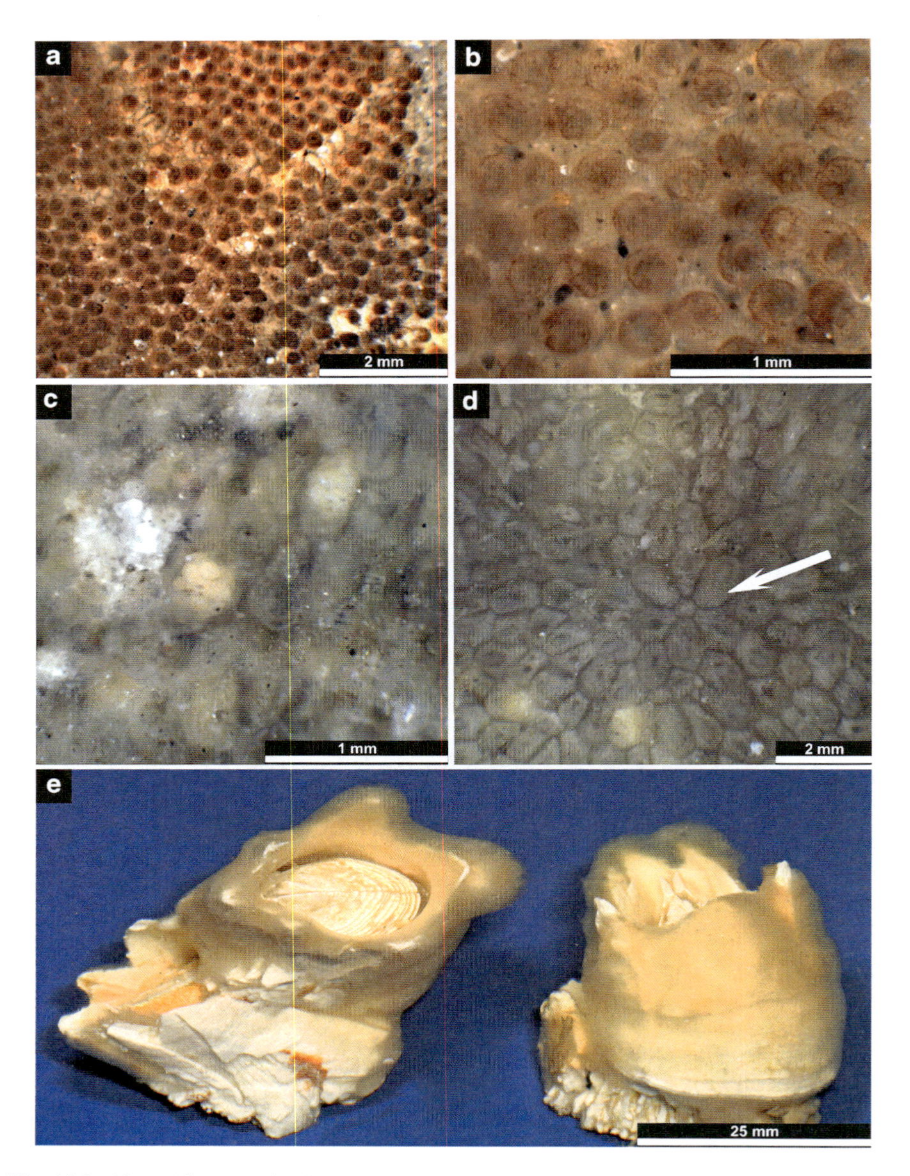


**Fig. 19.3** *Alcyonidium* species from the Pacific coast. (**a**, **b**) species #5; (**c**) zooid (*centre*) with a cluster of *pale pink* embryos, species #6; (**d**) early astogeny of a small colony, probably species #6; (**e**) species #7 overgrowing *Balanus nubilis*

The extent of this deposit must be expected to vary with age of the colony and time of year. The polypides have ~18 tentacles (higher than in co-occurring species) and will probably incorporate an ITO at certain times of the year, though none was seen in June–July.

Species #2 is very similar to, possibly identical with, European *A. parasiticum*. The main problem is that *A. parasiticum* is currently not properly defined and appears to comprise a complex of more than one similar species. For example, Levinsen (1894) and Marcus (1940) describe a species different from that of Hincks (1880), while we have seen specimens that conform to neither. Specimens from Elkhorn Slough and Coos Bay were noticed to have linear earthy stripes on the extended introvert (Fig. 19.1a), as described for *A. argyllaceum* by Castric-Fey (1971). Unfortunately neither she, nor any other European author, refers to presence or absence of such stripes in *A. parasiticum*. The American species cannot be named until the identity of *A. parasiticum* has been clarified.

Species #3 is distinctive (Fig. 19.2d–e), multilaminar, with the smooth surface raised into large rounded prominences (quite different from the mammillae of *A.* sp. #8 from Bamfield (Fig. 19.2f)). It formed large and conspicuous patches on the pilings of Santa Cruz wharf. An ITO was seen in June. *Alcyonidium* sp. #4 is problematic and is discussed below with sp. #6.

Species #5 (Fig. 19.3a–b) is unilaminar but recognizable by the variably raised peristome, though this feature is not developed to the extent seen, for example, in *A. mammillatum* (see Hincks (1880) or Hayward (1985)). Even when almost retracted, the peristome shows as puckering picked out with a circle of fine particles. The tentacle number (~18) is lower than in *A. mammillatum* (~21).

Smooth, unilaminar colonies, without the white zooid walls of sp. #1, have been seen on most of the rocky shores visited, their thin films coating the underside of boulders and rock faces. The first, from Pigeon Point, believed to be oviparous, was designated sp. #4. The ITO was subsequently seen in specimens (which, though extremely similar, may prove not be identical) collected in Coos Bay (Fig. 19.1b). When later examining specimens from Pigeon Point, however, a few zooids were found to contain pale pink embryos (Fig. 19.3c), and this larviparous species was designated #6. Earlier (1995), JSP had seen colonies with pink embryos at Carmel. Individual colonies lacking either embryos or an ITO cannot presently be attributed. All have about 15 tentacles. Our collections certainly contain colonies of different appearance – zooid shape, polypide morphology, and opacity of both matrix and (preserved) polypide, but these differences cannot yet be resolved into distinct species. The distinctive early astogeny of one colony, probably sp. #6, is shown in Fig. 19.3d.

The two remaining species are from northerly locations. Both have multilaminar colonies and lophophores with an ITO, but sp. #7 (Friday Harbor; Fig. 19.3e) has smooth, slightly translucent colonies, while #8 (Bamfield; Figs. 19.1c and 19.2f) has a mammillate colony recalling that of *A. nodosum* (Ryland 2001). In that species zooids in the mammillae were organized to exploit the location of exhalant chimneys to assist sperm discharge. An unsuccessful attempt was made to prepare histological sections of sp. #8; at some subsequent time the specimen was lost. Nothing resembling it has since been collected.

It is hoped that this brief preliminary account will both assist in the recognition of Pacific coast species and enable additional material and descriptive information to be obtained.

**Acknowledgements** We gratefully thank those who facilitated our collecting and study in the USA and Canada: Hank Chaney, Richard Emlett, Jeff Goddard, George Mackie, John and Vicky Pearse, Craig Young, Russel Zimmer and the late Christopher Reed.

## References

Banta WC, McKinney FK, Zimmer RL (1974) Bryozoan monticules: excurrent water outlets? Science NY 185:783–784

Bergey A, Denning D (1996) Phylum Bryozoa (Ectoprocta). In: Kozloff EN, Price LH (eds) Marine invertebrates of the Pacific Northwest. University of Washington Press, Seattle/London

Borg F (1930) Moostierchen oder Bryozoen (Ectoprocten). Tierwelt Deutschl 17:25–142

Cadman PS, Ryland JS (1996a) The characters, reproduction and growth of *Alcyonidium mytili* Dalyell, 1848 (Bryozoa: Ctenostomatida). In: Gordon DP, Smith A, Grant-Mackie J (eds) Bryozoans in space and time. NIWA, Wellington, pp 69–79

Cadman PS, Ryland JS (1996b) Redescription of *Alcyonidium mytili* Dalyell, 1848 (Bryozoa: Ctenostomatida). Zool J Linn Soc 116:437–450

Castric-Fey A (1971) Sur quelques bryozoaires de l'Archipel de Glenan (Sud-Finistère). Vie Milieu A 22:69–86

d'Hondt JL (1983) Tabular keys for the identification of the Recent ctenostomatous Bryozoa. Mém Inst océanogr (Monaco) 14:1–134

d'Hondt JL (2001) Flustrina versus Neocheilostomina (Bryozoa). Biosystematical remarks on supraspecific levels. Bull Soc Zool Fr 126:391–406

d'Hondt JL, Goyffon M (1989) New data on the intraspecific variability of *Alcyonidium polyoum* (Hassall, 1841), Bryozoa: Ctenostomida studied with gradient polyacrylamide gels. In: Ryland JS, Tyler PA (eds) Reproduction, genetics and distributions of marine organisms. Olsen & Olsen, Fredensborg, pp 273–282

de Putron S, Ryland JS (1998) Effects of the 'Sea Empress' oil spillage on reproduction and recruitment of *Alcyonidium* (Bryozoa) populations on *Fucus serratus*. In: Edwards R, Sime H (eds) The *Sea Empress* oil spill. T. Dalton & CIWEM, Lavenham, pp 457–466

Dick MH, Herrera-Cubilla A, Jackson JB (2003) Molecular phylogeny and phylogeography of free-living Bryozoa (Cupuladriidae) from both sides of the Isthmus of Panama. Mol Phylogenet Evol 27:355–371

Harmer SF (1915) The Polyzoa of the Siboga Expedition. Part 1. Entoprocta, Ctenostomata and Cyclostomata. Siboga Exped 28A:1–180

Hayward P (1985) Ctenostome bryozoans. In: Synopses of the British fauna, NS, vol 33. Linnean Society, London, pp 1–169

Hincks T (1880) A history of the British marine Polyzoa. van Voorst, London

Horowitz AS (1989) Listing of Recent ctenostome bryozoan species. Version of 23 July 1989. Circulated manuscript, p 32

Jang KH, Hwang UW (2009) Complete mitochondrial genome of *Bugula neritina* (Bryozoa, Gymnolaemata, Cheilostomata): phylogenetic position of Bryozoa and phylogeny of lophophorates within the Lophotrochozoa. BMC Genomics 10:167

Kluge GA (1962) Mshanki Severnykh Morei SSSR, vol 76, Opredeliteli po faune SSSR. Akademiia nauk SSSR, Moskva, pp 1–584 (In Russian)

Kluge GA (1975) Bryozoa of the northern seas of the USSR. Amerind, New Delhi

Le Brozec R (1955) Les *Alcyonidium* de Roscoff et leurs charactères distinctifs (Bryozoaires Ectoproctes). Trav Sta biol Roscoff 33 (4). Archs zool exp gén 93:35–50

Levinsen GMR (1894) Mosdyr. Zool Dan 4:1–105

Light SF (1941) Laboratory and field text in invertebrate zoology. Associated Students Store, University of California, Berkeley

Mackie JA, Keough MJ, Christidis L (2006) Invasion patterns inferred from cytochrome oxidase I sequences in three bryozoans, *Bugula neritina*, *Watersipora subtorquata*, and *Watersipora arcuata*. Mar Biol 149:285–295

Marcus E (1940) Mosdyr (Bryozoer eller Polyzoer). Danm Fauna 46:1–102

Nikulina EA, Hanel R, Schäfer P (2007) Cryptic speciation and paraphyly in the cosmopolitan bryozoan *Electra pilosa* – impact of the Tethys closing on species evolution. Mol Phylogenet Evol 45:765–776

O'Donoghue C, O'Donoghue E (1923) A preliminary list of Bryozoa (Polyzoa) from the Vancouver Island region. Contrib Can Biol Fish NS 1:143–201

O'Donoghue CH, O'Donoghue E (1925) Notes on certain Bryozoa in the collection of the University of Washington. Washington University Puget Sound Biological Station Publications 5:15–23

O'Donoghue CH, O'Donoghue E (1926) A second list of the Bryozoa (Polyzoa) from the Vancouver Island region. Contrib Can Biol Fish NS 3:49–131

Osburn RC, Soule JD (1953) Suborder Ctenostomata. In: Osburn RC (ed) Osburn: Bryozoa of the Pacific coast of America, part 3, Cyclostomata, Ctenostomata, Entoporocta and addenda. Allan Hancock Pacific expeditions. University of Southern California Press, Los Angeles, pp 726–758

Porter JS, Bloor P, Stokell BL, Ryland JS (2000) Intra- and inter-specific variation in tentacle number in the genus *Alcyonidium* (Bryozoa, Ctenostomatida). In: Herrera Cubilla A, Jackson JBC (eds) Proceedings of the 11th International Bryozoology Association conference. Smithsonian Tropical Research Institute, Balboa

Rattenbury JC, Smith RI (1954) Key to the more easily recognized Entoprocta and marine Bryozoa (Ectoprocta). In: Smith RI, Pitelka FA, Abbott DP, Weesner FM (eds) Intertidal invertebrates of the central California coast. University of California Press, Berkeley/Los Angeles

Reed CG (1991) Bryozoa. In: Giese AC, Pearse JS, Pearse VB (eds) Reproduction of marine invertebrates. Boxwood Press, Pacific Grove

Ricketts EF, Calvin J (1968) Between Pacific tides. Stanford University Press, Stanford

Robertson A (1900) Papers from the Harriman Alaska Expedition, VI. The Bryozoa. Proc Wash Acad Sci 2:315–340

Robertson A (1902) Some observations on *Asorhiza occidentalis* Fewkes, and related Alcyonidia. In: Proceedings of the Califorina Academy of Sciences, 3rd ser, vol 3, pp 99–108

Ryland JS (1963) Systematic and biological studies on Polyzoa (Bryozoa) from western Norway. Sarsia 14:1–60

Ryland JS (1969) A nomenclatural index to "A History of the British Marine Polyzoa" by T. Hincks (1880). Bull Brit Mus (Nat Hist) Zool 17:207–260

Ryland JS (2001) Convergent colonial organization and reproductive function in two bryozoan species epizoic on gastropod shells. J Nat Hist 35:1085–1101

Ryland JS (2002) Life history characters and ecology of some incrusting ctenostomate Bryozoa. In: Wyse Jackson P, Jones MS, Buttler C (eds) Bryozoan studies 2001. A.A. Balkema, Lisse

Ryland JS, Gordon DP (1977) Some New Zealand and British species of *Hippothoa* (Bryozoa Cheilostomata). Proc R Soc New Zealand 7:17–49

Ryland JS, Porter JS (2005) Variation in zooid size in two west European species of *Alcyonidium* (Ctenostomatida). In: HIM Moyano G, Cancino JM, Jackson PW (eds) Bryozoan studies 2004. A.A. Balkema, Leiden

Ryland JS, Porter JS (2006) The identification, distribution and biology of encrusting species of *Alcyonidium* (Bryozoa: Ctenostomatida) around the coasts of Ireland. Biol Environ Proc Roy Irish Acad 106B:19–33

Schwaninger HR (2008) Global mitochondrial DNA phylogeography and biogeographic history of the antitropically and longitudinally disjunct marine bryozoan *Membranipora membranacea* L. (Cheilostomata): another cryptic marine sibling species complex? Mol Phylogenet Evol 49:893–908

Silén L (1942) Carnosa and Stolonifera (Bryozoa) collected by Prof. Dr. Sixtén Bock's expedition to Japan and the Bonin Islands, 1914. Ark Zoo, Stockh 34A:1–33

Silén L (1943) Notes on Swedish Marine Bryozoa. Ark Zool Stockh 35A:1–16
Soule JD (1951) Two new species of incrusting ctenostomatous Bryozoa from the Pacific. J Wash Acad Sci 41:367–370
Soule JD, Soule DF, Pinter PA (1975) Phylum Ectoprocta (Bryozoa). In: Smith RI, Carlton JT (eds) Light's Manual: intertidal invertebrates of the central California coast. University of California Press, Berkeley/Los Angeles
Thorpe JP, Ryland JS (1979) Cryptic speciation detected by biochemical genetics in three ecologically important intertidal bryozoans. Estuar Coast Shelf Sci 8:395–398
Thorpe JP, Ryland JS (1987) Some theoretical limitations on the arrangement of zooids in encrusting Bryozoa. In: Ross JRP (ed) Bryozoa: present and past. Western Washington University, Bellingham, pp 277–283
Thorpe JP, Winston JE (1984) On the identity of *Alcyonidium gelatinosum* (Linnaeus, 1761) (Bryozoa, Ctenostomata). J Nat Hist 18:853–860
Waeschenbach A, Telford MJ, Porter JS, Littlewood DTJ (2006) The complete mitochondrial genome of *Flustrellidra hispida* and the phylogenetic position of Bryozoa among the Metazoa. Mol Phylogenet Evol 40:195–207
Waeschenbach A, Cox CJ, Littlewood DTJ, Porter JS, Taylor PD (2009) First molecular estimate of cyclostome bryozoan phylogeny confirms extensive homoplasy among skeletal characters used in traditional taxonomy. Mol Phylogenet Evol 52:241–251

# Chapter 20
# Distribution and Zoogeography of Cheilostomate Bryozoa Along the Pacific Coast of Panama: Comparison Between the Gulf of Panama and Gulf of Chiriquí

## Cheilostomes in Pacific Waters of Panama

**Priska Schäfer, Amalia Herrera Cubilla, and Beate Bader**

**Abstract** Along the Pacific coast of Panama, bryozoans are often a dominant component in epibenthic communities occupying more space than other taxa. Bryozoans dominate on carbonate substrates such as shells, coral rubble and stones around the islands scattered in the Gulf of Panama and Gulf of Chiriquí, shelves situated in very contrasting environmental settings. A quantitative survey of the encrusting cheilostome fauna occurring in both gulfs was undertaken to study distribution and diversity patterns. Species occurrences were analysed under the context of broader zoogeographic distribution patterns. Both the highest and lowest species richnesses (alpha) were found in the Gulf of Chiriquí whereas the Gulf of Panama showed a moderate richness. The bryozoan fauna in the Gulf of Chiriquí is more abundant and more diverse than in the Gulf of Panama probably due to a higher degree of habitat fragmentation, lower food level and a more stable environment. Q-mode cluster analyses of the 16 most abundant species revealed groupings between the two gulfs (presence/absence) and between shallow and deep sites (absolute abundances). R-Mode cluster analyses of species distribution revealed a Panama and Chiriquí cluster, in which the Chiriquí cluster showed a dominance of species with West Atlantic, Caribbean and Indo-Pacific distribution, whereas the Panama cluster had species known from Atlantic and East Pacific waters.

**Keywords** Bryozoa • Cheilostomata • Panama • Distribution • Ecology • Zoogeography

---

P. Schäfer (✉) • B. Bader
Institute of Geosciences, Kiel University, Ludewig-Meyn-Str. 10, Kiel D-24118, Germany
e-mail: ps@gpi.uni-kiel.de; bbader@online.no

A.H. Cubilla
Smithsonian Tropical Research Institute (STRI), MRC 0580-01 Apdo, Panama City 0843-03092, Republic of Panama
e-mail: herreraa@si.edu

A. Ernst, P. Schäfer, and J. Scholz (eds.), *Bryozoan Studies 2010*,
Lecture Notes in Earth System Sciences 143, DOI 10.1007/978-3-642-16411-8_20,

## Introduction

Bryozoans are often the dominant component in epibenthic communities along the Pacific coast of Panama occupying more space than other taxa. In contrast with the Caribbean Sea, little systematic work has been done on bryozoan faunas in the Pacific waters of Panama and the role of bryozoans in ecosystems is mostly unknown. Hastings (1930) conducted the first taxonomic study on bryozoans in Panama in the vicinity of the Panama Canal. Following this, Osburn (1950, 1952, 1953) subsequently described the bryozoan fauna of the Pacific coast of North America. He covered the full length of the coast from Alaska to Central America, and therefore his monographs lack much detail for the Panama region. Zoogeographic and evolutionary analyses in this area are hampered by: (1) an overall lack of detailed knowledge of tropical eastern Pacific bryozoan faunas, (2) failure in the past to correctly partition intraspecific versus interspecific variation in cheilostomes leading to taxonomic confusion (Jackson and Cheetham 1990; Soule et al. 2002; Dick and Mawatari 2004) and (3) a misbalance in the knowledge of extant and fossil faunas between the Caribbean and tropical eastern Pacific.

Bivalve shells obtained on two expeditions collecting surface carbonates on the Gulf of Panama and Gulf of Chiriquí shelves (Fortunato and Schäfer 2009; Schäfer et al. 2011; Reijmer et al. 2011), were studied with respect to bryozoan diversity patterns. The present study is a faunistic analysis of encrusting cheilostome bryozoans from both gulfs and from the ecological points of view. The final aim of the survey was to lay a basis for examining the zoogeographic relationships among regional bryozoan faunas of the Pacific side of Panama and between faunas on the Caribbean and Pacific sides.

## Geographic and Hydrographic Settings

The Pacific coast of Panama is divided into two parts, the Gulf of Panama in the east and the Gulf of Chiriquí in the west, separated by the Azuero Peninsula (Fig. 20.1). The Gulf of Panama is enclosed to the north, east and west forming a gradually deepening shelf ramp. This contrasts with the Gulf of Chiriquí, which is enclosed to the north and east but only semi-enclosed to the west forming a narrow, rapidly steepening but topographically structured shelf.

The Gulf of Panama and the Gulf of Chiriquí are characterized by distinctly different hydrographic settings (D'Croz and O'Dea 2007). The Gulf of Panama is prone to intense seasonal upwelling during the dry season (January to March) generated by the seasonal latitudinal shift of the Intertropical Convergence Zone (ITCZ) (Rodriguez-Rubio et al. 2003). During the dry season, strong NE trade winds blow across the isthmus bringing cold and saline and nutrient-rich waters to the sea surface (18–25°C, 33.3 33.6 PSU) and generating an intense primary production (chlorophyll-$\alpha$ values up to 1.44 mg $m^{-3}$). During the wet season

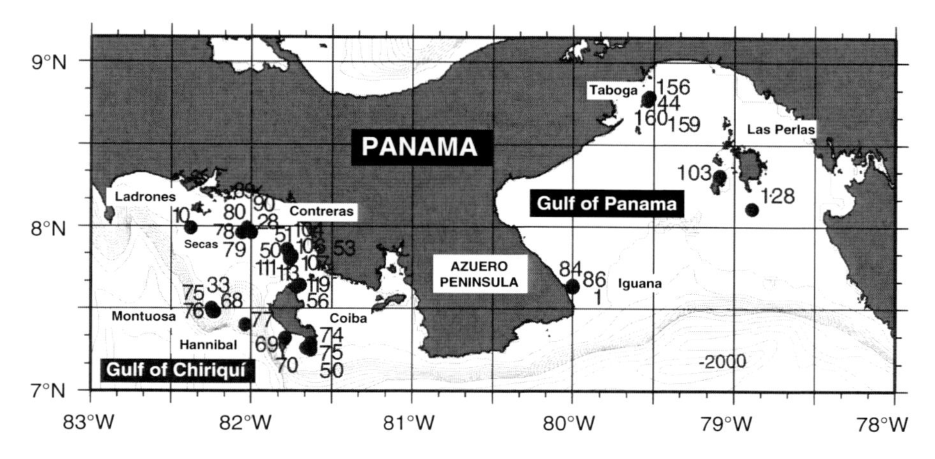


**Fig. 20.1** Bathymetric map of the Pacific coast of Panama including Gulf of Panama and Gulf of Chiriquí. Numbers refer to site localities included in the study

(April to December), sea surface salinity drops to 29.2–30.3.3 PSU (D'Croz and O'Dea 2007).

In contrast, the Gulf of Chiriquí is much less affected by seasonality due to changes in trade wind patterns. Here, the high Cordillera de Talamanca blocks off the NE trade winds. The largest seasonality results mostly from salinity changes between the dry and wet seasons (30.5–32.5 PSU), whereas sea surface temperatures are relatively stable (27.5–28.8°C) throughout the year. The range in chlorophyll-$\alpha$ is from 0.16–0.29 mg m$^{-3}$ in the dry season (D'Croz and O'Dea 2007). In the wet season, chlorophyll-$\alpha$ are equally low in both gulfs.

Benthic communities and carbonate sediments in the Gulf of Panama and Gulf of Chiriquí reflect the respective hydrographic conditions (Reijmer et al. 2011). Carbonate sediments dominated by bivalves and barnacles characterize the eutrophic Gulf of Panama whereas bivalves, coralline red algae, and occasionally scleractinians prevail in the mixotrophic Gulf of Chiriquí.

In both gulfs, carbonate-rich environments containing abundant bivalve shells, which serve as a substrate for bryozoans, are concentrated around the islands. In this study, we examined shells from around Isla Taboga, Isla Iguana, and Islas Perlas (Gulf of Panama) and from around Isla Coiba, Islas Contreras, Islas Ladrones, Isla Montuosa, and the 70 m-deep Banco Hannibal (Gulf of Chiriquí).

## Material and Methods

We conducted dredging expeditions aboard RV *Urraca* (Smithsonian Tropical Research Institute, STRI) in the Gulfs of Panama and Chriquí during May 2004 and February 2005 to sample the marine benthos and carbonate sediments. From bulk samples, we randomly selected 75 substrate samples, 30 from the Gulf of Panama and 45 from the Gulf of Chiriquí consisting of shells, stones, coral rubble,

**Table 20.1** Gulf of Panama and Gulf of Chiriquí station numbers, number of colonies and number of species, arranged for shallow ≤30 m and deep sites (≥30 m)

| | Shallow sites (≤30 m) | Sites | Colonies/ species | Deep sites (≥30 m) | Sites | Colonies/ species |
|---|---|---|---|---|---|---|
| Gulf of Panama | Isla Taboga | 44, 86, 159, 160 | 359/24 | – | – | – |
| | Isla Iguana | 1, 84, 86 | 254/31 | – | – | – |
| | Islas Perlas | 103, 128 | 90/25 | – | – | – |
| Gulf of Chiriqui | Islas Secas | 28, 80 | 142/20 | Islas Secas | 78, 79 | 142/20 |
| | Islas Contreras | 50, 51 | 369/39 | Islas Contreras | 104, 106, 107 | 264/32 |
| | Isla Montuosa | 33, 68 | 51/9 | Isla Montuosa | 75, 76 | 68/18 |
| | Isla Coiba | 56, 69, 70, 111, 113 | 388/24 | Isla Coiba | 50, 74, 75 | 27/13 |
| | – | – | – | Islas Ladrones | 10 | 27/7 |
| | – | – | – | Banco Hannibal | 77 | 44/8 |

or coralline red algae and examined them for encrusting cheilostome bryozoans. These samples come from 34 dredge stations at nine islands or sites, three are in the Gulf of Panama and six in the Gulf of Chiriquí. The depth range of the collections was 2–76 m. At shallow sites (≤3 m), we collected by snorkeling. All sites in the Gulf of Panama were shallower than 30 m depth whereas sites in the Gulf of Chiriquí were both shallow (above 30 m) and deeper (below 30 m) (Table 20.1).

## Data Acquisition

All substrate samples were outlined on paper and the location of every encrusting bryozoan colony was mapped onto the drawings by using a dissecting microscope. All encrusting colonies were counted and identified to species when possible following Jackson and Herrera-Cubilla (2000). For comparison, data were compiled from Soule et al. (1995) for Californian basins and from Dick et al. (2006) for Hawaii. Herein, the total abundance of cheilostomes is expressed as the number of discrete individuals (colonies).

## Data Analysis

*Species sampling adequacy*. The number of species was plotted against the number of colonies examined (Hayek and Buzas 1997). These cumulative species curves allowed us to judge the completeness of sampling in the gulf (Fig. 20.2).

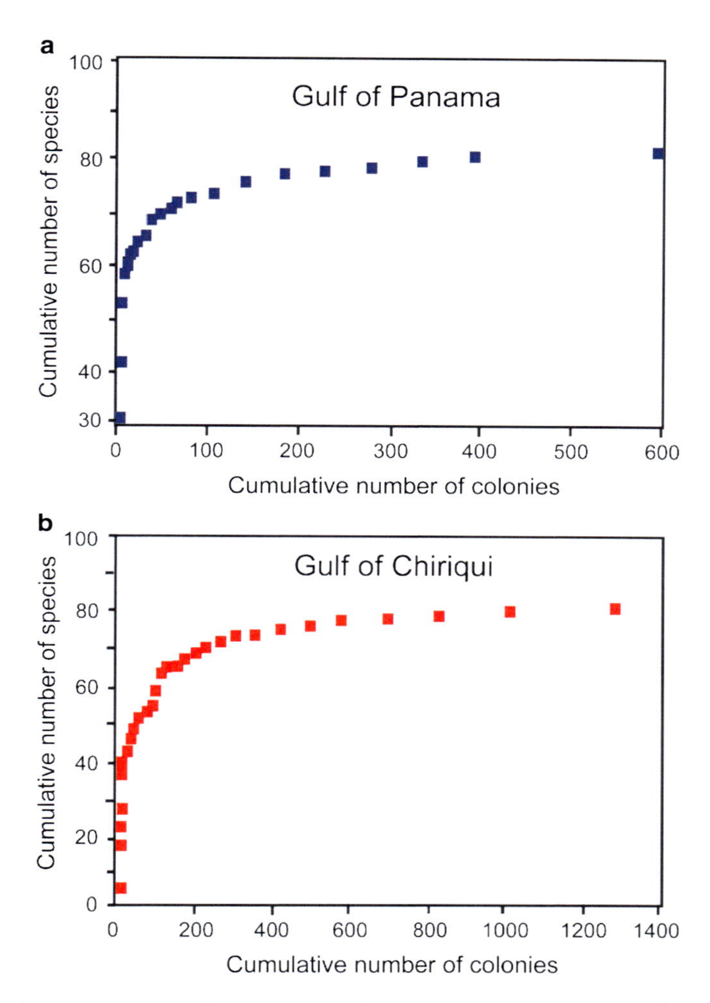


**Fig. 20.2** Cumulative curves of species. (**a**) Gulf of Panama; (**b**) Gulf of Chiriquí

*Species Diversity*. Diversity calculations broadly followed the recommendations of Magurran (1988). Sample species richness (the number of species found corrected for the number of individuals examined) was expressed using the log-series index (alpha), which has been widely recommended because it has good discriminant ability and low sensitivity to sample size.

*Similarity Indices*. Jaccard's, Sørensen's, and Montford's similarity indices were calculated as a means of comparing similarities in faunal composition between ecosystems (gulfs). Jaccard's and Sørensen's indices are the simplest of similarity comparisons between sample sets, but both indices are rather sensitive to differences in sample size. This problem is less in Montford's index of similarity (SPSS Inc. 2002).

*Comparison of community structures.* Quantitative methods based on the abundance of the 16 most common species co-occurring in both gulfs, including all species with ≥30 colonies in both gulfs (2,327 colonies in total) were used to compare community structures. The data matrix included 13 sites, with both shallow and deep sites in some localities. Sites in the Gulf of Panama include Taboga, Iguana, and Las Perlas, those in the Gulf of Chiriquí include Coiba (<30 m, >30 m), Contreras (<20 m, >30 m), Ladrones (50 m), Montuosa (<30 m, 40 m), Secas (<30 m, >30 m), and Banco Hannibal (70 m) (Table 20.1).

Clusters (R-mode) expressing the degree of association between species were generated based on the most common species from both gulfs. In addition, a "site – biofacies" cluster analysis (Q-mode) was conducted that grouped sites with similar species composition. The cluster analyses used the average linkage method in which distances between samples were expressed as Pearson correlations. Statistical analysis followed Rucker (1975) and Hughes and Jackson (1992) (SPSS Inc. 2002).

## Results

In all, 83 cheilostome species were found in both gulfs, representing 47 genera and including 23 anascans (+ cribrimorphs) and 60 ascophorans. Twenty-six genera were represented by one species, ten genera with two species, four genera by three species each, two genera by four and five species, respectively, and one genus (*Microporella*) by six species.

Fifty-one species were found in the Gulf of Panama and 77 species in the Gulf of Chiriquí. While 46 species were common to both gulfs, only five species were restricted to the Gulf of Panama compared to 31 restricted to the Gulf of Chiriquí. Jaccard's Index was 31.9% similarity, Sørensen's Index was 35.9% similarity, and Montford's Index was 71.8%. Species richness (alpha) was highest (> 8) in shallow and deep Coiba, followed by shallow and deep Contreras and Iguana. Species richness (alpha) was moderate (6–8) in shallow and deep Secas followed by Taboga, and lowest (3–5) in Las Perlas, deep Ladrones and Montuosa and on Banco Hannibal.

At shallow sites, species numbers in the Gulf of Chiriquí ranged from 9 (Montuosa) to 40 around Coiba (Fig. 20.3a). At shallow sites in the Gulf of Panama, species numbers were less variable and lower than in the Gulf of Chiriquí, ranging from 25 (Taboga and Las Perlas) to 32 (Iguana). Colony number at shallow sites ranged from 51 (Montuosa) to 388 (Coiba) in the Gulf of Chiriquí, and from 90 (Las Perlas) to 359 (Taboga) in the Gulf of Panama (Fig. 20.3b).

At deep sites, all in the Gulf of Chiriquí, species numbers ranged from 8 (Banco Hannibal) to 32 (Contreras) (Fig. 20.3c), and colony numbers ranged from 22 (Ladrones) to 264 (Contreras) (Fig. 20.3d).

In the Gulf of Panama, although bryozoan diversity and abundance are much lower than in the Gulf of Chiriquí, a higher percentage of colonies belonged to the 16 most abundant species, with the remaining 35 species showing distinctly lower abundances. In contrast, in the Gulf of Chiriquí, although the bryozoan diversity

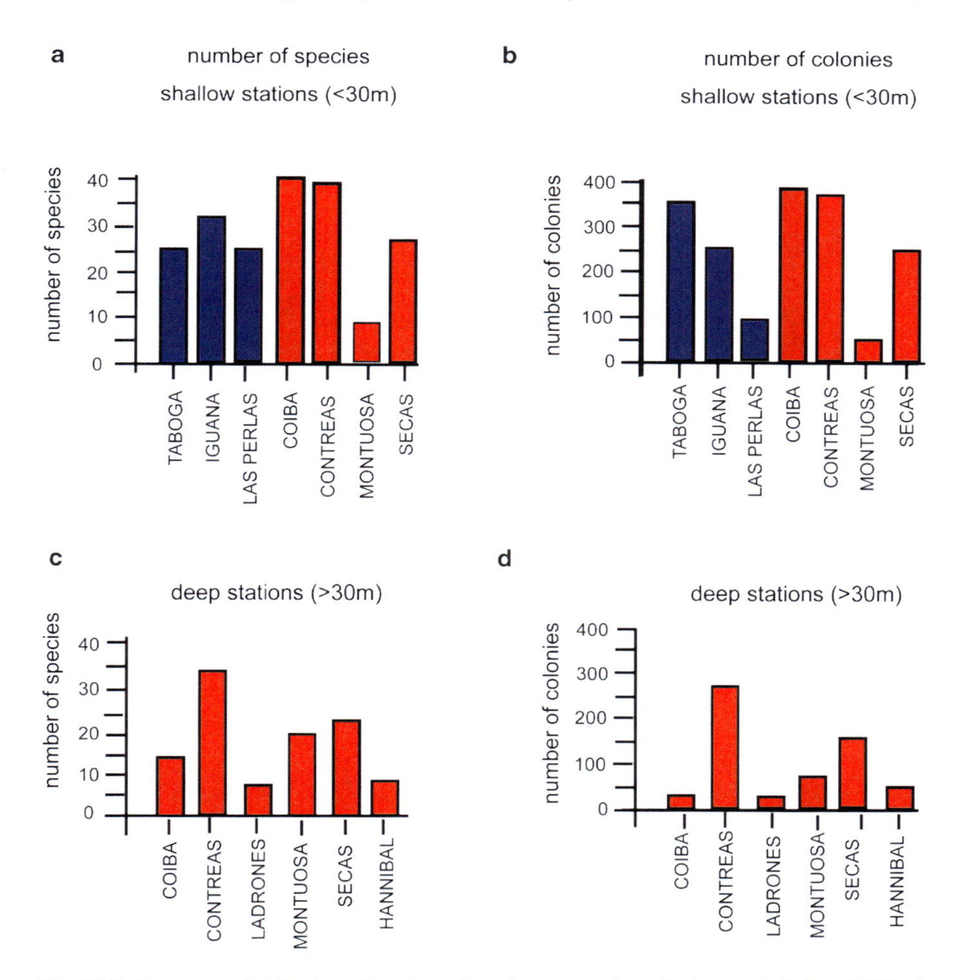


**Fig. 20.3** Frequency distribution of number of species and number of colonies at shallow (≤30 m) and deep sites (≥30 m)

and abundance were much higher, the percentage of colonies belonging to the 16 most common species was lower than in to the Gulf of Panama. In other words, more species in the Gulf of Chiriquí were represented by a few colonies.

The 16 most abundant species co-occurring in both gulfs (Fig. 20.4) contribute 76.5% of the total abundance of 2,327 colonies. The remaining 30 species in common thus represented less than one quarter of all colonies. In the Gulf of Panama, 81.6% (574 colonies) belonged to the 16 most common species, whereas the rest (18.8%) were distributed among the other 35 species. In the Gulf of Chiriquí (1,624 colonies in total), 74.15% of the colonies belonged to the 16 most common species, with the rest distributed among the remaining 61 species.

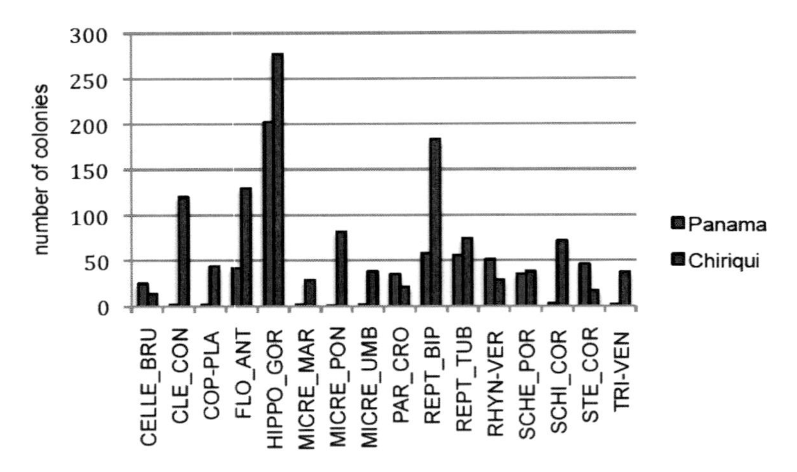


**Fig. 20.4** Frequency distribution of 16 most abundant species in the Gulf of Panama and Gulf of Chiriquí (same species as in Table 20.2)

The number of colonies of the 16 most abundant species in both gulfs varied considerably (Fig. 20.4). Twelve of these species showed more colonies in the Gulf of Chiriquí than in the Gulf of Panama. Most common in both gulfs are *Hippoporella gorgonensis* (479 colonies, 20.6%, 9 sites) and *Reptadeonella bipartita* (241 colonies, 10.4%, 10 sites) followed by *Floridina antiqua* (171 colonies, 7.35%, 8 sites) and *Reptadeonella tubulata* (130, 5.59%, 10 sites). *Hippoporina mexicana*, *Microporella tractabilis*, *Rhynchozoon* sp., *Antropora granulifera* and *Celleporaria* aff. *aperta* occurred as a few colonies and only at Gulf of Panama sites. Only *Parasmittina crosslandi*, *Rhynchozoon verruculatum* and *Steganoporella cornuta* had more colonies in the Gulf of Panama. *Hippoporella gorgonensis* was the most abundant species in both gulfs.

A plot of species richness against water depths (Fig. 20.5) shows an increase in the occurrence of species with decreasing depth. At 60 m depth, species numbers are low but individual species may occur with relatively high colony numbers. The opposite pattern occurs at 70 m water depth. At shallowest site (few meters), species numbers coincide with that in 10–40 m depth, whereas colony numbers in general are only moderately high. Although a negative trend is observable, a reliable statistical correlation was found neither for species numbers (Fig. 20.5) nor colony numbers (Fig. 20.6) versus water depth.

The Q-mode cluster analysis (16 species, abundances) used for sites/facies shows several clusters of which cluster I includes three both shallow and deep subclusters (both Panama and Chiriquí) and cluster II includes three deep Chiriquí sites (Fig. 20.7). If presence/absence data are considered, the Q-mode cluster analysis shows three main clusters of which cluster I includes two subclusters (Fig. 20.8).

The R-mode cluster analysis (16 species, abundances) used for species shows two distinct cluster groups (I and II) of which both include two subclusters (Fig. 20.9). The R-mode cluster analysis for presence/absence data shows two main clusters (I and II) of which cluster I includes several subclusters (Fig. 20.10).

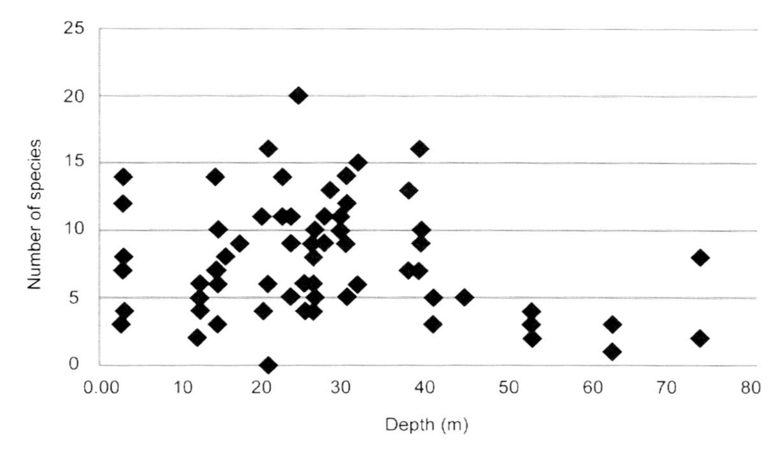


**Fig. 20.5** Species numbers versus water depths (all stations)

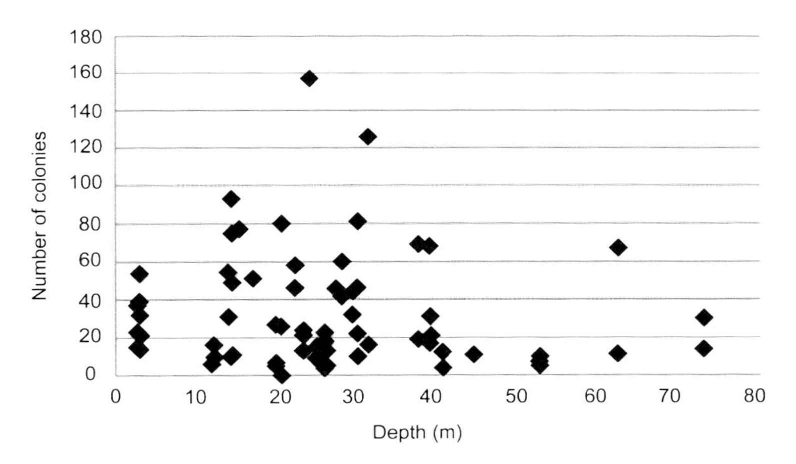


**Fig. 20.6** Colony numbers versus water depths (all stations)

# Discussion

The cheilostomate bryozoan fauna along the Pacific coast of Panama is distinctly less specific in the Gulf of Panama (only five species are restricted to this gulf) than in the Gulf of Chiriquí (with 32 species restricted here). This distinctiveness in faunal composition implies at first glance a less characteristic, more widespread bryozoan fauna (9.8% endemism) in the Gulf of Panama compared to the more specific faunas of the Gulf of Chiriquí (92.5% endemism).

*Species richness and environment.* Several reasons may account for the existence of higher species richness in the Gulf of Chiriquí. This could be due to the greater

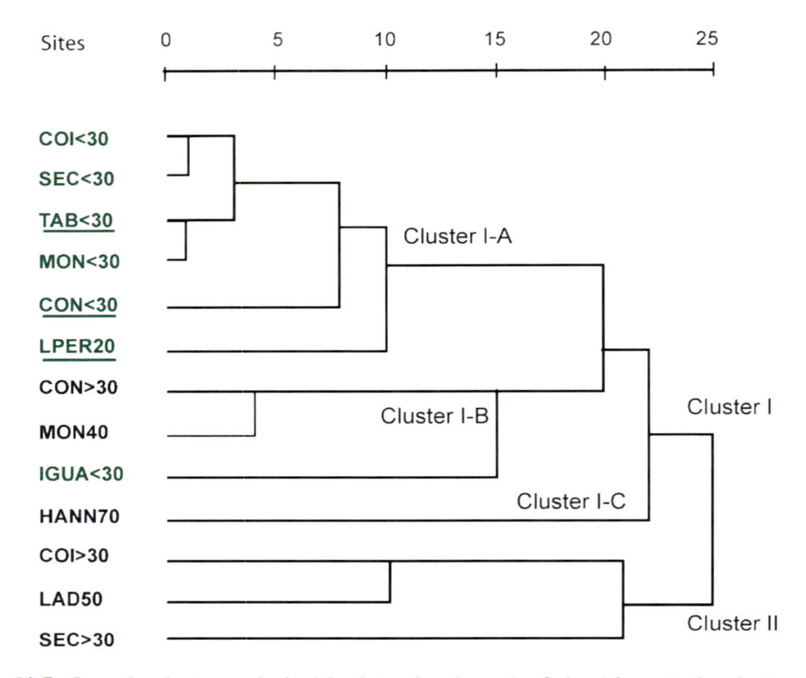


**Fig. 20.7** Q-mode cluster analysis (absolute abundances) of the 16 most abundant species; groupings according to sites (abbreviations see Table 20.1). *Green*: shallow-water sites; *black*: deep-water sites; *underlined* sites: Gulf of Panama

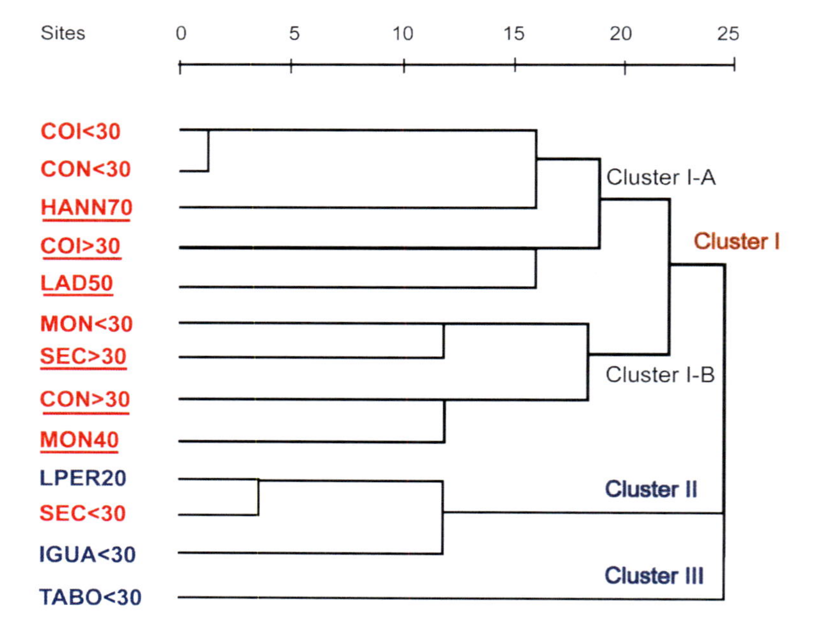


**Fig. 20.8** Q-mode cluster analysis (presence/absence) of the 16 most abundant species; groupings according to sites (abbreviations see Table 20.1). *Blue*: Panama sites; *red*: Chiriquí sites; *underlined* sites: stations >30 m depth

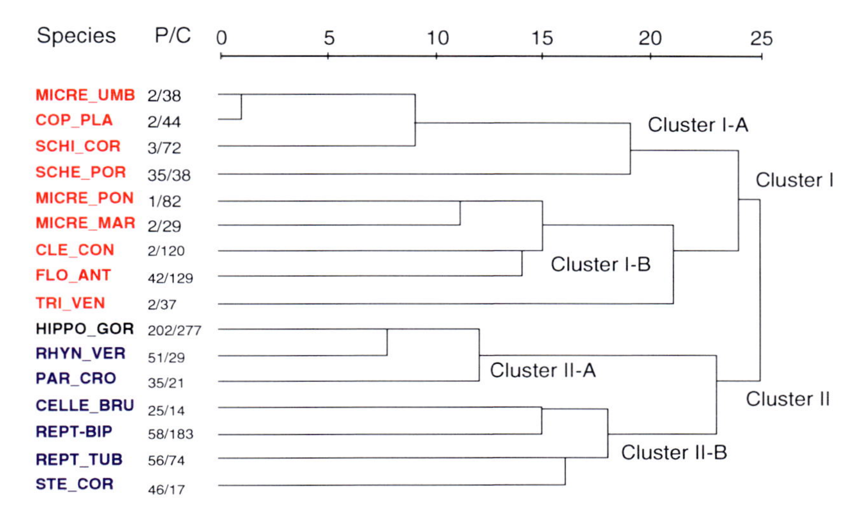


**Fig. 20.9** R-Mode cluster analysis (absolute abundances) of the 16 most abundant species; groupings according to species (abbreviations see Table 20.2). P/C: number of colonies at Panama and Chiriquí sites, respectively; *blue*: species with high abundance in the Gulf of Panama; *red*: species with high abundance in the Gulf of Chiriquí; black: species equally abundant in both gulfs

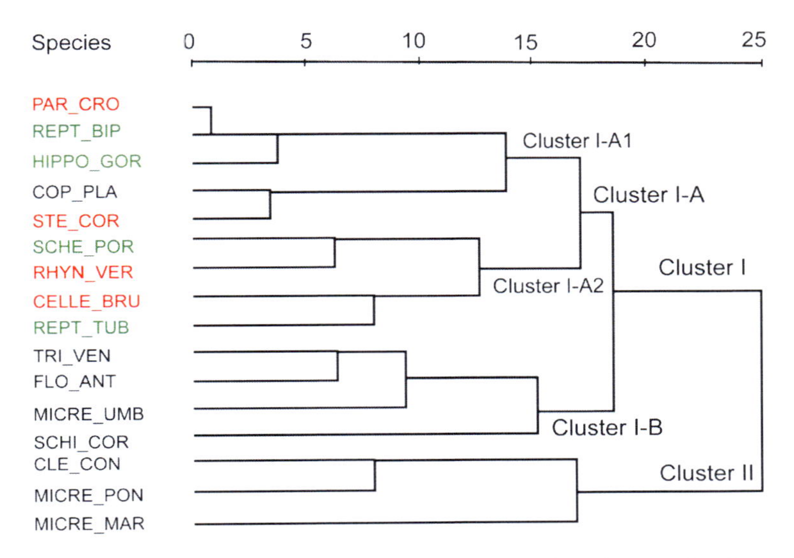


**Fig. 20.10** R-Mode cluster analysis (presence/absence) of the 16 most abundant species; groupings according to species (abbreviations see Table 20.2); *red*: species with higher abundance in the Gulf of Panama; *green*: species with higher abundances in the Gulf of Chiriquí

habitat fragmentation when compared with the Gulf of Panama. The Gulf of Panama has a broad, evenly steeping shelf the vast majority of which is covered with silts and only a few islands or island groups providing gravel substrate (stones and shells) suitable for larval settlement. In contrast, the Gulf of Chiriquí is partitioned into small-scale habitats due to a narrow shelf with many islands/island groups in which calm, near-coastal mangrove habitats, shell gravel lag sea-beds, coral reefs, and steep troughs with high accumulation rates of fine sediments are close together.

Nevertheless, the argument that limitation of suitable substrates causes low species diversity and abundances is weakened by the fact that balanid plates form the bulk of carbonate sediment particles around the islands in the Gulf of Panama (Reijmer et al. 2011) indicating that there is enough firm substrate (rock) for epibenthic larvae to settle. Balanids, however, are especially competitive in intertidal and shallow subtidal rocky habitats (Anderson 1994). Along the Pacific coast of Panama, layered, relatively stable substrates such as shells of epibenthic bivalves like species of *Spondylus, Pinctada, and Pteria* seem to be more suitable substrates for bryozoan larvae than the rock habitats where barnacles tend more to settle and be most competitive colonizers of rocky surfaces.

Contrasting with balanids, bryozoans are more common in the Gulf of Chiriquí than in the Gulf of Panama, probably due to the more intense fisheries in the latter. As a matter of fact, the fishing of the pearl oyster *Pinctada mazatlantica* since the sixteenth century by the Spaniards and its overexploitation during the early twentieth century contributed to the destruction of the extensive banks that had flourished in the Gulf of Panama in 1.5–18.0 m depth in rocky or gravelly areas (Cipriani et al. 2008) and that comprise the primary habitats of encrusting bryozoans. The living populations became then completely wiped out by the exceptionally strong red tides during the dry season 1938 (Fortunato pers. com.).

On the other hand, the much higher productivity in the Gulf of Panama does not seem to trigger an increase in species diversity and abundance. An important reason might be both the seasonality and instability of the environment resulting in a reduced success in larval settlement in the Gulf of Panama. Unpredictable environments although rich in food resources, result in low diversity and favour r-strategies (Valentine 1973). This hypothesis corresponds to the findings of Jackson and Herrera-Cubilla (2000), who compared ovicell area and size variation in cheilostomes between the Caribbean Sea and Eastern Pacific as well as between the Gulf of Panama and Gulf of Chiriquí. The largest differences in ovicell and zooid area ($mm^2$) were found between the Caribbean and the Gulf of Panama and were explained as adaptive strategies to differences in food availability, seasonality and environmental stability. This agrees with data concerning echinoid egg size (Lessios 1990), the proportions of strombinid gastropods with direct, lecitotrophic vs. planktotrophic development (Jackson et al. 1996), and gastropod development types (Fortunato 2004). According to Jackson and Herrera-Cubilla (2000), smaller differences in zooid and ovicell sizes exist between the Gulf of Panama and Gulf of Chiriquí, despite their trophic differences, indicating that local adaptation may be downplayed by gene flow in the eastern Pacific, at least on this spatial scale.

*Depth relationships.* Although negative relationships between water depth and benthic diversity are well known and have been described for many organisms along shelf-continental slope transects (Angel 1994; Gaston and Spicer 2007 and references therein), the picture is much more complex when shallow shelf benthos are considered. In general, highest bryozoan richness is known for 0–100 m water depths although a differentiated pattern is found if different colonial growth forms are considered (McKinney and Jackson 1989). In this study, the pattern of decreasing number of species and colonies with increasing depth is not much pronounced and might be best explained by the reduction of substrates available to colonisation. With depth, suitable substrates become rare, more frequently covered by fine sediment films probably preventing bryozoan larval settlement, whereas suspended sediment may additionally hamper nutrition of polypides. Overall food availability, instead, is not considered to play a major role here otherwise one would expect a much higher bryozoan diversity and abundance in the eutrophic Gulf of Panama, which is not the case.

*Community structure.* The Q-Mode cluster analyses show distinctly different results depending on whether absolute abundances or presence/absence of the 16 most abundant species is considered. Indeed, when absolute abundances are taken into account, one shallow-site (cluster I) and one deep-site cluster (cluster II) can be distinguished as well as two mixed, shallow/deep-site clusters (Fig. 20.7). Notice that all Panama sites are nested within the Chiriquí clusters while the deep Banco Hannibal forms a separate cluster. The existing correlation between clusters and water depths found for abundances may correspond with the inverse relationship between species richness and water depth. Certainly, any scatter in clustering depth might be an artefact of using a depth cut-off by distinction on water depths above and below 30 m ("shallow" vs. "deep sites").

The Q-mode cluster analysis based on presence/absence data shows three main (I–III) clusters, two of which include Panama sites (Fig. 20.8). Taboga forms a separate cluster (III), whereas Iguana and Las Perlas (both open Gulf of Panama) co-occur with shallow Secas (II) the latter probably indicating a mere sampling effect by chance, because the site has a high portion of species in common with the Gulf of Panama. In general, geographically neighbouring sites cluster together independently of the water depth. This contrasts with the absolute abundance clustering where a more pronounced distinction between shallow and deep sites is found independently of geographic closeness. The clear separation into a Panama and Chiriquí cluster may be due to differences in temperature, salinity, and nutrients triggering food availability, which may be equally important for species distribution as water depth.

The R-mode cluster analysis based on abundance data shows a cluster including species occurring with high abundances in the Gulf of Chiriquí (cluster I) and another including species dominating in the Gulf of Panama (cluster II) (Fig. 20.9). This distinct separation into a Chiriquí and Panama cluster may indicate a higher tolerance of the latter species to the highly changeable, much more seasonally fluctuating environment in the Gulf of Panama. The clusters, however, do not reflect the species richness of the genera to which they belong (Table 20.2). That both high- and low-diversity genera are equally distributed in the clusters reflects ecologic differences between the two bays and responses of the bryozoan fauna

**Table 20.2** Sixteen most abundant species in both bays. Alphabetic list, abbreviations, number of colonies in Panama and Chiriquí, number of species in the genus, stratigraphic range of genus, stratigraphic range of the species, and regional distribution of species (Data except for Gulf of Panama and Gulf of Chiriquí from http:bryozoa,net/and PPP databanks)

| Species | Abbreviation | Number of colonies Panama/ Chiriqui | Number of species in genus | Stratigraphic range of genus | First occurrence (ma)/ locality of species | Distribution of species (Cheetham et al. 2000; http:bryozoa.net/) |
|---|---|---|---|---|---|---|
| *Celleporaria brunnea* | CELLE_BRU | 25/14 | >100 | Palaeogene to Recent | 16.0/Dom. Rep. | Burica, Caribbean, NE-Pacific |
| *Cleidochasma (=Characodoma) contractum* | CLE_CON | 2/120 | >23 | Neogene to Recent | 16.2/Florida | ? Burica, Caribbean |
| *Copidozoum planum* | COP_PLA | 2/44 | >22 | Neogene to Recent | 4.3/Bocas del Toro | Burica, Australia, E-Pacific |
| *Floridina antiqua* | FLO_ANT | 42/129 | 48 | Cretaceous to Recent | 1.8/Burica | only recent species, W-Atlantic |
| *Hippoporella gorgonensis* | HIPPO_GOR | 202/277 | 27 | Neogene to Recent | 16.2/Florida | Burica, Caribbean, Atlantic |
| *Microporella marsupiata* | MICRE-MAR | 2/29 | >100 | Neogene to Recent | Recent/W-Atlantic | Only recent, W-Atlantic |
| *Microporella pontifica* | MICRE-PON | 1/82 | – | Neogene to Recent | Recent/Mexico | Only recent, Mexico, E-Pacific |
| *Microporella umbracula* | MICRE_UMB | 2/38 | – | Neogene to Recent | 5.7/Bocas del Toro | Burica, Caribbean, W-Atlantic, Indo-Pacific, Mediterranean |
| *Parasmittina crosslandi* | PAR_CRO | 35/183 | >100 | Palaeogene to Recent | 4.3/Bocas del Toro | Burica, Caribbean, E-Pacific, Central America |
| *Reptadeonella bipartita* | REPT_BIP | 58/183 | 30 | Neogene to Recent | 4.3/Bocas del Toro | Burica, Caribbean, W-Atlantic |
| *Reptadeonella tubulata* | REPT_TUB | 56/74 | – | Neogene to Recent | 3.1/Dom. Rep. | ? |
| *Rhynchozoon verruculatum* | RHYN_VER | 51/29 | 61 | Tertiary to Recent | 1.8/Burica | Burica, Caribbean, W-Atlantic |
| *Schedocleido-chasma porcellanum* | SCHE_POR | 35/38 | 4 | All Recent | 16.2/Florida | Burica, Caribbean, Indopazific, Mediterranean |
| *Schizoporella cornuta* | SCHI_COR | 3/72 | >100 | Palaeogene to Recent | 9.9/Canal Basin | Burica, Caribbean |
| *Steginoporella cornuta* | STE_COR | 46/17 | 59 | Palaeogene to Recent | 7.1/Dom. Rep. | Burica, Caribbean, tropical Pacific |
| *Trypostega venusta* | TRI_VEN | 9/37 | 14 | Eocene to Recent | 35.0/Coastal Plains (USA) | Burica, Caribbean, widespread |

rather than phylogenetic constraints. This separation between species with some being more abundant in one or the other bay is not seen in the R-mode presence/absence cluster diagram (Fig. 20.10). However, also here a more complex Panama cluster (IA) is separate from the two Chiriquí clusters (I-B and II). *Copidozoum planum* is the only species distinctly more abundant in Chiriquí included in the cluster I-A1 containing species more abundant in Panama. Although more abundant in the Gulf of Chriquí, *Reptadeonella bipartita* and *R. tubulata, Hippoporella gorgonensis*, and *Schedocleidochasma tubulata* occur also with raised abundances in the Gulf of Panama. Notice that both R-mode clusters distinguish between Chiriquí and Panama species.

*Zoogeographic connections and evolutionary implications.* According to Cheetham et al. (1999), only 13% species (10 of 132) from the Pacific side (Burica Peninsula, Early Quaternary, 1.8 Ma) do not also occur on the Caribbean side. Among these, *Smittoidea prolifera* and *Watersipora subovoidea* are considered to be "true" Pacific species not reported from Caribbean-Atlantic waters. Because of the strong correspondence between Pacific and Caribbean cheilostome faunas, the authors state that the closure of the isthmian seaway (3.5 Ma) had little evolutionary effect on the cheilostome fauna (see also Cheetham and Jackson 2000).

In our data set, 23.6% of species (15 of 56) have not been reported from the Caribbean side of Panama: *Hippoporina mexicana, Lagenicella hippocrepis, Microporella pontifica, M.* aff. *coriacea, M. tractabilis, Puellina californiensis, Reptadeonella tubulata, Rhynchozoum globosum, Stylopoma cornuta, Tecatia sinaloensis, Watersipora edmondsoni, W. subovoidea, Copidozoum planum, Floridina antiqua* and *Labioporella sinuosa*. These are twice as many as reported in the PPP database (Cheetham et al. 1999). Of these 15 modern Pacific species, four have been reported also from the early Pleistocene of Burica Peninsula (Pacific side of Panama): *Hippoporella mexicana, Watersipora subovoidea, Copidozoum planum* and *Floridina antiqua* (Cheetham et al. 1999).

Another interesting outcome from the R-mode cluster analyses (Fig. 20.9, less so Fig. 20.10) and comparison with the species distributions (Table 20.2, right column) is that species in the Panama cluster are reported to have a much wider distribution in the Atlantic and Eastern Pacific, in contrast to species in the Chiriquí cluster, which are more widely reported from the West Atlantic, Caribbean and Indo-Pacific oceans. If this pattern holds to be true, it would support the idea of the Gulf of Chiriquí as a more tropical, "Caribbean" environment compared to the Gulf of Panama with its distinctly colder, open-oceanic environment.

**Acknowledgements** This project was funded by the Deutsche Forschungsgemeinschaft and the Smithsonian Tropical Research Institute (STRI). We thank the Government of the Republic of Panama for the fieldwork and collecting permits; STRI for providing a scientific and logistic platform for the fieldwork; the master and crew of the RV URRACA (STRI) for nautical skills in collecting the samples during the cruises; Helena Fortunato (Kiel University) is thanked for valuable discussions and critically reading the manuscript; Ute Schuldt (SEM lab, Kiel University) for processing the SEM-photographs used for species identification; and Heidi Blaschek (Kiel University) for compiling taxa in the PPP-data and the IBA databanks. Maja Novosel and Matthew Dick are appreciated for their critical reviews of the manuscript and for many helpful suggestions.

## References


Anderson D (1994) Barnacles. Structure, function, development and evolution. Chapman and Hall, London

Angel MV (1994) Spatial distribution of marine organisms: pattern and processes. In: Edwards PJ, May RM, Webb NR (eds) Large-scale ecology and conservation biology. Blackwell Science, Oxford, pp 59–109

Cheetham AH, Jackson JBC, Sanner J (1999) Neogene cheilostome Bryozoa of tropical America: comparison and contrast between the Central American Isthmus (Panama, Costa Rica) and the North-Central Caribbean (Domenican Republic). Bull Am Paleontol 357:159–192

Cheetham AH, Jackson JBC (2000) Neogene history of cheilostome Bryozoa in tropical America. In: Proceedings of the 11th International Bryozoology Association conference 2000, Panama. pp 1–16

Cipriani R, Guzman HM, Lopez M (2008) Harvest history and current densities of the pearl oyster *Pinctada mazatlantica* (Bivalvia: Pteriidae) in Las Perlas and Coiba archipelagos, Panama. J Shellfish Res 27(4):691–700

D'Croz L, O'Dea A (2007) Variability in upwelling along the Pacific shelf of Panama and implications for the distribution of nutrients and chlorophyll. Estuar Coast Shelf Sci 2007:1–16

Dick MH and Mawatari SF (2004) Resolving taxonomic problems of North Pacific bryozoans. In: Mawatari SF, Okada H (eds) Integration of geoscience and biodiversity studies. In: Proceedings of the international symposium on Dawn of a new natural history-integration of geosciences and biodiversity studies, 5–6 Mar 2004, New Science of Natural History, Sapporo, pp 67–74

Dick MH, Tilbrook KJ, Mawatari SF (2006) Diversity and taxonomy of rocky-intertidal Bryozoa in the Island of Hawaii, USA. J Nat Hist 40(38–40):2197–2257

Fortunato H (2004) Reproductive strategies in gastropods across the Panama seaways. Invertebr Reprod Dev 46:139–148

Fortunato H, Schäfer P (2009) Coralline algae as carbonate producers and habitat providers on the Eastern Pacific coast of Panama: preliminary assessment. N Jahrb Geol Palaeont Abh 253:145–161

Gaston KJ, Spicer JI (2007) Biodiversity – an introduction. Blackwell Science, Oxford

Hastings AB (1930) Cheilostomatous Polyzoa from the vicinity of the Panama Canal collected by D. C. Crossland on the cruise of the S. Y. "St. George". Proc Zool Soc Lond 1929:670–740

Hayek LC, Buzas MA (1997) Surveying natural populations. Columbia University Press, New York http:bryozoa.net/iba.html

Hughes DJ, Jackson JBC (1992) Distribution and abundance of cheilostome bryozoans on the Caribbean reefs of Central Panama. Bull Mar Sci 51:443–465

Jackson JBC, Cheetham AH (1990) Evolutionary significance of morphospecies: a test with cheilostome Bryozoa. Science 248:579–583

Jackson JBC, Herrera Cubilla A (2000) Adaptation and constraint as determinants of zooid and ovicell size among encrusting ascophoran cheilostome Bryozoa from opposite sides of the Isthmus of Panama. In: Proceedings of the 11th International Bryozoology Association conference 2000, Panama. pp 249–258

Jackson JBC, Jung P, Fortunato H (1996) Paciphilia Revisited: transisthmian evolution of the *Strombina* group (Gastropoda. Columbellidae). In: Jackson JBC, Budd AF, Coates AG (eds) Evolution and environment in tropical America. Yale University Press, New Haven, pp 234–270

Lessios H (1990) Adaptation and phylogeny of egg size of echinoderms from the two sides of the Isthmus of Panama. Am Nat 135:1–13

Magurran AE (1988) Ecological diversity and its measurements. Princeton University Press, Princeton

McKinney FK, Jackson JBC (1989) Bryozoan evolution. Boston Unwin Hyman, London

Osburn RC (1950) Bryozoa of the Pacific coast of North America. Part 1, Cheilostomata Anasca. Allan Hancock Pacific Exped 14(1):1–269

Osburn RC (1952) Bryozoa of the Pacific coast of North America. Part 2, Cheilostomata Ascophora. Allan Hancock Pacific Exped 14(2):271–611

Osburn RC (1953) Bryozoa from the Pacific Coast of North America, Part 3, Cyclostomata, Ctenostomata, Entoprocta, and addenda. Allan Hancock Pacific Exped 14(3):613–841

Reijmer JJG, Bauch T, Schäfer P (2011) Carbonate facies patterns in surface sediments of upwelling and non-upwelling shelf environments (Panama, East Pacific). Sedimentol. doi:10.1111/j.1365-3091.2010.01214

Rodriguez-Rubio E, Schneider W, del Rio RA (2003) On the seasonal circulation within the Panama Bight derived from satellite observations of wind, altimetry and sea surface temperature. Geophys Res Lett 30:1410

Rucker JB (1975) Paleoecological analysis of cheilostome Bryozoa from Venezuela-British Guiana shelf sediment. Bull Mar Sci 17(4):787–839

Schäfer P, Fortunato H, Bader B, Liebetrau V, Bauch T, Reijmer J (2011) Facies, growth rates and carbonate production in upwelling and non-upwelling settings along the Pacific coast of Panama. Palaios 26(7):420–432

Soule DF, Soule JD, Chaney HW (1995) Taxonomic atlas of the benthic fauna of the Santa Maria basin and Western Santa Barbara channel, vol 2, Irene McCulloch foundation monograph series. Irene McCulloch foundation, Santa Barbara, pp 1–344

Soule DF, Soule JD, Morris PA (2002) Changing concepts in species diversity in the northeastern Pacific. In: Wyse Jackson PN, Buttler CJ, Spencer-Jones ME (eds) Bryozoan studies 2001. Swets and Zeitlinger, Lisse, pp 299–306

SPSS Inc. (2002) http://www.spss.com

Valentine JW (1973) Evolutionary paleoecology of the marine biosphere. Prentice-Hall, Englewood Cliffs/New York

# Chapter 21
# High Resolution Non-destructive Imaging Techniques for Internal Fine Structure of Bryozoan Skeletons

## Non-destructive Imaging of Bryozoans

**Rolf Schmidt**

**Abstract** Many aspects of skeletal morphological research in bryozoans have involved destructive methods, such as thin sectioning of Palaeozoic fossils or removal of basal walls to view internal structures in cheilostomes. Two relatively new technologies allow non-destructive visualisation of internal zooidal skeletal structures. Tomography is shown to be very effective for cheilostome zooid cavities as it allows resolution down to 1 μm, which can resolve features such as the morphological evidence of origins of frontal shields in ascophoran cheilostomes. It also generates a three-dimensional reconstruction of the whole structure, which can resolve complex internal structures like those of *Siphonicytara*, or how multilaminar colonies develop the communication between layers. The Synchrotron is best suited to image the internal structures of Palaeozoic bryozoans that have mineral-filled zooidal cavities, as it can better resolve differences in composition.

**Keywords** Morphology • X-ray tomography • Synchrotron • Non-destructive

## Introduction

Computed Tomography (CT) has been used for some time to create three-dimensional images of the internal structures of objects ranging from humans to machines, and even some large fossils. Bryozoans are too small for this technique, but X-ray micro-computed tomography (microCT) allows the scanning of much smaller structures, down to nanometre scale. As an imaging tool it plays an important role in many diverse applications such as bio-medical imaging (e.g. Weiss et al. 2005), metallurgy (Elmoutaouakkil et al. 2002), material science, food industry (van Dalen et al. 2003) and astronomy (Jurewicz et al. 2003).

R. Schmidt (✉)
Museum Victoria, PO Box 666, Melbourne VIC 3001, Australia
e-mail: rschmid@museum.vic.gov.au

A. Ernst, P. Schäfer, and J. Scholz (eds.), *Bryozoan Studies 2010*,
Lecture Notes in Earth System Sciences 143, DOI 10.1007/978-3-642-16411-8_21,

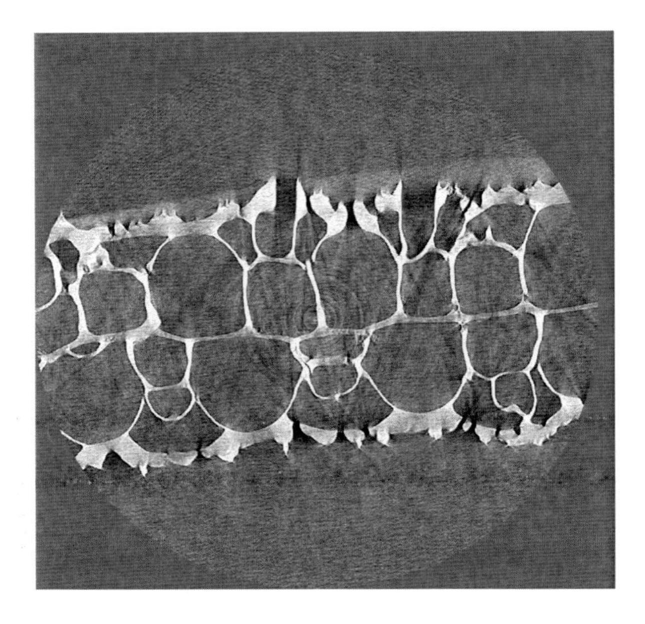

**Fig. 21.1** Single raw microCT image created by rotating the specimen of *Siphonicytara occidentalis* through the X-ray beam

This is a three-dimensional imaging technique for measuring and characterising internal structures as a non-destructive evaluation for materials with high resolution results (Ziegler et al. 2008, 2010).

Microtomographic imaging has been extensively used in various fossil groups, from Cambrian corals (Han et al. 2010) to dinosaur bone histology (Hieronymus et al. 2009). In recent years it has also been carried out on bryozoan fossils (Viskova and Pakhnevich 2010), on several species of cyclostomes (Mainwaring 2008; Taylor et al. 2008), and on bryozoan soft tissue (Metscher 2009).

## Methods

The tomography machine used for this initial work was an Xradia Inc. microXCT (micro X-ray Computed Tomography), using the following setup: Objective lens 10x magnification; Pixel size = 2.3 μm; Field of view ~ $(2.2\ \text{mm})^3$; Tube voltage 100 kV and power 9 W; source sample distance = 120 mm; Sample detector distance = 20 mm; Number of projection 721. This involved taking 1,024 image slices through the specimen as it rotated through the beam (Fig. 21.1)

The data was analysed using the software program 'drishti', developed at the Australian National University (ANU) (http://anusf.anu.edu.au/Vizlab/drishti/). This aspect of the imaging process can be difficult to master and very time consuming.

The specimen used for this study was a Recent colony of *Siphonicytara occidentalis* from a collection of the Southern Surveyor Voyage SS102005, Sample 112–16 off Western Australia held at Museum Victoria, Melbourne (NMV

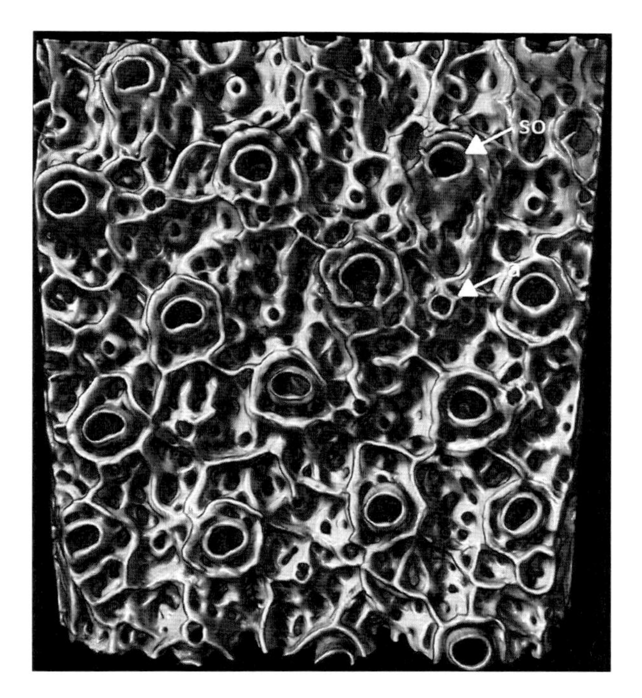


**Fig. 21.2** Frontal view of microCT 3D visualisation of a *Siphonicytara occidentalis* colony (NMV F109341.1); *so* secondary orifice, *a* ascopore

F109341.1). The colony fragment was bleached to allow for maximum contrast, as organic tissue can require significant digital processing to clean up the images to enhance the skeletal signal. It was scanned dry, though scanning wet (i.e. in 70% ethanol) is possible with many microCT machines.

## Results

Preliminary results using the Recent specimen of *Siphonicytara occidentalis* are very promising. It clearly shows numerous morphological features previously not seen, such as the peristomial brood chamber and marginal areolarpores (Figs. 21.2, 21.3, 21.4, and 21.5). With higher resolution scans, which are possible, even fine structure of the internal frontal wall could be imaged.

## Discussion

The three-dimensional reconstructions allow exploration of the complete interior of the colony, without need to re-image further samples. It also allows generation of 3D pixels (voxels), which can be utilised for very precise volume calculations. This is of particular value for use with type specimens or very rare material, where destructive analysis of the interior structures is unfeasible.

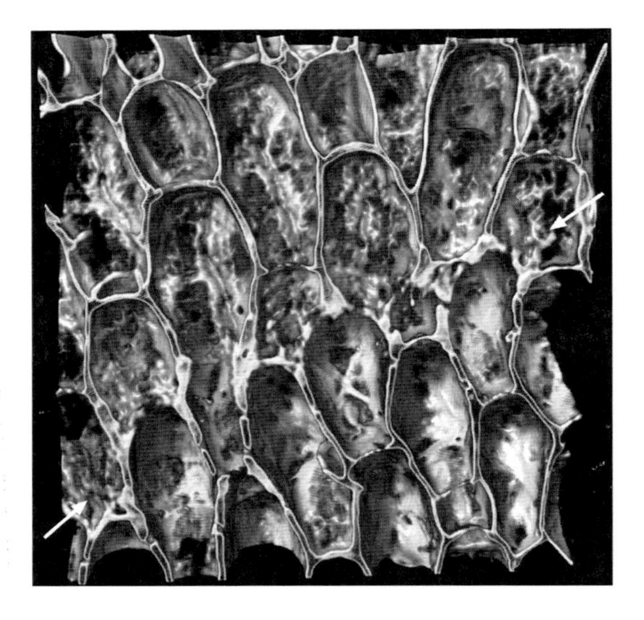

**Fig. 21.3** Slice through microCT 3D visualisation of a *Siphonicytara occidentalis* colony (NMV F109341.1) viewed from the frontal area, *arrows* indicate line where slice cuts through the basal wall between the bilaminar zooid layers, *above* the line the basal wall is visible, *below* the line the interior walls of the other zooid layer is visible

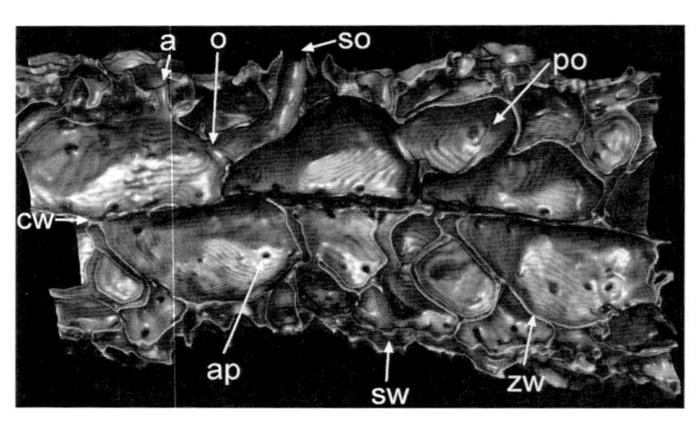


**Fig. 21.4** Lateral slice through microCT 3D visualisation of a *Siphonicytara occidentalis* colony (NMV F109341.1); *o* primary orifice, *so* secondary orifice, *a* ascopore, *po* peristomial ovicell, *ap* areolar pore, *cw* central wall, *zw* frontal zooid wall, *sw* secondary frontal wall

Results for tomography are best for high phase (density) contrast in the specimen (e.g., between soft-tissue and skeleton). Where the contrast is lower, such as in fossils where calcareous matrix is attached to, and fills the specimen, the results can be improved by sectioning a part of the specimen and using SEM scan data to improve the data clean-up by allowing it to be calibrated to the density contrasts.

Resolution here was only 2 μm, but future scans will aim to generate higher resolution images, well below 1 μm (maximum resolution is partly limited by the specimen size).

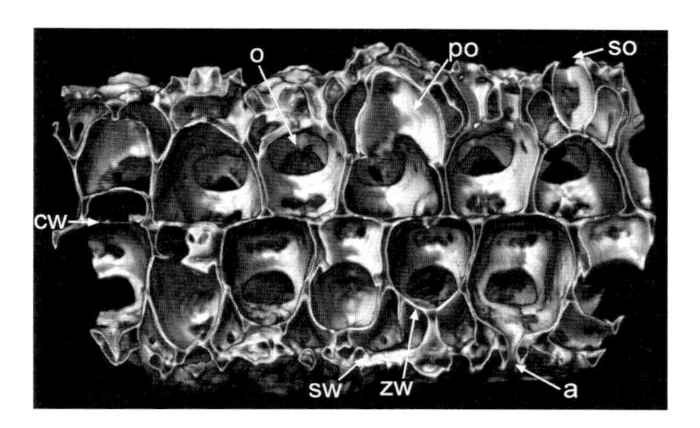


**Fig. 21.5** Slice through microCT 3D visualisation of a *Siphonicytara occidentalis* (NMV F109341.1) looking in the distal direction of the colony, showing primary orifices and central wall; *o* primary orifice, *so* secondary orifice, *a* ascopore, *po* peristomial ovicell, *cw* central wall, *zw* frontal zooid wall, *sw* secondary frontal wall

It is now the aim to also utilise the Australian Synchrotron facility. There are differences between the synchrotron as a source, and microcomputed tomography (Betz et al. 2007) which can enable better imaging.

The synchrotron produces many times more flux than a lab based instrument, so the exposure times are smaller, and therefore data collection is faster.

Synchrotron x-rays can be made monochromatic, and still be bright. So the quantification of parameters such as attenuation is far easier, allowing clearer 3D images.

The source of the x-rays is small just as in a microfocus tube, and the distances to the sample are large, 10s of metres. So coherence properties can be used over length scales of 100s of microns. This is often summarised in the phrase ‘phase contrast imaging’. Instead of using the attenuation to provide contrast, phase shifts of the x-rays as they pass through the object are also used. Phase contrast imaging is (a) more sensitive to density changes; and (b) can potentially be done at lower doses to the sample.

These properties may potentially allow the imaging of internal structures of more lithified Palaeozoic Bryozoa. It has already been utilised for other fossils (e.g. Heethoff et al. 2008). Efforts are currently underway to allow publication and sharing of the actual 3D data either by embedding in pdf files (Ruthensteiner and Hess 2008; Ruthensteiner et al. 2010) or by developing online databases (Martone et al. 2002).

**Acknowledgments** The tomography image was taken in the Department of Physics, La Trobe University, Melbourne, with the assistance of Drs. Benedicta Arhatari and Peter Kappen. Thanks to Tim Senden, Australian National University for assistance with further tomography. Chris Hall (Imaging and Medical Therapy (IMT) beamline at the Australian Synchrotron) for advice on analysing Palaeozoic Bryozoa. Tim Holland (Museum Victoria), Ajay Limaye (ANU), Drew Whitehouse (ANU) for assistance in processing that data. Thanks to Thomas Schwaha and Paul Taylor for greatly improving the article.

## References

Betz O, Wegst U, Weide D, Heethoff M, Helfen L, Lee W-K, Cloetens P (2007) Imaging applications of synchrotron X-ray phase-contrast microtomography in biological morphology and biomaterials science. I. General aspects of the technique and its advantages in the analysis of millimetre-sized arthropod structure. J Microsc 22:51–71

Elmoutaouakkil A, Salvo L, Maire E, Peix G (2002) 2D and 3D characterization of metal foams using X-ray tomography. Adv Eng Mater 4(10):803–807

Han J, Kubota S, Uchida H, Stanley GD Jr, Yao X, Shu D, Li Y, Yasui K (2010) Tiny sea anemone from the Lower Cambrian of China. PLoS One 5(10):e13276

Heethoff M, Helfen L, Cloetens P (2008) Non-invasive 3D-visualization of the internal organization of microarthropods using synchrotron X-ray-tomography with sub-micron resolution. J Vis Exp 15. http://www.jove.com/details.php?id=737. doi: 10.3791/737

Hieronymus TL, Witmer LM, Tanke DH, Currie PJ (2009) The facial integument of centrosaurine ceratopsids: morphological and histological correlates of novel skin structures. Anat Rec 292:1370–1396

Jurewicz AJG, Jones SM, Tsapin A, Mih DT, Connolly HC Jr, Graham GA (2003) Locating stardust-like particles in aerogel using X-ray techniques 34th lunar and planetary science conference, League

Mainwaring P (2008) Application of the Gatan X-ray ultramicroscope (XuM) to the investigation of material and biological samples. Microscopy Today 16:14–17

Martone ME, Gupta A, Wong M, Quian X, Sosinsky G, Ludäscher B, Ellisman MH (2002) A cell-centred database for electron tomographic data. J Struct Biol 138:145–155

Metscher BD (2009) MicroCT for comparative morphology: simple staining methods allow high-contrast 3D imaging of diverse non-mineralized animal tissues. BMC Physiol 9:11

Ruthensteiner B, Hess M (2008) Embedding 3D models of biological specimens in PDF publications. Microsc Res Tech 71:78–786

Ruthensteiner B, Bäumler N, Barnes DG (2010) Interactive volume rendering in biomedical publications. Micron 41:886.e1–886.e17

Taylor PD, Howard L, Gundrum B (2008) Microcomputed tomography X-ray microscopy of cyclostome bryozoan skeletons. In: Abstracts of the VIII Larwood Meeting 2008, University of Vienna

van Dalen G, Blonk H, van Aalst H, Hendriks CL (2003) 3-D imaging of foods using X-ray microtomography G.I.T. Imaging Microsc 3:18–21

Viskova LA, Pakhnevich AV (2010) A new boring bryozoan from the Middle Jurassic of the Moscow Region and its micro-CT research. Paleontol J 44(2):157–167

Weiss P, Le Nihouannen D, Rau C, Pilet P, Anguado E, Gauthier O, Jean A, Daculsi G (2005) Synchrotron and non synchrotron X-ray microtomography threedimensional representation of bone ingrowth in calcium phosphate biomaterials. Eur Cell Mater 9(Suppl 1):48–49

Ziegler A, Faber C, Mueller S, Bartolomaeus T (2008) Systematic comparison and reconstruction of sea urchin (Echinoidea) internal anatomy: a novel approach using magnetic resonance imaging. BMC Biol 6:33

Ziegler A, Ogurreck M, Steinke T, Beckmann F, Prohaska S, Ziegler A (2010) Opportunities and challenges for digital morphology. Biol Direct 6(5):45

# Chapter 22
# Being a Bimineralic Bryozoan in an Acidifying Ocean

## Ocean Acidification and Bryozoans

**Abigail M. Smith and Christopher J. Garden**

**Abstract** Strongly controlled calcification by bryozoans means that some species maintain complex skeletons formed of more than one mineral. Whether they are mainly intermediate-Mg calcitic with up to 50% aragonite, mainly aragonitic with small amounts of high-Mg calcite (>8 wt.% $MgCO_3$), or formed of both high- and low-Mg calcites, preservation of sediments formed of these bimineralic bryozoan skeletons may be more at risk from ocean acidification than the majority of bryozoan sediments formed of monomineralic skeletons. An acid-bath immersion experiment on seven species reveals that three (*Adeonella* sp., *Adeonella patagonica*, and *Adeonellopsis* sp.) are more resistant to dissolution than the other four. Skeletal carbonate mineralogy appears to influence dissolution history very little: the most soluble aragonite and high-Mg calcite species, *Adeonellopsis* sp., was more highly resistant to dissolution than species dominated by low-Mg calcite. In the context of ocean acidification, it is likely that bryozoan skeletons with high surface area and small delicate morphologies are at greatest risk of dissolution, irrespective of mineralogical composition.

**Keywords** Bryozoa • Ocean acidification • Experimental dissolution • Aragonite • Calcite

## Introduction

Bryozoans, particularly cheilostomes, are complex mineralisers, able to produce calcite (with a range of Mg contents), aragonite, and various mixtures of the two. While there are some broad phylogenetic (Smith et al. 2006) and evolutionary (Taylor et al. 2009) trends in skeletal carbonate mineralogy, bryozoan colonies are variable enough to require replicate tests (usually using x-ray diffractometry or

A.M. Smith (✉) • C.J. Garden
Department of Marine Science, University of Otago, P. O. Box 56, Dunedin 9054, New Zealand
e-mail: abby.smith@otago.ac.nz

A. Ernst, P. Schäfer, and J. Scholz (eds.), *Bryozoan Studies 2010*,
Lecture Notes in Earth System Sciences 143, DOI 10.1007/978-3-642-16411-8_22,

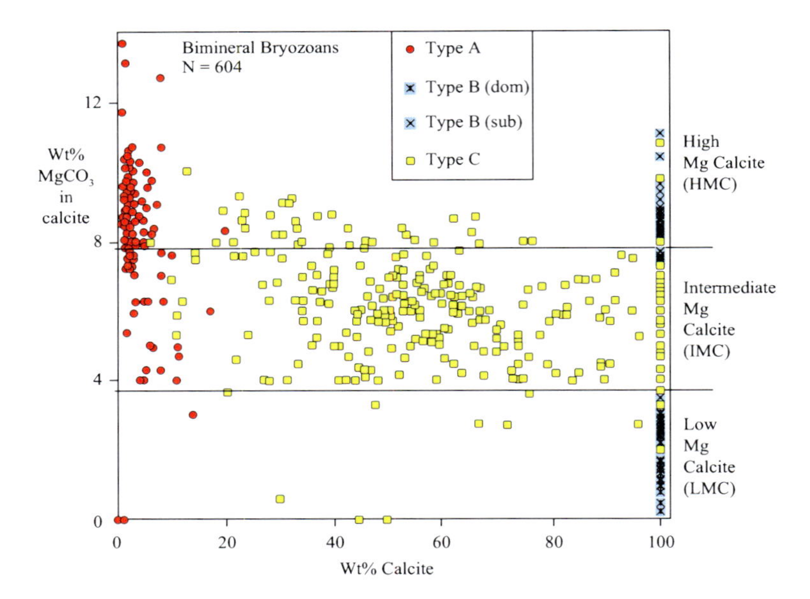


**Fig. 22.1** Skeletal carbonate mineralogy of 604 specimens of bryozoans, which form bimineralic skeletons (see Table 22.1). Type A (red) is mostly aragonitic, with 0–20% high-Mg calcite. Type B (blue) is formed of two discrete calcites, a dominant low-Mg calcite ('dom') and subdominant high-Mg calcite ('sub'). Type C (yellow) is formed primarily of intermediate-Mg calcite but has to a varying degree a secondary layer of aragonite, the thickness of which increases with age (Data are taken from Smith et al. 2006; Wejnert and Smith 2008; Taylor et al. 2009; Smith and Clark 2010; Smith and Girvan 2010; Smith and Lawton 2010)

XRD) within species to truly characterize biominerals present (Smith et al. 1998; Lombardi et al. 2008).

Recently published data on bimineralic bryozoans (Smith et al. 2006; Wejnert and Smith 2008; Taylor et al. 2009; Smith and Clark 2010; Smith and Girvan 2010; Smith and Lawton 2010) shows that they tend to fall into three main groups (Fig. 22.1; Table 22.1). Some bryozoan skeletons are formed mostly of aragonite (Type A in Fig. 22.1), e.g., *Adeonellopsis* sp. from New Zealand (Wejnert and Smith 2008). These species generally range from 0–20 wt.% calcite, usually at the high end of Mg content (6–12 wt.% $MgCO_3$).

A few species produce entirely calcitic skeletons, but a dual-peaked XRD curve suggests there are two discrete minerals present (called Type B in Fig. 22.1). The dominant mineral is usually low-Mg calcite with around 2–4 wt.% $MgCO_3$, with subdominant calcite higher: 8–12 wt.% $MgCO_3$. This mineralogy has only been described well in the Cellariidae (e.g., *Melicerita chathamensis*, Smith and Lawton in press), though since it is easy to miss when analyzing XRD peaks (Bone and James 1993), it may also occur in other taxa.

Other bimineralic cheilostomes are mineralogically more variable, though often formed mostly of calcite (Type C in Fig. 22.1), e.g., *Odontionella cyclops* (Smith and Girvan 2010). These species form a primary skeleton of calcite, which is gradually frosted with secondary aragonite (Smith and Girvan 2010). Calcite content

**Table 22.1** Skeletal carbonate mineralogy of some bimineralic bryozoan species. Mg content is re-calculated from that published to a standard unit of wt.% $MgCO_3$. Only species where two or more specimens have been reported are included here (Data from Smith et al. 2006; Wejnert and Smith 2008; Taylor et al. 2009; Smith and Clark 2010; Smith and Girvan 2010; Smith and Lawton 2010)

| **TYPE A: Mostly aragonite** Species | Family | No. of specimens | Mean Wt.% calcite (range) | Mean Wt.% $MgCO_3$ in calcite (range) |
|---|---|---|---|---|
| *Adeonellopsis* sp. (New Zealand) | Adeonidae | 150 | 3 (0–11) | 8.5 (4.3–10.7) |
| *Alderina solidula* (Hincks, 1860) | Calloporidae | 2 | 8 | 11.7 (10.7–12.7) |
| *Discoporella doma* (d'Orbigny, 1853) | Cupuladriidae | 2 | 6 (0–11) | 4.0 |
| *Discoporella umbellata* (DeFrance, 1823) | Cupuladriidae | 3 | 2 (0–4) | 5.0 (4.0–6.0) |
| *Gigantopora regularis* | Gigantoporidae | 3 | 5 (0–13) | 3.0 |
| *Odontoporella* cf. *adpressa* | Hippoporidridae | 2 | 14 (8–20) | 7.7 (7.0–8.3) |
| *Parasmittina spathulata* (Smitt, 1873) | Smittinidae | 3 | 6 (0–17) | 5.5 (5.0–6.0) |
| *Rhynchozoon digitatum* Gautier, 1962 | Phidoloporidae | 2 | 1 | 12.7 (11.7–13.7) |
| *Schizomavella arrogata* (Waters, 1879) | Bitectiporidae | 4 | 3 (2–3) | 7.9 (7.5–8.5) |
| *Schizomavella auriculata* (Hassall, 1842) | Bitectiporidae | 2 | 6 (1–10) | 7.6 |
| **Group A: 10 species** | **8 Familes** | **173** | **3 (0–20)** | **8.4 (3.0–13.7)** |

| **TYPE B: Dual calcite** Species | Family | No. of specimens | Mean Wt.% $MgCO_3$ in dominant calcite (range) | Mean Wt.% $MgCO_3$ in subdominant calcite (range) |
|---|---|---|---|---|
| *Cellaria immersa* (Tenison-Woods, 1880) | Cellariidae | 25 | 2.1 (0.3–4.3) | 9.1 (7.0–11.1) |
| *Cellaria tenuirostris* (Busk, 1853) | Cellariidae | 5 | 2.5 (0.5–6.0) | 8.4 (7.7–9.0) |
| *Melicerita chathamensis* Uttley and Bullivant, 1972 | Cellariidae | 92 | 2.1 (0.8–3.6) | 8.1 (6.6–9.7) |
| **Group B: 3 species** | **1 Family** | **122** | **2.1 (0.3–6.0)** | **8.3 (6.6–11.1)** |

| **TYPE C: Mostly calcite** Species | Family | No. of specimens | Mean Wt.% calcite (range) | Mean Wt.% $MgCO_3$ in calcite (range) |
|---|---|---|---|---|
| *Adeonella patagonica* Hayward, 1988 | Adeonellidae | 12 | 65 (6–100) | 7.7 (6.0–10.8) |
| *Adeonella* sp. (Chile) | Adeonellidae | 12 | 37 (18–77) | 7.4 (4.5–9.1) |

(continued)

**Table 22.1** (continued)

| **TYPE A: Mostly aragonite** Species | Family | No. of specimens | Mean Wt.% calcite (range) | Mean Wt.% $MgCO_3$ in calcite (range) |
|---|---|---|---|---|
| *Arachnopusia unicornis* (Hutton, 1873) | Arachnopusiidae | 11 | 90 (37–100) | 5.3 (4.0–7.0) |
| *Calpensia nobilis* (Esper, 1796) | Microporidae | 7 | 81 (73–90) | 4.3 (4.0–5.1) |
| *Chaperia granulosa* (Gordon, 1986) | Chaperiidae | 5 | 52 (11–100) | 4.1 (3.7–5.3) |
| *Cryptosula pallasiana* (Moll, 1803) | Cryptosulidae | 15 | 91 (50–100) | 5.1 (2.0–6.0) |
| *Escharella immersa* (Fleming, 1828) | Romancheinidae | 2 | 88 (85–91) | 4.4 (4.2–4.5) |
| *Escharoides coccinea* (Abildgaard, 1806) | Romancheinidae | 3 | 78 (74 to 83) | 5.6 (5.4–6.1) |
| *Hippomenella vellicata* (Hutton, 1873) | Schizoporellidae | 4 | 84 (68–100) | 5.9 (5.0–6.6) |
| *Hippomonavella flexuosa* (Hutton, 1873) | Bitectiporidae | 2 | 100 | 7 (6.9–7.1) |
| *Hippopetraliella marginata* (Canu and Bassler, 1928) | Petraliellidae | 2 | 95 (90–100) | 5.5 (5.0–6.0) |
| *Lepraliodes nordlandica* (Nordgaard, 1905) | Cyclicoporidae | 2 | 91 (90–92) | 5.2 (4.1–6.3) |
| *Metrarabdotos unguiculatum* Canu and Bassler, 1928 | Metrarabdotosidae | 6 | 59 (48–75) | 5.3 (4.0–8.0) |
| *Microporella* spp. | Microporellidae | 9 | 69 (45–92) | 6.4 (4.4–7.3) |
| *Odontionella cyclops* (Busk, 1854) | Calloporidae | 118 | 58 (27–100) | 6.2 (3.6–8.8) |
| *Oligotresium jacksonensis* (Canu and Bassler, 1920) | Lunulitidae | 2 | 48 (45–50) | 6.0 (4.0–8.0) |
| *Pachyegis princeps* (Norman, 1903) | Stomachetosellidae | 3 | 89 (83–93) | 6.5 (5.8–7.0) |
| *Parasmittina trispinosa* (Johnston, 1838) | Smittinidae | 4 | 85 (81–89) | 6.4 (5.8–6.9) |
| *Pentapora fascialis* (Pallas, 1766) | Bitectiporidae | 21 | 33 (13–56) | 7.8 (5.3–10.0) |
| *Pentapora foliacea* (Ellis and Solander, 1786) | Bitectiporidae | 7 | 41 (12–76) | 4.0 (4.8–6.7) |
| *Porella concinna* (Busk, 1854) | Bryocryptellidae | 3 | 57 (52–60) | 6.7 (6.5–7.3) |
| *Schizomavella linearis* (Hassall, 1841) | Bitectiporidae | 2 | 48 (39–57) | 7.2 (6.8–7.6) |
| *Schizoporella dunkeri* (Reuss, 1848) | Schizoporellidae | 2 | 30 (24–36) | 5.2 (5.0–5.3) |
| *Schizoporella errata* (Waters, 1878) | Schizoporellidae | 2 | 49 (40–68) | 6.2 (5.7–6.7) |

(continued)

**Table 22.1** (continued)

| **TYPE A: Mostly aragonite** Species | Family | No. of specimens | Mean Wt.% calcite (range) | Mean Wt.% $MgCO_3$ in calcite (range) |
|---|---|---|---|---|
| *Schizoporella unicornis* (Johnston in Wood, 1844) | Schizoporellidae | 24 | 45 (22–76) | 5.7 (0.6–8.0) |
| *Schizotheca serratimargo* (Hincks, 1886) | Phidoloporidae | 2 | 61 (53–70) | 5.8 (5.5–6.0) |
| *Smittina rigida* Lorenz, 1886, sensu Kluge | Smittinidae | 2 | 88 | 6.6 (6.3–6.9) |
| *Steginoporella perplexa* Livingstone, 1929 | Steginoporellidae | 7 | 80 (42–100) | 3.9 (2.7–4.8) |
| *Stylopoma* sp. | Schizoporellidae | 2 | 11–95 | 5.9–6 |
| **Group C: 29 species** | **19 Familes** | **293** | **60 (6–100)** | **6.1 (0.6–10.8)** |

may be as low as 30%, but at least some specimens within the species are close to 100% calcite. In general the calcite is intermediate in Mg content, 4–8 wt.% $MgCO_3$.

In the context of ongoing global acidification of surface seawaters (Raven et al. 2005), marine, calcifying organisms may find biomineralisation increasingly challenging (Orr et al. 2005). Equally, marine carbonate sediments formed from skeletal remains may dissolve far more rapidly than is now the case (Smith 2009), thus affecting the overall marine carbon budget. Experimental dissolution studies on bryozoan skeletons (and those of other invertebrates) have shown that surface area is probably the most important variable determining dissolution rate, but that when surface area is broadly similar, then mineralogical composition determines susceptibility to dissolution (Flessa and Brown 1983; Smith et al. 1992).

In general, sedimentologists would expect aragonite and high-Mg calcite (>8 wt.% $MgCO_3$) to dissolve earlier and more rapidly than calcite with lower Mg content, but what we do not know is how mixtures of different minerals might dissolve. Here we investigate rates of dissolution in bimineralic bryozoans of three types. Acid-bath studies are both efficient and effective, as they allow comparison of dissolution rates and features among a range of specimens. Understanding the ramifications of different bimineralic modes among bryozoans will enable us to project the effects of ocean acidification on important tracts of bryozoan carbonate sediments.

## Methods

Specimens of seven bimineralic bryozoans were collected from archived sediment samples from around southern New Zealand and Chile (Table 22.2). Samples of each species were cleaned of inorganic material in bleach and ten replicates of

**Table 22.2** Specimens used in this study, along with original weight and dissolution half-life (DHL). Means are given +/−standard error; in each case N = 10

| Species | Collected from | Mineralogy | Morphology | Mean original weight (g) | Mean DHL (min) |
|---|---|---|---|---|---|
| *Hippomenella vellicata* | Otago shelf, New Zealand | Mostly calcite (C) | Erect rigid foliose | 0.1829 | 41 |
| *Odontionella cyclops* | Snares platform, New Zealand | Mostly calcite (C) | Erect rigid robust branching to foliose | 0.2104 | 89 |
| *Adeonella patagonica* | Fiords, Chile | Mostly calcite (C) | Erect rigid robust branching | 0.2062 | 486 |
| *Adeonella* sp. | Fiords, Chile | Mostly calcite (C) | Erect rigid robust branching | 0.2637 | 481 |
| *Melicerita chathamensis* | Snares platform, New Zealand | Dual calcite (B) | Erect rigid sabre-shaped | 0.0592 | 58 |
| *Cellaria immersa* | Otago shelf, New Zealand | Dual calcite (B) | Erect flexible articulated | 0.0378 | 114 |
| *Adeonellopsis* sp. | Doubtful sound, New Zealand | Mostly aragonite (A) | Erect rigid robust branching | 0.3961 | 358 |

similar colony shape and size were selected, each of which was fresh looking and without encrusting organisms.

Each specimen was placed in a capped vial with ten 0.5 mm holes pierced in it to allow exchange of fluid (eight holes in sides of vial, one in top, one in bottom). The vials were weighed and carefully filled with 1% acetic acid (a concentration designed to cause total dissolution in a timeframe of days, see Smith et al. 1992) before being capped and submerged in one of ten acid baths under a fume hood, each containing 750 ml of 1% acetic acid. Additionally, an empty vial was placed in each bath as a blank, to check for any dissolution of the plastic throughout the course of the experiment.

After 15 min in acid the vials were removed, drained and gently rinsed in distilled water before being placed to dry at 75°C for a minimum of 3 h. Each specimen was weighed (in the vial) to an accuracy of +/−0.0001 g. The specimens were returned to the acid baths, and the process was repeated after cumulative immersion times of 30, 60, 120, 240, 480 and 960 min.

The pH of each acid bath was recorded using an immersion pH meter (+/−0.01 pH units) at the start and end of the experiment, and prior to each re-immersion of the specimens. Each bath was covered when the specimens were not immersed to prevent evaporation.

Dissolution curves showing percent of original weight remaining over time (less the vial) was calculated and graphed for each specimen (data available on request from the authors). Dissolution Half-Life (DHL), that is, the time it takes for half the

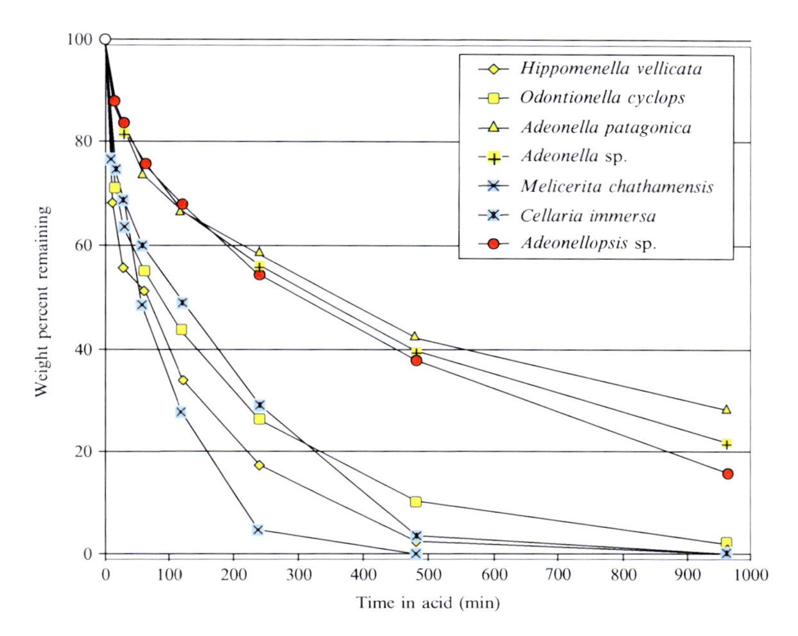


**Fig. 22.2** Mean dissolution curves for seven species of bimineralic bryozoans subjected to 1% acetic acid over the course of 960 min (ten replicates). Mineralogy type A: *red*. Type B: *blue*. Type C: *yellow*

original weight to dissolve, was calculated from each specimen's dissolution curve (Table 22.2), and mean DHL for each species calculated.

## Results

The pH of the ten acid baths ranged from 2.58 at the start of the experiment to a mean of 4.04 (range: 3.92–4.24, std dev = 0.09, N = 10) after 960 min. While there was some titration by dissolving carbonate, the pH remained low enough to ensure continuing dissolution throughout the experiment. There was no consistent difference among the acid baths in terms of pH change.

Empty tubes varied somewhat in weight over the course of the experiment, with six tubes losing weight and four gaining weight; the mean weight change was, however, 0.0000 g (range −0.00089 to +0.00120 g; std dev = 0.0005 g; N = 10). As a consequence of this variation, results were rounded to the nearest mg (0.001 g).

Mean dissolution half-life calculated from mean dissolution curves for each species (Fig. 22.2) ranged from 41.4 to 486.6 min (mean = 232.8 min; std dev = 186.6 min, N = 7) (Table 22.2). Three species (*Adeonellopsis* sp.,

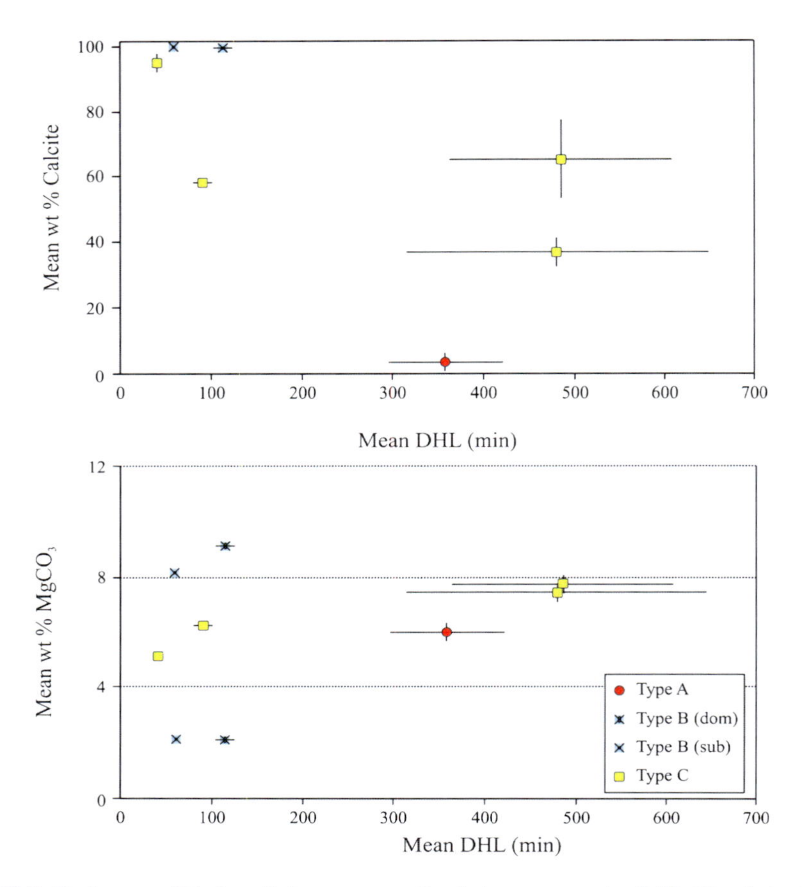


**Fig. 22.3** Resistance of bimineralic bryozoans to dissolution, expressed as DHL (dissolution half-life in minutes, x-error bars are standard error, N = 10) related to their mean skeletal carbonate mineralogy as reported in Table 22.1 (y-axis error bars are standard error, N = 11–118 depending on species). Mineralogy type A: *red*. Type B: *blue*. Type C: *yellow*

*Adeonella* sp. and *Adeonella patagonica*) were considerably more resistant to dissolution (mean DHL 358–486 min) than the other four (mean DHL 41–114 min). The more resistant species contained more aragonite, on average, and higher Mg in calcite than the less resistant species (Fig. 22.3).

## Discussion

If mineralogy were the only consideration in skeletal carbonate dissolution, we would expect high-Mg calcite to dissolve most readily, followed closely by aragonite. Low-Mg calcite, the least soluble polymorph, would last the longest (see, e.g.,

Kuklinski and Taylor 2009). Aragonite-dominated Type A bryozoans, then, would be most vulnerable to dissolution, with Type B's low-Mg-calcite-dominated skeletons perhaps the most resistant. Our acid-bath results, however, suggest the exact opposite.

The type A bryozoan *Adeonellopsis* sp. was resistant to dissolution, less so than the two *Adeonella* spp., but more resistant than any of the Type C bryozoans. This is, perhaps, the most geochemically surprising result (although it has been shown before by Smith et al. 1992). *Adeonellopsis* contains (on average) 97% aragonite, more than any other species in the study, and also has the highest mean Mg-content (8.5 wt.% $MgCO_3$). It has the same robust-branching shape as colonies of *Adeonella* and *Odontionella*. All other things being equal, we would expect this species to dissolve first. That it does not suggests that some other characteristic, perhaps structural or even ultra-structural in nature, enhances this species' resistance to dissolution.

Type B bryozoans *Melicerita chathamensis* and *Cellaria immersa*, formed of a combination of high-Mg and low-Mg calcite, dissolved rapidly. They show the two characteristic skeletal growth forms of the Cellariidae: *C. immersa* is an erect flexible articulated colony with thin cylindrical internodes less than 1 mm in diameter. *M. chathamensis* has a non-articulated blade-shaped colony, only two zooids thick. These colony forms have high surface-area-to-volume ratio, which increases their susceptibility to dissolution. In *M. chathamensis*, at least, while young parts of the colony may be entirely low-Mg calcite, secondary calcification with high-Mg calcite can result in parts of the colony with over 90% high-Mg calcite (Smith and Lawton 2010), again increasing vulnerability to dissolution.

Type C bryozoans varied in their dissolution history, in a way not obviously related to mineralogical content. *Adeonella* sp. and *A. patagonica* were among the most resistant to dissolution, whereas *Odontionella cyclops* and *Hippomenella vellicata* were less resistant, despite having a less soluble skeletal mineralogy with lower mean wt% aragonite and Mg in calcite. *H. vellicata* forms a high-surface-area-to-volume-ratio foliose colony, which may account for its faster dissolution (Fig. 22.4), but the other three species are all similar in colony form, with wide flat robust branches. *O. cyclops* was the fastest dissolving bryozoan found in a previous experiment (Smith et al. 1992). Perhaps its skeletal structure has an influence; it has been shown to add aragonite as secondary calcification in a discrete surface layer (Smith and Girvan 2010), which can readily be removed, either physically or chemically. If *Adeonella* spp. are not constructed in this bilaminar manner (e.g., see Taylor et al. 2008), that may account for their greater resistance to dissolution.

It appears that the role played by mineralogy in dissolution of bimineralic bryozoans is minimal. Whereas the importance of surface area is known, it has usually been assumed that mineralogy must play some role (Flessa and Brown 1983; Smith et al. 1992). We show here that, if it does, that role is more complex than simple relative solubility. Other possible factors, which may influence dissolution but have not been investigated here, include porosity and surface complexity (Chave 1964; Walter and Morse 1984), extent of organic matrix material present

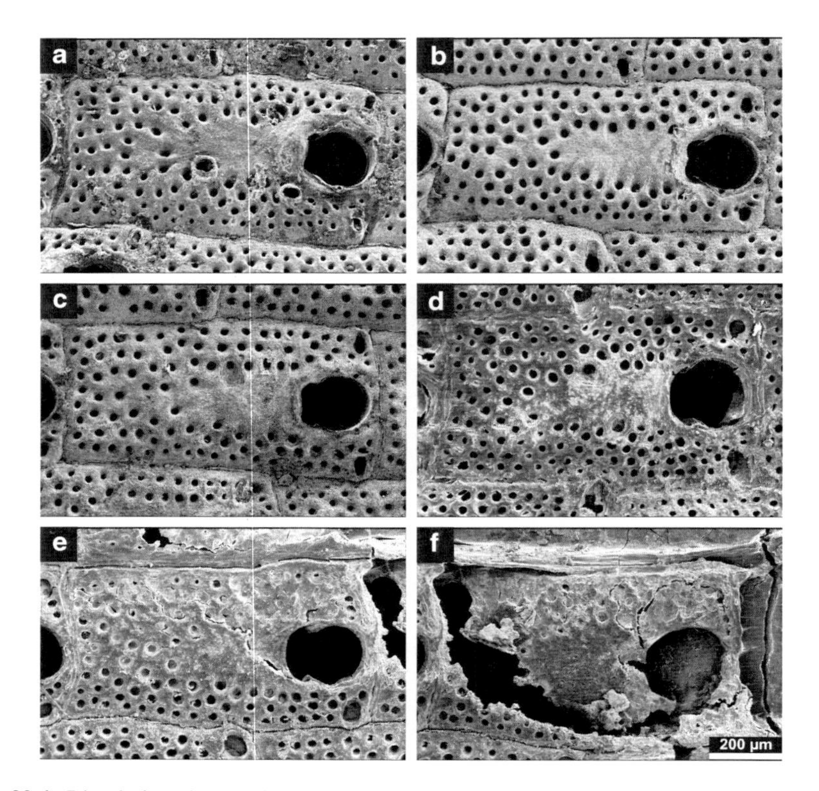


**Fig. 22.4** Dissolution characteristics of *Hippomenella vellicata* from the Otago Shelf. (**a**) – not dissolved; (**b, c**) – 7% dissolved (or 93% of pre-dissolution weight); (**d**) – 20% dissolved; (**e, f**) – 50% dissolved

(Glover and Kidwell 1993), and even possibly ultrastructure (Henrich and Wefer 1986).

In the context of increasing acidification of temperate seawater, it is likely that the shape, size, and skeletal structure of bryozoan sediments will affect their dissolution rate more than mineralogical characteristics. At-risk bryozoan sediments are those dominated by thin delicate species, or colonies like *Cellaria,* which break down into small fragments, irrespective of carbonate mineralogy.

**Acknowledgements** We thank Albert Zhou and Bev Dickson of the Portobello Marine Laboratory for assistance with laboratory work; Liz Girvan of the Otago Centre for Electron Microscopy for assistance with SEM photomicrography; Damian Walls of the Geology Department, University of Otago for support of XRD work. Sediment samples were mainly collected by the University of Otago's Research Vessel *Polaris II,* and we thank Bill Dickson and Phil Heseltine for their ongoing help in the field. This work was funded by the University of Otago Research Grant. We thank Prof. Marcus M. Key, Jr. and Prof. Priska Schäfer for their helpful reviews.

## References

Bone Y, James NP (1993) Bryozoans as carbonate sediment producers on the cool-water Lacepede Shelf, Southern Australia. Sediment Geol 86:247–271

Chave KE (1964) Skeletal durability and preservation. In: Imbrie J, Newell N (eds) Approaches to paleoecology. Wiley, New York

Flessa KW, Brown TJ (1983) Selective solution of macro-invertebrate calcareous hard parts: a laboratory study. Lethaia 16:193–205

Glover CP, Kidwell SM (1993) Influence of organic matrix on the postmortem destruction of molluscan shells. J Geol 101:729–747

Henrich R, Wefer G (1986) Dissolution of biogenic carbonates: effects of skeletal structure. Mar Geol 71:341–362

Kuklinski P, Taylor PD (2009) Mineralogy of Arctic bryozoan skeletons in a global context. Facies 55:489–500

Lombardi C, Cocito S, Hiscock K, Occhipinti-Ambrogi A, Setti M, Taylor PD (2008) Influence of sea water temperature on growth bands, mineralogy and carbonate production in a bioconstructional bryozoan. Facies 54:333–342

Orr JC, Fabry VJ, Aumont O, Bopp L, Doney SC et al (2005) Anthropogenic ocean acidification over the twenty-first century and its impact on calcifying organisms. Nature 437:681–686

Raven J, Caldeira K, Elderfield H, Hoegh-Guldberg O, Liss P et al (2005) Ocean acidification due to increasing atmospheric carbon dioxide, vol 12/05, Royal society policy document. The Royal Society, London

Smith AM (2009) Bryozoans as southern sentinels of ocean acidification: a major role for a minor phylum. Mar Freshw Res 60(5):475–482

Smith AM, Clark DE (2010) Skeletal carbonate mineralogy of bryozoans from Chile: an independent check of phylogenetic patterns. Palaios 25:229–233

Smith AM, Girvan E (2010) Understanding a bimineral bryozoan: skeletal structure and carbonate mineralogy of *Odontionella cyclops* (Foveolariidae: Cheilostomata: Bryozoa). Palaeogeogr Palaeoclimatol Palaeoecol 289:113–122

Smith AM, Lawton EI (2010) Growing up in the temperate zone: age, growth, calcification and carbonate mineralogy of *Melicerita chathamensis* (Bryozoa) in southern New Zealand. Palaeogeogr Palaeoclimatol Palaeoecol 298:271–277

Smith AM, Nelson CS, Danaher PJ (1992) Dissolution behavior of bryozoan sediments: taphonomic implications for non-tropical carbonates. Palaeogeogr Palaeoclimatol Palaeoecol 93:213–226

Smith AM, Nelson CS, Spencer HG (1998) Skeletal mineralogy of New Zealand bryozoans. Mar Geol 151:27–46

Smith AM, Key MM Jr, Gordon DP (2006) Skeletal mineralogy of bryozoans: taxonomic and temporal patterns. Earth Sci Rev 78:287–306

Taylor PD, Kudryavtsev AB, Schopf JW (2008) Calcite and aragonite distributions in the skeletons of bimineralic bryozoans as revealed by Raman spectroscopy. Invertbr Biol 127:87–97

Taylor PD, James NP, Bone Y, Kuklinski P, Kyser TK (2009) Evolving mineralogy of cheilostome bryozoans. Palaios 24:440–452

Walter LM, Morse JW (1984) Reactive surface area of skeletal carbonates during dissolution: effect of grain size. J Sediment Petrol 54:1081–1090

Wejnert KE, Smith AM (2008) Within-colony variation in skeletal mineralogy of *Adeonellopsis* sp. (Cheilostomata: Bryozoa) from New Zealand. N Z J Mar Freshw Res 42:389–395

# Chapter 23
# *Hornera striata* (Milne Edwards, 1838), a British Pliocene Cyclostome Bryozoan Incorrectly Recorded from New Zealand, with Notes on Some Non-fenestrate *Hornera* from the Coralline Crag

## *Hornera* from New Zealand and Britain

**Abigail M. Smith, Paul D. Taylor, and Rory Milne**

*Rory Milne died on 24th April 2011 while this article was in press.*

**Abstract** One of the most commonly reported species in the cyclostome bryozoan family Horneridae is *Hornera striata* Milne Edwards, 1838. First described from the Pliocene Coralline Crag of Suffolk in England, it has since been reported from many European fossil localities. It has also been recorded from the Tertiary and Recent of New Zealand. Here we revise *H. striata* and some other non-fenestrate species of *Hornera* described by Busk (1859) and Mongereau (1972) from the Coralline Crag. *Hornera lagaaiji* Mongereau, 1972 from the Coralline Crag has autozooidal apertures more distantly spaced than *H. striata* but is otherwise very similar to this species and is placed into synonymy, as are Coralline Crag specimens identified by Busk (1859) as *H. frondiculata* Lamouroux, 1821. Another Coralline Crag species, *H. humilis* Busk, 1859, is characterized by small colony and zooid size, and is also unusual in having autozooids opening on the outside of the cone of branches. No gonozooids have been found in any of the Coralline Crag species of *Hornera*, despite a high incidence of colony bases implying recruitment from larvae rather than through clonal fragmentation. Recent records of *H. striata* from New Zealand appear to represent misidentifications of *Hornera robusta* MacGillivray, 1883, while the first report of this British species from New Zealand, from the Miocene Orakei Greensand Member, is provisionally assigned to *H. lunularis* Stoliczka, 1865.

A.M. Smith
Department of Marine Science, University of Otago, P.O. Box 56, Dunedin, New Zealand

P.D. Taylor (✉) • R. Milne
Department of Earth Sciences, Natural History Museum, Cromwell Road, London, SW7 5BD, UK
e-mail: p.taylor@nhm.ac.uk

A. Ernst, P. Schäfer, and J. Scholz (eds.), *Bryozoan Studies 2010*,
Lecture Notes in Earth System Sciences 143, DOI 10.1007/978-3-642-16411-8_23,

**Keywords** Cyclostomata • Taxonomy • Cenozoic • England • New Zealand

## Introduction

*Hornera striata* is one of the most commonly reported species in the family Horneridae (Smith et al. 2008). First described from the Pliocene of England by Milne Edwards (1838) (Fig. 23.1a), it has since been recorded from many other European fossil localities. In 1865, the species was reported from the Tertiary of New Zealand by Stoliczka (Fig. 23.1c), and subsequently identified as still living in southern New Zealand by mollusc specialists. Many examples of bryozoan species with apparently pan-global distributions have been shown to have more restricted distributions following detailed taxonomic study (e.g. Kuklinski and Taylor 2008; Tompsett et al. 2009). Therefore, the identification of *H. striata* in New Zealand can be immediately questioned, especially in view of the low capacity for dispersal in species of *Hornera* with their short-lived larvae and erect, ramose colony not suited to rafting. This has prompted us to revise *H. striata*, based on topotypical material from the Coralline Crag, and to ascertain whether the New Zealand specimens really do belong to this species.

A survey of the non-fenestrate species of *Hornera* described from the Coralline Crag has proved to be desirable in order to gain a better appreciation of the morphological variation within and between species of this taxonomically challenging genus. Accordingly, we provide notes and illustrations of material belonging to the following species of *Hornera* recorded from the Coralline Crag: *H. striata* Milne Edwards, 1838, *H. lagaaiji* Mongereau, 1972, *H. frondiculata* Lamouroux, 1821, and *H. humilis* Busk, 1859. New Zealand records of apparent *H. striata* are shown mostly to comprise the austral species *H. robusta*, apart from Stoliczka's Orakei Greensand (Miocene) examples here provisionally assigned to *H. lunularis* Stoliczka, 1865.

## *Hornera striata*

### Systematics

Phylum Bryozoa Ehrenberg, 1831

Class Stenolaemata Borg, 1926

Order Cyclostomata Busk, 1852

Suborder Cancellata Gregory, 1896

Family Horneridae Smitt, 1866

Genus *Hornera* Lamouroux, 1821

*Hornera striata* Milne-Edwards, 1838

Figures 23.1a, 23.2, 23.3, 23.4, 23.5, and 23.6

1838 *Hornera striata* Milne Edwards, p. 213, pl. 11, fig. 1 [Fig. 23.1a herein].

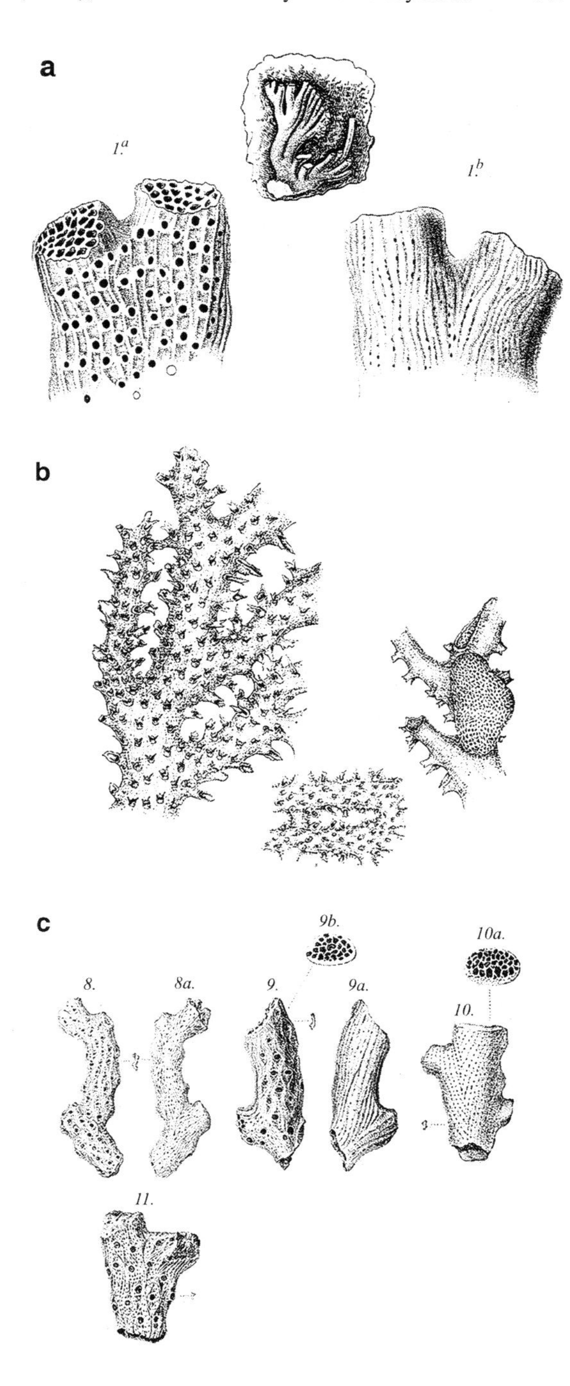


**Fig. 23.1** Original illustrations of: (**a**) *Hornera striata* Milne Edwards, 1838, pl. 11, fig. 1; (**b**) *Hornera robusta* MacGillivray, 1885, pl. 118, figs. 6–8; (**c**) *Hornera striata* Milne Edwards *sensu* Stoliczka, 1865, pl. 17, figs. 8–11

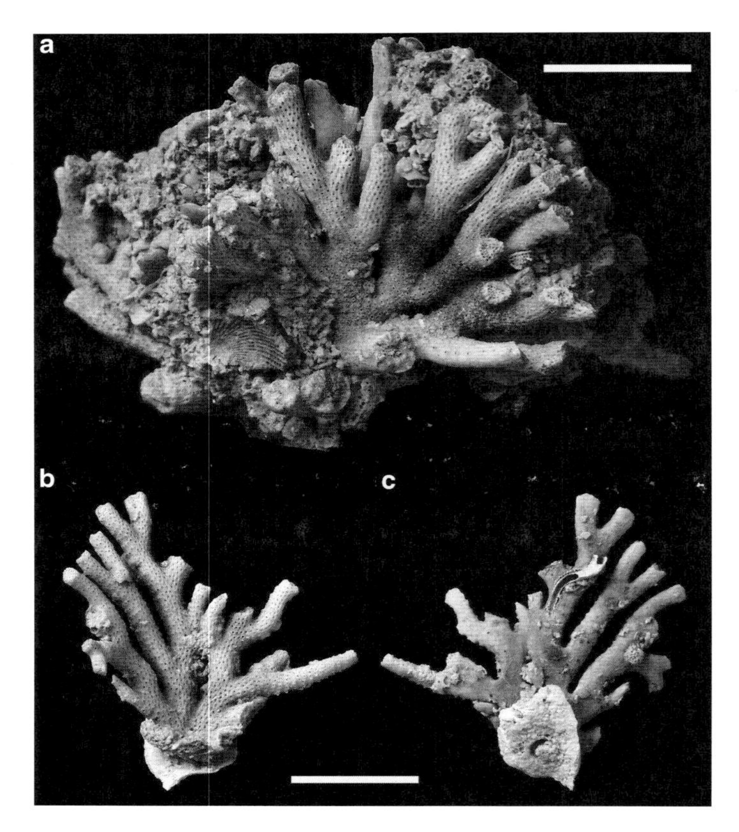


**Fig. 23.2** *Hornera striata* Milne Edwards, 1838, Pliocene, Coralline Crag, Suffolk. (**a**) photograph of Mongereau's (1972) neotype, S. Wood Collection, NHM D6897. (**b**, **c**) oblique frontal view and underside of another colony with a preserved base, NHM BZ 5819. *Scale bars*: 10 mm

1844 *Hornera striata* Milne Edwards: Wood, p. 16.

1859 *Hornera striata* Milne Edwards: Busk, p. 103, pl. 15, fig. 3, pl. 16, fig. 5.

1859 *Hornera frondiculata* Lamouroux: Busk, p. 102, pl. 15, figs. 1, 2, pl. 16, and 6 [Fig. 23.4 herein].

Non 1865 *Hornera striata* Milne Edwards: Stoliczka, p. 107, pl. 17, figs. 8–11 [Fig. 23.1c herein].

Non 1873 *Hornera striata* Milne Edwards: Hutton, p. 101.

Non 1880 *Hornera striata* Milne Edwards: Hutton, p. 196.

Non 1891 *Hornera striata* Milne Edwards: Hutton, p. 107.

Non 1898 *Hornera striata* Milne Edwards: Hamilton, p. 197.

Non 1904 *Hornera striata* Milne Edwards: Hutton, p. 299.

1952 *Hornera striata* Milne Edwards: Lagaaij, p 170, pl. 21, fig. 4a, b.

Non 1971 *Hornera striata* Milne Edwards: Fleming, p. 7.

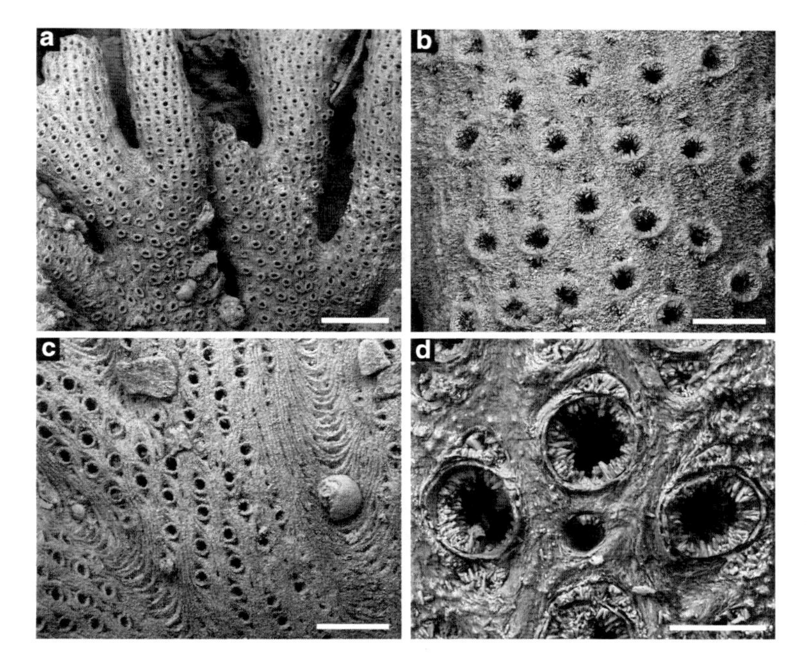


**Fig. 23.3** *Hornera striata* Milne Edwards, 1838, backscattered scanning electron micrographs of specimens from the Pliocene, Coralline Crag, Suffolk. (**a**, **b**) Mongereau's (1972) neotype, S. Wood Collection, NHM D6897; (**a**) oblique frontal view of bifurcating, subequal branches; (**b**) detail showing apertures and cancelli obscured by diagenetic cement. (**c**) basal branches of a colony from Gedgrave showing arcuate striae developed in branch bifurcations, NHM 23469(a). (**d**) detail of autozooidal apertures and cancelli with relatively little cement, NHM BZ 5820. *Scale bars*: a = 1 mm; b = 200 μm; c = 500 μm; d = 100 μm

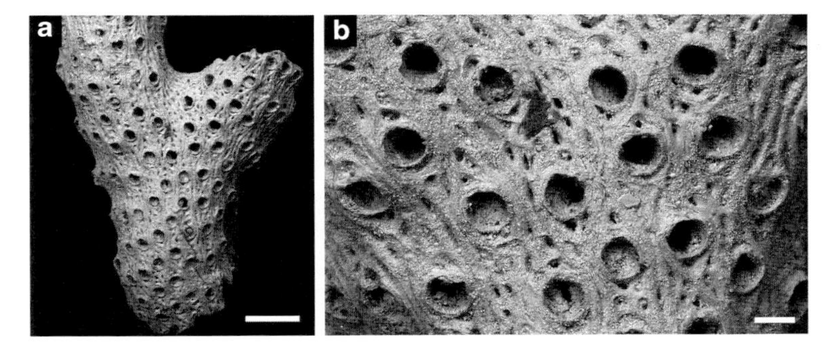


**Fig. 23.4** *Hornera striata* Milne Edwards, 1838, backscattered scanning electron micrographs of specimen figured by Busk (1859, pl. 16, fig. 6c) as *Hornera frondiculata* Lamouroux, Pliocene, Coralline Crag, Suffolk, S. Wood Collection, NHM B1615a. (**a**) branch frontal surface. (**b**) detail showing autozooids and cancelli. *Scale bars*: a = 500 μm; b = 100 μm

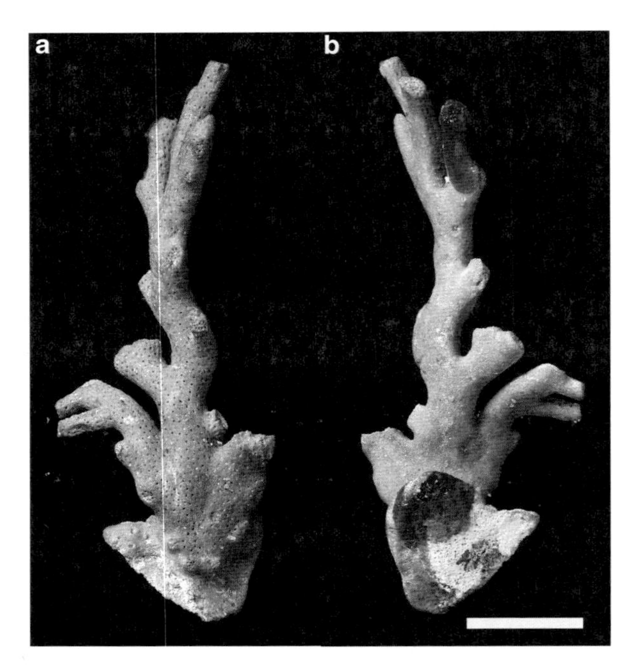


**Fig. 23.5** *Hornera striata* Milne Edwards, 1838. Photographs of frontal (**a**) and reverse (**b**) sides of the holotype of *Hornera lagaaiji* Mongereau, 1972, Pliocene, Coralline Crag, Suffolk, S. Wood Collection, NHM D38451. *Scale bars*: 5 mm

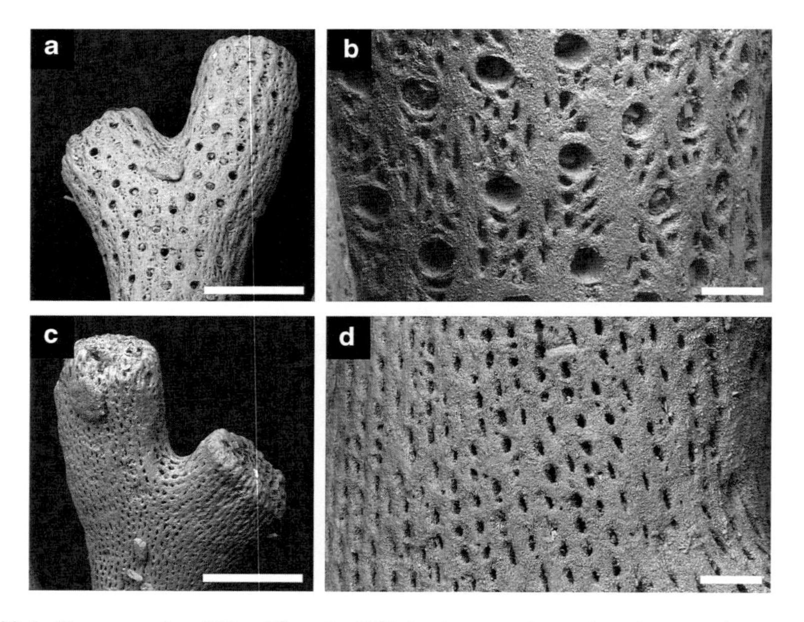


**Fig. 23.6** *Hornera striata* Milne Edwards, 1838, backscattered scanning electron micrographs of the holotype of *Hornera lagaaiji* Mongereau, 1972, Pliocene, Coralline Crag, Suffolk, S.Wood Collection, NHM D38451. (**a, b**) frontal surface showing autozooidal apertures and cancelli. (**c, d**) reverse surface with cancelli. *Scale bars*: a, c = 1 mm; b, d = 200 μm

1972 *Hornera striata* Milne Edwards: Mongereau, p. 345, pl. 5, figs. 8, 9, pl. 6, figs. 1–4, 6, pl. 7, figs. 5–8.

Note: this abbreviated synonymy only covers topotypical and putative New Zealand occurrences; the status of the numerous additional records of this species from Eocene–Recent deposits globally cannot be adequately evaluated without restudy of the relevant specimens, although most must be deemed questionable.

## Material

*Type*: A Pliocene fossil of this species from the Crag, Suffolk, England, was lodged in the Fischer Collection at the Museum National d'Histoire Naturelle, Paris (MNHN). Mongereau (1972) could not locate this specimen and thus proposed as the neotype Busk's (1859) figured specimen: NHM D6897 (incorrectly cited as B6897), Coralline Crag (Pliocene), Suffolk, England, S. V. Wood Collection (Fig. 23.2a and 23.3a, b). However, d'Hondt (2006, p. 28) subsequently rediscovered Milne Edwards' type of *H. striata* in the collections of the MNHN, where it is registered as No. 4417. The International Code of Zoological Nomenclature states that if, after the designation of a neotype, the type that was presumed lost is found still to exist, the rediscovered material again becomes the type and the neotype is set aside. Therefore, the holotype, by monotypy, of *H. striata* is No. 4417 Milne Edwards Collection, MNHN. Although not studied by the current authors, there is no reason to suspect that this specimen belongs to a different species than the old neotype, which is overwhelmingly the dominant non-fenestrate *Hornera* in the Coralline Crag.

*Other material:* NHM BZ 5819, 5820, Coralline Crag, Suffolk. NHM D49929-34, D49952, D50991-51000 [sample], D51180-3, Coralline Crag, Sudbourne Hall, Suffolk, Burrows Colln. NHM D50000-4, Coralline Crag, 'zone f', Pettistree Hall, Sutton, Suffolk.

## Description

Colony dendroid, the largest observed example 55 mm in width; branches generally 1–2 mm in transverse diameter, ovoidal in cross-section, the reverse surface relatively flat in young branches, becoming more convex in older branches. Primary branches arising in a circlet from an encrusting base (Fig. 23.2b, c) that may be flat, concave or tubular when moulded around a perished cylindrical substratum, up to 9 mm in diameter. Ramifications mostly dichotomous bifurcations at acute angles; lateral ramifications lacking. Anastomoses rare. Cancelli infilling areas between bifurcating branches in older, proximal parts of colonies, especially immediately above the base (Fig. 23.3c). Branches at the same level of subequal width, with the exception of occasional aborted branches, developed where space is restricted, these being narrow and tapering towards a rounded distal termination. Small colonies having branches arranged in a cone, larger colonies becoming more convoluted and three-dimensional through twisting, bending and reflexing of branches which may form semi-distinct palmate clusters at their distal ends. Growth tips rounded.

Autozooidal apertures small, more or less circular, arranged in semi-regular longitudinal rows (Fig. 23.4a), about 5–10 rows across the width of a branch. Thickened calcification forming a system of nervi winding between the apertures (Fig. 23.4b). Peristome may be slightly raised in well preserved specimens (Fig. 23.3d); cusps not evident. Frontal surface between apertures covered with cancelli varying in number and size ontogenetically and between colonies. Often a distinct cancellus is present above and below each aperture (Fig. 23.3d), giving a lozenge-shaped appearance.

Dorsal (reverse) surface densely covered with cancelli arranged in longitudinal rows, the cancelli elongate (Fig. 23.6d). Brood chamber unknown.

## Measurements

| | |
|---|---|
| Aperture longitudinal diameter | 0.06–0.10 mm (12 measurements from NHM D6897) |
| Aperture transverse diameter | 0.07–0.10 mm (12 measurements from NHM D6897) |
| Longitudinal apertural spacing | 0.20–0.43 mm (9 measurements from NHM D6897) |

## Remarks

Milne Edwards (1838, p. 213) characterised this Coralline Crag species as arborescent like *H. frondiculata* but with branches much less widely forked, and the apertures of the zooids very much closer and laid out in more-or-less regular longitudinal series, separated by small longitudinal ridges. Wood (1844) and Busk (1859) described further material from the Coralline Crag of Suffolk, England, noting that: colonies are densely tufted with cylindrical branches; apertures are small, round and organised in more-or-less longitudinal rows, with older ones developing a slightly thickened peristome, sometimes with 'an acute angle on one side' (Busk 1859, p. 103); pores [cancelli] occur above and below the apertures; smooth ridges form nearly diamond-shaped spaces around the apertures on the frontal surface, whereas the reverse side is smooth or subgranular with fine pores arranged longitudinally.

Michelin (1847) reported *H. striata* from two fossil localities in central France, remarking that the branches are close together and all end at about the same height with rounded tips. He referred to 'unequal pores', which are elongated parallel to branch length. Also in 1847, d'Archiac figured a specimen he called *Hornera hippolythus* Defrance (a complex species which has been misapplied regularly since then). According to Canu (1912), d'Archiac's specimen (then in the Ecole des Mines Collection, Paris) is also *H. striata*, having in the meantime been reassigned to *H. trabecularis* by Reuss (1869) and *H. frondiculata* by both Pergens (1887) and Waters (1877). Canu cited the size of the apertures, the regular occurrence of two 'vacuoles' next to each orifice, and anastomoses of the branches near the base as important features.

In his major revision of European *Hornera*, Mongereau (1972) noted the variation in morphology that can occur in a single specimen of *H. striata*, and commented on some characteristics of *H. frondiculata* he found to be developed close to the bases of specimens otherwise assignable to *H. striata*. For this reason he re-assigned

*H. striata* (which he erroneously ascribed to Busk 1859) to '*Hornera frondiculata* Auct. forma *striata* Busk, 1859', arguing that *H. frondiculata* is not a true biological species but an 'ensemble paléontologique' (Mongereau 1972, p. 350). In contrast, Lagaaij (1952) considered *H. striata* and *H. frondiculata* to be truly different species, though easily confused. He noted that *H. striata* has narrower branches, smaller and more regular apertures (0.05–0.06 mm vs. 0.06–0.10 mm), and fewer cancelli (vacuoles) between the apertures on the frontal surface.

Supposed examples of the extant *H. frondiculata* Lamouroux, 1821 recorded from the Coralline Crag (Fig. 23.4) are re-determined here as *H. striata*. Busk (1859) acknowledged some differences between the material he identified as *H. frondiculata* from the Coralline Crag and Recent specimens of this species but explained these in terms of '.. the protean habits of this perplexing genus..' (Busk 1859, p. 103). The characteristics he regarded as important for identifying *H. frondiculata* were: (1) branch ramifications in the same or more or less the same plane; (2) emargination of the border of the aperture ('mouth'), giving it a bifid appearance; and (3) coarse perforation of the dorsal surface. Character (1) is a feature of most species of *Hornera*, and characters (2) and (3) may be dependent on ontogeny and, in fossils, preservation. Recent material of *H. frondiculata* from Marseille in the NHM collections has apertural diameters of about 0.12 mm, very slightly larger than putative *H. frondiculata* from the Coralline Crag, which has longitudinal and transverse apertural diameters both measuring 0.09–0.10 mm (5 measurements from B1615a) and a longitudinal apertural spacing of 0.26–0.31 mm (3 measurements from B1615a). The Coralline Crag material does not show bifid apertural margins, although this may be due to taphonomic loss.

A Coralline Crag species described as *Hornera frondiculata* Auct., forme *lagaaiji* Mongereau, 1972 (p. 339, pl. 9, figs. 7, 8) is also here placed in synonymy with *H. striata*. The type specimen of this species (NHM D37807; Figs. 23.5 and 23.6) is a fragment preserving the colony base but stripped of most of the branches. It has slightly more distantly spaced autozooidal apertures than is typical for *H. striata* but this seems to represent an end-member in a population showing high levels of variability as all intermediates can be found among material from the Coralline Crag.

Putative *Hornera striata* has since been reported frequently from the Tertiary of Europe, including southern Italy (Waters 1877; Seguenza 1879; De Stefani 1884; Neviani 1896a,b, 1900; Scotti 1936), Rhodes (Pergens 1887), France (Julien 1940; Roger and Buge 1946; Buge 1957), Belgium (Lagaaij 1952), Austria (Gillard 1938; Vávra 1979), and Slovakia (Zágorsek and Hudácková 2000). Some of the specimens from the Miocene of Austria and Hungary described by Reuss (1848) as *H. hippolythus* Defrance were later re-identified as *H. striata* by Manzoni (1877), an opinion supported by Canu and Bassler (1924) and Bobies (1958). Roemer's (1863) new species *H. punctata* and *H. sulcato-punctata* from the Oligocene of northern Germany were also placed in synonymy with *H. striata* by Mongereau (1972), as was *H. radians* Defrance.

Canu (1908) reported *H. striata* in the Miocene of Argentina, and the species has been recorded several times from the Miocene–Recent of New Zealand. The enormous geographical separation alone makes it unlikely that southern hemisphere records of *H. striata* really do represent this species.

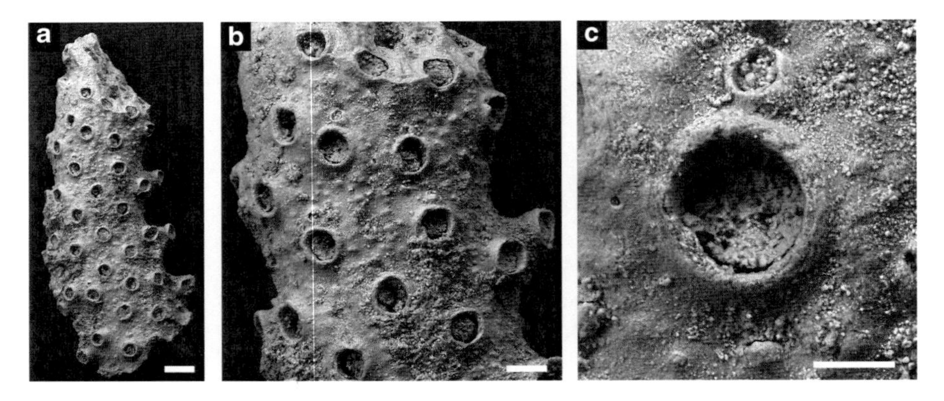


**Fig. 23.7** *Hornera lunularis* Stoliczka, 1865, backscattered scanning electron micrographs of branch from the Miocene Orakei Greensand of Hobson Bay, Auckland, New Zealand, NHM BZ 5821. (**a**) branch frontal. (**b**) detail showing autozooidal apertures. (**c**) autozooidal aperture with a small cancellus above. *Scale bars*: a = 200 μm; b = 100 μm; c = 50 μm

Reports of *H. striata* from New Zealand began with Stoliczka (1865) and were perpetuated by several later authors. The supposed *H. striata* described by Stoliczka (Fig. 23.1c) is from the greensand of Orakei Bay in Auckland, a deposit now known as the Orakei Greensand Member (Ballance 1976), which forms part of the Waitemata Group, East Coast Bays Formation of early Miocene age. Stoliczka recorded three species of *Hornera* from Orakei Bay – *H. striata* and two new species, *H. lunularis* and *H. pacifica* – which, along with the other 29 species he described from Orakei, are conserved in the fossil collections of the Naturhistorisches Museum, Vienna under catalogue numbers 1865.49.0005, 1865.49.0006 and 1865.49.0007, respectively. Optical microscopic study of the three species, all represented by small, infertile branch fragments, revealed no significant differences between them, and the Orakei species is here referred provisionally to *H. lunularis* Stoliczka, 1865 (*H. lunularis* has page and figure priority over *H. pacifica*). Based on scanning electron microscopy of a branch in the NHM collections (Fig. 23.7) that was matched with Stoliczka's types, *H. lunularis* has proportionally larger area of skeleton between the autozooidal apertures and a smaller area occupied by cancelli (Fig. 23.7b, c). The species was refigured and briefly redescribed by Mongereau (1972, p. 337, pl. 4, figs. 7, 8).

According to Neviani (1900), the stratigraphical range of *H. striata* in Europe is Eocene to Recent. Buge (1957) revised this range to Aquitanian (lower Miocene) to Quaternary, subsequently extended back into the Eocene by Malecki (1963) and the lower Eocene by Mongereau (1972; as *Hornera frondiculata* forme *striata*). However, records of *H. striata* older or younger than Pliocene require careful re-evaluation before they can be accepted as belonging to this species.

There were no gonozooids on the Coralline Crag specimens of *H. striata* described by Milne Edwards (1838) or Busk (1859). Later authors from Pergens (1887) to Lagaaij (1952) also noted a lack of gonozooids in their specimens. Indeed, no gonozooids were described in *H. striata* until 1972 when Mongereau identified examples in putative *H. striata* from the Burdigalian of the Rhône Basin.

**Fig. 23.8** Photograph of a typical colony of *Hornera robusta* MacGillivray, 1883 from the Otago Shelf, New Zealand, NHM 2010.7.23.1. Note the pinnate branching pattern with narrow secondary branches arising from the sides of broader primary branches (cf. *H. striata*, Fig. 23.2a). *Scale bar*: 10 mm

He characterised these as being variable and much like those of *H. frondiculata*. We have been unable to find gonozooids among the large number of topotypical specimens of *H. striata* in the collections of the NHM. This is puzzling given that bases diagnostic of colonies produced by larvae are common and there is no evidence that any of the colonies were the result of fragmentation of older colonies.

## *Hornera robusta*

### Systematics

*Hornera robusta* MacGillivray, 1883

Figures 23.8, 23.9, and 23.10

1873 *Hornera striata* Milne Edwards: Hutton, p. 101.

1880 *Hornera striata* Milne Edwards: Hutton, p. 196.

1883 *Hornera robusta* MacGillivray, p. 291, pl. 1, fig. 1.

1885 *Hornera robusta* MacGillivray: MacGillivray in McCoy, p. 72, pl. 118, figs. 6–8.

1891 *Hornera striata* Milne Edwards: Hutton, p. 107.

1904 *Hornera striata* Milne Edwards: Hutton, p. 299.

1971 *Hornera striata* Milne Edwards: Fleming, p. 7.

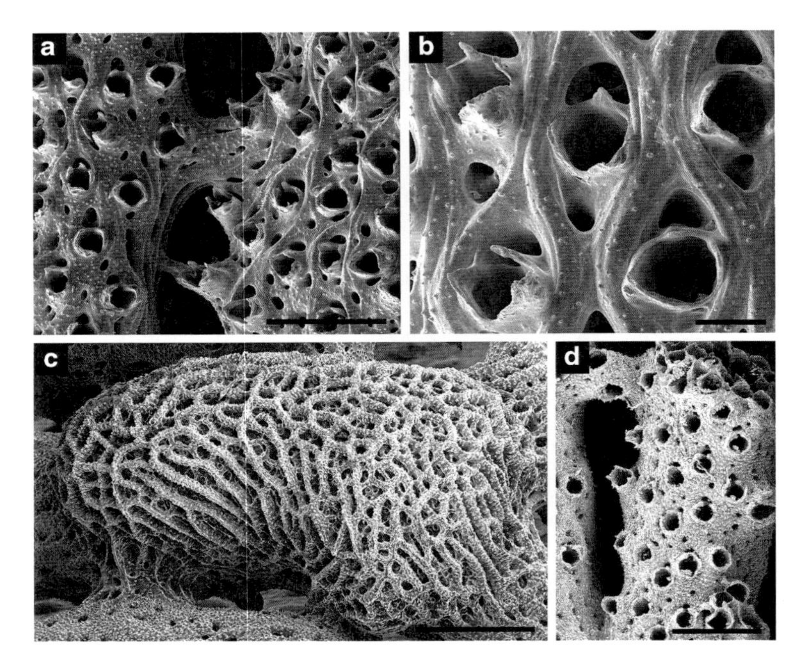


**Fig. 23.9** *Hornera robusta* MacGillivray, 1883, secondary scanning electron micrographs of coated Recent specimens. (**a–c**), syntype, Museum Victoria, Melbourne, F110130; (**a**) branches connected by an anastomosis; (**b**) cuspate apertures and cancelli; (**c**) gonozooid. (**d**) part of a New Zealand colony from Stewart Island misidentified by F. W. Hutton as *Hornera striata*, GNS2. *Scale bars*: a, c, d = 500 μm; b = 100 μm

## Material

GNS2, a large specimen and 2 fragments labelled '*Hornera striata* Stolizka (ident. F W Hutton) (fragments from Cant Mus specimen), Stewart Island, Recent.' (Fig. 23.9d). GNS3, four small fragments in a cardboard slide labelled '*Hornera striata*, Foveaux St'. GNS4, small fragment in a glass slide labelled 'Polyzoa, *Hornera striata* M. Edwards, Stewart Island'. GNS5, small colony in a cardboard slide labelled '*H. striata*, Fov. St.' GNS6, colony in a cardboard box labelled '*Hornera striata* M. Edw., Stewart Id., Recent'. NHM 2010.7.23.1-3, Otago Shelf, New Zealand.

## Remarks

This species, originally described from the Recent of Port Phillip Heads, Victoria, Australia, appears to be the dominant non-fenestrate *Hornera* found in waters around New Zealand at the present day. Molecular comparisons between Australian and New Zealand populations are, however, needed to show whether they are truly conspecific. Specimens of *H. robusta* from both Australia (Port Phillip, Victoria)

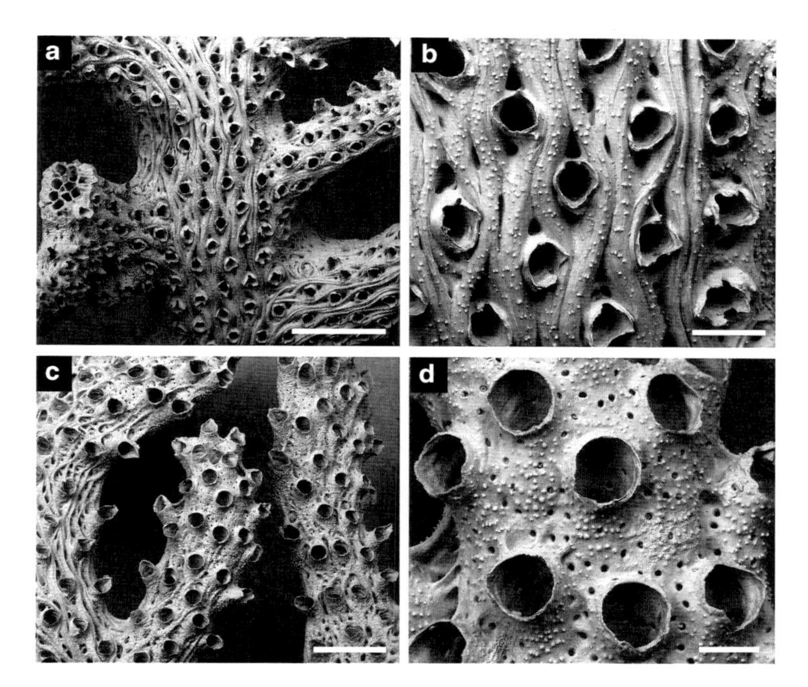


**Fig. 23.10** *Hornera robusta* MacGillivray, 1883, backscattered scanning electron micrographs of bleached Recent specimens. (**a**, **b**) NHM 1934.2.20.6 (a), Port Phillip, Victoria, Australia; (**a**) frontal view of main branch with smaller side branches; (**b**) detail of a mature branch showing apertures and cancelli with thick calcification forming sinuous striae. (**c**, **d**) NHM 2010.7.23.2, Otago Shelf, New Zealand, 82 m; (**c**) branching pattern; (**d**) immature branch showing pustules and pores on the interior frontal walls between the autozooidal apertures. *Scale bars*: a = 1 mm; b = 200 μm; c = 500 μm; d = 100 μm

and New Zealand (Otago Shelf) have autozooidal apertures that are consistently 0.10 mm in diameter, and spaced longitudinally between about 0.3 and 0.5 mm apart. The general appearance of colonies from the two regions is also very similar (Fig. 23.10).

It is clear that most, if not all, Recent records of *H. striata* from New Zealand represent misidentified specimens of *H. robusta*. Notable features of *H. robusta* enabling its distinction from *H. striata* are the pinnate colonies (Fig. 23.8) with anastomosing minor branches (Fig. 23.9a), typically tricuspate peristomes (Fig. 23.9b), and lateral zooids with elongated peristomes (Figs. 23.9d and 23.10c). The pinnate growth pattern of *H. robusta* means that branches vary appreciably in diameter at any given level within the colony, whereas branches from the same level in *H. striata* colonies have approximately equal diameters. Gonozooids of *H. robusta* are ridged and beaded (Fig. 23.9c) and have a characteristic frontally oriented ooeciostome. As noted above, the gonozooid has not been found in unequivocal *H. striata*.

## *Hornera humilis*

### Systematics

*Hornera humilis* Busk, 1859
Figure 23.11

1859 *Hornera humilis* Busk, p. 100, pl. 14, fig. 5 only.

1952 *Hornera humilis* Busk: Lagaaij, p. 171, pl. 20, figs. 5, 6.

?1952 *Hornera humilis* Busk: Malecki, p. 217, pl. 14, fig. 15.

1972 *Hornera humilis* Busk: Mongereau, p. 334, pl. 4, fig. 9, pl. 8, figs. 3, 4.

### Material

*Lectotype*: NHM D6896, Coralline Crag, Suffolk, S. Wood Collection, chosen by Lagaaij (1952).

*Other material:* NHM D49460, Coralline Crag ('zone f'), Pettistree Hall, Sutton, Suffolk, Burrows Collection. NHM 60198a, b, 60317a, b, Coralline Crag, no locality details, Suffolk. NHM D50974, Coralline Crag, Broom Hill, Suffolk.

### Description

Colony small, erect, dichotomously branched, branches narrow, ranging from 0.6–0.8 mm in diameter, branch anastomoses not observed. Attached base about 1 mm in diameter, giving rise to three primary branches diverging at 120° and initiating a broad, cone-shaped colony (Fig. 23.11a, c, d). Branch surfaces immediately above base covered by cancelli, lacking autozooidal apertures (Fig. 23.11a, c). Apertures of autozooids opening on outside of cone (Fig. 23.11d); cancellate branch reverse surfaces oriented towards interior of cone.

Autozooids with small, subcircular apertures arranged in poorly defined longitudinal rows separated by sinuous low ridges, about six rows across the width of a branch (Fig. 23.11e). Apertural spines and peristomes lacking. Calcification between apertures smooth, without pustules. Cancelli distributed between autozooidal apertures on branch frontal surface, about 3–4 times more abundant than autozooids, circular to longitudinally elliptical (Fig. 23.11f). On branch reverse surfaces cancelli aligned in uniserial longitudinal rows separated by slightly raised areas of granular calcification (Fig. 23.11b).

Gonozooids unknown.

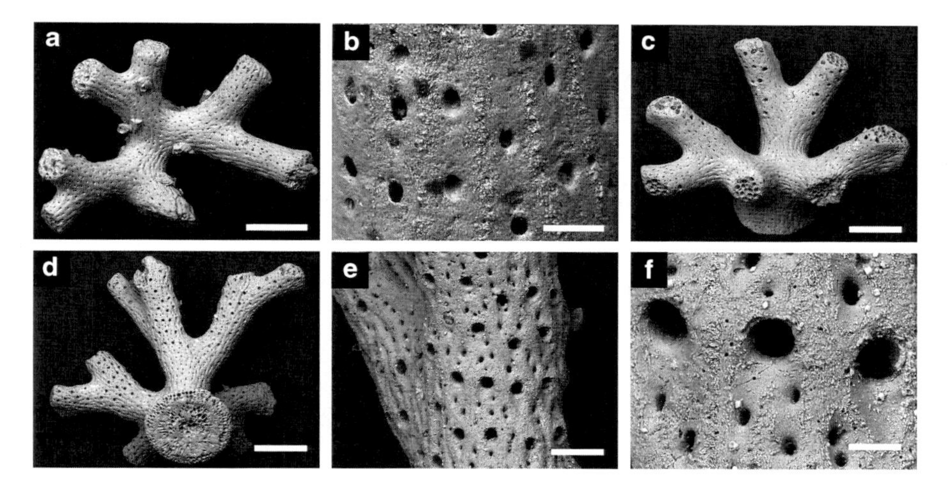


**Fig. 23.11** *Hornera humilis* Busk, 1859, Pliocene, Coralline Crag, Suffolk; backscattered scanning electron micrographs of uncoated specimens. (**a**, **b**), lectotype, NHM D6896; (**a**) colony from above showing reverse sides of branches; (**b**) detail of cancelli on branch reverse. (**c**–**f**), NHM D49459; (**c**) colony in oblique profile; (**d**) colony from the underside showing base and autozooidal apertures on frontal sides of branches above basal expansion; (**e**, **f**) dorsal branch surface showing autozooidal apertures and cancelli. *Scale bars*: a, c, d = 1 mm; b, f = 100 μm; e = 500 μm

## Measurements

| | |
|---|---|
| Aperture longitudinal diameter | 0.05–0.08 mm (5 measurements from D49459) |
| Aperture transverse diameter | 0.05–0.08 mm (5 measurements from D49459) |
| Longitudinal apertural spacing | 0.21–0.34 mm (5 measurements from D49459) |
| Dorsal cancelli longitudinal diameter | 0.02–0.03 mm (5 measurements from D6896) |
| Dorsal cancelli transverse diameter | 0.02–0.03 mm (5 measurements from D6896) |
| Longitudinal dorsal cancelli spacing | 0.11–0.15 mm (4 measurements from D6896) |

## Remarks

This Coralline Crag species is readily distinguished from the much commoner *H. striata*. Not only are the autozooidal apertures significantly smaller, but the tiny colonies are very unusual in showing what Lagaaij (1952) referred to as an inversion of the frontal and dorsal surfaces. In most species of *Hornera*, the frontal surface bearing autozooidal apertures lies on the inside of the initial cone formed by the branches of the young colony. However, in *H. humilis*, the frontal surface is on the outside of the cone. Aside from implying a significant difference in bud orientations during early astogeny, this inversion would have resulted in a reversal of the colony-wide current system. Normally, in *Hornera* incurrents can be inferred to enter through the open end of the cone formed by the young colonies, with excurrents passing centifugally outwards between the branches. In *H. humilis*, however, incurrents would have approached the cone centripetally, with excurrents passing between the branches for venting out through the open top of the cone.

Lagaaij (1952) recorded *H. humilis* also from Scaldisian localities in Belgium. It seems to be a rare species in the Coralline Crag, although this apparent rarity may in part be because the colony base must be present for the inversion that is diagnostic of the species to be seen.

## Discussion

By far the most abundant hornerid cyclostome in the Pliocene Coralline Crag of Suffolk, England is *Hornera striata* Milne Edwards. Subsequent reports of this species from different aged deposits and from beyond the North Sea Basin either require confirmation or are wrong. In particular, records of *H. striata* from New Zealand are misidentifications: most represent *H. robusta* but some can be assigned to *H. lunularis*, a species originally described from the Miocene Orakei Greensand by Stoliczka (1865) in the same paper that first cited *H. striata* in New Zealand.

A second, far less common species of *Hornera* found in the Coralline Crag is *H. humilis*. The small colonies of this species are peculiar in having an inverted branch polarity, with the autozooids opening on the outside rather than the inside of the cone formed by the early branches.

**Acknowledgements** Andreas Kroh (Vienna) kindly allowed one of us (PDT) to study material of *Hornera* and other bryozoans in the Stoliczka Collection. John Simes and Chris Hollis allowed examination and SEM (by AMS) of *Hornera* from the GNS Science collections in Lower Hutt, New Zealand. We also acknowledge assistance from Liz Girvan (Otago Centre for Electron Microscopy), and Mary Spencer Jones and Vicki Holmes (Department of Zoology, NHM).

## References

d'Archiac A (1847) Description des fossils du groupe Nummulitique aux environs de Bayonne et de Dax. Mém Soc Géol France 2(3):397–456

Ballance PF (1976) Stratigraphy and bibliography of the Waitemata Group of Auckland, New Zealand. NZ J Geol Geophys 19:897–932

Bobies CA (1958) Bryozoenstudien III/2 Die Horneridae (Bryozoa) des Tortons im Wiener und Eisenstädter Becken. Sitz Österreich Akad Wiss, Mathem.-naturw. Kl., Wien, Abstract I, Bd. 167(3–4):119–137

Borg F (1926) Studies on Recent cyclostomatous Bryozoa. Zool Bidr Upps 10:181–507

Buge E (1957) Les Bryozoaires du Néogène de l'ouest de la France et leur signification stratigraphique et paléobiologique. Mém Mus Natn Hist Nat, Série C 6:1–435

Busk G (1859) A monograph of the fossil Polyzoa of the Crag. Palaeontogr Soc Monogr 14:1–136, 22 pls

Canu F (1908) Les Bryozoaires fossiles des terrains du Sud-Ouest de la France. II Lutétien. Bull Soc Géol Fr, Ser 4, 8:382–390

Canu F (1912) Les Bryozoaires fossiles des terrains du Sud-Ouest de la France. VI Bartonien: Auversien (fin). Bull Soc Géol Fr, Ser 4, 12: 623–630

Canu F, Bassler RS (1924) Contribution a l'étude des Bryozoaires d'Autriche et de Hongrie. Bull Soc Géol Fr, Ser 4, 24:672–690

De Stefani C (1884) Escursione scientifica nella Calabria (1877–78), Jejo, Montalto e Capo Vaticano. Mem. R. Accad. D. Lincei, Studio Geologico, Ser 3, 18,1–292

Ehrenberg CG (1831) Symbolae physicae, seu Icones et descriptiones corporum naturalium novorum aut minus cognitorum. . . Pars zoologica. Dec. I. Ex officina Academica, Berlin

Fleming CA (1971) New Zealand Bryozoa. Catalogue of specimens in the N.Z. Geological Survey. NZ Geol Surv Rep 50:1–43

Gillard PA (1938) Sur les Bryozoaires helvétiens des Faluns de la Vienne. C R Somm Seances Soc Géol Fr 9:153–155

Gregory JW (1896) Catalogue of the fossil Bryozoa in the Department of Geology, British Museum (Natural History). The Jurassic Bryozoa. Trustees of the British Museum (Natural History), London

Hamilton A (1898) A list of Recent and fossil Bryozoa collected in various parts of New Zealand. Trans Proc NZ Inst 30:192–199

d'Hondt J-L (2006) The Henri Milne Edwards' (1800–1885) collection of recent and fossil Bryozoa. Linzer Biol Beiträge 38:25–38

Hutton FW (1873) Catalogue of the marine Molluscs of New Zealand. Colonial Museum and Geological Survey Department, Wellington

Hutton FW (1880) Manual of the New Zealand Mollusca. Colonial Museum and Geological Survey Department, Wellington

Hutton FW (1891) Revised list of the marine Bryozoa of New Zealand. Trans NZ Inst 1890:102–107

Hutton FW (1904) Polyzoa. In: Index Faunae Novazealandiae. Dulau, London, pp 293–299

Julien M (1940) Révision de la faune vindobonienne de Saint-Fons (Rhône). Trav Lab Géol Fac Sci Lyon 38:1–60

Kuklinski P, Taylor PD (2008) Arctic species of the cheilostome bryozoan *Microporella*, with a redescription of the type species. J Nat Hist 42:1893–1906

Lagaaij R (1952) The Pliocene Bryozoa of the Low Countries and their bearing on the marine stratigraphy of the North Sea region. Mededel Geol Sticht, Serie C, 5(5):1–233

Lamouroux JVF (1821) Exposition méthodique des genres de l'ordre des polypiers, avec leur description et celles des principales espèces figures dans 84 planches; les 63 premiers appartenant a l'histoire naturelle des zoophytes d'Ellis et Solander.V. Agasse, Paris

MacGillivray PH (1883) Descriptions of new or little-known Polyzoa, Part 1. Trans Proc Roy Soc Vict 19:287–293

MacGillivray PH (1885) [*Hornera foliacea* and *Hornera robusta*]. In: McCoy F (ed) Prodromus of the zoology of Victoria, decade 11, pp 71–73, pl. 118

Malecki J (1952) Les Bryozoaires des sables – *Heterostegina* aux environs de Cracovie et miechow. Ann Soc Geol Pologne 21:181–234

Malecki J (1963) Bryozoa from the Eocene of the central Carpathians between Grybow and Dukla. Polska Akad Prace Geol 16:1–151

Manzoni A (1877) I Briozoi fossili del Miocene d'Austria ed Ungharia. Denks K Akad Wiss Wien 38(2):1–24

Michelin H (1847) Iconographie Zoophytologique. Bertrand, Paris

Milne Edwards MH (1838) Mémoire sure les Crisies, les Hornères et plusieurs autres Polypes vivans ou fossils dont l'organisation est analogue à celle des Tubulipores. Ann Sci nat Zool Biol anim 2(9):193–238

Mongereau N (1972) Le genre *Hornera* Lamouroux, 1821, en Europe (Bryozoa – Cyclostomata). Ann Naturhist Mus Wien 76:311–373

Neviani A (1896a) Briozoi neozoici di alcune località d'Italia. Parte Terza: Briozoi postpliocenici di Spilinga (Catanzaro). Boll Soc Romana Studi Zoologii 5:114–120

Neviani A (1896b) Briozoi postpliocenici di Spilinga (Calabria). Atti Accad Gioenia Sci Nat 9:1–66

Neviani A (1900) Briozoi neogenici delle Calabrie. Palaeontogr It 6:115–265

Pergens E (1887) Pliocäne Bryozoen von Rhodos. Ann K K Naturhist Hofmus Wien 2:1–34

Reuss AE (1848) Die fossilen Polyparien des Wiener Tertiärbeckens. Haidingers' Naturwiss Abh 2:1–109

Reuss AE (1869) Paläontologische Studien uber die älteren Tertiärschichten der Alpen, II, Die fossilen Anthozoen und Bryozoen der Schichtengruppe von Crosara. Denk K Akad Wiss Wien 29:215–298

Roemer FA (1863) Beschreibung der norddeutschen tertiären Polyparien. Palaeontolog 9:199–245

Roger J, Buge E (1946) Les Bryozoaires du Redonien. Bull Soc Geol Fr, Ser 5, 16:217–230

Scotti P (1936) Briozoi fossili del miocene della Collina di Torino (Collezione Rovasenda). Atti Reale Accad Sci Torino 71(1):402–431

Seguenza G (1879) Le formazioni terziarie nella provincia di Reggio. Atti Accad Lincei, Mem Classe Sci Fis 3/6:1–445

Smith AM, Taylor PD, Spencer HG (2008) Resolution of taxonomic issues in the Horneridae (Bryozoa: Cyclostomata). Ann Bryozool 2:359–412

Stoliczka F (1865) Fossile Bryozoen aus dem tertiären Grünsandstein der Orakei-Bay be Auckland. Reise der Österreichischen Fregatte "Novara" um die Erde in den Jahren 1857, 1858, 1859. Geologischer Teil 1:89–158

Tompsett S, Porter JS, Taylor PD (2009) Taxonomy of the fouling cheilostome bryozoans *Schizoporella unicornis* (Johnston) and *Schizoporella errata* (Waters). J Nat Hist 43: 2227–2243

Vávra N (1979) Bryozoa from the Miocene of Styria (Austria). In: Larwood GP, Abbott MB (eds) Advances in Bryozoology. Academic, London, pp 585–610

Waters AW (1877) Remarks on the Recent geology of Italy. Trans Manchester Geol Soc 14: 251–282

Wood SV (1844) Descriptive catalogue of the zoophytes from the Crag. Ann Mag Nat Hist 13:10–21

Zágorsek K, Hudácková N (2000) Ottnangian Bryozoa and Foraminifera from the Vienna Basin (Slovakia). Slovak Geol Mag 6(2–3):110–115

# Chapter 24
# *Schizomavella grandiporosa* and *Schizomavella sarniensis*: Two Cryptic Species

**Javier Souto, Oscar Reverter-Gil, and Eugenio Fernández-Pulpeiro**

**Abstract** The type specimen of *Schizomavella grandiporosa* Canu and Bassler, 1925, has been located in the MNHN, Paris, and reveals that this species is closely similar to *Schizomavella sarniensis* Hayward and Thorpe, 1995. All available material of both species has been studied, including SEM-photographies. Morphometric analysis has been performed using n-MDS and tested with ANOSIM. Results confirm the separation of the two species which are, however, morphologically very similar. The geographic distributions overlap with *S. grandiporosa* being more southern (northwest Spain–Morocco–Algeria) than *S. sarniensis* (English Channel–southern Portugal).

**Keywords** Bryozoa • Cheilostomata • Bitectiporidae • *Schizomavella* • Cryptic species

## Introduction

In 1925, Canu and Bassler described *Schizomavella grandiporosa* from a single colony collected at Fedhala (now Mohammedia), at the Atlantic coast of Morocco. The original description of this species is extremely vague and their figures are not very informative. This is perhaps the main reason why this species seems to have been unnoticed since its original description.

Seventy years later, Hayward and Thorpe (1995) described *Schizomavella sarniensis* from the Bay of Biscay, which is a well characterised species but not much reported since then.

J. Souto (✉) • O. Reverter-Gil • E. Fernández-Pulpeiro
Departamento de Zooloxía e Antropoloxía Física, Facultade de Bioloxía,
Universidade de Santiago de Compostela, 15782 Santiago de Compostela, Spain
e-mail: javier.souto@usc.es; oscar.reverter@usc.es; eugenio.fernandez.pulpeiro@usc.es

A. Ernst, P. Schäfer, and J. Scholz (eds.), *Bryozoan Studies 2010*,
Lecture Notes in Earth System Sciences 143, DOI 10.1007/978-3-642-16411-8_24,

The Laboratoire de Paléontologie of the Muséum National d'Histoire Naturelle, Paris, keeps numerous samples of the Canu Collection partly stored in a long row of files, with fossil and Recent specimens mixed up, and not very well labelled. However, in a lateral corridor there are two drawers where the specimens figured by Canu and Bassler in their monographs on the Bryozoans from Morocco (Canu and Bassler 1925, 1928) and Tunisia (Canu and Bassler 1930) are correctly classified, though not numbered. Among them we found the type specimen of *S. grandiporosa*.

At first sight under the stereomicroscope, this species is similar to *S. sarniensis*, originally described from the South of England and the North of Spain, and later reported from the English Channel, Ferrol (NW Spain), and Algeria (Reverter Gil and Fernández Pulpeiro 1995), from Vigo (NW Spain) (Soto García et al. 2002) and from the South of Portugal (Souto et al. 2010).

The close resemblance of the two species leads us to wonder if all the material might correspond to a single species. Therefore, all available material of both species was revised and compared using morphometric data and statistical techniques in order to test the differences between the colonies (e.g. Garraffoni and Camargo 2006, 2007; Carter et al. 2010).

## Material and Methods

Besides of the type specimen of *S. grandiporosa*, all material previously identified as *S. sarniensis* was also studied. These samples are currently stored in the Natural History Museum, London (NHMUK) and in the Muséum National d'Histoire Naturelle, Paris (MNHN) as well as in our own collection. The examined material came from 13 stations in the Atlantic-Mediterranean region. All samples were dry, and some colonies were bleached. Almost all colonies were examined using SEM photographs. Previous identifications of all studied colonies were ignored for the visual examination.

Seven characters were measured from digital SEM images using the ImageJ open source software (http://rsbweb.nih.gov/ij/). Measurements used are: primary orifice length and width, sinus depth and width, condyles length and width, avicularia width (Fig. 24.1). Length and width of autozooids were not considered, as they are extremely variable due to frontal budding.

### *Statistical analyses*

Depending on the size of the colony, between 8 and 20 measurements from each colony were done using ImageJ, Measurements for every character were averaged and a single value for each of them per colony was obtained. Each variable was then $\log_{10}$ transformed and standardised by dividing it by the sinus width (Gardner 2004; Carter et al. 2010).

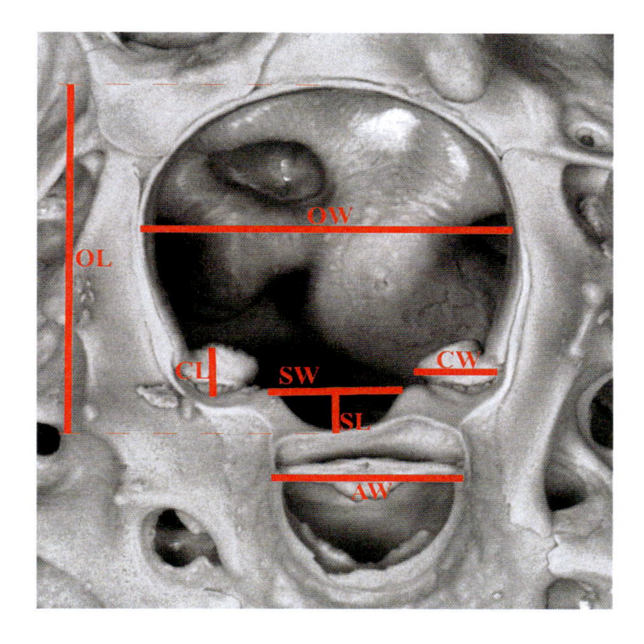


**Fig. 24.1** Orifice view with morphological variables annotated: *OL* orifice length, *OW* orifice width, *SL* sinus length, *SW* sinus width, *CL* condyle length, *CW* condyle width, *AW* avicularia width

Euclidian distances were used to create a similarity matrix. This similarity matrix was subsequently used with non-parametric multidimensional scaling (MDS) and an analysis of similarities (ANOSIM) to statistically test the null hypothesis that there are no significant morphometric differences between the colonies. Both MDS and ANOSIM analyses were carried out using PRIMER v.6 statistical package (Clarke and Gorley 2006).

## Results

*Schizomavella grandiporosa* and *S. sarniensis* are quite similar species, and it is difficult to differentiate them morphologically. Both species show similar colony development, with autozooids of similar shape and frontal shield. Ovicell shape and development are also similar, being initially smooth and prominent, and covered afterwards by a thick and rough secondary calcification.

However, a detailed study of all available material shows several fine differences, which as a whole allow us to define two clearly different groupings.

The group including the type material of *S. sarniensis* (Fig. 24.2, Table 24.1) shows a high inter- and intra-colonial variability. Indeed, the same colony has both primary orifices with a wide sinus, and others with a much narrower sinus (Fig. 24.3). The primary orifice is somewhat narrow proximally, with lateral walls converging proximally. The condyles have a denticulate distal edge. The suboral avicularium is proportionally larger, although this character shows a high variability. Finally, in some colonies the avicularium may be occasionally enlarged, occupying most of the frontal surface of the autozooid.

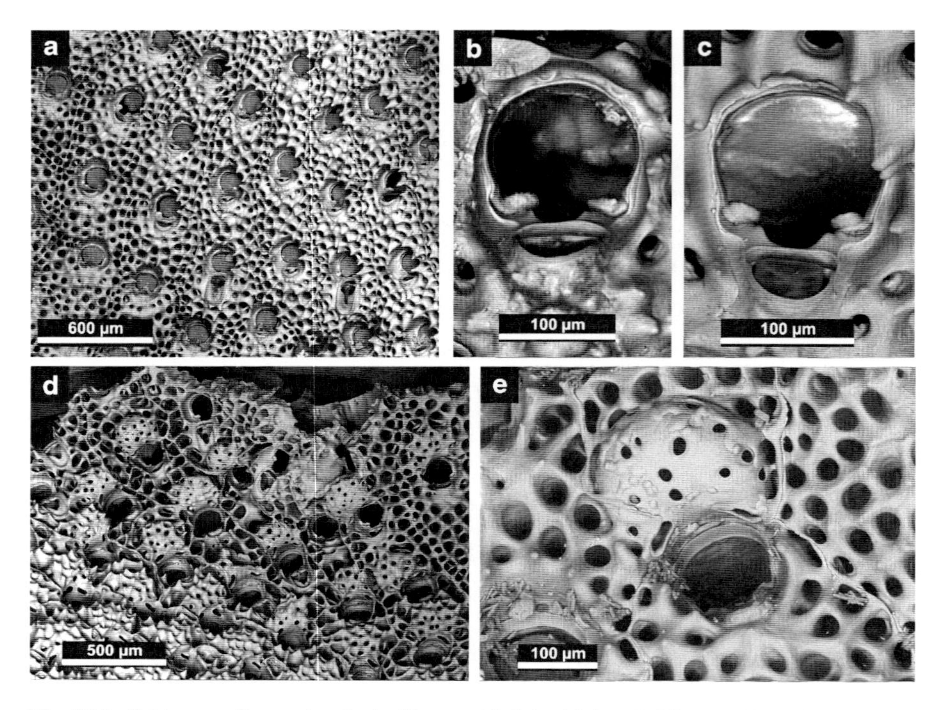


**Fig. 24.2** *Schizomavella sarniensis*. (**a**–**b**) type. (**c**) Saint Malo. (**d**) Vigo. (**e**) Roscoff

**Table 24.1** Measurements of *S. sarniensis*, type: *ZL* zooid length, *ZW* zooid width, *OL* orifice length, *OW* orifice width, *SL* sinus length, *SW* sinus width, *CL* condyle length, *CW* condyle width, *AW* avicularia width

| | ZL | ZW | OL | OW | SL | SW | CL | CW | AW |
|---|---|---|---|---|---|---|---|---|---|
| **Mean** | 0.491 | 0.362 | 0.133 | 0.143 | 0.017 | 0.055 | 0.018 | 0.035 | 0.078 |
| **SD** | 0.038 | 0.054 | 0.008 | 0.011 | 0.002 | 0.007 | 0.003 | 0.004 | 0.007 |
| **Max** | 0.568 | 0.463 | 0.146 | 0.158 | 0.019 | 0.065 | 0.022 | 0.042 | 0.087 |
| **Min** | 0.425 | 0.265 | 0.117 | 0.125 | 0.014 | 0.045 | 0.014 | 0.027 | 0.061 |
| **N** | 12 | 12 | 11 | 11 | 11 | 11 | 11 | 11 | 11 |

The group including the type material of *S. grandiporosa* (Fig. 24.4, Table 24.2) shows a limited variability of the sinus width, which is wider than in *S. sarniensis*. The suboral avicularia is narrower and never enlarged, whereas the primary orifice is parallel sided in its proximal part, and the condyles are smooth.

Both groups are distinctly shown in the MDS (Fig. 24.5), which has a very low stress value (minimum stress: 0.01), considered to be a sign of high resolution (Clarke and Warwick 1994). The MDS analysis clearly separates the colonies into two clusters confirmed by the ANOSIM permutation test, showing that these are significantly different ($p < 0.05$; global R: 0,798).

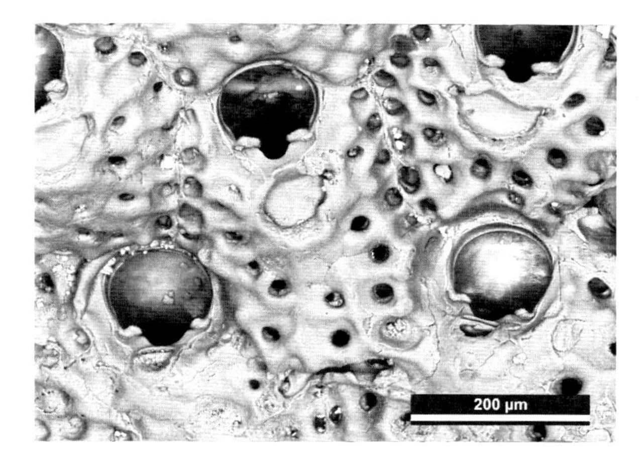


**Fig. 24.3** Intracolonial variability in *S. sarniensis* (Laredo Beach: NHM-1965.1.4.4)

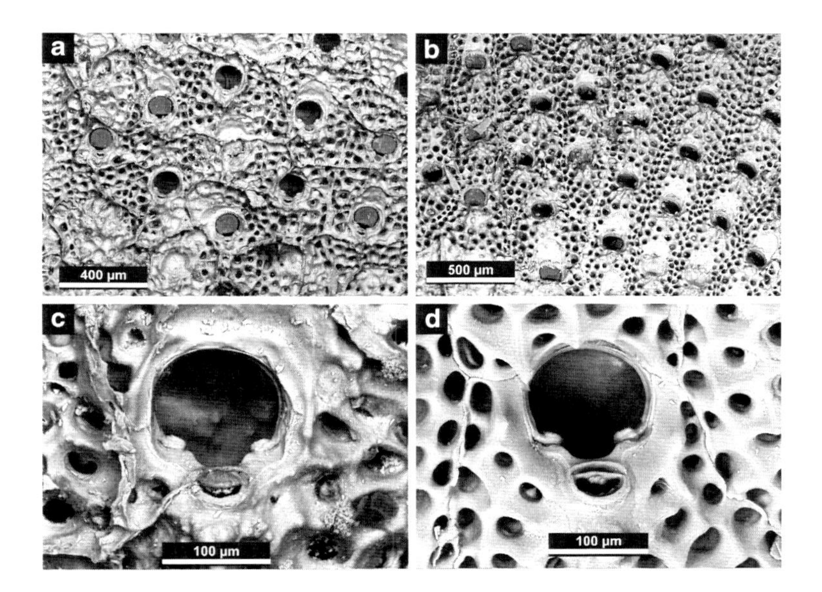


**Fig. 24.4** *Schizomavella grandiporosa*. (**a–c**) type. (**d**) sample from Ferrol

As a conclusion, we consider that *S. grandiporosa* and *S. sarniensis* are different species, though very closely related making it very difficult to identify a single colony under the stereomicroscope with certainty. A re-description of *S. grandiporosa* is included here as the original one is quite ambiguous.

*Schizomavella grandiporosa* Canu and Bassler, 1925

*Schizomavella grandiporosa* Canu and Bassler, 1925: 26, pl. 3, fig. 1.
*Schizomavella auriculata* (Hassall): Canu and Bassler, 1925: 24 (part: material from Fedhala).

**Table 24.2** Measurements of *S. grandiporosa*, type: *ZL* zooid length, *ZW* zooid width, *OL* orifice length, *OW* orifice width, *SL* sinus length, *SW* sinus width, *CL* condyle length, *CW* condyle width, *AW* avicularia width

| | ZL | ZW | OL | OW | SL | SW | CL | CW | AW |
|---|---|---|---|---|---|---|---|---|---|
| **Mean** | 0.420 | 0.372 | 0.113 | 0.117 | 0.014 | 0.050 | 0.013 | 0.024 | 0.048 |
| **SD** | 0.052 | 0.054 | 0.004 | 0.004 | 0.002 | 0.004 | 0.002 | 0.002 | 0.002 |
| **Max** | 0.577 | 0.500 | 0.120 | 0.123 | 0.016 | 0.056 | 0.015 | 0.027 | 0.053 |
| **Min** | 0.327 | 0.296 | 0.108 | 0.113 | 0.010 | 0.042 | 0.010 | 0.020 | 0.045 |
| **N** | 20 | 20 | 8 | 8 | 8 | 8 | 8 | 8 | 8 |

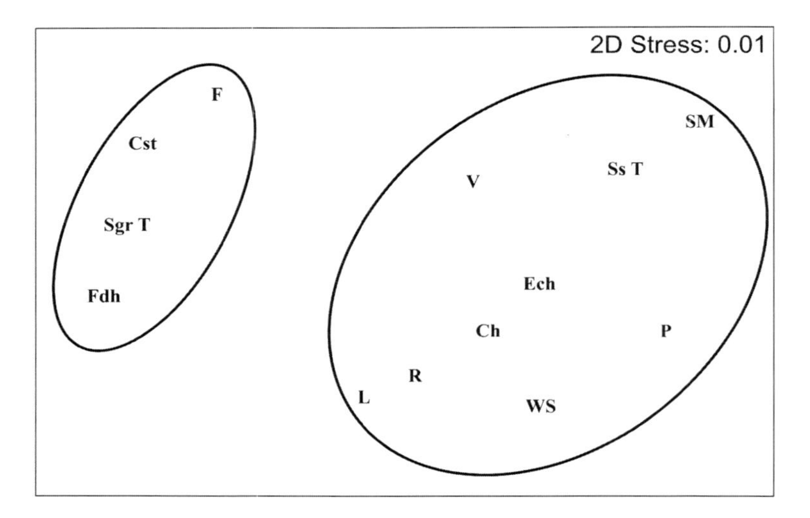


**Fig. 24.5** Non-parametric MDS ordination plots for morphometrics among localities. *Cst* Castiglione, *Ch* Cherbourg, *Ech* English Channel, *F* Ferrol, *Fdh* Fedhala, *L* Laredo beach, *P* South of Portugal, *R* Roscoff, *Sgr T* Type of *S. grandiporosa*, *SM* Saint Malo, *Ss T* Type of *S. sarniensis*, *V* Vigo, *WS* West of Spain

*Schizomavella sarniensis* Hayward and Thorpe: Reverter Gil and Fernández Pulpeiro 1995: 265, figs. 2, 3d-e (only material from Ferrol and Algeria); Reverter-Gil and Fernández-Pulpeiro 2001: 116 (only material from Ferrol).

### *Material Examined*

Holotype: *Schizomavella grandiporosa* Canu and Bassler, 1925: MNHN-Paléontologie (no registration number), Fedhala (Morocco).

Other material: *Schizomavella grandiporosa*: MNHN-4358: Fedhala. Dét. Canu as *Schizomavella auriculata* (Hassall). MNHN-17042: Gabés (Tunisia) or Castiglione (now Bou Ismail, Algeria), Coll. Dienzeide as *Schizoporella auriculata*? var.? MNHN-17101: Castiglione (now Bou Ismail), 30 m, Dét. Prenant as *Schizoporella auriculata* (Hassall). MNHN-17102: Castiglione (now Bou Ismail), 30 m, Dét. Prenant as *Schizoporella auriculata* (Hassall). Personal Collection: Ría de Ferrol.

*Schizomavella sarniensis*: NHMUK-1911.10.1.1543: Salcombe. Holotype and Paratype. NHMUK-1993.11.16.1, English Chanel, 49º29,25′N, 2º11.4′W, 44 m, 22/07/1961; NHM-1872.2.3.150, west coast of Spain; NHMUK-1965.1.4.4, Laredo Beach Santander; MNHN-1621: Rade de Cherbourg, 12-7-1881, 12 m. Coll. Jullien as *Lepralia auriculata* Hassall. MNHN-6583: Roscoff, Coll. d'Hondt as *Schizomavella linearis*. MNHN-14724: N de St. Malo, 12–14 m as *Schizomavella linearis*. Personal Collection: Armação de Pêra Bay (southern Portugal), 19–21 m, 11-12/XII/2007. Ría de Vigo.

*Description*

Colony unilaminar to multilaminar, developing extensive nodular sheets. Autozooids in linear series at growing edges of new laminae, losing orientation in subsequent layers. Autozooids quadrangular to polygonal, separated by distinct sutures. Frontal wall thickly calcified, perforated by numerous large rounded pores; coarsely granular, becoming prickly in later ontogeny; typically developing a short suboral umbo.

Primary orifice slightly longer than wide, surrounded by a peristomial rim that produces short pointed denticles. Proximal border with a shallow, U-shaped sinus occupying about one half of its width. A pair of smooth oval condyles, conspicuous, hardly reaching the sinus. Young autozooids with three distal spines.

Avicularium suboral, small, immediately proximal to the sinus, almost perpendicular to the plane of orifice, with a short semi-oval mandible. Apertural bar with a poorly developed denticle.

Ovicells initially hemispheric, smooth, but soon immersed by a secondary calcification and covered by a nodular imperforate sheet, except for the proximal semielliptical area.

## Geographic Distribution

*Schizomavella grandiporosa* is present in the Atlantic coast of Morocco, ranging through the North and NW of the Iberian Peninsula (Ría de Ferrol), and East to the Mediterranean, reaching Algeria. The distribution of *S. sarniensis* in the NE Atlantic ranges from the English Chanel to the Cantabrian coast of Spain, NW of the Iberian Peninsula (Ría de Vigo) and South of Portugal. Therefore, the geographic distributions of both species overlap in the western coast of the Iberian Peninsula (Fig. 24.6).

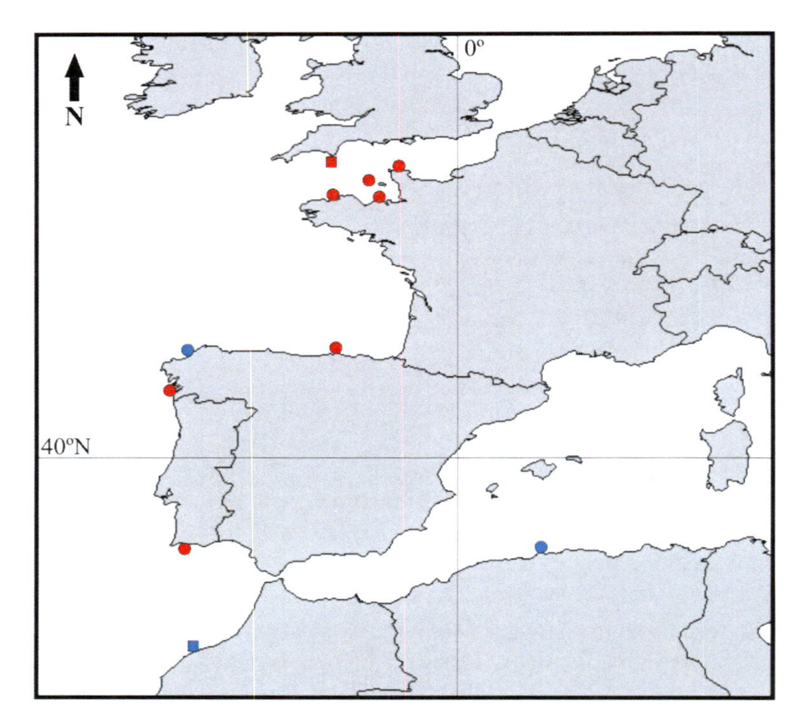


**Fig. 24.6** Distribution of the two species. In *blue S. grandiporosa*. In *red S. sarniensis*. *Squares* show the type localities

**Acknowledgements** We are grateful to P. Lozouet, J.-L. d'Hondt, J. P. Saint Martin and N. Améziane (MNHN) as well as M. Spencer Jones (NHMUK) for their valuable assistance during our visits and the loan of material. We also wish to thank M. Barreiro (Servizo de Microscopía Electrónica, Universidade de Santiago de Compostela) for the SEM photographs. Finally, to the two referees.

This work was supported by the project '*Fauna Ibérica: Briozoos I (Ctenostomados y Queilostomados Anascos)*' (CGL2006-04167), co-financed by the Ministerio de Ciencia e Innovación, Spanish Government and FEDER.

JS gratefully acknowledges a Synthesys grant (FR-TAF-3988) to visit the MNHN. The synthesys Project (http://www.synthesys.info/) is financed by European Community Research Infrastructure Action under the FP6 'Structuring the European Research Area' Programme.

## References

Canu F, Bassler RS (1925) Les Bryozoaires du Maroc et de Mauritanie (1er Mémoire). Mém Soc Sci Nat 10:1–79

Canu F, Bassler RS (1928) Les Bryozoaires du Maroc et de Mauritanie (2e Mémoire). Mém Soc Sci Nat 18:1–85

Canu F, Bassler RS (1930) Bryozoaires marins de Tunisie. Stat Océanogr Salammbô 5:1–91

Carter MC, Gordon DP, Gardner JPA (2010) Polymorphism and variation in modular animals: morphometric and density analyses of bryozoan avicularia. Mar Ecol Prog Ser 399:117–130

Clarke KR, Gorley RN (2006) PRIMER v6: user manual/tutorial. PRIMER-E, Plymouth

Clarke KR, Warwick RM (1994) Changes in marine communities: an approach to statistical analysis and interpretation. Plymouth Marine Laboratory, Plymouth, UK

Gardner JPA (2004) A historical perspective of the genus *Mytilus* (Bivalvia: Mollusca) in New Zealand: multivariate morphometric analyses of fossil, midden and contemporary blue mussels. Biol J Linn Soc 82:329–344

Garraffoni ARS, Camargo MD (2006) First application of morphometrics in a study of variations in uncinial shape present within the Terebellidae (Polychaeta). Zool Stud 45(1):75–80

Garraffoni ARS, Camargo MD (2007) A new application of morphometric in a study of variation in uncinial shape present within the Terebellidae (Polychaeta): a re-evaluation from digital images. Cah Biol Mar 48:229–240

Hayward PJ, Thorpe JP (1995) Some British species of *Schizomavella* (Bryozoa: Cheilostomatida). J Zool London 235:661–676

Reverter Gil O, Fernández Pulpeiro E (1995) Some species of *Schizomavella* (Bryozoa, Cheilostomatida) from the Atlanto-Mediterranean region. Cah Biol Mar 36:259–275

Reverter-Gil O, Fernández-Pulpeiro E (2001) Inventario y cartografía de los Briozoos marinos de Galicia (N.O. de España). Nova Acta Científica Compostelana, Santiago de Compostela

Soto García E, Fernández Pulpeiro E, Ramil Blanco F (2002) briozoos infralitorales de la Ría de Vigo (España). Bol R Soc Esp Hist Nat (Sec Biol) 97 (1–4):85–96

Souto J, Reverter-Gil O, Fernández-Pulpeiro E (2010) Gymnolaemate bryozoans from the Algarve (S. Portugal). New species and biogeographical considerations. J Mar Biol Assoc UK 90 (7):1417–1439

# Chapter 25
# A Diverse Bryozoan Fauna from Pleistocene Marine Gravels at Kuromatsunai, Hokkaido, Japan

## Bryozoan Diversity in a Japanese Pleistocene Gravel

**Paul D. Taylor, Matthew H. Dick, Diana Clements, and Shunsuke F. Mawatari**

**Abstract** A remarkable locality, nicknamed 'Kokemushi Paradise', at the town of Kuromatsunai in Hokkaido, northern Japan, exposes 6.5 m of the Nakasato Conglomerate Member of the Pleistocene Setana Formation, dated at about 1.2–1.0 Ma. The sediment consists of marine gravels containing abundant shells, pebbles and cobbles, most densely encrusted by bryozoans that are typically well preserved. Broken branches of erect bryozoans, especially *Myriapora*, are common in the matrix. Several bryozoan genera are recorded for the first time in the fossil record, notably the cold-water genera *Septentriopora, Harmeria, Doryporella* and *Rodinopora*. Up to 25 bryozoan species can be found on a single clast, and 120 bryozoan species have been provisionally identified, dominated by ascophoran cheilostomes. Although this species richness is high compared to other fossil bryozoan assemblages from marine gravels, viewed in the context of bryozoan diversities in modern rocky subtidal habitats in the northern Pacific region it may be unexceptional. Lack of calcareous algae and presence of stylasterid hydrocorals suggest a subtidal environment, possibly in a fast-flowing channel between the Sea of Japan and the Pacific Ocean.

**Keywords** Bryozoans • Pleistocene • Japan • Species richness • Palaeoecology • Palaeoenvironment

P.D. Taylor (✉) • D. Clements
Department of Earth Sciences, Natural History Museum, Cromwell Road, London SW7 5BD, UK
e-mail: p.taylor@nhm.ac.uk; d.clements@nhm.ac.uk

M.H. Dick • S.F. Mawatari
Natural History Sciences, Hokkaido University, North 10 West 8, Kita-Ku, Sapporo 060-0810, Japan
e-mail: mhdick@mail.sci.hokudai.ac.jp; shunfm@sci.hokudai.ac.jp

A. Ernst, P. Schäfer, and J. Scholz (eds.), *Bryozoan Studies 2010*,
Lecture Notes in Earth System Sciences 143, DOI 10.1007/978-3-642-16411-8_25,

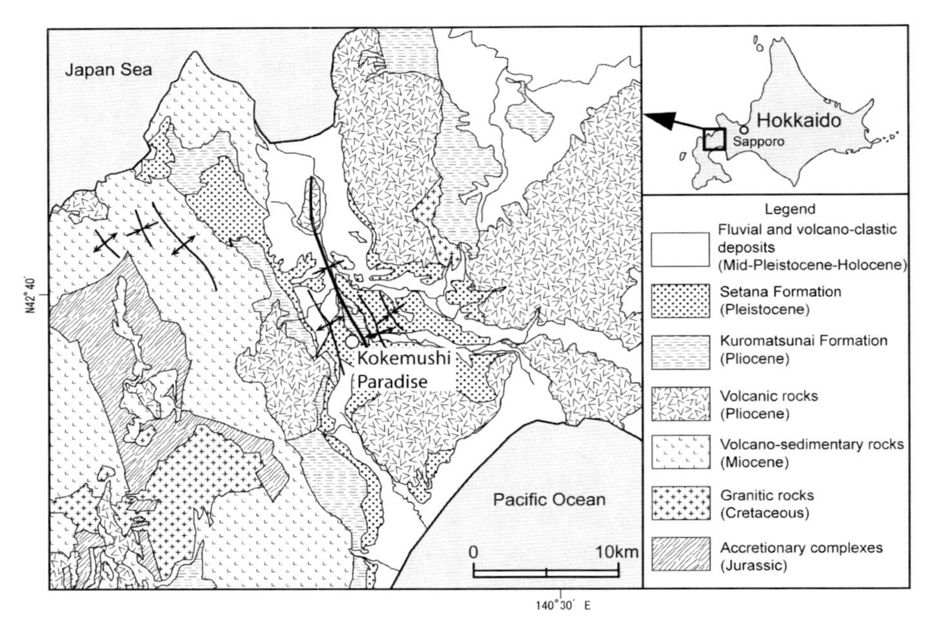


**Fig. 25.1** Map showing the location of Kokemushi Paradise (From Takashima et al. 2008)

## Introduction

In August 2005, an outstanding site for fossil bryozoans was discovered at the town of Kuromatsunai on Hokkaido, the northern island of Japan (Fig. 25.1). A small quarry in a forest exposes marine gravels belonging to the Pleistocene Setana Formation. It is immediately obvious from the white patches of encrusting bryozoans on the surfaces of the stones used to construct the trackway leading into the quarry that this is an exceptional bryozoan locality. Indeed, the great majority of the pebble- and cobble-sized clasts constituting the gravel are encrusted by bryozoan colonies, most excellently preserved in view of the coarseness of the sediment. We are unaware of any other marine gravel deposit of any age in the geological record containing such a rich bryozoan fauna. Additional fragments of erect bryozoans occur abundantly in the matrix between the pebbles and cobbles (Fig. 25.2a). The locality was dubbed 'Kokemushi Paradise', 'kokemushi' being the Japanese word for bryozoans (Dick et al. 2008). Aside from the bryozoans, there are abundant stylasterid hydrocorals and bivalves, with occasional barnacles, serpulid worms (including spirorbids) and encrusting foraminifera. Like the lithoclasts, the stylasterids and bivalves (Fig. 25.2b) are also typically heavily encrusted with bryozoans. Unfortunately, quarrying operations have ceased and the main face of the quarry has been graded. This occurred before detailed studies could be made of the distribution of bryozoans through the 6.5-m section and most of the clasts collected are not stratigraphically localized, nor are their original orientations known. Nevertheless, over 700 encrusted lithoclasts are now available for study

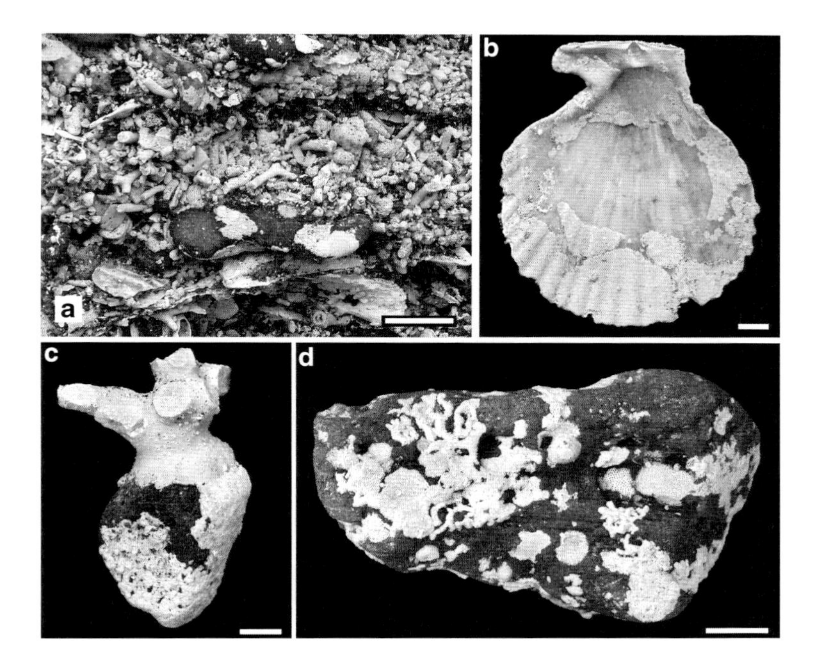


**Fig. 25.2** Matrix and clasts from the Nakasoto Conglomerate Member at Kokemushi Paradise. (**a**) field photograph showing encrusted lithoclasts in a matrix comprising branches of mostly *Myriapora subgracilis* (d'Orbigny, 1852). (**b**) bryozoans encrusting the interior surface of a pectinid valve, NHM BZ 5478. (**c**) stylasterid attached to a lithoclast also encrusted by bryozoans, NHM BZ 5474. (**d**) diverse bryozoans encrusting a lithoclast, NHM BZ 5476. *Scale bars*: a = 2 cm; b–d = 10 mm

in the collections of the Natural History Museum, London (NHM), with additional specimens at the University of Hokkaido in Sapporo. Many of these have been digitally photographed and/or examined using a LEO 1455VP low vacuum scanning electron microscope at the NHM.

Here we introduce the geological setting and palaeoenvironment at Kokemushi Paradise, and comment on the preservation and diversity of the bryozoan fauna, highlighting the first fossil examples of some genera previously known only from the Recent.

## Geological Setting

Kokemushi Paradise, otherwise known as the Utasai Section, is a small quarry in a forest close to the Buna Center, Kuromatsunai (N46°38′56″, E140°18′36″). About 6.5 m of the upper part of the Nakasato Conglomerate Member of the Setana Formation are exposed here (Takashima et al. 2008). The sediment comprises weakly cemented, coarse bioclastic sandstone containing beds of clast- or matrix-supported conglomerate (Fig. 25.2a). There is an absence of fine-grained mud. Clasts are of pebble to cobble grade, well rounded, and mostly consist of various types of volcanic rocks. An aragonitic fauna is generally lacking (but see below), presumably due to leaching from groundwater passing through the porous sediment. The dominant macrofossils apart from bryozoans are bivalves, including *Mizuhopecten yesoensis*, *Swiftopecten swiftii*, *Chlamys islandicus* and *Acirsa ochotensis*, all previously recorded from the Setana Formation (Suzuki 1989).

The age of the Nakasato Conglomerate Member, which forms the basal part of the Setana Formation, is estimated to be 1.2–1.0 Ma (Takashima et al. 2008). The member is located at the overlap between the ranges of the planktonic foraminifers *Neogloboquadrina pachyderma* and *N. incompta*, and within the CN13b calcareous nannofossil zone.

## Palaeoenvironment

Evidence from the bivalves has been taken to indicate a shoreface depositional setting for the Nakasato Conglomerate Member (Suzuki 1991). However, the presence of stylasterid hydrocorals (Fig. 25.2c) and absence of calcareous algae both favour a more offshore setting. Lindner et al. (2008) stated that 89% of modern stylasterid species are deep water (>50 m), while the lack of calcareous algal fossils at Kuromatsunai suggests deposition beneath the photic zone. In contrast, the slightly younger Soebetsu Sandstone Member of the Setana Formation exposed at the type locality in the Soebetsu River (see Takashima et al. 2008) close to Kokemushi Paradise contains abundant encrusting and branching calcareous algae, implying shallower deposition within the photic zone.

This alternative, deeper water model for deposition of the Nakasato Conglomerate Member at Kokemushi Paradise envisages a channel between the Sea of Japan to the northwest and the Pacific Ocean to the southeast. Kuromatsunai is at a relatively low altitude at present, and palaeogeographical maps for the Pleistocene (Kano et al. 1991) show that it was the location of a narrow strait at the time.

Cold-water conditions dominated during deposition of the Nakasoto Conglomerate Member (Suzuki 1989; Nojo et al. 1999). This is indicated by the molluscan fauna, including *Yabepecten tokunagai*, *Chlamys islandica* and *Acirsa ochotensis*, that characterises the Subarctic Province and is indicative of sea surface temperatures of 12–20°C in the summer and 0–2°C in the winter, according to Suzuki and Akamatsu (1994). The bryozoan fauna also indicates cold conditions, as it contains several genera typical now of Arctic or northern Boreal latitudes (see below). Today, the seas around northern Japan are mainly influenced by two currents, the warm-water Tsushima Current, which flows northwards through the Sea of Japan, and the cold-water Oyashio Current flowing southwards in the Pacific Ocean (Kitamura and Kimoto 2006). Given palaeontological evidence suggesting a cold climate, the Oyashio Current may have prevailed during the interglacial highstand when the Nakasoto Conglomerate Member was deposited.

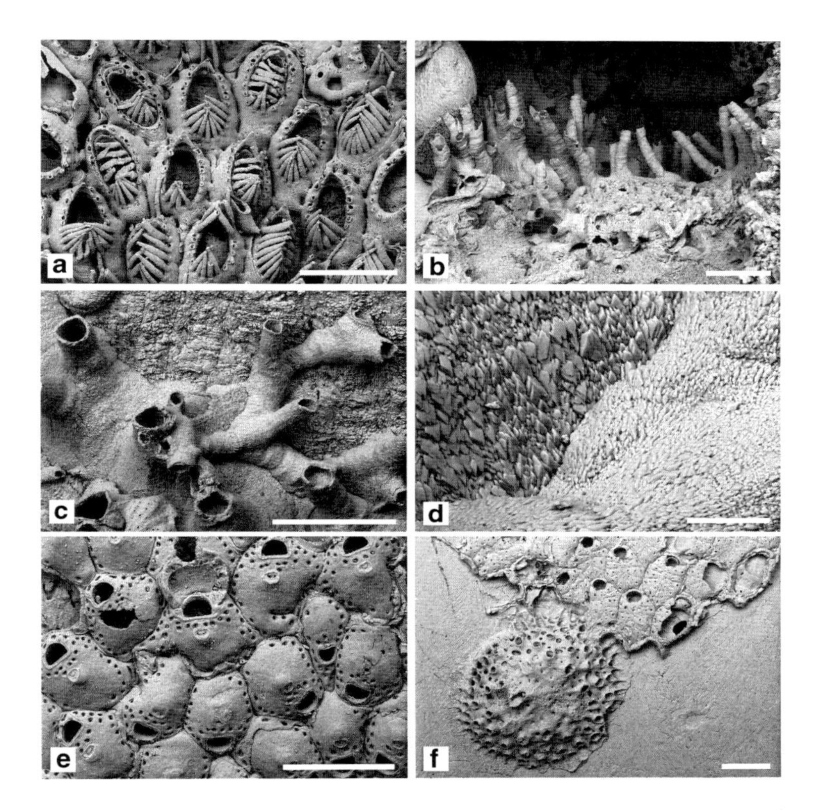


**Fig. 25.3** Preservation of bryozoans from the Nakasoto Conglomerate Member at Kokemushi Paradise. (**a**) exceptionally preserved colony of *Cauloramphus disjunctus* Canu and Bassler, 1929, with intact articulated spines and pedunculate avicularia, NHM BZ 5823. (**b**) profile of a cyclostome colony in a recess showing preservation of long peristomes, NHM BZ 5824. (**c, d**) colony of '*Stomatopora*' (*right*) overgrowing *Oncousoecia* (*left*), with an enlarged view of the central region to show the difference in the size of the epitaxial cement crystals on the frontal walls; the crystals are small in the former (*lower right*) and large in the latter (*upper left*); NHM BZ 5536. (**e**) reparative growth in a colony of *Fenestrulina*, NHM BZ 5825. (**f**) colony of *Disporella* (*lower left*) recruited onto a patch of substrate formerly covered by a broken colony of *Hayamiellina constans* Grischenko and Gordon, 2004 (*top*), NHM BZ 5569. *Scale bars*: a, c, e, f = 500 μm; b = 1 mm; d = 50 μm

Notwithstanding the generally cold climate, MART analysis from two bryozoan species (*Porella concinna* (Busk, 1854) = *P. hanaishiensis* Hayami, 1975, and *Escharoides hataii* Hayami, 1975) indicates a highly seasonal climatic regime, with a calculated temperature range averaging 11.7°C (Dick et al. 2008). This may point to alternate flushing of the inferred depositional channel by the cold-water Oyashio Current in the winter and the warm-water Tsushima Current in the summer.

Although the erect bryozoans in the matrix of the conglomerate are broken and often abraded, the encrusters on the clasts generally have very well preserved surfaces (Fig. 25.2d), suggesting colonization of the clasts in situ rather than colonization elsewhere followed by transportation to the site of deposition at Kokemushi Paradise. While some of the preservation of delicate structures such as cheilostome spines (Fig. 25.3a) and cyclostome peristomes (Fig. 25.3b) can be explained by colonies having been protected as a result of growing in cavities on clast surfaces, the excellent preservation on more exposed surfaces cannot be easily explained using a model of pre-transportational encrustation, as collisions between the heavy clasts would have caused substantial damage to the bryozoan colonies. Therefore, the pebbles and cobbles were likely colonized at or very close to their sites of eventual burial. Clasts are commonly encrusted on all sides, and occasionally smaller clasts are totally enveloped by bryozoans. Although this may signify periodic overturning, it is equally possible that there was sufficient space for water to circulate beneath the clasts and allow colonization of their undersides (cf. Spjeldnaes 2000) in the coarse sediment of the Nakasoto Conglomerate Member. Such cryptic encrustation is certainly conceivable for the thin encrusting bryozoans if not the large, erect stylasterids. The latter evidently grew concurrently with the bryozoans (examples can be found of bryozoans overgrowing stylasterids and vice-versa) and must have occupied the upper surfaces of clasts with free space above into which the branching colonies could grow (Fig. 25.2c). The occurrence of these stylasterids, which were probably slow growing, points to instances of clast stability.

Episodic influxes of new clasts would have buried the cobbles and boulders previously carpeting the sea floor, causing widespread mortality of the sclerobiotic communities but providing fresh surfaces for colonization. Given the tectonic instability of the region, which today has several active volcanoes, submarine landslides triggered by earthquakes provide a viable mechanism for transporting clasts from a shallow source area into the hypothesized channel at Kuromatsunai. For example, Neogene marine gravel deposits interpreted to have been formed by tsunamis have been described from elsewhere in Japan (Soh 1989; Yamazaki et al. 1989; Shiki and Yamazaki 1996) and, although these make no mention of encrusting communities on the clasts, it is conceivable that tsunamis were responsible for transportation of the lithoclasts at Kuromatsunai.

## Bryozoan Preservation

As already mentioned, bryozoan preservation at Kokemushi Paradise is generally excellent. Nowhere is this better seen than in recesses in the clasts where it is common to find preserved very long peristomes of cyclostomes (Fig. 25.3b) as well as in situ, basally articulated spines and stalked avicularia of the cheilostome *Cauloramphus disjunctus* Canu and Bassler, 1929 (Fig. 25.3a). Growth of carbonate cement and pressure solution from contacting grains may locally obscure or destroy some of the surface detail. Cement growth on bryozoan surfaces is epitaxial and probably syntaxial, as different bryozoan species can show different cement fabrics (Fig. 25.3c, d). Colonies sometimes flake off clast surfaces during collection or cleaning (Fig. 25.2b). While this has the advantage of revealing the undersides of colonies and any overgrown organisms, it compromises studies of hard substrate palaeoecology by destroying parts of the original sclerobiotic community.

Evidence is occasionally found that bryozoan colonies were locally or more extensively damaged on the seabed. For example, reparative growth from broken proximal edges is sometimes observed in encrusting colonies (Fig. 25.3e), as is intramural budding following damage or death of individual zooids. Removal of parts of colonies from the substrate during the development of the sclerobiotic community can be demonstrated when other colonies foul the newly available patches of substrate (Fig. 25.3f).

The absence of aragonite-shelled molluscs at Kokemushi Paradise is indicative of diagenetic solution of the aragonitic biota, cf. the Soebetsu River section where the aragonitic biota of the Setana Formation is preserved in a siltstone. Dissolved molluscan aragonite undoubtedly contributed to the growth of cement on the surfaces of colonies at Kokemushi Paradise. A significant proportion of fossil and Recent cheilostome bryozoans are known to utilize aragonite for their skeletons, either alone or in combination with calcite to form a bimineralic skeleton (Taylor et al. 2009). While completely aragonitic bryozoans are lacking at high latitudes (e. g. Kuklinski and Taylor 2009), bimineralic species with basal calcitic skeletons overgrown by aragonite do occur in low to moderate frequencies. Despite the typical loss of aragonite at Kokemushi Paradise, a few of the ascophoran species do preserve remnants of aragonite. This aragonite is visible under an optical microscope as patches of white skeleton overlying the surface of darker, more vitreous skeleton. Laser Raman spectroscopy (see Taylor et al. 2008) of a colony (Fig. 25.4a, b) of *Stomachetosella* cf. *collaris* (Kluge, 1946) preserved in this way proved that the white surficial material was aragonite and the darker underlying material calcite.

The effect of routine dissolution of aragonite from the skeletons of bimineralic species at Kokemushi Paradise deserves comment. Aragonite loss causes thinning of the frontal shield in ascophorans, leaving boundary walls, orificial rims and adventitious avicularia standing at a higher elevation than in colonies with complete preservation of the skeleton. Ornament may also be lost from the surface of the frontal shield, as in a species of *Celleporaria* with a finely reticulate patterning on pristine frontal shields that disappears after aragonite dissolution (Fig. 25.4c–f). Caution must be exercised when attempting to identify bimineralic species exhibiting such diagenetic artifacts resulting from aragonite dissolution.

## Bryozoan Diversity

Bryozoan species richness at Kokemushi Paradise totals 120 species according to our current (July 2010) estimate (Appendix A). Of the taxa recognised, 29 have been identified to species level, 37 are close to known species and are categorized as 'cf.', 'aff.' or '?', and 54 have yet to be identified as or unequivocally allied to

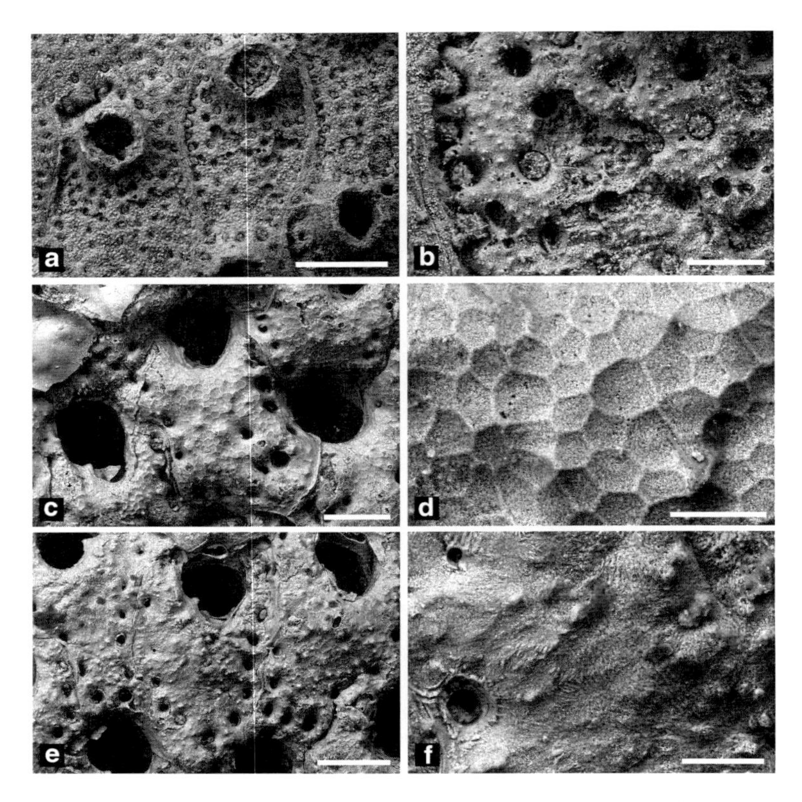


**Fig. 25.4** Effects of partial dissolution of the frontal shield in ascophorans from the Nakasoto Conglomerate Member at Kokemushi Paradise. (**a, b**) incomplete loss of the pustulose aragonitic surface layer in *Stomachetosella* cf. *collaris* (Kluge, 1946), NHM BZ 5600. (**c–f**) zooids with (**c**, **d**) and without (**e**, **f**) a frontal layer (?aragonitic) with reticulate ornamentation in *Celleporaria* sp., NHM BZ 5462. *Scale bars*: a = 500 μm; b = 100 μm; c, e = 200 μm; d, f = 50 μm

particular species. Uncertainties in identification are not due to the quality of the fossil bryozoans but instead largely reflect deficiencies in the literature. Very few descriptions of Recent and fossil bryozoans from the Japan region are accompanied by scanning electron micrographs (Dick et al. 2008), a prerequisite for precise identification. Fossil species described by, for example, Sakakura (1935), Kataoka (1961) and Hayami (1975), are difficult to evaluate, as are some of the species described by Ortmann (1890) in his classic monograph of Recent bryozoans from Japan. Similar problems afflict the use of the pre-SEM Russian references on the taxonomy of northern Pacific bryozoans. Cyclostome identification is particularly problematical owing to the invariable lack of information about gonozooid and pseudopore morphology (cf. Taylor and Zatoń 2008) in the literature.

Encrusting species dominate at Kokemushi Paradise, accounting for 111 (93%) of the species found. Erect cheilostomes number only three species (*Microporina* sp., *Myriapora subgracilis* (d'Orbigny, 1852) and *Reteporella*), whereas a quarter of the cyclostome taxa have erect colonies.

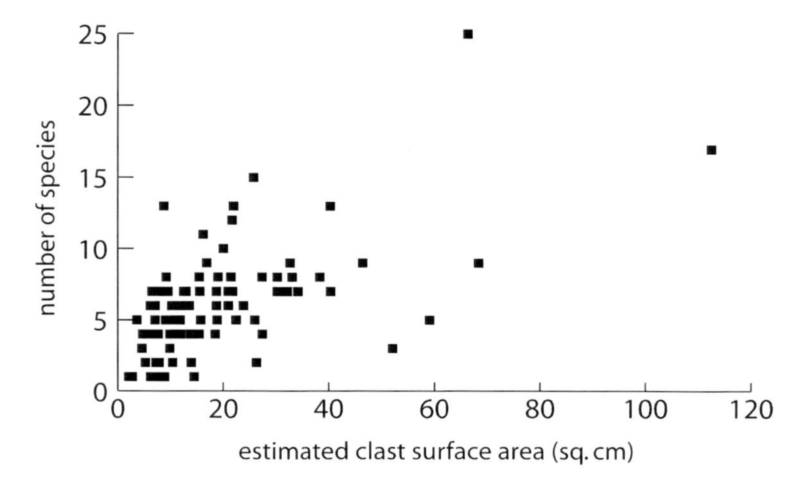


**Fig. 25.5** Correlation of estimated clast surface area and number of encrusting bryozoan species in a sample of 85 lithoclasts from the Nakasato Conglomerate Member at Kukemushi Paradise

The great majority of the encrusting species possess sheet-like colonies, although there are a few uniserial runners belonging to Cyclostomata ('*Stomatopora*') and Cheilostomata (*Aetea*, *Hippothoa*), as well as ribbon-like encrusting cyclostomes (*Platonea*, *Oncousoecia*), the dome-like cheilostome *Celleporina*, and bases of the erect cyclostome *Crisia*. In terms of species diversity of higher taxa, the bryozoan fauna is dominated by ascophorans, which constitute 77 of the species (64%), followed by cyclostomes with 24 species (20%) and anascans with 19 species (16%). Some genera are represented by multiple species at Kokemushi Paradise, notably *Porella* (8 spp.) and *Microporella* (6 spp.), but most have only one or two species.

No quantitative assessment has been made of the abundance of species in terms of number of colonies or areal coverage. However, visual impressions suggest that ascophorans also dominate in these two metrics. Some ascophoran species, notably species of *Celleporaria*, *Smittina*, *Parasmittina*, *Myriozoella* and *Stomachetosella*, may occur as large colonies that cover the entire surface of smaller clasts. Individual clasts are more often encrusted by multiple species, and this is always the case for the larger, cobble-sized clasts. A survey of 85 catalogued lithoclasts showed the maximum number of species on an individual clast to be 25, although most of the surveyed clasts were encrusted by eight or fewer species (Fig. 25.5). The anticipated correlation between clast size and number of encrusting bryozoan species is present but is less strong than expected.

Overgrowths between colonies of the same and different species are frequently observed. However, multilayering of bryozoans is uncommon.

## First Generic Fossil Records

Four bryozoan genera previously unknown as fossils have been found at Kokemushi Paradise: *Septentriopora*, *Doryporella*, *Harmeria* and *Rodinopora*.

## Septentriopora *Kuklinski and Taylor, 2006a*

This genus was introduced for calloporids lacking pore chambers, and having a reduced ovicell and a pair of small interzooidal avicularia distolateral of the opesia that face proximally or proximolaterally. The type species is *Tegella nigrans* Hincks, 1882. Another species of *Septentriopora*, *S. karasi* Kuklinski and Taylor, 2006a, occurs at Kokemushi Paradise (Fig. 25.6a), although it is very rare. This species was originally described from the Recent of Spitsbergen, Franz Josef Land, Arctic Canada, Alaska and the Kara Sea, i.e. exclusively cold-water localities. Indeed, the other two species of *Septentripora* are also Arctic or Boreal species.

## Doryporella *Norman, 1903*

*Doryporella* is an unusual anascan with an extensive, yet thin, cryptocyst ornamented by polygonal reticulations. Four species have been recognized, all Arctic or Boreal in distribution (see Grischenko et al. 2000, 2004). A few fragmentary colonies of *Doryporella* cf. *alcicornis* (O'Donoghue and O'Donoghue, 1923) have been found at Kokemushi Paradise (Fig. 25.6b), differing from Recent *D. alcicornis* in having opesiae with a straighter proximal edge. Today, *D. alcicornis* is a Pacific, high-Boreal species, recorded from British Columbia northwards in the eastern Pacific and the Commander Islands and Sea of Okhotsk in the western Pacific, at depths ranging from 11 to 152 m.

## Harmeria *Norman, 1903*

Assigned to the ascophoran family Cryptosulidae, this genus is represented only by *H. scutulata* (Busk, 1855), a distinctive species with an inner zone of large autozooids surrounded by an outer ring of small brooding zooids (Kuklinski and Taylor 2006b). The specimens of *H. scutulata* from Kuromatsunai (Fig. 25.6c) are very fragile but sufficiently well preserved to show the diagnostic zooidal dimorphism. At present, *H. scutulata* has been recorded from depths of up to 37 m and has a circumpolar Arctic distribution (Kuklinski and Taylor 2006b), extending southward to south-central Alaska in the eastern Pacific (Dick and Ross 1988), which provides further evidence for cold-water deposition of the Nakasato Conglomerate Member.

## Rodinopora *Taylor and Grischenko, 1999*

The type species of this monospecific cyclostome genus is *R. magnifica* Taylor and Grischenko, 1999, originally described from the Sea of Okhotsk at 75–88 m depth.

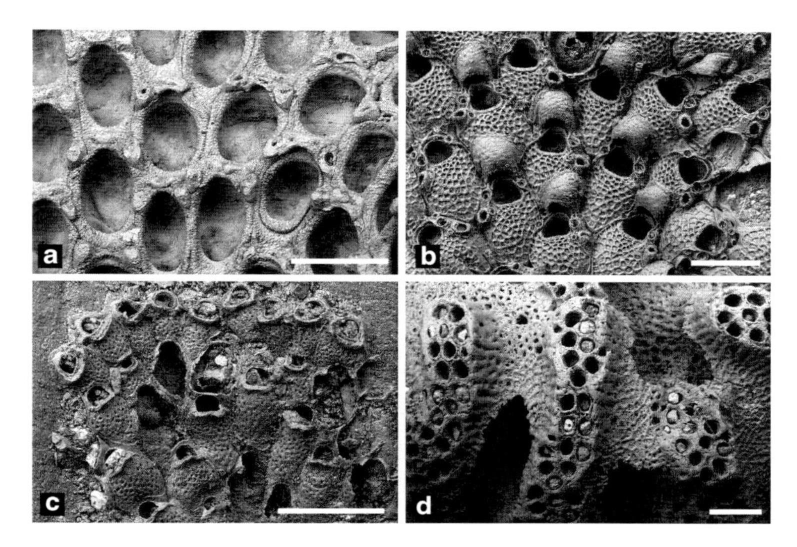


**Fig. 25.6** Some genera recorded as fossils for the first time in the Nakasato Conglomerate Member at Kuromatsunai; all are from Kokemushi Paradise, with the exception of d, which comes from the locality at Neppu. (**a**), *Septentriopora karasii* Kuklinski and Taylor, 2006, NHM BZ 5582. (**b**), *Doryporella* cf. *alcicornis* (O'Donoghue and O'Donoghue, 1923), NHM BZ 5826. (**c**), *Harmeria scutulata* (Busk, 1855); note the small brooding zooids at the *top* and larger autozooids *below*, NHM BZ 5599(1). (**d**), *Rodinopora magnifica* Taylor and Grischenko, 1999, autozooidal fascicles with broken interior-walled roof of gonozooid stretching between them, NHM BZ 5598. *Scale bars*: a, c, d = 500 μm; b = 200 μm

The fungiform colonies have autozooids opening in radial, multiserial rows on the top, separated by free-walled kenozooids. Lack of gonozooids in the original material placed a question mark over the subordinal affiliation of *Rodinopora*. Material of *R. magnifica* from Kuromatsunai (Figs 25.6d and 25.7) shows the gonozooid to be roofed by interior walled calcification (Fig. 25.6d), implying affiliation with the suborder Rectangulata. The species is more common at a second locality in Kuromatsunai, an abandoned quarry at 34 Neppu (also Lower Setana Formation), than it is at Kokemushi Paradise.

## Discussion

Kokemushi Paradise contains a uniquely diverse assemblage of fossil bryozoans associated with an ancient marine gravel. The Nakasato Conglomerate Member at Kokemushi Paradise is interpreted to have accumulated in a current-swept submarine channel between the Sea of Japan and the Pacific in a region where periods of tectonic instability transported pebble- and cobble-sized clasts from nearshore into a deeper setting where they became colonized by bryozoans, stylasterid corals and occasional serpulids, barnacles, and foraminifera.

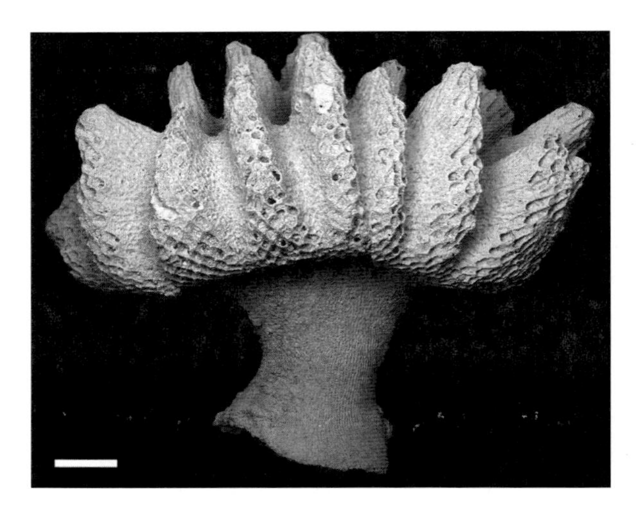

**Fig. 25.7** Profile view of a fungiform colony of *Rodinopora magnifica* Taylor and Grischenko, 1999, from the Nakasato Conglomerate Member at Kokemushi Paradise, NHM BZ 5585. *Scale bar*: 1 mm

The bryozoan-dominated encrusting fauna at Kokemushi Paradise offers unprecedented opportunities for studying the dynamics of an ancient hard-substrate community, including overgrowth competition for substrate space between encrusters, patterns of fouling, short-term ecological succession on clasts, microhabitat differentiation, and the influence of clast type, size and shape on the encrusters present. However, in addition to the loss of soft-bodied and fully aragonitic components of the sclerobiotic community, flaking of encrusters from clast surfaces is an issue when analyzing hard substrate palaeoecology at Kokemushi Paradise.

At well over 100 species, the diversity of encrusting bryozoans at Kokemushi Paradise appears to be exceptionally high for a non-tropical locality. However, this must be viewed in the context of high bryozoan diversity throughout the northern Pacific. Dick and Ross (1988), for example, found 57 cheilostome species and roughly 74 bryozoan species overall (including cyclostomes and a few ctenostomes) in rocky intertidal habitats of layered cobble, shells, and slabs at Kodiak, Alaska; subtidal diversity in the nearshore shelf zone around Kodiak is undoubtedly much higher (see Grischenko et al. 2007: 1151–1152). The diversity of bryozoans encrusting pebbles and bivalve shells is also high in the western Aleutian Islands, Alaska, at depths of 50–350 m, where at least 120 mostly cheilostome species have been detected in samples from only two trawl hauls (MD, unpublished data).

Palaeoenvironmental information potentially of value in understanding climatic and oceanographical conditions in northern Japan approximately one million years ago can be obtained from the bryozoans at Kokemushi Paradise. The bryozoan fauna overwhelmingly comprises species of cold-water aspect, matching the molluscs, and includes some characteristically Arctic species recorded as fossils for the first time, for example, *Harmeria scutulata* and *Septentriopora karasii*. A seasonal climatic regime with a mean annual range in temperature of about 12°C has been calculated previously from variance in zooid size (Dick et al. 2008).

**Acknowledgements** This work was supported in part by the COE (Centers of Excellence) Program 'Neo-Science of Natural History' at Hokkaido University, funded by MEXT (Ministry of Education, Culture, Sports, Science and Technology, Japan), and by grants from the JSPS (Japan Society for the Promotion of Science) and by Natural History Museum Special Funds to PDT and DC. We thank Piotr Kuklinski, Andrei Grischenko, Andrew Ostrovsky, Reishi Takashima, Masato Hirose, Luella Taranto and James Taylor for collecting assistance in the field; Dennis Gordon for taxonomic advice; and Bill Schopf and Anatoliy Kudryavtsev for laser Raman spectroscopy at UCLA (University of California, Los Angeles). Patrick Clements and Anna Taylor helped with estimating clast surface areas, and Phil Crabb is thanked for specimen photography.

# Appendix A: Bryozoan Species Recorded from the Nakasato Conglomerate Member, Setana Formation, at the Kokemushi Paradise Locality in Kuromatsunai, Hokkaido, Japan

| | |
|---|---|
| Cheilostomata: Anasca | *Aetea* aff. *truncata* (Landsborough, 1852) |
| | cf. *Ammatophora nodulosa* (Hincks, 1880) |
| | *Amphiblestrum* cf. *canui* Sakakura, 1935 |
| | *Amphiblestrum* aff. *denticulatum* Canu and Bassler, 1926 |
| | *Antropora* aff. *japonica* (Canu and Bassler, 1929) |
| | *Callopora craticula* (Alder, 1856) |
| | *Cauloramphus disjunctus* Canu and Bassler, 1929 |
| | *Copidozoum* aff. *japonica* (Ortmann, 1890) |
| | *Doryporella* cf. *alcicornis* (O'Donoghue and O'Donoghue, 1923) |
| | *Ellisina* aff. *andense* (Hayami, 1975) |
| | *Flustrellaria* cf. *microecia* (Kataoka, 1961) |
| | *Gontarella*? *spatulata* (Kataoka, 1961) |
| | *Microporina* sp. |
| | *Monoporella* sp. |
| | *Septentriopora karasi* Kuklinski and Taylor, 2006 |
| | *Tegella* aff. *aquilirostris* (O'Donoghue and O'Donoghue, 1923) |
| | *Tegella* aff. *arctica* (d'Orbigny, 1851) |
| | *Tegella* sp. A |
| N = 19 | *Tegella* sp. B |
| Cheilostomata: Ascophora | *Aimulosia palliolata* (Canu & Bassler, 1928) |
| | ?*Arthropoma* sp. A |
| | ?*Arthropoma* sp. B |
| | *Buffonellaria biaperta* sensu Hayami, 1975, non Michelin |
| | *Calyptotheca* sp. |
| | *Celleporaria*? *bidentata* (Androsova, 1958) |
| | *Celleporaria*? cf. *distincta* (Sakakura, 1936) |
| | *Celleporaria* sp. A |
| | *Celleporaria* sp. B |

(continued)

| | |
|---|---|
| | *Celleporella hyalina* (L., 1767) |
| | *Celleporella* sp. A |
| | *Celleporina minima* Grischenko et al., 2007 |
| | *Cleidochasma tuberculata* Osburn, 1952 |
| | *Coleopora*? *tsugaruensis* Kataoka, 1957 |
| | ?*Collarina* sp. |
| | *Cribrilaria caesia* (Dick et al., 2005) |
| | *Cryptosula*? *pirikaensis* (Hayami, 1975) |
| | *Cyclocolposa* sp. |
| | *Cylindroporella* sp. |
| | *Elleschara* cf. *rylandi* (Soule et al., 1995) |
| | *Escharella* cf. *klugei* Hayward, 1978 |
| | *Escharella* sp. |
| | *Escharoides hataii* Hayami, 1975 |
| | *Exochella* sp. A |
| | *Exochella* sp. B |
| | *Fenestrulina* sp. |
| | *Harmeria scutulata* (Busk, 1855) |
| | *Hayamiellina constans* Grischenko and Gordon, 2004 |
| | *Hincksipora* cf. *spinulifera* (Hincks, 1889) |
| | ?*Hippopodinella* sp. |
| | *Hippoporella* aff. *hippopus* (Smitt, 1868) |
| | ?*Hippoporella huziokai* Hayami, 1975 |
| | *Hippoporella* aff. *kurilensis* (Gontar, 1979) |
| | ?*Hippoporella* sp. |
| | *Hippothoa divaricata* Lamouroux, 1821 |
| | *Hippothoa* sp. A |
| | *Integripelta japonica* Gordon et al., 2002 |
| | *Lagenipora* sp. |
| | *Lepraliella* aff. *contigua* (Smitt, 1868) |
| | *Metacleidochasma* sp. |
| | *Microporella* ?*alaskana* Dick and Ross, 1988 |
| | *Microporella* aff. *californica* (Busk, 1856) |
| | *Microporella* cf. *formosa* Suwa and Mawatari, 1998 |
| | *Microporella* aff. *stellata* (Verrill, 1879) |
| | *Microporella* sp. A |
| | *Microporella* sp. B |
| | *Myriapora subgracilis* (d'Orbigny, 1852) |
| | *Myriozoella plana* (Dawson, 1859) |
| | *Odontoporella* sp. |
| | *Pachyegis* cf. *princeps* (Norman, 1903) |
| | *Parasmittina jeffreysi* (Norman, 1876) |
| | *Parasmittina* cf. *triangularis* (Mawatari, 1952) |
| | *Parasmittina* aff. *trispinosa* (Johnston, 1838) |
| | *Parasmittina* sp. |
| | *Petraliella*? *pirikaensis* Hayami, 1975 |

(continued)

| | |
|---|---|
| | *Porella* aff. *acutirostris* Smitt, 1868 |
| | *Porella* cf. *columbiana* O'Donoghue and O'Donoghue, 1923 |
| | *Porella hanaishiensis* Hayami, 1975 |
| | *Porella* sp. A |
| | *Porella* sp. B |
| | *Porella* sp. C |
| | *Porella* sp. D |
| | *Porella* sp. E |
| | *Reginella* cf. *kokubuensis* Hayami, 1975 |
| | *Reteporella* sp. |
| | *Rhamphostomella scabra* (Fabricius, 1824) |
| | *Rhamphostomella* sp. |
| | ?*Robertsonidra* sp. |
| | *Schizomavella* cf. *magniporata* Nordgaard, 1906 |
| | *Schizomavella* sp. A |
| | *Schizomavella* sp. B |
| | *Smittina hanaishiensis* Hayami, 1975 |
| | *Stomachetosella* cf. *collaris* (Kluge, 1946) |
| | *Stomachetosella perforata* (Canu and Bassler, 1929) |
| | *Stomachetosella* cf. *producta* (Packard, 1863) |
| | *Stomachetosella* sp. |
| N = 77 | Gen. nov. aff. *Kubaninella* sp. |
| Cyclostomata | *Crisia* sp. A |
| | *Crisia* sp. B |
| | ?*Densipora* sp. |
| | *Desmeplagioecia pastiliformis* Gontar, 2009 |
| | *Diplosolen* sp. |
| | *Disporella* sp. |
| | ?*Entalophoroecia* sp. |
| | *Fasciculiporoides* cf. *simplex* (Ortmann, 1889) |
| | *Favosipora* sp. |
| | *Filifascigera* cf. *grandiosa* Sakakura, 1935 |
| | *Heteropora* sp. |
| | '*Hyporosopora*' sp. |
| | aff. *Liripora* sp. |
| | *Oncousoecia* cf. *palmata* (Wood, 1844) sensu Smitt, 1867 |
| | *Oncousoecia* sp. |
| | *Pencilleta* sp. |
| | *Plagioecia* cf. *sarniensis* (Norman, 1864) |
| | *Platonea* cf. *murmanica* (Kluge, 1915) |
| | *Platonea* sp. A |
| | *Platonea* sp. B |
| | *Prosthenoecia* sp. |
| | *Rodinopora magnifica* Taylor and Grischenko, 1999 |
| | '*Stomatopora*' sp. |
| N = 24 | *Tubulipora* cf. *pulchra* MacGillivray, 1885 |

# References

Busk G (1855) Zoophytology. Q J Microscop Sci 3:253–256
Canu F, Bassler RS (1929) Bryozoa of the Philippine region. Bull US Natn Mus 100(9):1–685
Dick MH, Ross JRP (1988) Intertidal Bryozoa (Cheilostomata) of the Kodiak vicinity, Alaska. Cent Pac NW Stud Occas Pap 23:1–133
Dick MH, Takashima R, Komatsu T, Kaneko N, Mawatari SF (2008) Overview of Pleistocene bryozoans in Japan. In: Okada H, Mawatari SF, Suzuki N, Gautam P (eds) Origin and evolution of natural diversity. Hokkaido University, Sapporo, pp 83–91
Grischenko AV, Dick MH, Mawatari SF (2007) Diversity and taxonomy of intertidal Bryozoa (Cheilostomata) at Akkeshi Bay, Hokkaido, Japan. J Nat Hist 41:1047–1161
Grischenko A, Mawatari SF, Taylor PD (2000) Systematics and phylogeny of the cheilostome bryozoan *Doryporella*. Zool Scripta 29:247–264
Grischenko AV, Taylor PD, Mawatari SF (2004) *Doryporella smirnovi* sp. nov. (Bryozoa: Cheilostomata) and its impact on phylogeny and classification. Zool Sci 21:327–332
Hayami T (1975) Neogene Bryozoa from northern Japan. Sci Rep Tohoku Univ Geol 45:83–126
Hincks T (1882) Report on the Polyzoa of the Queen Charlotte Islands. Ann Mag Nat Hist 10:459–471
Kano K, Kato H, Yanagisawa Y, Yoshida F (1991) Stratigraphy and geological history of the Cenozoic of Japan. Rep Geol Surv Japan 274:1–114
Kataoka J (1961) Bryozoan fauna from the "Ryukyu Limestone" of Kikai-jima, Kagoshima Prefecture, Japan. Sci Rep Tohoku Univ Geol 32:213–272
Kitamura A, Kimoto K (2006) History of the inflow of the warm Tsushima Current into the Sea of Japan between 3.5 and 0.8 Ma. Palaeogeogr Palaeoecol 236:355–366
Kluge GA (1946) New and less known bryozoans from the Arctic Ocean. In: Bujnitskiy VH (ed) Trudy dreifuyuschei ekspeditsii glavsevmorputi na ledokolnom parachodie "G.Sedov" 1937–1940, vol 3. Izdatelstvo Glavsevmorputi, Moscow, pp 194–223 (In Russian)
Kuklinski P, Taylor PD (2006a) A new genus and some cryptic species of Arctic and boreal calloporid cheilostome bryozoans. J Mar Biol Assoc UK 86:1035–1046
Kuklinski P, Taylor PD (2006b) Unique life history strategy in a successful Arctic bryozoan, *Harmeria scutulata*. J Mar Biol Assoc UK 86:1305–1314
Kuklinski P, Taylor PD (2009) Mineralogy of Arctic bryozoan skeletons in a global context. Facies 55:489–500
Lindner A, Cairns SD, Cunningham CW (2008) From offshore to onshore: multiple origins of shallow-water corals from deep-sea ancestors. PLoS One 3(6):1–6
Nojo A, Hasegawa S, Okada H, Togo Y, Suzuki A, Matsuda T (1999) Interregional lithostratigraphy and biostratigraphy of the Pleistocene Setana Formation, southwestern Hokkaido, Japan. J Geol Soc Japan 105:370–388
Norman AM (1903) Notes on the natural history of East Finmark. Polyzoa. Ann Mag Nat Hist 7:567–598
O'Donoghue CH, O'Donoghue E (1923) A preliminary list of the Bryozoa from the Vancouver Island region. Contrib Can Biol Fish, NS 1:143–201
Orbigny AD d' (1851–1854) Paleontologie Francaise. Descriptions des Mollusques et rayonnes 432 fossiles. Terrains Crétacés, V. Bryozoaires. Victor Masson, Paris, p 1191
Ortmann A (1890) Die japanische Bryozoenfauna. Arch Naturgesch 54:1–74
Sakakura K (1935) Pliocene and Pleistocene Bryozoa from the Boso Peninsula. (1). J Fac Sci Imp Univ Tokyo Sec 2 4:1—48
Shiki T, Yamazaki T (1996) Tsunami-induced conglomerates in Miocene upper bathyal deposits, Chita Peninsula, central Japan. Sed Geol 104:175–188
Soh W (1989) Coarse clast dominant submarine debrite, the Mio-Pliocene Fujikawa Group, central Japan. In: Taira A, Masuda F (eds) Sedimentary facies on active plate margin. Terra Scientific Publishing, Tokyo, pp 495–510

Spjeldnaes N (2000) Cryptic bryozoans from West Africa. In: Herrera Cubilla A, Jackson JBC (eds) Proceedings of the 11th International Bryozoology Association Conference. Smithsonian Tropical Research Institute, Balboa Republic of Panama pp 385–391

Suzuki A (1989) Molluscan fauna from the Setana Formation in the Kuromatsunai district, southwestern Hokkaido, Japan. Earth Science (Chikyu-Kagaku) 43:277–289

Suzuki A (1991) Sedimentary environment of the Setana Formation in the Pirika-Hanaishi area, southwestern Hokkaido. J Geol Soc Japan 97:329–344

Suzuki A, Akamatsu M (1994) Post-Miocene cold-water molluscan faunas from Hokkaido, Northern Japan. Palaeogeogr Palaeclimatol Palaeoecol 108:353–367

Takashima R, Dick MH, Nishi H, Mawatari SF, Nojo A, Hirose M, Gautam P, Nakamura K, Tanaka T (2008) Geology and sedimentary environments of the Pleistocene Setana Formation in the Kuromatsunai district, southwestern Hokkaido, Japan. In: Okada H, Mawatari SF, Suzuki N, Gautam P (eds) Origin and evolution of natural diversity. Hokkaido University, Sapporo, pp 75–82

Taylor PD, Grischenko AV (1999) *Rodinopora* gen. nov. and the taxonomy of fungiform cyclostome bryozoans. Species Divers 4:9–33

Taylor PD, James NP, Bone Y, Kuklinski P, Kyser TK (2009) Evolving mineralogy of cheilostome bryozoans. Palaios 24:440–452

Taylor PD, Kudryavtsev AB, Schopf JW (2008) Calcite and aragonite distributions in the skeletons of bimineralic bryozoans as revealed by Raman spectroscopy. Invert Biol 127:87–97

Taylor PD, Zatoń M (2008) Taxonomy of the bryozoan genera *Oncousoecia*, *Microeciella* and *Eurystrotos* (Cyclostomata: Oncousoeciidae). J Nat Hist 42:2557–2574

Yamazaki T, Yamaoka M, Shiki T (1989) Miocene offshore tractive current-worked conglomerates – Tsubutegaura, Chita Peninsula, central Japan. In: Taira A, Masuda F (eds) Sedimentary facies on active plate margin. Terra Scientific, Tokyo, pp 483–494

# Chapter 26
# Early Carboniferous Bryozoans from Western Siberia, Russia

## Carboniferous Bryozoans from Siberia

**Zoya Tolokonnikova**

**Abstract** A diverse bryozoan fauna is described from two boreholes from the Kurgan Region, Russia, containing rocks of Lower Carboniferous (Tournaisian-Viséan) age. The bryozoan assemblage includes 13 fenestellids and 5 rhabdomesines, of which one species is new: *Pseudonematopora sibirica* n. sp. The studied fauna shows the most similarity to the fauna of the southwest of Western Siberia and Kazakhstan, the Kuznetsk Basin of Russia, and the United States of America.

**Keywords** Bryozoa • Fenestellida • Rhabdomesina • Carboniferous • Mississippian • Western Siberia • Russia

## Introduction

Carboniferous deposits are widespread in the southwestern region of the Western Siberian plain, Russia. They are covered by Mesozoic-Cenozoic deposits and are best seen in a series of boreholes. Carboniferous deposits include various but poorly investigated fossils. Tournaisian-Viséan bryozoans from the target area have been earlier described by Mezentseva (2007), who reported the following taxa of the order Fenestellida and suborder Rhabdomesina (order Cryptostomida) from boreholes C-310 and East-Kurganskaya 44: *Laxifenestella* cf. *exigua* (Ulrich, 1890), *Rectifenestella* sp., *Alternifenestella* sp., *Polypora burlingtonensis* (Ulrich, 1890), *Rectifenestella* cf. *bukhtarmensis* (Nekhoroshev, 1956), *Rectifenestella* sp., *Spinofenestella* sp., and *Hemitrypa* sp. In the present study abundant bryozoans were found in core samples from the boreholes Kurgan-Uspenskaya 1 and East-Kurganskaya 44. The aim of the study is to describe the faunal composition, to

Z. Tolokonnikova (✉)
Kuzbass State Pedagogical Academy, Kuznetsov Str., 6, 654041 Novokuznetsk, Russia
e-mail: zalatoi@yandex.ru

A. Ernst, P. Schäfer, and J. Scholz (eds.), *Bryozoan Studies 2010*,
Lecture Notes in Earth System Sciences 143, DOI 10.1007/978-3-642-16411-8_26,

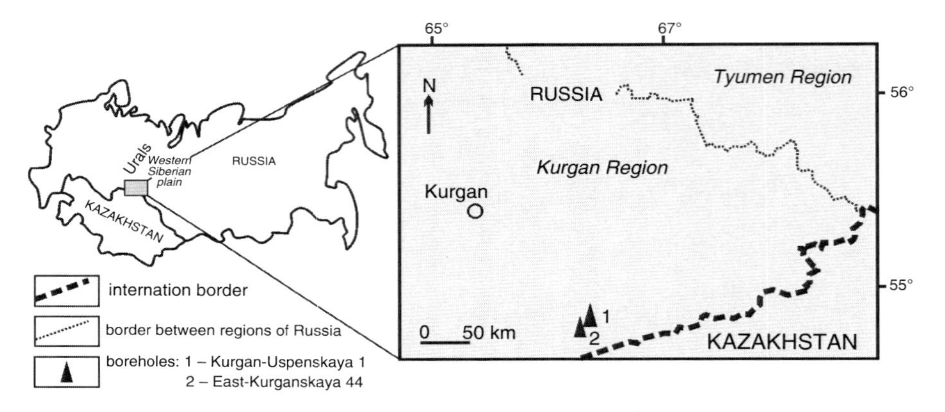


**Fig. 26.1** Location of boreholes

establish palaeobiogeographic connections, and to assess the possible biostratigraphic value of the bryozoan fauna.

## Geological Setting

The boreholes East-Kurganskaya 44 and Kurgan-Uspenskaya 1 are located in the southeast of the Kurgan Region (southern Russia) in the joint zone of Urals and Kazakhstan tectonic structures (Fig. 26.1). The basement of the Western Siberian plains includes pre-Devonian formations, and is overlain by Middle Palaeozoic to Middle Triassic formations, as well as Mesozoic–Cenozoic rocks (Ivlev 2008). Pre-Jurassic deposits were disclosed in a series of boreholes in the 1970s–1990s and later in the 2000s. Discovery of iron ore deposits and oil-and-gas prospects determines the ongoing interest in Carboniferous deposits (Pumpyanski 1992, 1999). The biostratigraphy of the Western Siberian plains is based on foraminifers and brachiopods (Bogush 1985; Pumpyanski 1988, 1990, 1992, 1999).

During the Early Carboniferous, Western Siberia was a shelf bordering a closing oceanic basin (Mizens 2002). Mizens (2002) characterized the deposits to indicate periodical changes of the hydrodynamic regime. The early Tournaisian underwent a shoaling of the basin; and in some places lagoons developed (Mizens and Kokshina 2010). Widening of the basin and its deepening to 100–150 m occurred in the late Tournaisian. Deposition of predominantly carbonates indicates normal salinity and a standard marine regime. These conditions were favorable for a prosperous marine benthic fauna. In the Viséan, the oceanic basin closed and the region was characterized by an increase of volcanic activity and supply of terrigenous materials from continental heights (Puchkov 2000; Mizens 2002).

## Material and Methods

Borehole Kurgan-Uspenskaya 1 was drilled in 2007 and penetrated Upper Devonian-Lower Carboniferous deposits (upper Famennian–lower Viseán) (Stepanova et al. 2010a; Stepanova et al. 2011). Borehole East-Kurganskaya 44 was drilled in the second part of the twentieth century and penetrated upper Tournaisian deposits only (Pumpyanski 1990; Stepanova et al. 2010b). Selected core samples from these boreholes were given to the author in 2009 for age determination. Core samples were chosen from several intervals with visible bryozoan accumulations (Fig. 26.2). Most of bryozoan remains are embedded in floatstones, packstones, and grainstones, in which recrystallization and deformation have taken place. The bryozoans occur together with foraminifers, algae, brachiopods, crinoids, corals, and ostracods.

Laboratory techniques include preparation of thin sections and their study with the use of a microscope. The present study describes bryozoans of the order Fenestellida and the suborder Rhabdomesina (order Cryptostomata) as defined by Boardman et al. (1983). In total, 142 thin sections were studied. Eighteen species belonging to 9 genera were recognized in the studied association. Three taxa were placed in open nomenclature due to the fragmentary nature of the material. Some taxa were identified to genus level only. One new species, *Pseudonematopora sibirica* sp. nov., was discovered among rhabdomesine bryozoans. The new taxon was described based on measurements of skeletal structures of the colony following Boardman et al. (1983) and Gorjunova (1985). Numerical statistics (mean, standard deviation, variation coefficient, and minimum/maximum values) were applied according to Köhler et al. (1996).

Results of a palaeobiogeographic comparison are considered in a cluster analysis. Cluster analysis used the unweighted pair-group average method (UPGMA), in which clusters are joined based on the average distance between all members in the two groups (Hammer et al. 2001). Similarity between regions was measured using the Jaccard coefficient for absence-presence data (Jaccard 1901) defined as the proportion of characters that match between samples, excluding those that are missing in both samples (the size of the intersection divided by the size of the union of the sample sets). Analyses were performed using PAST software (version 1.97; Hammer et al. 2001).

Trepostome and cystoporid bryozoans, although equally abundant in the core samples, will be described in a later publication. The studied material is stored at the Kuzbass State Pedagogical Academy (Novokuznetsk, Russia; collection number 4).

## Bryozoan Fauna

Tournaisian-Viséan deposits from southwest of the Western Siberian plain contain a bryozoan fauna, which shares many species with contemporary deposits of Kazakhstan (73.7%), the Kuznetsk Basin (57.4%), and the United States of America (47.4%) (Ulrich 1890; Nekhoroshev 1953, 1956; Trizna 1958; Snyder 1991). There is less similarity with Mongolia (26.3%), Uzbekistan (21%), Northeastern Russia (15.8%), and Eastern Transbaikalia (10.5%) (Mikhno and Balakin 1975; Morozova 1981; Popeko 2000;

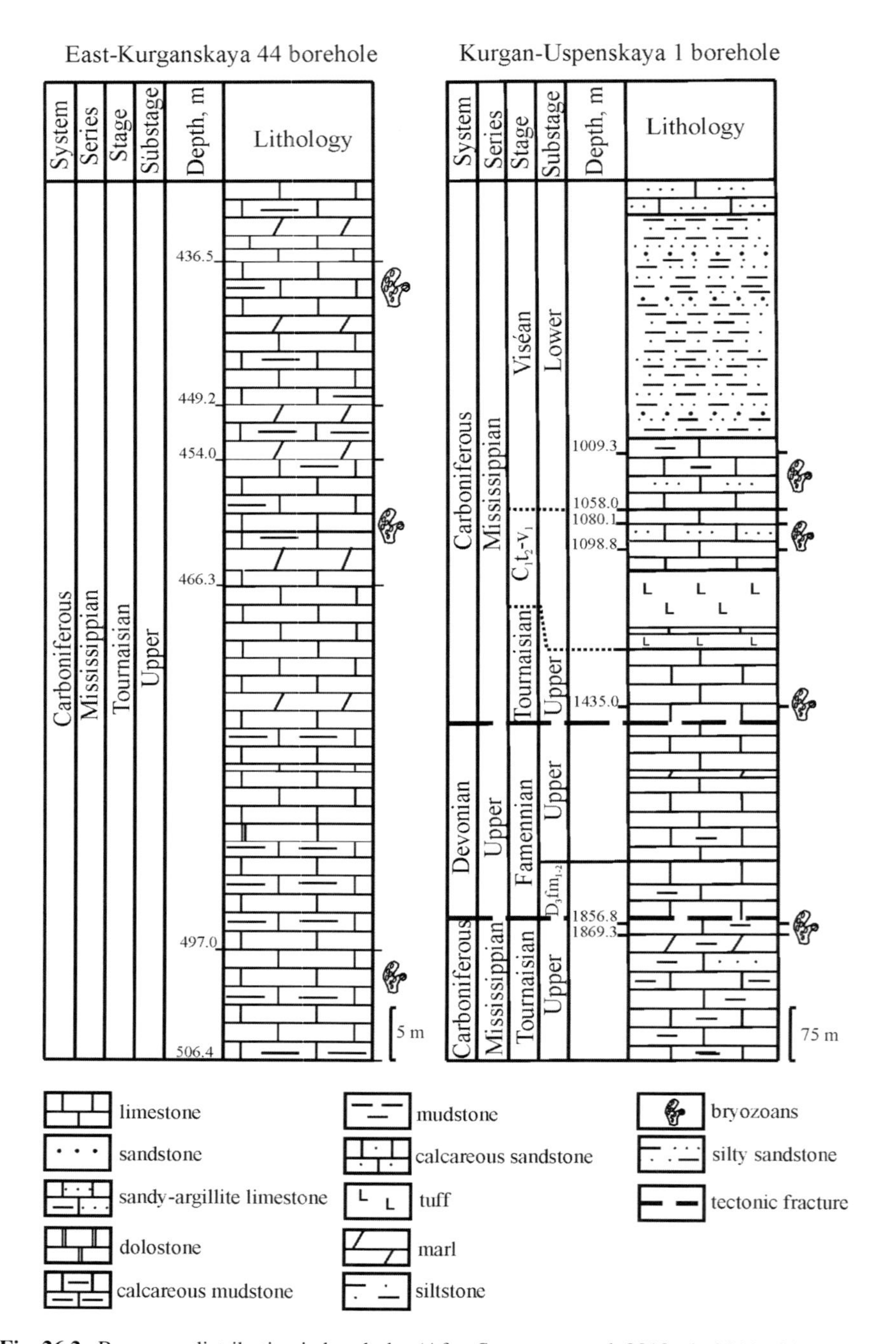


**Fig. 26.2** Bryozoan distribution in boreholes (After Stepanova et al. 2010a, b, 2011 with changes)

Morozova et al. 2003). Fenestellid bryozoans were dominant in that assemblage representing 14 of 18 identified species. In general, the Carboniferous was the time of a peak in the development of fenestellids (Ross 1981; Snyder 1991; Gorjunova et al. 2004). This tendency is confirmed by their abundance in the studied assemblage.

The fenestellid fauna contains 5 species with a wide geographical distribution occurring also in regions other than Kazakhstan, the Kuznetsk Basin, and the United States of America. Thus, *Polyporella* cf. *spininodata* (Ulrich, 1890) is also known from Uzbekistan and Eastern Transbaikalia (Mikhno and Balakin 1975; Popeko 2000). *Rectifenestella* aff. *nododorsalis* (Ulrich, 1890) occurs in the Mississippian of Mongolia, and Eastern Transbaikalia, and *Rectifenestella triserialis* (Ulrich, 1890) is known from the Mississippian of Turkmenistan, Northeastern Russia, and Uzbekistan. *Rectifenestella cesteriensis* (Ulrich, 1890) was previously reported from the Mississippian of Turkmenistan, and *Polypora maccoyana* Ulrich, 1890 from the Mississppian of Mongolia (Nikiforova 1938; Mikhno and Balakin 1975; Gorjunova and Morozova 1979; Morozova 1981; Popeko 2000; Morozova et al. 2003). All fenestellid genera of the studied assemblage are cosmopolitan, except the genus *Pseudopolypora* (Morozova 2001). The majority of these genera appeared in the Lower Devonian, and most of them disappeared in the Late Permian (Morozova 2001). The genus *Hemitrypa* disappeared in the Mississippian (Morozova 2001; Gorjunova et al. 2004).

Rhabdomesine bryozoans had small geographic ranges and only occurred in one or two regions outside the southwest of the Western Siberian plain. *Nikiforovella multipitata* Trizna, 1958 and *Rhombopora novitia* Trizna, 1958 also occur in the Kuznetsk Basin, and *Nikiforovella ulbensis* Nekhoroshev, 1956 in Kazakhstan and Northeastern Russia (Nekhoroshev 1956; Trizna 1958; Morozova 1981). The species *Nicklesopora simulatrix* (Ulrich, 1884) was distributed in the territory of Mongolia, the Kuznetsk Basin, and the Unites States of America during the Early Carboniferous (Ulrich 1884; Trizna 1958; Gorjunova and Morozova 1979).

Most of the rhabdomesine genera appeared in the Late Devonian (Famennian) or Early Carboniferous (Mississippian) (Gorjunova 1985), except for the genus *Orthopora* that ranged from the Silurian to Carboniferous. *Rhabdomeson*, *Rhombopora*, *Nicklesopora* are cosmopolitan, whereas *Nikiforovella*, *Orthopora*, *Streblotrypa*, *Pseudonematopora* have a wide geographical distribution (Eurasia, North America) (Nikiforova 1948; Nekhoroshev 1953, 1956; Balakin 1974; Gorjunova 1985, 1988, 2001; Wyse Jackson 1996; Xia 1997; Ariunchimeg 2005; Ernst 2005; Ernst and Nakrem 2005). The genus *Klaucena* was found in the Kuznetsk Basin and China (Trizna 1958; Yang et al. 1988).

The fenestellid fauna is characterized by delicate colonies with thin branches and dissepiments. These bryozoans probably had reticulate or funnel-shaped habits (Fig. 26.3). Colonies thickness varies from 0.35 to 0.75 mm. Ramose rhabdomesines are 0.6–2.0 mm in diameter (Fig. 26.4). Although the bryozoans are fragmented, there is little sign of abrasion. It is concluded that colonies have not been transported over extensive distances. The bryozoan fauna reflects a normal-marine environment with moderate to low-water energy. This interpretation is also confirmed by sedimentological data (Mizens and Kokshina 2010).

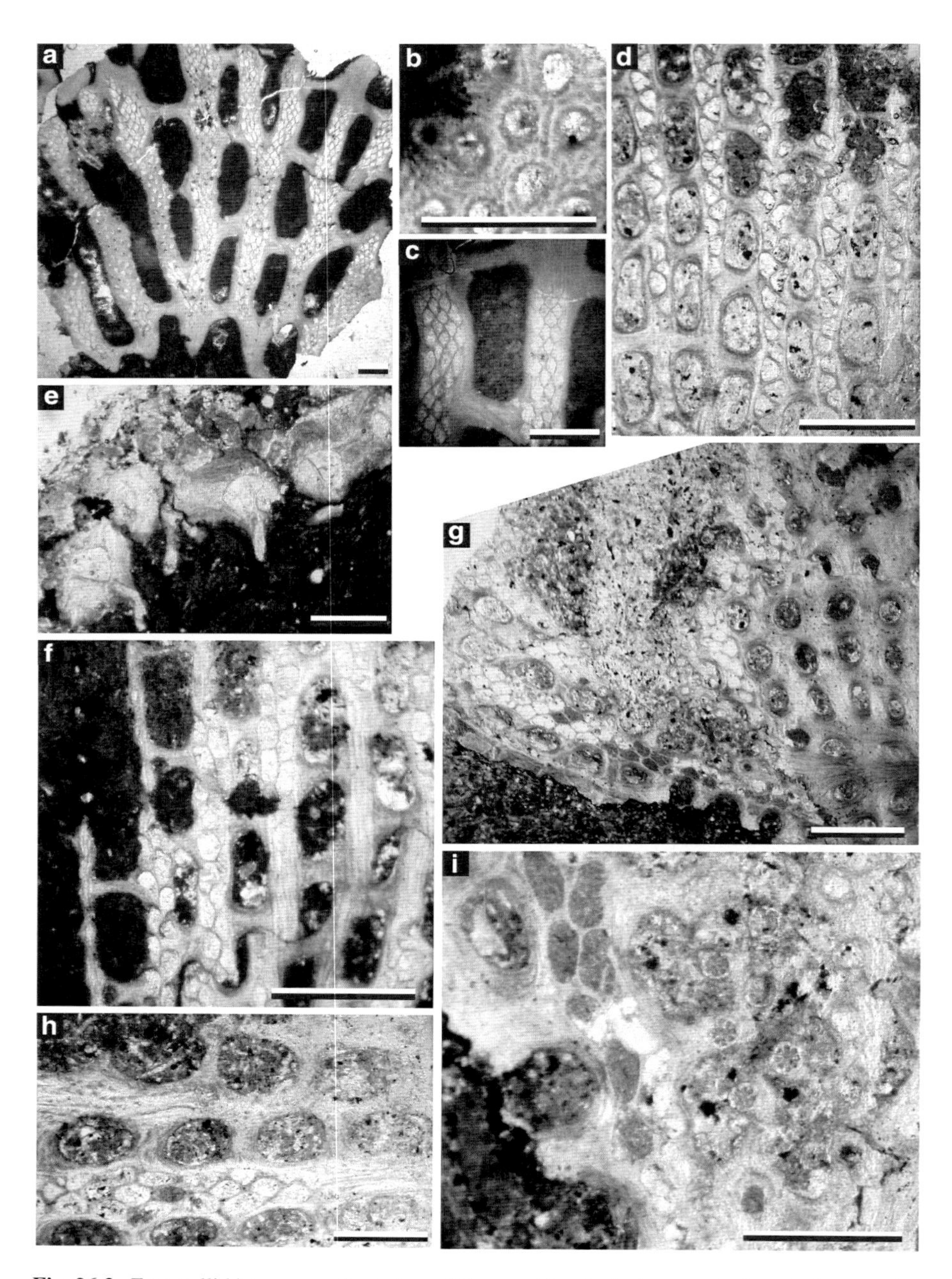

**Fig. 26.3** Fenestellid bryozoans. *Scale bars*: 0.5 mm – (**d**, **e**, **h**, **i**); 1 mm. – (**a**, **b**, **c**, **f**, **g**). (**a**, **b**, **c**) – *Polypora maccoyana* Ulrich, 1890, 4/31.1. (**d**) *Alternifenestella triangularis* (Nekhoroshev, 1956), 4/32.1. (**e**, **f**) *Fenestella cesteriensiformis* Nekhoroshev 1956, 4/33.1.(**g**, **h**, **i**) *Polyporella radialis* (Ulrich, 1890), 4/34.1

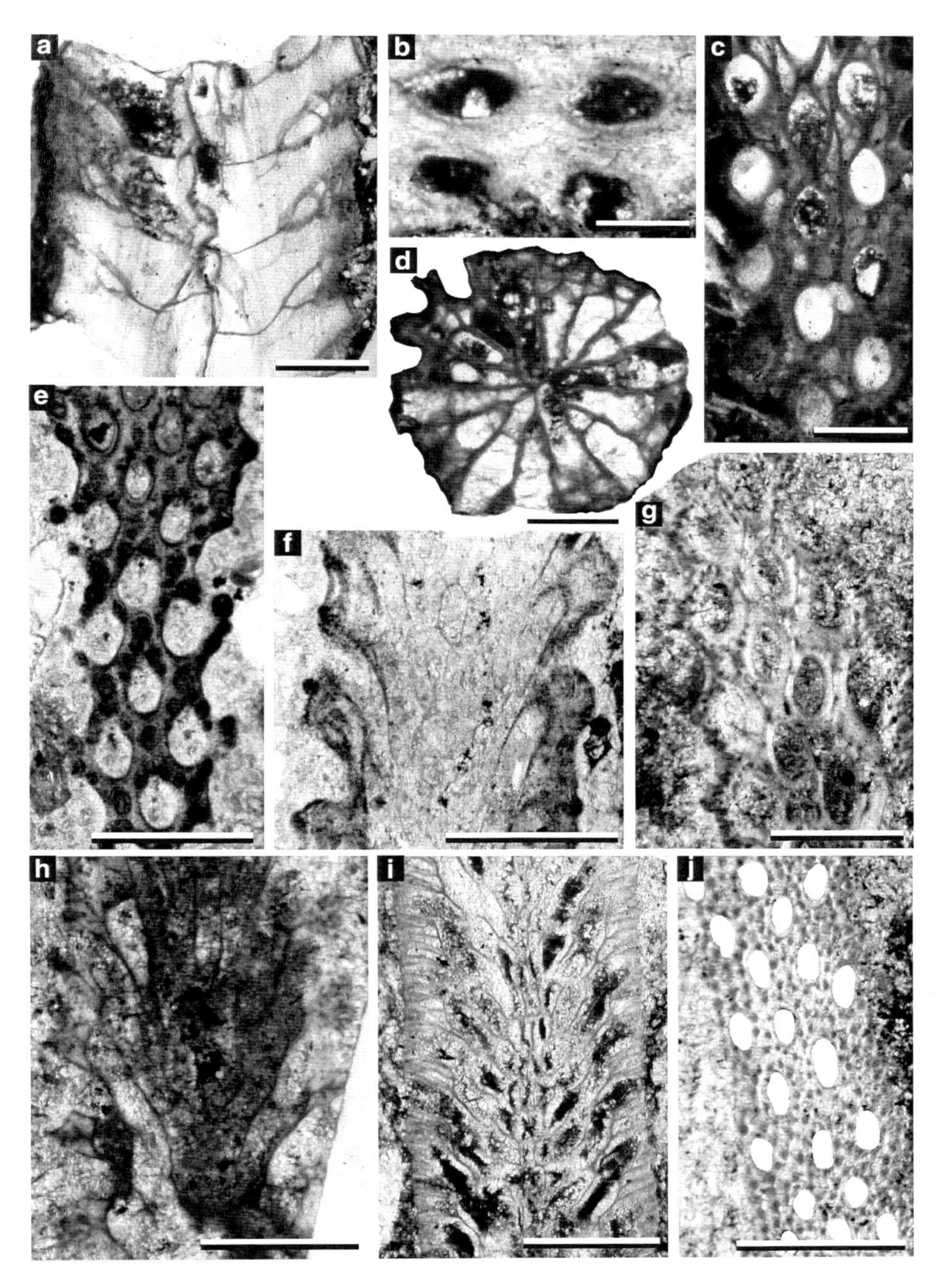

**Fig. 26.4** Rhabdomesine bryozoans. All *scale bars*: 0.5 mm. (**a–d**) *Pseudonematopora sibirica* sp. nov. (**a**) – longitudinal section, 4/27.1. (**b**) – tangential section, 4/27.2. (**c**) – tangential section, 4/27.1. (**d**) – transverse section, 4/27.3, (**e, f**) *Rhombopora novitia* Trizna, 1958, 4/28.1. (**g, h**) – *Nicklesopora simulatrix* (Ulrich, 1884), 4/29.1. (**i, j**) *Nikiforovella multipitata* Trizna, 1958, 4/30.1

**Table 26.1** Measurements of *Pseudonematopora sibirica* sp. nov. N – number of measurements, X – mean, SD – standard deviation, CV – coefficient of variation, MIN – minimal value, MAX – maximal value

| | N | X | SD | CV | MIN | MAX |
|---|---|---|---|---|---|---|
| Branch width, mm | 5 | 1.68 | 0.294 | 17.45 | 1.45 | 2.12 |
| Endozone width, mm | 5 | 0.49 | 0.057 | 11.71 | 0.40 | 0.56 |
| Exozone width, mm | 5 | 0.58 | 0.105 | 17.98 | 0.50 | 0.75 |
| Aperture width, mm | 10 | 0.19 | 0.015 | 8.13 | 0.17 | 0.22 |
| Aperture length, mm | 10 | 0.29 | 0.021 | 7.37 | 0.27 | 0.32 |
| Aperture spacing along branch, mm | 10 | 0.18 | 0.021 | 11.77 | 0.15 | 0.20 |
| Aperture spacing across branch, mm | 10 | 0.17 | 0.028 | 16.64 | 0.12 | 0.20 |

Phylum Bryozoa Ehrenberg, 1831

Class Stenolaemata Borg, 1926

Order Cryptostomata Vine, 1884

Suborder Rhabdomesina Astrova and Morozova, 1956

Family Arthrostylidae Ulrich, 1882

Genus *Pseudonematopora* Balakin, 1974

Type species: *Nematopora*? *turkestanica* (Nikiforova, 1948). Carboniferous, Mississippian, upper Tournaisian; Kazakhstan.

Diagnosis: Thin-branched colonies. Branches are circular or semicircular in transverse section. Autozooecia occur in 6–16 longitudinal rows. Autozooecial apertures are round to oval, with peristomes, and originate from a central axis or a narrow mesotheca. Terminal diaphragms are developed in some species. Tektitozooecia may be present in the exozone in varying numbers. Other heterozooecia are absent.

Occurrence: China; Upper Devonian (Famennian); Ireland, Germany, Kazakhstan, Turkmenistan, Russia; Carboniferous, Mississippian (Tournaisian, Viséan); Mongolia; Pennsylvanian (Bashkirian, Moscovian); Canadian Arctic Islands; Lower Permian.

Discussion: The genus *Pseudonematopora* Balakin, 1974 differs from the genus *Nematopora* Ulrich, 1888 by the presence of tektitozooecia and mesotheca.

*Pseudonematopora sibirica* sp. nov.

Figure 26.4a–d, Table 26.1

Etymology: Referring to Siberia where this species was found.

Holotype: 4/27.1; paratypes: 4/27.2-4/27.12.

Type locality: Kurgan Region, East-Kurganskaya 44 borehole, depths 454.0–460.0 m and 497.0–506.4 m.

Type horizon: Carboniferous, Mississippian, upper Tournaisian.

Diagnosis: Branched colonies with a narrow mesotheca. Autozooecial apertures are oval in shape, with a peristome, and occur in 6–8 longitudinal rows. Tektitozooecia are abundant.

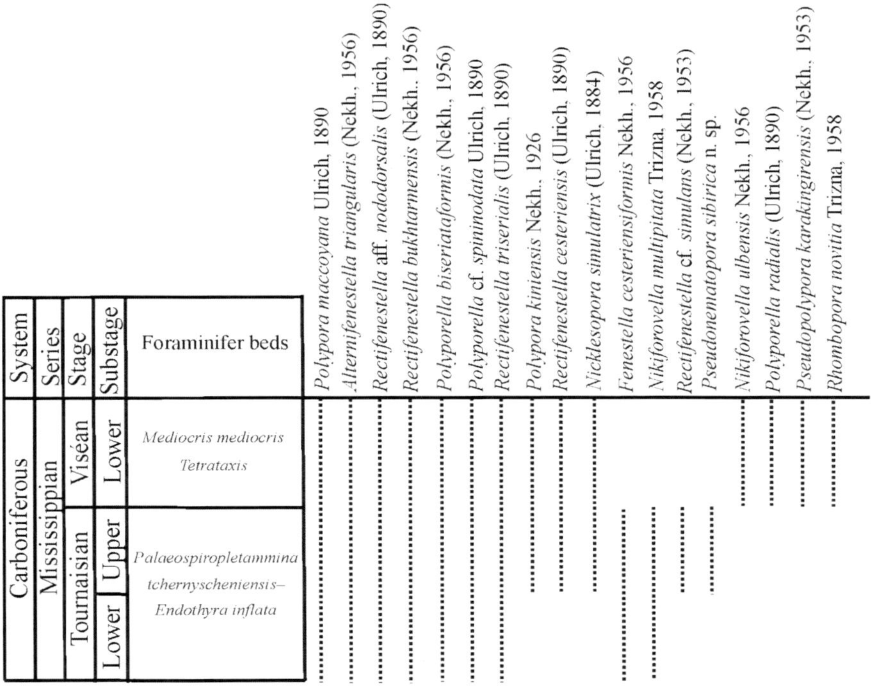

| System | Series | Stage | Substage | Foraminifer beds |
|---|---|---|---|---|
| Carboniferous | Mississippian | Viséan | Lower | *Mediocris mediocris* *Tetrataxis* |
| | | Tournaisian | Upper | *Palaeospiropletammina tchernyschenìensis–Endothyra inflata* |
| | | | Lower | |

*Polypora maccoyana* Ulrich, 1890
*Alternifenestella triangularis* (Nekh., 1956)
*Rectifenestella* aff. *nododorsalis* (Ulrich, 1890)
*Rectifenestella bukhtarmensis* (Nekh., 1956)
*Polyporella biseriataformis* (Nekh., 1956)
*Polyporella* cf. *spininodata* Ulrich, 1890
*Rectifenestella triserialis* (Ulrich, 1890)
*Polypora kiniensis* Nekh., 1926
*Rectifenestella cesteriensis* (Ulrich, 1890)
*Nicklesopora simulatrix* (Ulrich, 1884)
*Fenestella cesteriensiformis* Nekh., 1956
*Nikiforovella multipitata* Trizna, 1958
*Rectifenestella* cf. *simulans* (Nekh., 1953)
*Pseudonematopora sibirica* n. sp.
*Nikiforovella ulbensis* Nekh., 1956
*Polyporella radialis* (Ulrich, 1890)
*Pseudopolypora karakingirensis* (Nekh., 1953)
*Rhombopora novitia* Trizna, 1958

**Fig. 26.5** Stratigraphic distribution of bryozoans

Description: Branched colonies. Mesotheca narrow, length 0.05–0.75 mm and 0.014 mm thick. Autozooecial apertures are ovally shaped. Apertures have a peristome 0.025–0.042 mm thick. Four apertures spaced in 2 mm along and across colony. Four tektitozooecia are arranged around each aperture, and contain thin diaphragms. Tektitozooecia covered by calcareous skeletal material on the colony surface. Rare tubules 0.014 mm in diameter occur.

Discussion: *Pseudonematopora sibirica* sp. nov. differs from *Pseudonematopora turkestanica* (Nikiforova, 1948) from the Tournaisian-Viséan of Kazakhstan and Turkmenistan in the size and number of tubules (rare, 0.014 mm in diameter versus abundant, 0.015–0.045 mm in diameter in *P. turkestanica*), the number of rows around colony (6–8 vs. 8–16 in *P. turkestanica*), the number of apertures in 2 mm across colony (4 vs. 6 in *P. turkestanica*) and the constant number of tektitozooecia (constant 4 vs. 1–5 in *P. turkestanica*).

## Biostratigraphy

According to specialists, bryozoans may be of considerable biostratigraphic value (Ross 1981; Bancroft 1987; Ross and Ross 1990). Unfortunately, most taxa from the southwest Western Siberian plain have long stratigraphic ranges

(Fig. 26.5). Most of the fenestellids occur in lower Tournaisian–lower Viséan rocks. *Polypora kiniensis* Nekh., 1926 and *Rectifenestella cesteriensis* (Ulrich, 1890) characterize upper Tournaisian–lower Viséan deposits. *Fenestella cesteriensiformis* Nekhoroshev, 1956 is restricted to the Tournaisian. The species *Rectifenestella* cf. *simulans* (Nekhoroshev, 1953) was present in the upper Tournaisian only. *Polyporella radialis* (Ulrich, 1890) and *Pseudopolypora karakingirensis* only appeared in the lower Viséan (Nekhoroshev, 1953).

In contrast to the former, rhabdomesines are characterized by shorter stratigraphic ranges (Fig. 26.5). *Nicklesopora simulatris* (Ulrich, 1884) was present in the upper Tournaisian–lower Viséan. *Nikiforovella multipitata* Trizna, 1958 existed during the Tournaisian age (Trizna, 1958). *Nikiforovella ulbensis* Nekhoroshev, 1956 and *Rhombopora novitia* Trizna, 1958 are characteristic for the lower Viséan (Trizna 1958; Nekhoroshev 1956).

The distribution of bryozoans in the boreholes studied is shown in Table 26.2. Bryozoans from the upper Tournaisian of Kurgan-Uspenskaya 1 and East-Kurganskaya 44 boreholes are different at species and genera levels. *Fenestella cesteriensiformis* Nekhoroshev 1956 and *Alternifenestella triangularis* (Nekhoroshev 1956) were found in both boreholes but in different stages.

*Polyporella spininodata* (Ulrich, 1890), *Rectifenestella nododorsalis* (Ulrich, 1890), *Rectifenestella triserialis* (Ulrich, 1890), *Rectifenestella cesteriensis* (Ulrich, 1890), and *Polypora maccoyana* Ulrich, 1890 may serve as biostratigraphic markers for the upper Tournaisian–lower Viséan (Ulrich 1890; Nikiforova 1938; Nekhoroshev 1953, 1956; Trizna 1958; Mikhno and Balakin 1975; Gorjunova and Morozova 1979; Morozova 1981; Snyder 1991; Popeko 2000; Morozova et al. 2003). They also display wide geographical distribution making them valuable for biostratigraphy.

Upper Tournaisian–lower Viséan deposits from the southwest Western Siberian plain are correlated with contemporaneous deposits of Kazakhstan and Kuznetsk Basin of Russia by bryozoans. The bryozoan assemblage studied here shows similar characteristics to the Rusakovski horizon (upper Tournaisian), Ishimski horizon and lower part of Yagovkinski horizon (lower Viséan) of Kazakhstan; Phominski horizon (upper Tournaisian), Pod'yakovski horizon (lower Viséan) of Kuznetsk Basin (Nekhoroshev 1953, 1956; Trizna 1958; Decisions of III Kazakhstan Stratigraphic meeting by Precambrian and Phanerozoic 1991). Faunal similarity between the southwest of the Western Siberian plains and Kazakhstan, the Kuznetsk Basin, the Middle and South Urals corresponds to that of foraminifers, brachiopods, and algae (Stepanova et al. 2010).

At the present time, the Tournaisian stage of the Western Siberian plain is based on the *Palaeospiropletammina tchernyscheninsis – Endothyra inflata* Foraminifers beds, and the Viséan stage on the *Mediocris mediocris, Tetraxis* Foraminifers beds (Decisions of Interdepartmental meeting about examination and acceptance of regional Stratigraphic scheme of Palaeozoic deposits of Western Siberian plate 1999). New data obtained from different fossils including bryozoans will be useful for interregional correlation.

**Table 26.2** Bryozoan distribution in the studied boreholes

| Borehole | Depth, m | Stage (After Stepanova et al. 2010) | Species |
|---|---|---|---|
| East-Kurganskaya 44 | 436.5–449.2 m | upper Tournaisian | *Polyporella* cf. *spininodata* (Ulrich, 1890), *Rectifenestella bukhtarmensis* (Nekhoroshev, 1956), *Polyporella biseriataformis* (Nekhoroshev, 1956) |
| | 454.0–466.3 m | | *Pseudonematopora sibirica* sp. nov., *Nikiforovella ulbensis* Nekhoroshev, 1956, *Polypora maccoyana* (Ulrich, 1890) |
| | 497.0–506.4 m | | *Polyporella biseriataformis* (Nekhoroshev, 1956), *Pseudonematopora sibirica* sp. nov., *Rectifenestella bukhtarmensis* (Nekhoroshev, 1956), *Fenestella* sp., *Rectifenestella* aff. *nododorsalis* (Ulrich, 1890), *Fenestella cesteriensiformis* (Nekhoroshev, 1956), *Alternifenestella triangularis* (Nekhoroshev, 1956) |
| Kurgan-Uspenskaya 1 | 1009.3–1023.1 m<br>1040.4–1058.5 m | Viséan | *Rhombopora novitia* (Trizna, 1958), *Pseudopolypora karakingirensis* (Nekhoroshev, 1953), *Rectifenestella* cf. *simulans* (Nekhoroshev, 1953), *Rhabdomeson* sp., *Orthopora* sp., *Hemitrypa* sp., *Nikiforovella* sp., *Fenestella cesteriensiformis* (Nekhoroshev, 1956), *Rectifenestella cesteriensis* (Ulrich, 1890), *Rectifenestella triserialis* (Ulrich, 1890) |
| | 1080.1–1098.8 m | Tournaisian-Viséan undivided | *Polyporella radialis* (Ulrich, 1890), *Alternifenestella triangularis* (Nekhoroshev, 1956), *Nikiforovella multipitata* Trizna, 1958, *Polypora kiniensis* Nekhoroshev, 1956 |
| | 1434.0–1435.5 m<br>1856.8–1869.3 m | upper Tournaisian | *Klaucena* sp., *Nicklesopora simulatrix* (Ulrich, 1884), *Rectifenestella* sp., *Orthopora* sp., *Streblotrypa* sp. |

## Palaeobiogeography

The present record of Mississippian bryozoans from the southwest of the Western Siberian plain shows an important faunal link between faunas from the Kazakhstan and Kuznetsk Basins. Results of a palaeobiogeographic comparison are considered

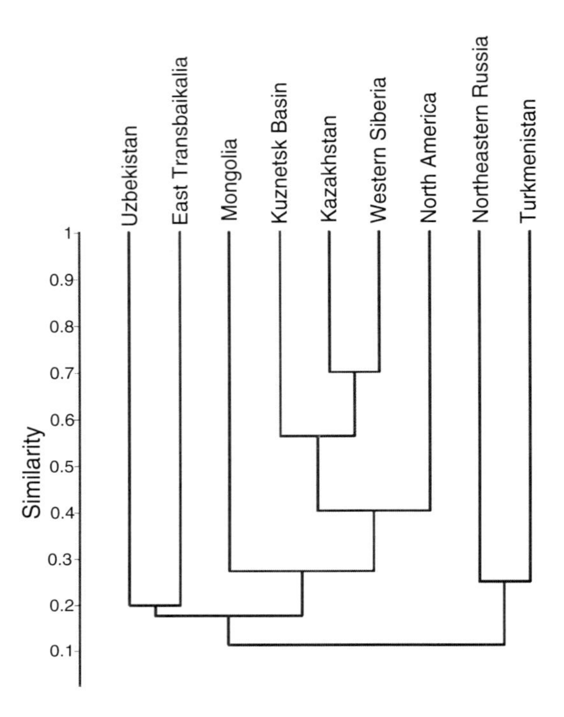


**Fig. 26.6** Cluster diagrams showing similarities between bryozoan faunas of different areas during the Mississippian (UMGMA, Unweighted Pair Group Method with Arithmetic Mean, Jaccard similarity coefficient, PAST ver. 1.81, Hammer et al. 2001)

in a cluster analysis (Fig. 26.6). The late Tournaisian–early Viséan bryozoan faunas of Western Siberia and Kazakhstan form a distinct cluster, and are closely associated with the Kuznetsk Basin. Bryzoan faunas of the United States of America are more similar to the above regions than to those elsewhere. Mongolian, Uzbekistan, and Eastern Transbaikalian faunas are also close to these regions but at lower levels of similarity. Northeastern Russia and Turkmenistan display very low similarity with the other regions.

## Conclusions

Twenty species of fenestellid and rhabdomesine bryozoans were found in upper Tournaisian–lower Viséan deposits from the southwest of the Western Siberian plain.

*Polyporella spininodata* (Ulrich, 1890), *Rectifenestella nododorsalis* (Ulrich, 1890), *Rectifenestella triserialis* (Ulrich, 1890), *Rectifenestella cesteriensis* (Ulrich, 1890), *Polypora maccoyana* (Ulrich, 1890) are suggested as markers for the upper Tournaisian–lower Viséan stages. In the Mississippian, these species have a wide geographic distribution on the shelves of Laurussia (5–7 habitats).

The Early Carboniferous was the time of dominance of fenestellids and the beginning of the expansion of rhabdomesines. Fenestellids developed abundant cosmopolitan genera.

Bryozoan species distribution during the Mississippian revealed an important link between faunas from the southwest Western Siberian plain and Kazakhstan, the Kuznetsk Basin, and the United States of America. Faunas from Turkmenistan and northeastern Russia show little similarity with other faunas during the late Tournaisian–lower Viséan.

Upper Tournaisian-lower Viséan deposits from the southwest Western Siberian plain are correlated by bryozoans with the Rusakovski horizon (upper Tournaisian), Ishimski horizon and lower part of Yagovkinski horizon (lower Viséan) of Kazakhstan; Phominski horizon (upper Tournaisian), Pod'yakovski horizon (lower Viséan) of Kuznetsk Basin of Russia.

**Acknowledgements** The author is grateful to Dr. G. Mizens, Yekaterinburg for providing core samples and consultations on the regional geology. Many thanks to Dr. D. Ruban, Rostov-on-Don for constructive suggestions to improve the manuscript. Part of this study was funded by the Sepkoski Grant-2010 from the Paleontological Society. Dr. A. Ernst, Kiel and Dr. C. Reid, Christchurch are thanked for critically reading of the manuscript and for helpful comments.

## References

Ariunchimeg Ya (2005) New Carboniferous bryozoans from Mongolia. Palaeontol J 3:264–271

Astrova GG, Morozova IP (1956) About systematics of the order Cryptostomata. Dokl Akad Nauk SSSR 110(4):661–664

Balakin GV (1974) *Pseudonematopora*, a new Early Carboniferous bryozoan genus. Palaeontol J 8:557–559

Bancroft AJ (1987) Biostratigraphical potential of Carboniferous Bryozoa. Courier Forsch-Inst Senckenberg 98:193–197

Boardman RS, Cheetham AH, Blake DB, Cook PL (1983) Treatise on invertebrate paleontology. Part. G (1): Bryozoa (revised). In: Robinson RA (ed) Geological Society of America and University of Kansas Press, Kansas

Bogush OI (1985) Foraminiferu i stratigraphiya paleozoya nizhnego karbona Zapadno-Sibirskoi plitu [Foraminifers and stratigraphy of Lower Carboniferous of Western Siberian plate]. (In Russian) In: Dubatolov VN et al. (eds) Biostratigraphy of Palaeozoic of Western Siberia, Novosibisrsk, Nauka

Borg F (1926) Studies of Recent cyclostomatous Bryozoa. Zool Bidr Uppsala 10:181–507

Ehrenberg CG (1831) Animalia invertebrata exclusis insects. Symbolae Physicae, seu Icones et descriptiones Corporum Naturalium novorum aut minus cognitorum. Pars Zoologica 4. Mittler, Berlin

Ernst A (2005) Lower Carboniferous Bryozoa from some localities in Sauerland, Germany. In: Moyano GI, Cancino JM, Wyse Jackson PN (eds) Bryozoan studies 2004. Proceedings of the 13th International Bryozoology Association conference, Balkema, London, pp 49–62

Ernst A, Nakrem HA (2005) Bryozoans from the Artinskian (Lower Permian) Great Bear Cape Formation, Ellesmere Island (Canadian Arctic). In: Moyano GI, Cancino JM, Wyse Jackson PN (eds) Bryozoan Studies 2004 – Proceedings of the 13th International Bryozoology Association conference. Balkema, London, pp 63–68

Decisions of III Kazakhstan Stratigraphic meeting by Precambrian and Phanerozoic (1991) (In Russian). Alma-Ata

Decisions of Interdepartmental meeting about examination and acceptance of regional Stratigraphic scheme of Palaeozoic deposits of Western Siberian plate (1999) (In Russian). Novosibirsk

Gorjunova RV, Morozova IP (1979) Pozdnepaleozoiskie mshanki Mongolii [Late Palaeozoic bryozoans of Mongolia] (In Russian). Trudy Sovmestnoi Sovetsko-Mongol'skoi Paleontologicheskoi Ekspedistii 9:1–140
Gorjunova RV (1985) Morphologia, klassifikatsiya i philogeniya mshanok (otryad Rhabdomesida) [The morphology, classification and phylogeny of the bryozoans (Order Rhabdomesida)] (In Russian). Akad Nauk SSSR 208:1–152
Gorjunova RV (1988) Novye kamennoygol'nye mshanki Gobiiskogo Altaya [New Carboniferous bryozoans of the Gobi Altai] (In Russian). Trudy Sovmestnoi Sovetsko-Mongol'skoi Paleontologicheskoi Ekspeditsii 33:10–29
Gorjunova RV (2001) Mshanki [Bryozoa]. In: Alekseev AS, Shik SM (ed) Srednii karbon Moskovskoi sineklizy [Middle Carboniferous of Moscow syncline (southern part)] (In Russian). Nauchnyi mir, Moscow
Gorjunova RV, Markov AB, Naimark EB (2004) Evolutsiya i biogeographiya paleozoiskich mshanok: rezultat kolichestvennogo analiza [Evolution and biogeography of the Palaeozoic Bryozoa: results of the numerical analysis] (In Russian). GEOS, Moscow
Hammer Ø, Harper DAT, Ryan PD (2001) PAST: paleontological statistics software package for education and data analysis. Palaeontologia Electronica 4(1):9 http://palaeo-electronica.org/2001_1/past/issue1_01.htm
Ivlev AI (2008) Magmatizm i geodinamika oblasti sochleneniya Urala i Kazakhstana [Magmatism and geodynamic of joint zone of Urals and Kazakhstan]. (In Russian) Rydnui, Kostanai
Jaccard P (1901) Étude comparative de la distribution florale dans une portion des Alpes et des Jura. Bull del la Société Vaudoise des Sciences Naturelles 37:547–579
Köhler W, Schachtel G, Voleske P (1996) Biostatistik. Einführung in die Biometrie für Biologen und Agrarwissenschaftler. Springer Berlin Heidelberg
Mezentseva OP (2007) O nizhnekamennougol'nych mshankach iz yugozapadnoi okrainy Zapadno-Sibirskoi plity [About Early Carboniferous bryozoans from southwest edge of Western Siberian plate] (In Russian). In: All-Russian conference. Upper Paleozoic of Russia: stratigraphy and palaeogeography, Kazan, Russia pp 208–210
Mikhno NM, Balakin GV (1975) Foraminifery i mshanki nizhnego karbona Chatkal'skich gor [Foraminifers and bryozoans from Lower Carboniferous of Chatkalsk Mountains] (In Russia). FAN, Tashkent
Mizens GA (2002) Sedimentatsionnue basseinu i geodinamicheskie obstanovki v pozhdnem devone-rannei permi yuga Urala [Sedimentation basins and geodynamic situations in Late Devonian-Early Permian of south of Urals] (In Russian) Yekaterinburg
Mizens GA, Kokshina LV (2010) Usloviya osadkonakopleniya v srednepaleozoiskich baseinach na yugo-zapade Zapadnoi Sibiri (zona sochleneniya Uralskich i kazakhstanskich struktur) [Sedimentation conditions in Middle Palaeozoic basins from southwest of Western Siberia (joint zone of Urals and Kazakhstan structures)] (In Russian), In: All-Russian conference. Tyumen, pp 111–113
Morozova IP (1981) Pozdnepaleozoiskii mshanki severo-vostoka SSSR [Late Palaeozoic bryozoans from northeast of SSSR] Tr Paleontologicheskogo Inst Akad Nauk SSSR 188:1–104
Morozova IP (2001) Mshanki otryada Fenestellida (morphologiya, systema, istoricheskoe razvitie) [Bryozoa of order Fenestellida (morphology, system, historical development)] (In Russian). Tr Paleontologicheskogo Inst Akad Nauk SSSR 227:1–177
Morozova IP, Gorjunova RV, Ariunchimeg Ya (2003) Mshanki [Bryozoa] (In Russian). In: Morozova IP (ed) Palaeontology of Mongolia Nauka, Moscow
Nekhoroshev VP (1953) Nizhnekamennougol'nye mshanki Kazakhstana [Lower Carboniferous Bryozoa of Kazakhstan] (In Russian). Trudy VSEGEI 11:1–236
Nekhoroshev VP (1956) Nizhnekamennougol'nye mshanki Altaya i Sibiri [Lower Carboniferous Bryozoa of Altai and Siberia] (In Russian). Trudy VSEGEI NS 13:1–420
Nikiforova AI (1938) Tipy kamennoygol'nych mshanok evropeiskoi chasti SSSR [Types of Carboniferous bryozoans from the European part of SSSR] (In Russian). AN SSSR, Moscow-Leningrad
Nikiforova AI (1948) Nizhnekamennougol'nye mshanki Karatau [Lower Carboniferous bryozoans of Karatau] (In Russian). Alama-Ata

Paleozoi yugo-vostoka Zapadno-Sibirskoi plitu [Palaeozoic of southeast of Western Siberian plate] (1984) (In Russian) Novosivirsk

Popeko LI (2000) Karbon Mongolo-Ochotskogo orogennogo poyasa [Carboniferous of the Mongol-Okhotsk orogenic belt] (In Russian). Dalnauka, Vladivostok

Puchkov VN (2000) Paleogeodinamika Yuzhnogo i Srednego Urala [Palaeogeodynamic of Southern and Middle Urals] (In Russian) Dauria

Pumpyanski AM (1988) Kamennoygol'nue otlozheniya Tyumensko-Kustanaiskogo progiba [Carboniferous from Tyumen-Kustanai inflection] Biostratigraphy and lithology of Upper Palaeozoic of Urals. (In Russian). Sverdlovsk

Pumpyanski AM (1990) Devonskie otlozheniya dourskogo fundamenta yujnoi chaste Zapadno-Sibirskoi plitu [Devonian of the pre-Jurassic basement of southern part of the Western Siberian plate]. (In Russian). All-Russian conference, Sverdlovsk, pp 49–58

Pumpyanski AM (1992) Stratigraphiya kamennoygol'nuch otlojenii severnoi chasti Tyumen-Kustanaiskogo progiba [Stratigraphy of Carboniferous deposits of the northern part of Tyumen-Kustanai inflection]. (In Russian). Toporkovskii chteniya 1:25–32

Pumpyanski AM (1999) Kamennoygol'nue otlojeniya Kurganskogo Zaural'ya [Carboniferous of Kurgan Trans-Urals]. (In Russian). Toporkovskii chteniya 4:55–62

Ross JF (1981) Biogeography of Carboniferous ectoproct Bryozoa. Palaeontol 24(2):313–341

Ross JRP, Ross CA (1990) Late Palaeozoic bryozoan biogeography. In: McKerrow WS, Scotese CR (eds) Palaeozoic palaeogeography and biogeography, vol 12, Geological Society Memoir. The Geological Society, London, pp 353–362

Snyder EM (1991) Revised taxonomic procedures and paleoecological applications for some North American Mississippian Fenestellidae and Polyporidae (Bryozoa). Palaeontogr Amer 57:1–275

Stepanova TI, Kucheva NA, Mizens LI, Tolokonnikova ZA (2010a) Novue dannye po stratigraphii verchnepaleozoiskich otlojenii vskrytykh skvazhinoi Kurgan-Uspenskaya 1 [New data about stratigraphy Upper Palaeozoic opens Kurgan-Uspenskaya 1 borehole]. (In Russian). In: All-Russian conference, Tyumen, pp 155–158

Stepanova TI, Kucheva NA, Mizens GA (2010b) O vozraste i usloviyach obrazovaniya terrigenno-karbonatnuch otlozenii, vskrytych skvazinoi VK-44 v Vagai-Ischimskoi vpadine (ug Zapadnoi Sibiri) [About age and formation conditions of terrigenous-carbonate deposits from borehole VK-44 in Vagai-Ischimskoi depression (south of Western Siberia)] (In Russian). Yearbook-2009 157:83–87

Stepanova TI, Kucheva NA, Mizens GA, Ivanova RI, Mizens LI, Tolokonnikova ZA. Rylkov SA (2011) Stratigraphiya paleozoiskogo razreza vskrytogo parametricheskoi skvazhinoi Kurgan-Uspenskaya-1 (ugo-zapadnaya okraina Zapadnoi Sibiri) [Stratigraphy of Palaeozoic section uncovered by Kurgan-Uspenskaya-1 key borehole (Western Siberia southwest margin)] (In Russian). Lithosphera 3:3–21

Trizna VB (1958) Rannekamennougolnye mshanki Kuznetzkoi kotloviny [Early Carboniferous bryozoans of the Kuznetzk depression] (In Russian). Trudy VNIGRI 122:1–433

Ulrich EO (1882) American Paleozoic Bryozoa. J Cincinnati Soc Nat Hist 5:233–257

Ulrich EO (1884) American Paleozoic Bryozoa. J Cinci Soc Nat Hist 7:24–51

Ulrich EO (1888): *Sceptropora*, a new genus of Bryozoa, with some remarks on *Helopora* Hall, and other genera of that type. The American Geologist, 1(4):228-234

Ulrich EO (1890) Paleozoic Bryozoa. Bull Geol Surv Illinois 8:283–688

Vine GR (1884) Fourth report of the Committee appointed for the purpose of reporting on fossil Polyzoa. Report of the 53th meeting of the British association for the advancement of science, pp 161–209

Wyse Jackson PNW (1996) Bryozoa from the Lower Carboniferous (Visean) of County Fermanagh, Ireland. Bull Nat His Mus London (Geol) 52(2):119–171

Xia FS (1997) Marine microfaunas (bryozoans, conodonts and microvertebrate remains) from the Frasnian-Famennian interval in northwestern Junggar Basin of Xinjiang in China. Beitr Paläont 22:91–207

Yang KC, Hu ZX. Xia F (1988) Bryozoans from Late Devonian and Early Carboniferous of Central Hunan. Palaeontol Sinica 174, NS B 23:1–197

# Chapter 27
# The Use of Early Miocene Bryozoan Faunal Affinities in the Central Paratethys for Inferring Climatic Change and Seaway Connections

## Early Miocene Bryozoans in the Central Paratethys

**Norbert Vávra**

**Abstract** The bryozoan faunas from a few important localities of the early Miocene of the Austrian part of the Central Paratethys are briefly described. All bryozoan faunas belong to the late Eggenburgian or early Ottnangian except for two localities yielding only a few rather poor fragments of bryozoans. Faunal migrations from the Western Mediterranean via the Rhône Basin into the Alpine Foredeep and the Central Paratethys best explain the geographic distribution of a number of bryozoan genera in the faunas of the Eggenburgian. A short climatic deterioration during this time-span may explain the occurrence of bryozoan-rich strata at this time characterized by extended temperate-water sedimentation. This climatic event could also explain why some genera are either extremely rare (e.g. *Steginoporella*) or even completely absent.

**Keywords** Bryozoa • Miocene • Paratethys • Palaeoclimatology • Palaeogeography

## Introduction

### *Historical Review*

Whereas bryozoan faunas from the middle Miocene ('Badenian' according to the regional stage concept of the Paratethys) have been described in extensive publications since the nineteenth century (e.g., Reuss 1847, 1874; Manzoni 1877, 1878), bryozoans from the early Miocene ('Eggenburgian', Molasse Zone) have remained understudied until today. Suess (1866) was the first author who mentioned

---

N. Vávra (✉)
Department of Palaeontology, Geozentrum, University of Vienna, Althanstrasse 14, Vienna A-1090, Austria
e-mail: norbert.vavra@univie.ac.at

A. Ernst, P. Schäfer, and J. Scholz (eds.), *Bryozoan Studies 2010*,
Lecture Notes in Earth System Sciences 143, DOI 10.1007/978-3-642-16411-8_27,

**Fig. 27.1** Bryozoan localities of the Eggenburgian in lower Austria; *asterisks* indicate localities

Bryozoa from the early Miocene of the Paratethys. Stoliczka (in Suess 1866) identified only a few specimens from three localities: Burgschleinitz, Oberdürnbach, and Maissau (Lower Austria, see Fig. 27.1). This material became lost and was never revised. In fact, Reuss and Manzoni largely overlooked this early report, mentioning in the publications above five species from four early Miocene localities. These few data gave the only information until Kühn's description of 22 species of Bryozoa, seven of them being new (Kühn 1925). A few specimens of this material are still available in the collections of the 'Geologische Bundesanstalt' (Austrian Geological Survey, Vienna). Thirty years later, the same author published a study of Bryozoa from the early Miocene 'Retzer Sands' (Kühn 1955). This fauna has been restudied recently (Vávra 2008). Except for two contributions dealing with special aspects of the bryozoan faunas from the Eggenburgian of the Paratethys (Vávra 1981, 1987), a comprehensive study is still lacking.

## *Aims of the Present Research*

After collecting at a number of Eggenburgian localities in Lower Austria for more than 35 years, the present author has yielded a large collection of bryozoan material, which

**Table 27.1** Stratigraphy of the Miocene of the Molasse Zone in Lower Austria (according to Wessely (2006)). (1) Fels-Formation, (2) Zogelsdorf-Fm., Retz-Fm., Brugg, Grübern, Limberg, Oberdürnbach

| Epochs | Age (in Ma) | Stages | Stages of the Central Paratethys | Formations | Planktonic Foraminifera | Calcareous Nanno-plankton |
|---|---|---|---|---|---|---|
| ↑ Middle Miocene | 16,4 | Langhian | ↑ Badenian | | ↑ M5 (N8) | NN4 |
| Early Miocene | 17,2 | Burdigalian | Karpatian | | M4 (N7) | |
| | 18,8 | | Ottnangian | ■ (2) | M3 (N6) | NN3 |
| | 20,5 | | Eggenburgian | ■ (1) | M2 (N5) | NN2 |
| | 23,8 | Aquitanian | Egerian ↓ | Melk-Fm. | | |

is the basis of a detailed taxonomic study currently in preparation. A few preliminary results and their palaeobiogeographic and palaeoclimatologic implications are given here. Several new answers to old questions can be given: Does the Austrian Miocene have a rather uniform bryozoan fauna or are there any differences between the early and the middle Miocene faunas? Moisette et al (2006) discussed biogeographical aspects of middle Miocene bryozoan faunas of Hungary. Among these Hungarian faunas, a considerable number of species of Eastern Atlantic/Mediterranean relationship, species endemic to the Mediterranean or the Paratethys, and also some species with Indo–Pacific affinities were mentioned. These data need further studies, however, and will therefore not be discussed in detail here.

As will be shown below, the diverse faunas from the Neogene of Western Europe (especially France) are important for determining the exact timing of possible bryozoan migrations from Western Europe into the Central Paratethys.

## *Localities and Stratigraphy*

The bryozoan material under study was collected at various localities, most of them rather close to the city of Eggenburg, situated in the NW part of Lower Austria (see Fig. 27.1). In this area, sediments of early Miocene age lie transgressively on top of crystalline rocks of the Bohemian Massif. Right in this area are the stratotypes for the 'Eggenburgian' as part of the stage concept of the Paratethys (Steininger and Seneš 1971). The exact stratigraphic correlation is based on planktonic foraminifera and calcareous nannoplankton (Table 27.1). A considerable variety of local biostratigraphic facies exist. The present state of knowledge concerning the lithostratigraphic units is summarized in Wessely (2006).

Nearly all Eggenburgian bryozoan localities belong to the late Eggenburgian, perhaps partly even to the early Ottnangian. Only two localities (Fels and Gösing) expose the Fels Formation, which offers a chance to identify possible immigrants from the Indo–Pacific. Unfortunately, these two faunas have such a low biodiversity due to poor preservation that any species determination of specimens is uncertain (for details see below).

## *Bryozoan Faunas*

For many years, the main problem in studying bryozoan material from the Eggenburgian had been the low number of specimens available. Kühn (1925) identified 22 species in his description of Bryozoa from the Eggenburg area, among them seven new species and one new genus. In addition to the low number of specimens, many of them have been lost. Even in Kühn's (1955) study of the Bryozoa from the Retz Formation, he could only identify 10 taxa, each of them on the basis of only a few specimens. Within the last 35 years, the situation has changed due to the discovery of new bryozoan localities by field mapping geologists (Prof. Steininger, Eggenburg; Dr. Roetzl, Geological Survey, Vienna) and intensified collecting by the author of this paper. As a result, an enormous amount of bryozoan material is now available for study. Realizing that this material includes a number of difficult taxonomic problems (e.g. celleporids, cerioporids, lichenoporids), the present study will certainly not be the last.

## *Brugg, Austria*

This outcrop is situated in a ditch of a water-reservoir, close to the road connecting Sigmundsherberg with Brugg. A detailed description of the section, consisting mainly of fine-grained sands, has been given in Steininger and Vávra (1983: 27–28). The pectinid assemblage found here is typical for the Eggenburg Formation. The outcrop has yielded one of the richest and best-preserved bryozoan faunas from the Eggenburgian of the Paratethys.

## *Grübern, Austria*

An abandoned sandpit ('Sandpit J. Fiedler'), situated about 700 m south of Grübern, has yielded a bryozoan fauna dominated by large, globular celleporids. The sediment consists of coarse sands with pebbles and frequent skeletal clasts (pectinids, balanids, brachiopods). Larger bulk-samples yielded a bryozoan fauna of more than 40 taxa. The sediments represent a high-energy environment of late Eggenburgian to Ottnangian age, which was also confirmed by K. Holcová (Charles University, Praha, pers. commun.) on the basis of its foraminifer assemblage.

## *Limberg, Austria*

This quarry ('Steinbruch Hengl') is situated about 3.5 km NE of Maissau (see Fig. 27.1); it exposes sediments of the Zogelsdorf Formation, which overlies granites of the Thaya pluton. The section starts with a transgressive horizon and ends with fine sands containing bryozoans, barnacles and coralline algae (Nebelsick et al. 1991). The increasing number of bryozoans from base to top has been attributed to decreasing hydrodynamic energy and continued transgression. From this outcrop, about 28 different bryozoan taxa have been identified, a few of them being helpful in resolving biogeographic problems: *Frondipora verrucosa, Mesenteripora meandrina, Myriapora truncata, Onychocella demarcqi, Pseudofrondipora davidi, "Schizoporella" geminipora*, and *Smittina cervicornis.*

## *Oberdürnbach, Austria*

Outcrops in the alleyway 'Kellergasse' within the area of the village Oberdürnbach have been mentioned already by Schaffer (1913: 76–77). These fine-grained sands have yielded a well preserved bryozoan fauna, with more than 60 taxa including a number of species that are rare or unknown from the Eggenburgian (e.g. *Steginoporella cucullata*). The sediments are late Eggenburgian and/or early Ottnangian. Studies of foraminiferal assemblages confirmed the stratigraphic range of 'Eggenburgian-Ottnangian' (Holcová, pers. comm.).

## *Zogelsdorf, Austria*

In the surroundings of Zogelsdorf, bryozoan-rich limestones have been mined from early Medieval Ages until the twentieth century. The 'Zogelsdorfer Sandstein' or 'Weißer Stein von Eggenburg' has been quarried for buildings and sculptured stone. The sediments are described by Nebelsick (1989) as one of the rare examples of a temperate water carbonate facies from the Paratethys. Though bryozoans are common (especially in one of the three main facies types), the biodiversity is low: only 18 taxa have been identified. Bryozoan growth forms have been used for paleoecological interpretations at this locality (Nebelsick 1989).

## *Notable Occurrences and Absences*

Unfortunately, the database does not allow a statistically reliable comparison of the faunas. Nevertheless, a number of striking occurrences and absences yield useful results.

In the following, a number of bryozoan taxa will be discussed in detail including possible candidates for immigration from Western Europe, prominent absences, and genera of use in comparing Eggenburgian and Badenian faunas. Only the most important examples will be given here; for the Cheilostomata the 'Interim Classification' as given by Gordon (2011) is used.

## *Cheilostomata*

### *Copidozoum rectirostre* David et al., 1972

David et al. (1972) established this taxon on the basis of Neogene (Burdigalian) material of Mus (Gard), Rhône Basin. It was also identified by Vávra (1987) in sediments from the Eggenburgian of Oberdürnbach.

### *Amphiblestrum appendiculatum* (Reuss, 1847)

This species has been identified in the Neogene of SW Spain, France, Germany, Hungary, Poland, and Czech Republic (Zágoršek et al. 2007). In Austria, it has been confirmed for the Badenian (middle Miocene) of Baden, Eisenstadt and Mödling. Moissette et al. (2006) listed it among the endemic Mediterranean species of the Badenian. Realizing that most species recently described as *Ramphonotus minax* are probably conspecific with *A. appendiculatum* (Berning 2006: 19–20), its biogeographic distribution is far from certain. David et al. (1972) described *A. appendiculatum* from the Burdigalian of Mus (Gard). The occurrence of this species in Oberdürnbach may therefore be regarded as the result of immigration.

### *Cupuladria* sp.

Small fragments and poorly preserved zoaria, impossible to identify to the species level, have been found at Gösing (early Eggenburgian) and possibly also at Gauderndorf. Although this genus has no value to solve biogeographic problems, its occurrence is remarkable with respect to sedimentology: the zoaria could even survive in areas with fine grained sediments.

### *Scrupocellaria* sp.

This genus is common in the Badenian (middle Miocene) of Austria; it has been reported from more than 12 different localities (Vávra 1977). The material described is traditionally summarized as *Scrupocellaria elliptica* (Reuss, 1847), although it is seen as a complex comprising a number of different species

(Schmid 1989; Berning 2006: 32ff). From the Eggenburgian, until now it is known from Oberdürnbach (Vávra 1987) but so far missing from other localities.

### *Onychocella angulosa* (Reuss, 1847)

This species is common in the Badenian (middle Miocene) of Austria (15 localities in Burgenland, Lower Austria, Styria, and Vienna). In the Eggenburgian, it has been identified in material from Brugg and Klein-Meiseldorf. In France, it occurs in the Aquitaine Basin, in the Loire and Rhône valleys. These occurrences also place it in the Burdigalian (David et al. 1972). Possibly, this taxon represents one more immigrant from the Western Mediterranean in the late Eggenburgian.

### *Onychocella demarcqi* David et al., 1970

The typical bifoliate, flat, and branching zoaria of this species are among the dominant faunal elements in the sand facies of the late Eggenburgian at Brugg (Fig. 27.2d). Their correspondence with material from France and importance when discussing migration routes was already pointed out by Vávra (1987). Meanwhile, its occurrence has been confirmed for two more localities in the Eggenburgian: Unternalb ('Retzer Sands') and Grübern. It has not been found at any of the localities from the Badenian of Austria, Hungary, Czech Republic, Poland, etc. However, erect, broadly bilaminar branches of bryozoan zoaria described as *Onychocella* sp. have been reported from the Neogene (early Ottnangian) of Gurlan (Lower Bavaria, Germany) (Schneider et al. 2009: 82). They are described as conspecific with the material from the Eggenburgian of Austria. Smaller opesia and a better-developed cryptocyst possibly confirms the establishment of a new taxon for both, however, further studies are required.

### *Steginoporella cucullata* (Reuss, 1847)

This is a widespread species. It has been listed among the endemic Mediterranean taxa by Moisette et al. (2006), Sefian et al. (1999) and Berning (2006: 43) mention it from 'Late Miocene Atlantic waters', (Fig. 27.2b). So far, it has been reported from at least 12 different localities from the Badenian (Middle Miocene) of Austria (Burgenland, Lower Austria, Styria),, SW Spain and from the Czech Republic, Hungary, and Poland. For the Eggenburgian, it has been tentatively identified by Vávra (1987), an occurrence confirmed here. Realizing that the genus *Steginoporella* is rare in the Eggenburgian in general, it may be important in connection with possible migration.

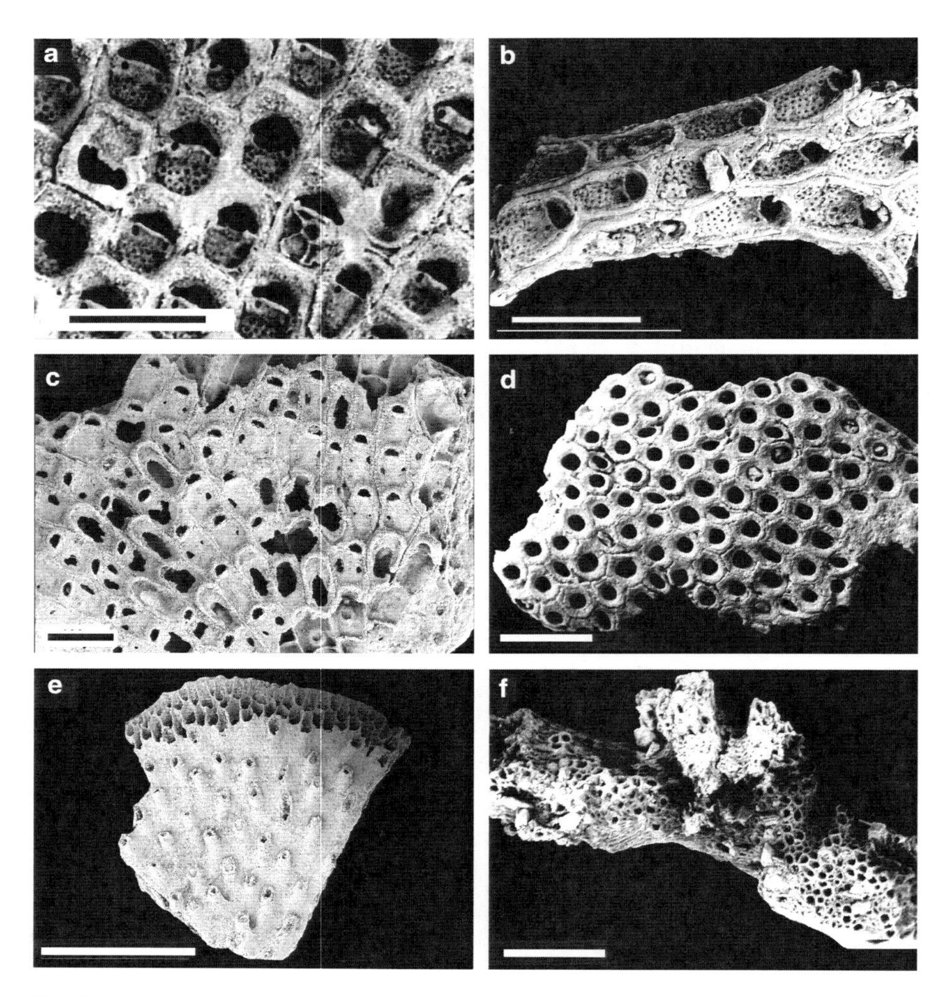


**Fig. 27.2** Bryozoan taxa that have migrated from the western Mediterranean into the Central Paratethys. (**a**) *Steginoporella rhodanica* Buge and David, 1967, Grübern, bar: 1 mm, (**b**) *Steginoporella cucullata* (Reuss, 1847), Oberdürnbach, bar: 1 mm, (**c**) *Thalamoporella* cf. *neogenica* Buge, 1950, Brugg, bar: 0,1 mm, (**d**) *Onychocella demarcqi* David et al., 1970, Brugg, bar: 1 mm, (**e**) *Mesenteripora meandrina* (Wood, 1844), Brugg, bar: 1 mm, (**f**) *Pseudofrondipora davidi* Mongereau, 1970, Oberdürnbach, bar: 1 mm

### *Steginoporella rhodanica* Buge and David, 1967

This species, first described by Buge and David (1967) from the Rhône Basin, was discovered in the Eggenburgian of Grübern, and its biogeographical connections were discussed by Vávra (1987), (Fig. 27.2a). It was also found in the early Miocene of Switzerland (St. Croix, Molasse Zone), as was a species of close morphological relationship – *Steginoporella elegans* – also in the Molasse Zone of Bavaria (Höch, Vávra 1981). Furthermore, *Steginoporella rhodanica* itself was recently reported from the early Ottnangian of Bavaria (Schneider et al. 2009: 82, Fig. 8c). These finds indicate a faunal migration from the Western Mediterranean via the Rhône Basin into the Alpine Foredeep, and finally into the Central Paratethys of the Eggenburg area. This agrees with the opinion of Pouyet and David (1979), who suggested a center of radiation for *Steginoporella* in Europe, presumably Western Europe, in the Miocene.

### *Thalamoporella* cf. *neogenica* Buge, 1950

This taxon is here reported from the Eggenburgian. A number of zoaria from Brugg is identical with – or at least very close to – *Thalamoporella neogenica*, which has been reported from the Aquitaine basins, the Loire Valley, and other localities from the Burdigalian (Fig. 27.2c). In addition, the occurrence of *Thalamoporella spathulata* David, 1949 from the Burdigalian of Mus (Gard), a species showing a rather similar morphology, should be mentioned (David et al. 1972). Finds at Brugg may represent one more possible hint for faunal migration during the late Eggenburgian.

### *Adeonella polystomella* (Reuss, 1847)

This taxon is well known from at least eight different localities from the Badenian (middle Miocene) of Austria. It has been also reported from the Neogene of the Czech Republic, France, Italy, Poland, and other countries (Vávra 1977). In the Eggenburgian, it had been found at Oberdürnbach (Vávra 1987); meanwhile, additional material from the Eggenburgian has been identified: Brugg – and possibly also Grübern and Limberg – have to be added to the list of localities. Realizing that this species has been reported also from the Burdigalian of France (e.g. Mus, David et al. 1972), its occurrence in the Eggenburgian of the Central Paratethys can be regarded as one more hint for possible faunal migration during this time.

### *Escharoides* sp.

This genus is common at a few of the Badenian (middle Miocene) localities in Austria, but it had been missing from the Eggenburgian. Now, a few specimens have been found at Brugg (late Eggenburgian). One of them can be attributed tentatively to *Escharoides megalota* (Reuss, 1847), a species, listed by Moissette et al. (2006) among the endemic Mediterranean taxa. Because the latter taxon has been reported also from the late Miocene of Spain (Berning 2006: 72), further studies are needed before any statement concerning possible migration of the genus *Escharoides* can be made.

### *Metrarabdotos* sp.

*Metrarabdotos maleckii* Cheetham, 1968 is one of the most common representatives of Cheilostomata in many bryozoan faunas from the Badenian (middle Miocene) of Austria. Even more so, its (nearly) complete absence from the faunas of the Eggenburgian is noteworthy. Though reported from at least 12 different localities of the Badenian (situated in Burgenland, Lower Austria, Styria, and Vienna), its occurrence in the Eggenburgian faunas still remains uncertain. Kühn (1925) reported a few specimens of *Metrarabdotos moniliferum* (possibly identical with *Metrarabdotos maleckii*?) from two localities of Eggenburgian age (Eggenburg-railway station and Kühnring). He stated, however, that this material was poorly preserved. Meanwhile, it has been lost and never revised. After more than 35 years of collecting, the author has seen only a few tiny undeterminable fragments, which may be close to *Metrarabdotos*. Knowing, that *Metrarabdotos* is regarded as an almost ideal subject for studies of evolutionary mode and tempo (Cheetham et al. 2007), and realizing its palaeoclimatic significance (e.g. Cheetham 1967), the complete absence of this genus in the Eggenburgian requires an explanation. If we assume a (short) deterioration of climatic conditions for the late Eggenburgian and early Ottnangian – as discussed above – the absence of *Metrarabdotos* makes sense. This climatic deterioration is supported by the rarity of *Steginoporella* during this time.

### "*Schizoporella*" *geminipora* (Reuss, 1847)

This species was reported by Kühn (1925) from two localities in the Eggenburgian: Klein-Meiseldorf and Kremserberg near Eggenburg. Vávra (1981, 1987) confirmed its occurrence in Brugg, Grübern, and Oberdürnbach, meanwhile it has been found also at Limberg, Pulkau, and Zogelsdorf. Based on its occurrence also in the Burdigalian of France (e.g. David et al. 1972) as well as on its role as an endemic Mediterranean taxon, it can be regarded as another species, which invaded the Central Paratethys during the late Eggenburgian – and possibly once more again during the Badenian.

### *Margaretta cereoides* (Ellis and Solander, 1786)

Comparing bryozoan faunas from the Eggenburgian and the Badenian of the Central Paratethys, the complete absence of the genus *Margaretta* in the Eggenburgian is one of the most striking differences. Colonies of this genus can be identified easily; even as tiny, poorly preserved fragments – they can hardly be overlooked making this genus an ideal object for paleobiogeographic comparison. Moissette et al. (2006) listed *Margaretta cereoides* as an Indo–Pacific species. *Margaretta* is still common in Recent faunas in the Indo–Pacific. Harmer (1957) described nine species, five of which were found in the SW Pacific. Lu (1991) mentioned four species from the Nansha Sea, and Tilbrook (2006) described two species from the Solomon Islands. As one of the fossil records, the occurrence of *Margaretta cereoides* in the Aquitanian of Qoum (125 km S of Tehran, Iran) is worth mentioning in this connection (Furon and Balavoine 1960). The Recent genus *Margaretta* is not restricted to the Pacific, but also occurs in the tropical and subtropical parts of the Atlantic, Red Sea, and Mediterranean including the Adriatic Sea. *Margaretta cereoides* has even been regarded as 'possibly endemic to the Mediterranean' (Hayward and McKinney 2002: 74). Its complete absence from the faunas of the Eggenburgian of Austria may perhaps be due to climatic conditions, as already discussed for other taxa.

### *Myriapora truncata* (Pallas, 1766)

This is – except for the celleporids – one of the most common bryozoans in the faunas of the Eggenburgian. It has been found at about 14 different localities: Brugg, Burgschleinitz (Roßberg), Eggenburg (Brunnstube, Kremserberg, Schindergraben), Klein-Meiseldorf, Kühnring, Limberg, Maissau (Schloßberg), Maissauer-Straße, Oberdürnbach, Obernalb, Pulkau, and Zogelsdorf. Realizing that it is known, among others, also from the Burdigalian of France (e.g. David et al. 1972) and that it is even regarded as an endemic Mediterranean bryozoan species (Moissette et al. 2006), but otherwise reported to occur also in the Atlantic (Berning 2006: 92), two immigrations into the Central Paratethys may have occurred: the first one during the late Eggenburgian and a second one during the Badenian.

### Celleporids

Large globular zoaria of this group are among the most common bryozoans found in the Eggenburgian. As taxonomic studies of this group have not yet been finished, at present time any generalization would still include a rather high degree of uncertainty.

## *Cyclostomata*

### *Ceriopora tumulifera* Canu and Lecointre, 1934

This species is rather common at different bryozoan localities in the Miocene of France. It has recently been described as a rare faunal element from the Badenian of the Moravian part of the Vienna Basin (Zágoršek et al. 2007). Meanwhile, one specimen in the surroundings of Oberdürnbach has been discovered – being the first specimen from the Eggenburgian of the Central Paratethys. It may refer to possible faunal migration from Western Europe into the Eggenburg area.

### *Frondipora verrucosa* (Lamouroux, 1821)

This species has been found in the Eggenburgian from Brugg, Grübern, Limberg, Maissau, Oberdürnbach, Pulkau, and Zogelsdorf. It represents one more Western European faunal element in the Eggenburgian.

### *Lichenopora* sp.

A detailed taxonomic study of the lichenoporids of the Eggenburgian has not yet been done. A general aspect already stated earlier (Vávra 1981), however, can now be confirmed on the basis of rich material: zoaria from the Eggenburgian show only uniserial fascicles. This contrasts the Badenian faunas, where bi- and triserial fascicles in lichenoporids are also common.

### *Mesenteripora meandrina* (Wood, 1844)

This species has been found in a number of faunas from the Eggenburgian (Brugg, Limberg, Oberdürnbach, Pulkau, and Zogelsdorf), which suggests correspondence with bryozoan faunas from Western Europe (Fig. 27.2e).

### *Polyascosoeciella* sp. – *Crisidmonea* sp.

According to Moissette et al. (2006), *Polyascosoeciella coronopus* (Canu and Bassler, 1922) is an Indo-Pacific bryozoan species. Tiny fragments of this zoarial type from the Fels Formation (early Eggenburgian) can be regarded as the result of immigration through an open Tethyan seaway from the Indo–Pacific at that time.

### *Pseudofrondipora davidi* Mongereau, 1970

This species established on the basis of material from the Burdigalian of France (Roua, Taulignan – Drôme), occurred at a number of localities of Badenian age, but also in the Eggenburgian: Limberg, Oberdürnbach, and Pulkau, (Fig. 27.2f). It may therefore represent one more faunal affinity with Western European bryozoan faunas.

## *Biogeographical Aspects*

During the late Eocene, the vanishing Tethys Ocean had given birth to two new marine realms at its former western end, the Eurasian Paratethys basin and the Mediterranean. The complex palaeogeographic evolution, the development and closure of various seaways, and the consequences for marine and terrestrial faunal development have been the subject of numerous studies (e.g. Rögl and Steininger 1983; Rögl 1998; Harzhauser and Piller 2007). During the late Oligocene and early Miocene (late Chattian – early Burdigalian), a maximum connection with the Indian Ocean existed, the upper NN2 nannoplankton zone being characterized by extensive Indo–Pacific connections. It is therefore not surprising, that molluscan faunas of this time-span from the Mediterranean, the Paratethys, and the Iranian Qoum Basin show strong relationships (Harzhauser et al. 2002). The possibility that many taxa had their origin in the Paratethys and migrated from there towards the east has been suggested by Harzhauser and Piller (2007), the Paratethys having possibly been a 'biodiversity hotspot' (Renema et al. 2008) during the Oligocene. As far as Bryozoa are concerned, only low-diversity faunas have been discovered from this time (e.g. Gösing, Fels am Wagram) in the Central Paratethys (this study). During the time of nannoplankton zones NN3 to lower NN4, the seaway connection between the Mediterranean and Indian Ocean was closed, and a continental migration bridge between Africa and Eurasia formed, resulting in terrestrial mammal migrations. For the marine faunas of the late Eggenburgian (and early Ottnangian), a seaway existed from the Western Mediterranean through the Rhône Basin into the Alpine Foredeep and the Central Paratethys at this time (Fig. 27.3, after Kroh and Harzhauser 1999). For bryozoan faunas, a few examples of possible immigration have been discussed earlier (Vávra 1981, 1987); meanwhile, faunal migration has been confirmed also for echinoderms (Kroh and Harzhauser 1999).

Since then, continuous bryozoan studies of faunas from the Eggenburgian have resulted in more detailed informations, thus providing a longer list of possible immigrants and presenting a more detailed picture of these faunal migrations. There are faunal elements, which obviously experienced quite acceptable ecological conditions in the Central Paratethys and consequently became more widespread. Other genera, however, were already near to their ecological limits in the Central Paratethys; the short cooling in the late Eggenburgian resulted in the accumulation of bryozoan-rich sediments but may also explain why a few genera (e.g. *Steginoporella*) remained rare during this time span.

Another faunal migration from the Mediterranean happened at the beginning of the middle Miocene following the far reaching transgression into the circum-Mediterranean area. This transgression brought bryozoan faunas to the Central Paratethys in which subtropical to tropical elements were common (Vávra 1980). This is not surprising, however, because this transgression coincided with a global warming, resulting in the spreading of warm water elements into higher latitudes in general.

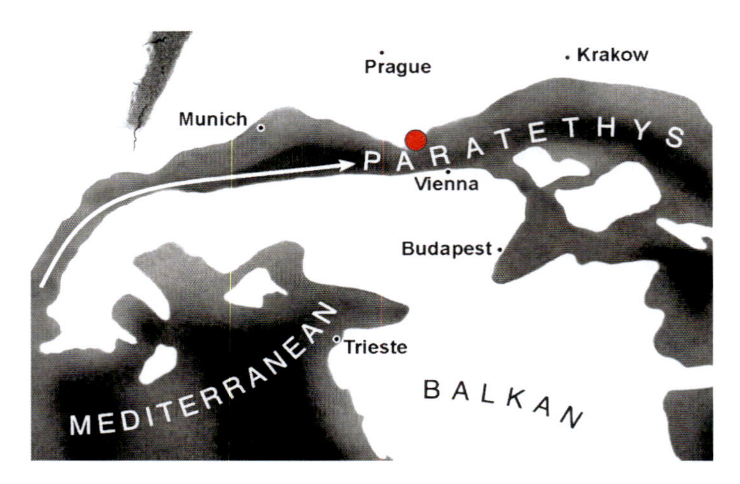


**Fig. 27.3** Faunal migration pattern from the Mediterranean via the Rhône Basin and Alpine foredeep into the Central Paratethys (palaeogeography after Kroh and Harzhauser (1999), with kind permission of the authors). *Red dot*: Eggenburg

## *Palaeoclimatological Background*

A number of publications have recently contributed to a renewal of the discussion of climatic conditions during the middle Burdigalian (Ottnangian according to the local stage concept of the Paratethys area). Palaeontological investigations yield data supporting the idea of a short, cooler period during that time (Nebelsick 1989; Grunert et al. 2009 and references given therein). Rich occurrences of diatoms, silicoflagellates, and radiolarians as well as a small-sized fauna of planktonic foraminifera indicate rather low temperatures (15°C to 17°C) of surface waters for the early Ottnangian (Bachmann 1973; Rupp and Haunold-Jenke 2003). Benthic foraminifera and echinoderm faunas indicate cooling of the deeper-water body (Kroh 2007). The palaeontological data have recently been confirmed by palaeotemperature reconstructions based on oxygen isotopes of shark teeth from the Paratethys (Kocsis et al. 2009). The closure of the connection of the Paratethys with the Indo–Pacific seems to have been an important factor for climate evolution – considerable changes of the flow pattern of marine currents and the absence of warm Indo–Pacific waters may have contributed essentially to a cooling in the Mediterranean and the Paratethys area (Grunert et al. 2009).

For the early Ottnangian, temperate-carbonate sediments rich in Bryozoa, and the absence of any hermatypic corals, has already been emphasized by Nebelsick (1989). Temperate-water carbonate facies differ considerably from tropical carbonates (Nelson 1978; Scoffin 1987). In contrast to the global scale, fossil examples of temperate-water carbonates from the Central Paratethys are rare, resulting in an increased interest in the Zogelsdorf Formation and related sediments. Bryozoan skeletons have been repeatedly discussed as a dominant constituent of cool-water sediments in the Cenozoic (e.g. Hageman et al. 2000); this adds an additional, more general aspect increasing the interest in the study of these rich, bryozoan-bearing sediments at the turn from the Eggenburgian Stage to the Ottnangian Stage.

**Acknowledgements** Our bryozoan studies have been supported by the Fonds zur Förderung der wissenschaftlichen Forschung (Austria), FWF-Project P19337-B17 ('Biodiversity and faunal interchange: Bryozoans of the Paratethys') and P15600-B06 ('Bryozoan sediments in Cenozoic tropical environments'), a support, which is gratefully acknowledged. The author is also greatly indebted to A. Ostrovsky (University of Vienna) for a number of stereoscan pictures, to K. Holcová (Charles University, Praha) for biostratigraphical and ecological studies of foraminifera, R. Roetzel (Austrian Geological Survey) for information concerning biostratigraphy, and M. Vávra (Vienna) for graphical work. A number of valuable suggestions by the reviewers (B. Berning, K. Zágoršek, and M. Key) and by the editor (P. Schäfer) are gratefully acknowledged.

## References

Bachmann A (1973) Die Silicoflagellaten aus dem Stratotypus des Ottnangien. In: Papp A, Rögl F, Seneš J (eds) Miozän M2 – Ottnangien. Die Innviertler, Salgotarjaner, Bantapusztaer Schichtengruppe und die Rzehakia Formation. Chronostratigraphie und Neostratotypen, Miozän der zentralen Paratethys 3. Vydavatelstvo Slovenskej akadémie vied, Bratislava

Berning B (2006) The cheilostome bryozoan fauna from the Late Miocene of Niebla (Guadalquivir Basin, SW Spain): environmental and biogeographic implications. Mitt Geol-Paläont Inst Univ Hamburg 90:7–156

Buge E (1950) Note sur la synonymie de trois éspèces des bryozoaires: *Diastopora latomarginata* d'Orbigny 1852, *Eschara andegavensis* Michelin 1847 et *Obelia disticha* Michelin 1847. Bull soc géol France 5. sér. 20 (7–9):459–465

Buge E, David L (1967) Révision des éspèces de *Steginoporella* (Bryozoa-Cheilostomata) du Néogène Français. Trav Lab Géol Univ Lyon NS 14:7–27

Canu F, Bassler RS (1922) Studies on the cyclostomatous Bryozoa. Proc US Nat Mus 61 (22):1–160

Canu F, Lecointre G (1934) Les Bryozoaires cyclostomes des Faluns de Touraine et d'Anjou. Mém Soc Géol France (ns) 4:131–215

Cheetham AH (1967) Paleoclimatic significance of the bryozoan *Metrarabdotos*. Trans Gulf Coast Ass Geol Soc 17:400–407

Cheetham AH (1968) Morphology and systematics of the bryozoan Genus *Metrarabdotos*. Smithsonian Misc Coll 153(1):121

Cheetham AH, Sanner J, Jackson JBC (2007) *Metrarabdotos* and related genera (Bryozoa: Cheilostomata) in the Late Paleogene and Neogene of Tropical America. J Paleont/Paleont Soc Mem 81/67:1–96

David L (1949) Quelques bryozoaires nouveaux du Miocène du Gard et de l'Hérault. Bull Soc Géol France 5.sér. 19 (7–9):539–544

David L, Mongereau N, Pouyet S (1970) Bryozoaires du Néogène du bassin du Rhône. Gisements burdigaliens de Taulignan (Drôme). Docum Lab Géol Fac Sci Lyon 40:97–175

David L, Mongereau N, Pouyet S (1972) Bryozoaires du Néogène du bassin du Rhône. Gisements burdigaliens de Mus (Gard). Docum Lab Géol Fac Sci Lyon 52:1–118

Ellis J, Solander D (1786) The natural history of many curious and uncommon zoophytes, collected from various parts of the globe. White, London

Furon R, Balavoine P (1960) Les Bryozoaires Aquitaniens de Qoum (Iran). Bull soc géol France, 7 1(3):294–302
Gordon DP (2011) Genera & Subgenera of Cheilostomata. Interim Classification (Working Classification for Treatise)
Grunert P, Harzhauser M, Piller WE (2009) Temperate Klimabedingungen im Mittleren Burdigalium Mitteleuropas – globales oder lokales Phänomen? Ber Geol Bundesanst 81:13–14
Hageman SJ, James NP, Bone Y (2000) Cool-water carbonate production from epizoic bryozoans on ephemeral substrates. Palaios 15:33–48
Harmer SF (1957) The Polyzoa of the Siboga expedition. Part IV: Cheilostomata Ascophora, II (Ascophora except Reteporidae, with additions to part II, Anasca). Rep Siboga Exped 28d:642–1147
Harzhauser M, Piller WE (2007) Benchmark data of a changing sea – palaeogeography, palaeobiogeography and events in the Central Paratethys during the Miocene. Palaeogeogr Palaeoclimat Palaeoecol 253:8–31
Harzhauser M, Piller WE, Steininger FF (2002) Circum-Mediterranean Oligo-Miocene biogeographic evolution – the gastropods' point of view. Palaeogeogr Palaeoclimat Palaeoecol 183:103–133
Hayward PJ, McKinney FK (2002) Northern Adriatic Bryozoa from the vicinity of Rovinj, Croatia. Bull Am Mus Nat Hist 270:1–139
Kocsis L, Vennemann TW et al (2009) Constraints on Miocene oceanography and climate in the Western and Central Paratethys: O-, Sr-, and Nd-isotope compositions of marine fish and mammal remains. Palaeogeogr Palaeoclimat Palaeoecol 271:117–129
Kroh A (2007) Climatic changes in the Early to Middle Miocene of the Central Paratethys and the origin of its echinoderm fauna. Palaeogeogr Palaeoclimat Palaeoecol 253:169–207
Kroh A, Harzhauser M (1999) An echinoderm fauna from the Lower Miocene of Austria: paleoecology and implications for Central Paratethys paleobiogeography. Ann Naturhist Mus Wien 101A:145–191
Kühn O (1925) Die Bryozoen des Miozäns von Eggenburg. Abh kk Geol Reichsanst 22(3):21–39
Kühn O (1955) Die Bryozoen der Retzer Sande. Sitz-Ber Österr Akad Wiss Math-Naturwiss Kl Abt I 164(4,5):231–248
Lamouroux JVF (1821) Exposition méthodique des genres de l'ordre des Polypiers, avec leur description et celles des principales espèces, figurées fans 84 planches: les 63 premières appartenant à l´Histoire naturelle des Zoophytes d´Ellis et Solander, Paris
Lu L (1991) Holocene bryozoans from the Nansha sea area. In: Team M (ed) Quaternary biological groups of the Nansha Islands and the neighbouring waters. Zhongshan University Publishing House, Guangzhou
Manzoni A (1877) I Briozoi fossili del Miocene d'Austria ed Ungheria II. Denkschr K Akad Wiss Math-Naturwiss Cl 37:49–78
Manzoni A (1878) I Briozoi fossili del Miocene d'Austria ed Ungheria III. Denkschr K Akad Wiss Math-Naturwiss Cl 38:1–24
Moissette P, Dulai A, Müller P (2006) Bryozoan faunas in the Middle Miocene of Hungary: biodiversity and biogeography. Palaeogeogr Palaeoclimat Palaeoecol 233:300–314
Mongereau N (1970) Les bryozoaires cyclostomes branchus du Miocène du Bassin du Rhône (France): Données nouvelles. Géobios 3(1):29–42
Nebelsick JH (1989) Temperate water carbonate facies of the Early Miocene Paratethys (Zogelsdorf Formation, Lower Austria). Facies 21:11–40
Nebelsick JH, Steininger FF et al (1991) F/11: Limberg, Steinbruch Hengl. In: Steininger FF Roetzel R, Rögl F (eds) Einführung. Geologische Grundlagen, Lithostratigraphie, Biostratigraphie und chronostratigraphische Korrelation und Paläogeographie der Molassesedimente am Ostrand der Böhmischen Masse. In: Roetzel R, Nagel D (eds) Exkursionen im Tertiär Österreichs. Molassezone. Waschbergzone. Korneuburger Becken. Wiener Becken. Eisenstädter Becken. Österreichische Paläontologische Gesellschaft, Wien

Nelson CS (1978) Temperate shelf carbonate sediments in the Cenozoic of New Zealand. Sedimentology 25:737–771
Pallas PS (1766) Elenchus Zoophytorum. Hagae-Comitum
Pouyet S, David L (1979) Revision of the Genus *Steginoporella* (Bryozoa, Cheilostomata). In: Larwood GP, Abbott MB (eds) Advances in bryozoology. Academic Press, London
Renema W, Bellwood DR, Braga JC, Bromfield K, Hall R, Johnson KG, Lunt P, Meyer CP, McMonagle LB, Morley RJ, O'Dea A, Todd JA, Wesselingh FP, Wilson MEJ, Pandolfi JM (2008) Hopping hotspots: global shifts in marine diversity. Science 321:654–657
Reuss AE (1847) Die fossilen Polyparien des Wiener Tertiärbeckens. Naturwiss Abh 2:1–109
Reuss AE (1874) Die fossilen Bryozoen des österreichisch-ungarischen Miocäns. Denkschr K Akad Wiss Math-Naturwiss Kl 33:141–190
Rögl F (1998) Palaeogeographic considerations for Mediterranean and Paratethys seaways (Oligocene to Miocene). Ann Naturhist Mus Wien 99A:279–310
Rögl F, Steininger FF (1983) Vom Zerfall der Tethys zu Mediterran und Paratethys. Die neogene Paläogeographie und Palinspastik des zirkum-mediterranen Raumes. Ann Naturhist Mus Wien 85A:135–163
Rupp C, Haunold-Jenke Y (2003) Untermiozäne Foraminiferenfaunen aus dem oberösterreichischen Zentralraum. Jb Geol Bundesanst 143(2):227–302
Schaffer FX (1913) Geologischer Führer für Exkursionen im Wiener Becken. III. Teil. Nebst einer Einführung in die Kenntnis der Fauna der ersten Mediterranstufe (= Sammlung geologischer Führer 18). Borntraeger, Berlin
Schmid B (1989) Cheilostome Bryozoen aus dem Badenien (Miozän) von Nussdorf (Wien) Beitr Paläont Österreich 15:1–101
Schneider S, Berning B, Bitner MA, Carriol R-P, Jäger M, Kriwet J, Kroh A, Werner W (2009) A parautochthonous shallow marine fauna from the Late Burdigalian (early Ottnangian) of Gurlan (Lower Bavaria, SE Germany): macrofaunal inventory and paleoecology. N Jb Geol Paläont, Abh 254(1–2):63–103
Scoffin TP (1987) An introduction to carbonate sediments and rocks. Blackie, Glasgow-London
Sefian NL, Pouyet S, El Hajjaji K (1999) Bryozoaires du Miocène supérieur de Charf el Akab (NW Maroc). Rev Paléobiol 18(1):221–252
Steininger F, Seneš J (1971) $M_1$ Eggenburgien. Die Eggenburger Schichtengruppe und ihre Stratigraphie. Chronostratigraphie und Neostratotypen, Miozän der zentralen Paratethys 2, Vydavatelstvo Slovenskej akadémie vied, Bratislava
Steininger F, Vávra N (1983) International Bryozoology Association. In: Sixth international conference. Palaeontological field meeting, University of Vienna, Vienna, 12–15 July 1983
Suess E (1866) Untersuchungen über den Charakter der österreichischen Tertiärablagerungen. I. Über die Gliederung der tertiären Bildungen zwischen dem Mannhart, der Donau und dem äußeren Saume des Hochgebirges. Sber Akad Wiss Wien Math-Naturwiss Kl Abt I 54:87–152
Tilbrook KJ (2006) Cheilostomatous Bryozoa from the Solomon Islands. Santa Barbara Mus Nat Hist Monogr 4:1–385
Vávra N (1977) Bryozoa tertiaria. In: Zapfe H (ed) Catalogus Fossilium Austriae, Vb/3, Österreichische Akademie der Wissenschaften, Wien
Vávra N (1980) Tropische Faunenelemente in den Bryozoenfaunen des Badenien (Mittelmiozän) der Zentralen Paratethys. Sber Österr Akad Wiss Math-Naturwiss Kl Abt I 189 (1–3):49–63
Vávra N (1981) Bryozoa from the Eggenburgian (Lower Miocene, Central Paratethys) of Austria. In: Larwood GP, Nielsen C (eds) Recent and fossil Bryozoa. Olsen & Olsen, Fredensborg
Vávra N (1987) Bryozoa from the early Miocene of the central Paratethys: biogeographical and biostratigraphical aspects. In: Ross JRP (ed) Bryozoa: present and past. Western Washington University, Bellingham
Vávra N (2008) Bryozoans of the Retz-Formation (Early Miocene, Austria) – A high-energy environment case study. In: Hageman SJ, Key jr. MM, Winston JE (eds) Bryozoan studies 2007.

Proceedings of the 14th International Bryozoology Association conference, Boone, 1–8 July 2007 (=Virginia Museum of Natural History, Special Publications No. 15). Martinsville, 2008

Wessely G (2006) Geologie der österreichischen Bundesländer. Niederösterreich. Geologische Bundesanstalt, Wien 1–416

Wood SV (1844) Descriptive catalogue of the Zoophytes from the Crag. Ann Mag Nat Hist 13:10–21

Zágoršek K, Vávra N, Holcová K (2007) New and unusual Bryozoa from the Badenian (Middle Miocene) of the Moravian part of the Vienna Basin (Central Paratethys, Czech Republic). N Jb Geol Paläont Abh 243(2):201–215

# Chapter 28
# Palaeoecology, Preservation and Taxonomy of Encrusting Ctenostome Bryozoans Inhabiting Ammonite Body Chambers in the Late Cretaceous Pierre Shale of Wyoming and South Dakota, USA

## Cretaceous Ctenostome Bryozoans of Wyoming and South Dakota

**Mark A. Wilson and Paul D. Taylor**

**Abstract** The straight ammonite *Baculites* is locally abundant in the Pierre Shale deposited across the Western Interior Seaway (WIS) of North America during the Late Cretaceous (Campanian-Maastrichtian). Fossils of *Baculites* are commonly preserved with body chambers filled by fine-grained sediment. Removal of the ammonite shell occasionally exposes the undersurfaces of runner-like bryozoan colonies embedded within this sediment. These bryozoans encrusted inner shell surfaces of empty body chambers of ammonites that apparently floated for some interval after death, before dropping to the seafloor where they were grazed by chitons, crustaceans and limpets and further encrusted by brachiopods and other bryozoans. SEM studies show that the runner-like bryozoans, previously identified in the literature as 'pyriporoid' or 'pyriporid' cheilostomes, are in fact ctenostomes belonging to a new genus and species, *Pierrella larsoni*. The ctenostome identity of these Pierre Shale bryozoans is evident from not only distorted zooid shapes and the lack of a calcareous skeleton, but also the remarkable preservation of some zooids with setigerous, or pleated, collars. The mode of preservation of these ctenostomes is enigmatic, although epibiont shadowing due to dissolution of surrounding shell played a role in most instances. As there is only minor overgrowth by organisms with mineralized skeletons, it is clear that the colonies are not preserved by bioimmuration. Most or all of the encrusted ammonites are concretionary so it is

M.A. Wilson (✉)
Department of Geology, The College of Wooster, Wooster, OH 44691, USA
e-mail: mwilson@wooster.edu

P.D. Taylor
Department of Earth Sciences, Natural History Museum, Cromwell Road, London SW7 5BD, UK
e-mail: p.taylor@nhm.ac.uk

A. Ernst, P. Schäfer, and J. Scholz (eds.), *Bryozoan Studies 2010*,
Lecture Notes in Earth System Sciences 143, DOI 10.1007/978-3-642-16411-8_28,

possible that rapid growth of authigenic calcite, which has been shown to be present by laser Raman spectroscopy, preserved the ctenostome zooids through a process of lithoimmuration before complete decay of their body walls.

**Keywords** Ctenostome bryozoans • Cretaceous • Western Interior Seaway • Lithoimmuration

## Introduction

By the Campanian and Maastrichtian stages of the Cretaceous, cheilostome-dominated bryozoan faunas of moderate to high diversity had become established in many regions of the world (e.g. India (Guha and Nathan 1996), South Africa (Brood 1977), southeastern USA (Taylor and McKinney 2006)) beyond their earlier diversity focus in Europe. However, some regions known to be rich in other shallow marine invertebrates have yet to produce diverse bryozoan faunas. Among these is the Western Interior Seaway of North America with sparse records of Late Cretaceous bryozoans (Cuffey et al. 1981; Taylor and Cuffey 1992; Cuffey 1994; Roberts et al. 2008).

One of the thickest and most fossiliferous units of the Western Interior is the Campanian-Maastrichtian Pierre Shale. This deposit has a broad outcrop across Colorado, Wyoming, Montana, South Dakota and North Dakota. Bryozoans have never been formally described from the Pierre Shale but they have been mentioned (e.g. Cuffey 1990, 1994). Most notably, Gill and Cobban (1966), in a paper on the palaeoecology of the Pierre Shale at Red Bird, Wyoming, depicted runner-like bryozoans that had encrusted the interiors of the body chambers of dead baculitid ammonites. Most of these bryozoans were identified as 'pyriporoid' cheilostomes. Our aim in this paper is to describe these supposed pyriporoid cheilostomes from the Pierre Shale, show them to be ctenostomes, and discuss their palaeoecology and mode of preservation.

## Geological Setting

The bryozoan fauna in this study was collected from the Pierre Shale, a thick and lithologically diverse unit that was deposited in the Western Interior Seaway (WIS) of North America during the Late Cretaceous (Campanian-Maastrichtian). The WIS extended northward from the Gulf of Mexico to the Arctic Ocean (almost 5,000 km) and was about 1,600 km at its widest (Kauffman 1977). The deepest part of the seaway, the Western Interior Basin, was the foreland basin of the Sevier Orogeny to the west. The orogenic highlands supplied most of the fine-grained clastic sediment that filled the western parts of the seaway (Molenaar and Rice 1988; Hicks et al. 1999). The complex stratigraphy within the WIS is the product of five transgressive-regressive cycles, the Pierre Shale deposited during the last two (Molenaar and Rice

1988). The rocks are correlated by macrofossil biostratigraphy and radiometric dates from numerous bentonite clay layers (summarized in Cobban et al. 2006).

The Pierre Shale is a fossiliferous, dark grey shale and claystone with interbeds of siltstones and sandstones, and occasional carbonate lenses ('teepee buttes') associated with cold seeps. It is exposed over most of the Great Plains of North America and consequently has numerous local members. The Red Bird section (Niobrara County, Wyoming) is the best-known complete exposure of the Pierre Shale and has been thoroughly described by Gill and Cobban (1966) and correlated by Hicks et al. (1999).

## Collecting Locations

Specimens from two primary locations were used for this study:

*Heart Tail Ranch, South Dakota* – Extensive exposures of the Pierre Shale are found here across rangeland in Butte County, South Dakota. Two sites at this locality, separated by 5.6 km, are: C/W-300 (N 43.78325°, W 102.89597°) and C/W-301 (N 43.76366°, W 102.86224°). Both sites are in the upper Gammon Ferruginous Member above the Groat Sandstone Bed, which is latest early Campanian in age (Larson et al. 1997).

*Red Bird, Montana* – Material from the USGS collections of Gill and Cobban (1966) made at Red Bird, Montana, was also examined for this study. These specimens came from the Red Bird Silty Member (upper Campanian), the Lower Unnamed Member (upper Campanian) and the Upper Unnamed Member (lower Maastrichtian) of the Pierre Shale.

## Methods and Materials

In the summer of 2008, we collected baculitid conchs and internal moulds from the Pierre Shale in the Black Hills region of South Dakota, Wyoming, and Montana. Most of the specimens were found eroding from low outcrops of shale and silty claystone, with a few removed from carbonate concretions. Our collections were supplemented by specimens donated or loaned by the Black Hills Institute of Geological Research in Hill City, South Dakota. We also borrowed specimens collected by Gill and Cobban (1966) from the Red Bird section in Wyoming and now housed in the repository of the United States Geological Survey (USGS) at Boulder, Colorado.

Specimens were studied in the laboratories of the College of Wooster, Ohio, and the Natural History Museum, London, with light microscopy, thin-sections, and low-vacuum scanning electron microscopy using a LEO 1455-VP. Some specimens were further examined with laser Raman spectroscopy in the Department of Earth and Space Sciences at the University of California, Los Angeles. The majority of specimens were not prepared in any way other than cleaning in water. A few baculitid conchs with remnant aragonite shell could be carefully exfoliated to expose fresh cryptic attachment surfaces and trace fossils in the sediment filling body chambers beneath.

## Cryptic Fauna within Baculitid Ammonite Conchs

*Baculites* is a Late Cretaceous heteromorphic ammonite with a broadly cyrtoconic and distally expanding, aragonitic shell that is almost orthoconic in its distal sections. It can reach up to 2 m in length. The shell has an elliptical to ovate cross-section and is either smooth or has weak ribbing and/or shallow nodes as ornamentation. Henderson et al. (2002) provided a review of baculitid functional morphology. Our study used the following described species: *Baculites baculus* Meek and Hayden, 1861; *B. gregoryiensis* Cobban, 1951; *B. crickmayi* Williams, 1930; and *B. rugosus* Cobban, 1962. In addition, we studied specimens of the undescribed species *Baculites* sp. (smooth) and *B.* sp. (weak flank ribs) noted in Cobban (1993) (see also Larson et al. 1997). Despite this taxonomic diversity, we believe that all the baculitids in our collection shared essentially the same life modes and environments, and the taphonomic controls on their preservation were close to identical.

Sclerobiont colonization of the baculitid conch interiors apparently started soon after the host cephalopod had died and the decay of the carcass removed it from the shell (Fig. 28.1). The internal moulds of body chambers of well-preserved conchs commonly show abundant nacreous ‘spots’, which on closer examination prove to be circular pedestals of baculite aragonite (about 0.5 mm in diameter) standing in positive relief from the original shell interior surface (Figs. 28.1c and 28.2a, b). We interpret these as the sites of attachment of unmineralized organisms on the shell interiors that are preserved by the process of ‘epibiont shadowing’ (Palmer et al. 1993). We agree with Gill and Cobban (1966) in interpreting the pedestals as areas of aragonite shell protected from the dissolution that etched the surrounding shell. When present, the pedestals are evenly distributed in a band around the inner circumference of the body chambers. This distribution favours a model of colonization when the conchs were still suspended in the water column before sinking to the sea floor when infilling sediment could prevent colonization of the lowermost sector of the body chamber. Other sclerobiont and trace fossils inside the baculitid conchs postdate these pedestals, as indicated by cross-cutting and overgrowth relationships. The identity of the organism/s producing the pedestals is unclear. However, their resemblance to supposed gastropod eggs recorded attached to the interiors of the living chambers of some Jurassic ammonites (Kaiser and Voigt 1983) suggests that they may represent the sites of attachment of mollusc eggs, possibly even baculite eggs (the embryonic shells of baculites described by Landman (1982) average 0.7 mm in diameter and are therefore similar in size to the pedestals).

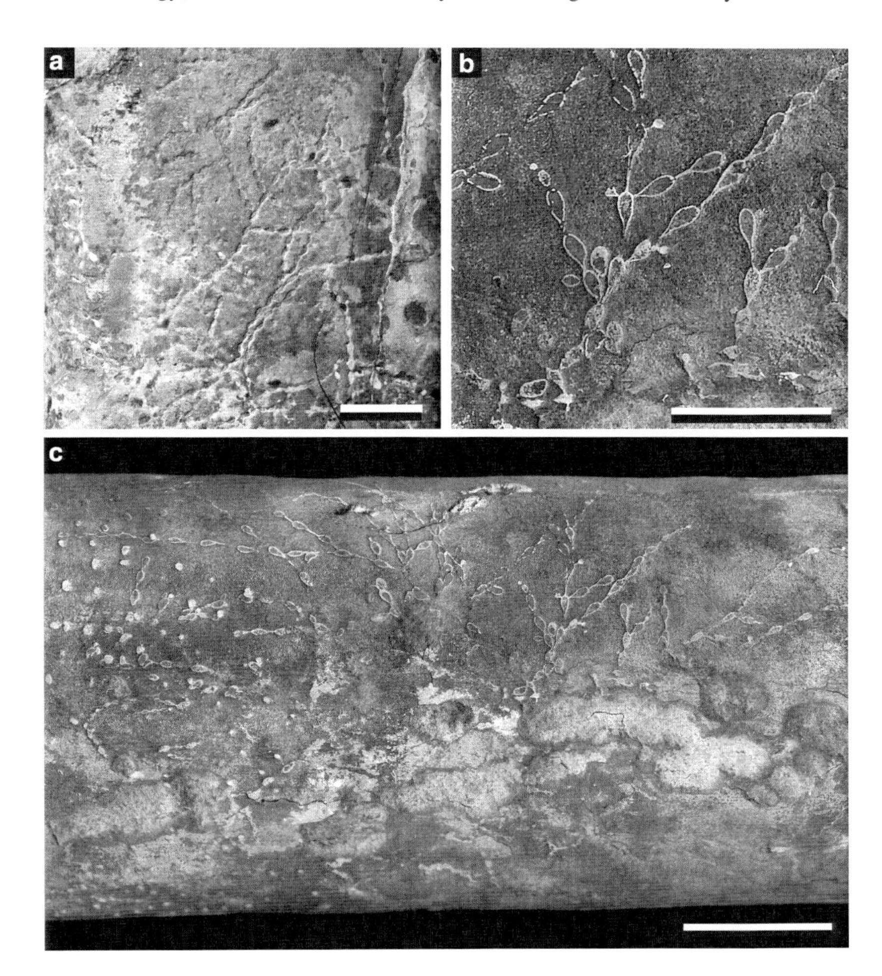


**Fig. 28.1** Photographs of baculite steinkerns showing undersides of colonies of *Pierrella larsoni* gen. et sp. nov. from the Pierre Shale. (**a**) corroded steinkern, Red Bird, Wyoming, USNM 542450. (**b**, **c**) well preserved steinkern from Heart Tail Ranch, South Dakota, NHM BZ 5815; in addition to *P. larsoni*, spots of an unknown epibiont are visible in c on the left. *Scale bars*: a, c = 10 mm; b = 5 mm

The runner-like bryozoans found inside the baculitid conchs are described in detail below. They spread throughout the body chambers, being most common in the posterior portions (Fig. 28.2c, d). During growth the bryozoan colonies were deflected around obstacles, most notably the pedestals described above. Most of the colonies expanded along the axis of the baculitid conch, either posteriorly or anteriorly. They did not grow around the entire circumference of the conch interior. As with the spot-like pedestals mentioned above, the bryozoan zooids are invariably marked by raised, pyriform patches of shell, suggesting a similar role for epibiont shadowing in their preservation. Several of the baculitid conch internal moulds have scraping trace fossils that cut across the pedestals and the bryozoans.

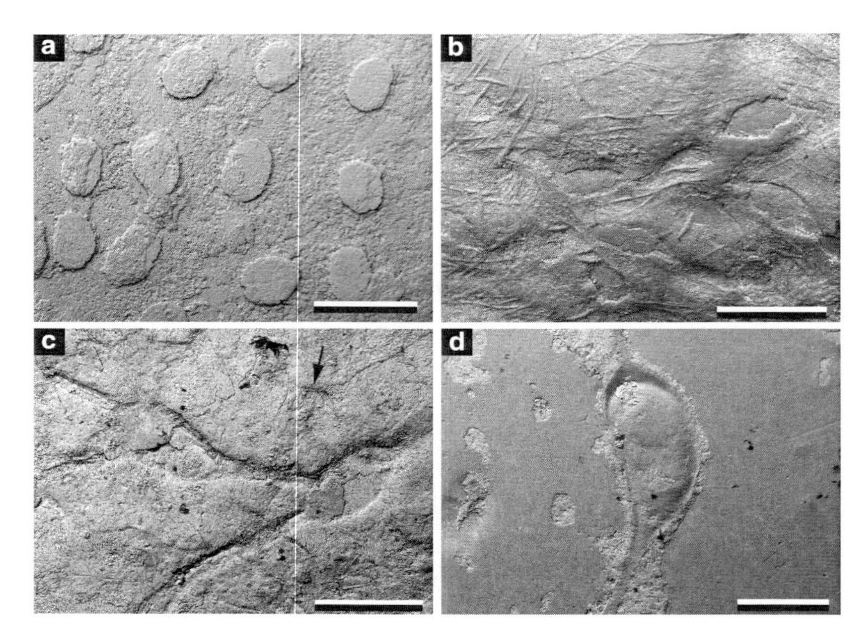

**Fig. 28.2** Scanning electron micrographs of baculite sclerobionts from the Pierre Shale. (**a**) internal surface of fragment from the body chamber of a baculite showing pedestals (epibiont shadows) indicating the original locations of non-preserved organisms (eggs?) attached to the shell, Heart Tail Ranch, South Dakota, NHM BZ 5818. (**b**) baculite steinkern with moulds of the grazing trace *Radulichnus* ichnosp. 1 and poorly preserved, pedestal-like epibiont shadows of *Pierrella larsoni* gen. et sp. nov., Red Bird, Wyoming, USNM 542451. (**c**) baculite steinkern with zooids of *Pierrella larsoni* gen. et sp. nov. bioimmured by a multiserial cheilostome which is difficult to see having infilled zooids (arrow points to a transverse wall), Red Bird, Wyoming, USNM 542542. (**d**) zooid of *P. larsoni* on baculite steinkern showing collapsed frontal wall, Heart Tail Ranch, South Dakota, NHM BZ 5816. *Scale bars*: a, b = 1 mm; c, d = 500 μm

There are three types: (1) *Radulichnus* ichnosp. 1 (Fig. 28.2b) is a series of long arcuate parallel ridges briefly described and photographed by Gill and Cobban (1966, pl. 8, fig. 6) and is apparently the grazing trace of a polyplacophoran; (2) *Radulichnus* ichnosp. 2 is a bilaterally-symmetrical set of inwardly directed long scratches which overlap to form a cross-hatched pattern and probably represents crustacean scratch marks; and (3) *Radulichnus inopinatus* Voigt, 1977, also figured by Gill and Cobban (1966, pl. 11, figs. 1, 2), comprises modular units of four to six parallel short striae likely made by the doccoglossan radula of a limpet.

## Bryozoan Occurrence

All of the studied bryozoans associated with Pierre Shale baculites encrusted the interiors of body chambers (Fig. 28.1); we have found no examples of bryozoans encrusting outer shell surfaces of baculites. Likewise, Henderson et al. (2002) noted that none of the roughly 2,000 baculitids they studied showed any clear indications of having been colonized by epizoans during the life of the ammonite. Although pristine outer shell surfaces are not often visible because outermost shell layers readily exfoliate, the location of the bryozoans on shell interiors is unlikely to represent an artefact of preservation but instead indicates preferential colonization of this cryptic habitat.

As the body chambers of the baculites are usually filled with hard, lithified sediment, encrusted inner surfaces of their shells are not generally exposed. However, in many baculite specimens used in this study the entire thickness of the shells have flaked away from the sediment that infilled the body chambers, revealing the undersides of the encrusting bryozoans on the internal moulds of the body chambers.

Mesozoic bryozoans are invariably identified on the basis of characters of the frontal surfaces of zooids. Therefore, the style of preservation of the baculite encrusters presents a taxonomic challenge. In some instances, the bryozoan zooidal chambers have themselves been filled with sediment and/or diagenetic minerals resembling infillings of the baculite body chamber, whereas in others the zooids are 'epibiont shadows' marked by raised pedestals of baculite shell. The outline shapes of the zooids, and the budding and branching pattern of the colony, are the only morphological characters available in these cases. Much more useful are examples of zooids that were not filled with sediment or minerals and in which the undersides of the frontal surfaces of the zooids can be seen.

## Identity of the Bryozoans

Previous works identified the baculite-encrusting bryozoans from the Pierre Shale as 'pyriporoid' or 'pyriporid' cheilostomes (e.g. Gill and Cobban 1966; Cuffey 1994). The implication is that they are closely related to such runner-like anascans as *Pyripora*, *Rhammatopora* and *Herpetopora*, all of which have been recorded from the Cretaceous (e.g. Larwood 1973; Taylor 1988, 2010; Taylor and McKinney 2006). These malacostegine genera have uniserial colonies consisting of tear-shaped to pyriform zooids, differing from one another in characters of the frontal wall: *Pyripora* has a pustulose cryptocyst and no spines or oral spine bases only; *Rhammatopora* possesses numerous small spine bases encircling the mural rim; and *Herpetopora* lacks both cryptocyst and spine bases. A characteristic of all three genera are secondary astogenetic gradients of increasing zooid length along new distolaterally budded branches, particularly well developed in *Herpetopora* (Taylor 1988). This character is also evident in the Pierre Shale bryozoans (Fig. 28.3c, d).

Before giving reasons why we believe that the Pierre Shale bryozoans are not 'pyriporoid' cheilostomes, it is necessary to consider alternative possibilities. The combination of pyriform zooids and a uniserial colony growth pattern with cruciate branching is unknown in post-Palaeozoic cyclostome bryozoans, although it can be found in the Ordovician-Permian genus *Corynotrypa* (Taylor and Wilson 1994). Stomatoporid cyclostomes, which are common in the Cretaceous, never produce a distal and paired distolateral buds to give a cruciate pattern of branching, except for some colonies of *Voigtopora* which have broad, non-pyriform zooids with small distal apertures quite unlike the Pierre Shale bryozoan. Aside from the malacostegine 'pyriporoids' mentioned above, other uniserial cheilostomes recorded from the Cretaceous are the calloporids *Allantopora*, *Pyriflustrina* and *Unidistelopora*, the cribrimorph *Andriopora* and the hippothoid *Tecatia*, but none of these genera show the steep secondary gradient of astogenetic change seen in the Pierre Shale bryozoan. Characteristics such as spinocystal (*Andriopora*) or gymnocystal (*Tecatia*) frontal shields, and ovicells (all four genera) provide further distinction.

While sharing zooids of similar shape and colonies with comparable growth patterns, several lines of evidence show the Pierre Shale bryozoans not to be 'pyriporoids' or indeed any other group of cheilostomes:

1. All cheilostomes have a mineralized skeleton, composed of calcite, aragonite or a combination of these two calcium carbonate polymorphs. There is no indication of a mineralized skeleton in any of the Pierre Shale 'pyriporoids'. White patches forming the bases of or outlining the zooids in some colonies (Fig. 28.1b, c) are remnants of baculite shell that did not flake away from the steinkerns. While it is not unusual for the skeletons of aragonitic cheilostomes to be dissolved by leaching during fossilization (e.g. Taylor et al. 2009), diagenetic loss of aragonite is unlikely here because of the preservation of the skeletons of other aragonitic fossils, notably the encrusted baculites themselves, which have been shown by laser Raman spectroscopy to consist of aragonite. Furthermore, there is no mouldic space where a leached aragonitic skeleton might once have been present. One of the 'pyriporoid'-encrusted Red Bird baculites is also encrusted by a small multiserial cheilostome with skeletal walls intact (Fig. 28.2c), proving the potential for bryozoan skeletal carbonate to be preserved in this depositional and diagenetic setting. The lack of a mineralized skeleton provides the first indication that the Pierre Shale 'pyriporoids' are ctenostomes rather than cheilostomes.
2. Many of the zooids are bent or distorted in a way suggesting that they were soft-bodied. This can be seen in places where zooids grew around obstructions on the substrate, such as the pedestals, or encountered other bryozoan branches (Fig. 28.3b). Unfilled zooids reveal that the undersides of the frontal surfaces are typically crushed inwards to varying degrees (Figs. 28.2d and 28.4c, d), which would not be expected if the zooid was supported by mineralized vertical walls but is explicable if it was entirely soft bodied.
3. Partly formed new buds are visible at the tips of some branches (Fig. 28.4b). These take the form of bulbous outgrowths with rounded ends that are smaller than mature zooids. In contrast, partly formed buds in 'pyriporoid' cheilostomes are seldom observed as calcification of the zooid appears to happen rapidly once the bud has attained an adult size.
4. 'Pyriporoid' cheilostomes generally have kenozooids that are characteristically subtriangular with gently concave sides (e.g. Taylor 1988). No such kenozooids have been observed in the Pierre Shale 'pyriporoids'.

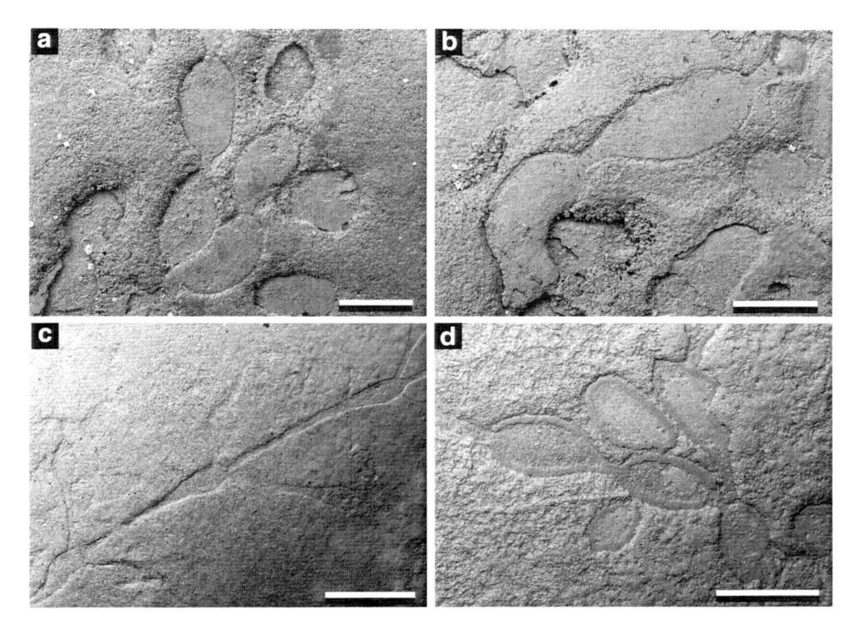


**Fig. 28.3** Scanning electron micrographs of baculite steinkerns showing undersides of colonies of *Pierrella larsoni* gen. et sp. nov. from the Pierre Shale of Red Bird, Wyoming (**a**–**c**) and the Heart Tail Ranch, South Dakota (**d**). (a, b) USNM 542453; (**a**) short pyriform zooids from two parallel branches; (**b**) branch deflected around an obstruction (*lower mid-left*). (**c**) narrow, elongate zooids showing astogenetic increase in length along a branch and cruciform budding pattern, USNM 542452. (**d**) short zooids with distinct peripheral rims, NHM BZ 5815. *Scale bars*: a, b = 500 μm; c, d = 1 mm

5. In some zooids an orifice is clearly visible (Fig. 28.4c, d). Instead of having a semicircular shape consistent with the presence of a closing operculum, as in most cheilostomes, it has a puckered appearance with about 8 radial rays (Fig. 28.4d). This structure can be interpreted as a pleated collar. In Recent bryozoans pleated collars are inferred to function to protect the opening lophophore from surficial debris that they push aside (McKinney and Dewel 2002). They are present in many ctenostomes but just a handful of primitive cheilostomes, namely *Aetea*, *Scruparia* and *Conopeum chesapeakensis* (Banta et al. 1995). Ctenostomes with pleated collars include several Recent genera resembling the Pierre Shale bryozoan (see below), whereas the few cheilostomes with this structure are morphologically quite dissimilar to the Pierre Shale 'pyriporoids'.

## Systematics

Order Ctenostomata Busk, 1852

Family Arachnidiidae Hincks, 1880

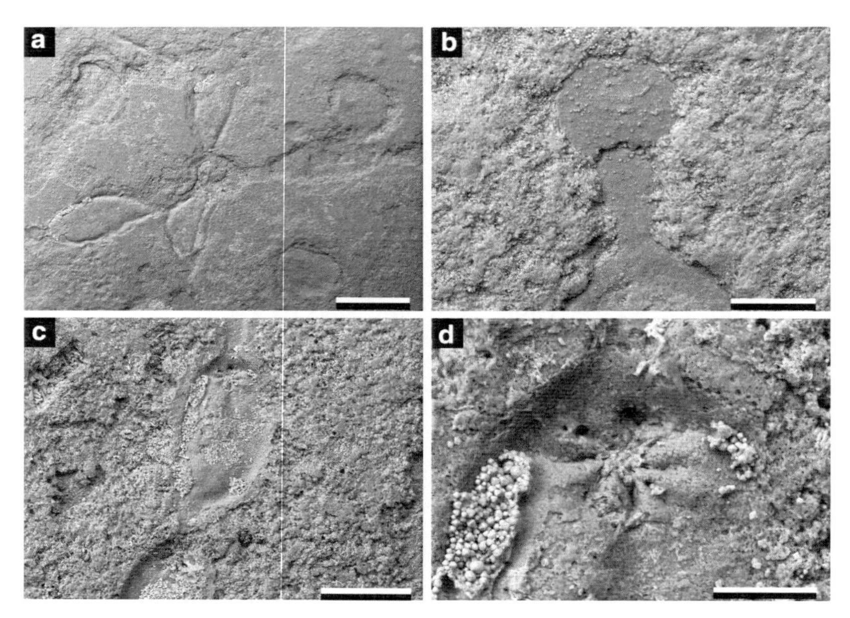

**Fig. 28.4** Scanning electron micrographs of baculite steinkerns showing undersides of colonies of *Pierrella larsoni* gen. et sp. nov. from the Pierre Shale of the Heart Tail Ranch, South Dakota. (**a**) apparent colony origin showing almost circular ancestrula (?) with what are interpreted as proximally (*right*) and distolaterally budded zooids, NHM BZ 5817. (**b**) new, incomplete bud at the end of a branch, NHM BZ 5815. (**c, d**) undersides of zooids showing collapsed frontal surfaces and orifices, NHM BZ 5814 (holotype); (**c**) zooids mostly unfilled apart from some pyrite framboids; (**d**) detail of an orifice showing setigerous collar moulded from the underside. *Scale bars*: a = 500 μm; b = 100 μm; c = 200 μm; d = 50 μm

*Pierrella* gen. nov.

Type species: *Pierrella larsoni* sp. nov.

Etymology: From its occurrence in the Pierre Shale.

Diagnosis: Colony adnate, uniserial, branching pattern cruciate, new branches originating from left and right distolateral buds, successive zooids showing astogenetic increase in size; no anastomoses. Ancestrula (?) ovoidal, budding three zooids, seemingly two distolaterally and one proximally. Zooids monomorphic, tear-shaped to pyriform, the orifice subterminal, closed by inwardly tapering radial processes of the setigerous collar. Pore chambers not observed. Subtriangular kenozooids lacking.

Remarks: As with all non-boring ctenostomes found in the fossil record, only a limited suite of morphological characters is available and comparisons with Recent genera can be difficult. Furthermore, comparisons between fossils are complicated by variations in their taphonomy and modes of preservation. In the case of the Pierre Shale bryozoan, the principal characters are the uniserial branches with a cruciate branching pattern, tear-shaped to pyriform zooids increasing in length along the branches, and orifice with a collar comprising 8 'setae'. Similar branching pattern and zooidal shape is found in several Recent and fossil carnosan ctenostomes, including *Paludicella* Gervais, *Arachnidium* Hincks, *Hislopia* Carter, and *Arachnoidea* Moore (see d'Hondt 1983; Hayward 1985). The two Mesozoic *Simplicidium* Todd et al., 1997 and *Cardoarachnidium* Taylor, 1990 differ from *Pierrella* in respectively having a simple orifice without setae and an orifice with a D-shaped operculum. The new genus also apparently lacks pore chambers, unlike both *Simplicidium* and *Cardoarachnidium*. None of the unpublished fossil genera named in the thesis of Todd (1993) exhibit the conspicuous setigerous collar that is evident in specimens of *Pierrella* from both the Red Bird section in Wyoming and the Heart Tail Ranch section in South Dakota. However, it must be acknowledged that collar visibility in fossil ctenostomes may be just as much a function of preservational mode as it is of taxonomy in view of the ubiquity of setigerous collars in modern ctenostomes. Assignment to Arachnidiidae is provisional.

Distribution: Lower Campanian-early Maastrichtian, Pierre Shale Formation, Wyoming and South Dakota, USA.

*Pierrella larsoni* sp. nov

Material: Holotype: NHM BZ 5814, Lower Campanian, Pierre Shale Formation, Gammon Ferruginous Member, Heart Tail Ranch, Butte County, South Dakota, USA, (Figs. 28.1, 28.2, 28.3, and 28.4). Paratypes: NHM BZ 5815-7, details as for holotype. USNM 542450, Lower unnamed shale member, Pierre Shale Formation, upper Campanian, tan-weathering silty calcareous concretions, east side of US Highway 85, 1.0 mile southwest of bridge over Mule Creek, Red Bird, Niobrara County, Wyoming, USA, NW1/4SW1/4 sec.13, T. 39 N., R. 61 W. USNM 542451, Pierre Shale Formation, Red Bird, Niobrara County, Wyoming, USA. USNM 542452, 542453, Pierre Shale Formation, Lower Maastrichtian, brown-weathering calcareous concretions, 364–394 feet below top, about one mile southeast of US Highway 85, NW1/4 sec.25, T. 39 N., R. 62 W, Red Bird, Niobrara County, Wyoming, USA.

Etymology: In honour of Neal Larson (Black Hills Institute of Geological Research, Hill City, South Dakota) who advised us about Pierre Shale localities, provided assistance in the field, and allowed us to borrow material from the collections of his institute.

Description: Colony adnate, uniserial, comprising ramifying branches, new branches originating as distolateral buds from parent zooids, often paired, giving a cruciform pattern. Branching angles range from 22° to 105°, averaging 77° (n = 378). Secondary gradients of astogenetic change marked by the budding of zooids of increasing length along branches, distal-most zooids reaching lengths of over 2 mm. Branch anastomoses lacking. Possible colony origin comprising an ovoidal zooid (putative ancestrula), 0.34 mm long by 0.29 mm wide, apparently with a pair of distolateral buds and one proximal bud, although polarity is uncertain. Pore chambers lacking or not preserved.

Zooids monomorphic, tear-shaped to pyriform, the proximal part (cauda) varying in length, straight or sinuous when overgrowing another zooid or growing around an obstruction; typically just over two times longer than wide; length 531–1,584 microns, averaging 1,013 microns (n = 104); width 280–556 microns, averaging 411 microns (n = 104). Most zooids infilled with minerals and/or sediment and visible only as moulds or raised patches of baculite shell, some bearing a distinct peripheral rim. Unfilled zooids showing variable degrees of frontal membrane collapse; orifice subterminal, circular, ca 0.06 mm in diameter, closed by an average of 8 inwardly tapering radial processes forming the setigerous collar. Bulb-shaped new buds visible at distal ends of growing branches. Kenozooids lacking.

Remarks and distribution: As for genus.

## Mode of Bryozoan Preservation

Non-boring ctenostome bryozoans can be preserved as fossils by two processes: bioimmuration and epibiont shadowing. Bioimmuration involves overgrowth by organisms with mineralized skeletons (reviewed by Taylor and Todd 2001). This process cannot account for the preservation of the ctenostome *Pierrella larsoni* as the great majority of specimens show no sign of having been overgrown; a rare exception is part of a colony overgrown by a multiserial cheilostome that grew together with the ctenostome on the inside of the body chamber of a baculite from Red Bird (Fig. 28.2c).

Epibiont shadowing provides an adequate explanation for the preservation of some of the colonies of *P. larsoni*, as well as the enigmatic 'spots' that are common on many baculite steinkerns. This process entails protection of the surface of the substrate immediately beneath the epibiont from corrosion and infestation by microendoliths (Palmer et al. 1993). Sometimes corrosional shadows stand up above the general level of the substrate as 'perched shadows', as is evidently also the case in the Pierre Shale. Palmer et al. (1993, fig. 6) illustrated a Jurassic arachnidiid ctenostome in which the epibiont shadow had a distinct rim like that often seen in *P. larsoni* (Fig. 28.3d). Rims are interpreted to represent the location of the thickened bases of the vertical zooidal walls that took longer to decay and therefore protected the substrate beneath them from corrosion for a longer period of time.

Epibiont shadowing alone is insufficient to explain all aspects of the preservation of *Pierrella larsoni* as this process preserves only the basal outlines of zooids and cannot account for preservation of the frontal membrane and orifice seen in some zooids (Fig. 28.2b). Many soft-bodied organisms in the fossil record are preserved by early diagenetic mineralization of their tissues, usually by phosphate minerals but also sometimes by pyrite or carbonates (Wilby et al. 1996; Briggs 2003). However, SEM studies coupled with EDX analysis failed to detect any evidence of permineralization of the frontal wall of the bryozoan zooids; instead, the frontal surface is preserved only in the form of a mould impressed into the sediment forming the steinkerns of the baculites.

When Vialov (1961) introduced the concept of bioimmuration he also coined the term 'lithoimmuration' for immuration caused by inorganic processes, citing as an example the envelopment of hibernating snakes by calcareous tufa. We hypothesize that lithoimmuration was involved in the preservation of *Pierrella larsoni*. The baculites hosting *P. larsoni* come from carbonate concretions in the Pierre Shale (Gautier 1982). It is well known that such concretions can grow very rapidly after burial. For example, carbonate concretions can form around organic nuclei on the marshes of north Norfolk within a year or so (Allison and Pye 1994). Authigenic carbonate is clearly associated with *P. larsoni* in the Pierre Shale: infillings of the bryozoan zooids may comprise a mixture of bladed crystals of calcite and subsidiary pyrite, as demonstrated by laser Raman spectroscopy. Early precipitation of calcite in conjunction with concretionary growth is hypothesized to have been capable of lithoimmuring zooids of *P. larsoni*, especially if colonies were rapidly buried in an environment where decay was inhibited. This allowed the bryozoan zooids to be preserved as imprints on the surface of the baculite steinkerns. While this hypothesis provides the most plausible explanation for the kind of preservation seen, for example, in Fig. 28.4c, d, it needs to be tested by geochemical and other means that are beyond the scope of this paper.

**Acknowledgements** We are grateful to Neal Larson (Black Hills Institute of Geological Research) for arranging fieldwork in South Dakota and allowing us to borrow specimens from his collection. John Sime (The College of Wooster) assisted during fieldwork and searched for material in the USGS collections, while Luke Larson, Jamie Brezina and Mike Ross also facilitated fieldwork. Bill Schopf and Anatoliy Kudryavtsev provided access to the laser Raman spectroscope in the Department of Earth and Planetary Sciences, UCLA. This work was also supported in part by the Luce and Wengerd Funds at The College of Wooster.

## References

Allison PA, Pye K (1994) Early diagenetic mineralization and fossil preservation in modern carbonate concretions. Palaios 9:561–575

Banta WC, Perez FM, Santagata S (1995) A setigerous collar in *Membranipora chesapeakensis* n. sp. (Bryozoa): implications for the evolution of cheilostomes from ctenostomes. Invertebr Biol 114:83–88

Briggs DEG (2003) The role of decay and mineralization in the preservation of soft-bodied fossils. Ann Rev Earth Planet Sci 31:275–301

Brood K (1977) Upper Cretaceous Bryozoa from Need's Camp, South Africa. Palaeontol Africana 20:65–82

Busk G (1852) An account of the Polyzoa and sertularian zoophytes collected in the voyage of the "Rattlesnake" on the coast of Australia and the Louisiade Archipelago. In: MacGillivray J (ed) Narrative of the voyage of *H.M.S. Rattlesnake* commanded by the late Captain O. Stanley during the years 1846–1850. Boone, London, pp 343–402

Cobban WA (1951) New species of *Baculites* from the Upper Cretaceous of Montana and South Dakota. J Paleontol 25:817–821

Cobban WA (1962) *Baculites* from the lower part of the Pierre Shale and equivalent rocks in the Western Interior. J Paleontol 36:704–718

Cobban WA (1993) Diversity and distribution of Late Cretaceous ammonites, Western Interior, United States. In: Caldwell WGE, Kauffman EG (eds) Evolution of the Western Interior basin. Geol Assoc Can Sp Pap 39:435–451

Cobban WA, Walaszczyk I, Obradovich JD, McKinney KC (2006) A USGS zonal table for the Upper Cretaceous middle Cenomanian-Maastrichtian of the Western Interior of the United States based on ammonites, inoceramids, and radiometric ages. US Geol Surv Open-File Rep 2006–1250:1–46

Cuffey RJ (1990) Cretaceous bryozoans on Baculite Mesa, Colorado. Geol Soc Am Abstr 22(6):7

Cuffey RJ (1994) Cretaceous bryozoan faunas of North America – preliminary generalizations. In: Hayward PJ, Ryland JS, Taylor PD (eds) Biology and palaeobiology of bryozoans. Olsen and Olsen, Fredensborg, pp 55–56

Cuffey RJ, Feldmann RM, Pholable KE (1981) New Bryozoa from the Fox Hills Sandstone (Upper Cretaceous, Maastrichtian) of north Dakota. J Paleontol 55:401–409

Gautier DL (1982) Siderite concretions: indicators of early diagenesis in the Gammon Shale (Cretaceous). J Sed Petrol 52:859–871

Gill JR, Cobban WA (1966) The Red Bird section of the Upper Cretaceous Pierre Shale in Wyoming, with a section on a new echinoid from the Cretaceous Pierre Shale of eastern Wyoming by Porter M. Kier. US Geol Surv Prof Pap 393:1–73

Guha AK, Nathan DS (1996) Bryozoan fauna of the Ariyalur Group (Late Cretaceous). TamilNadu and Pondicherry, India. Palaeontol Indica (ns) 49:1–217

Hayward PJ (1985) Ctenostome bryozoans. Synop British Fauna (ns) 33:1–169

Henderson RA, Kennedy WJ, Cobban WA (2002) Perspectives of ammonite paleobiology from shell abnormalities in the genus *Baculites*. Lethaia 35:215–230

d'Hondt J-L (1983) Tabular keys for identification of the Recent ctenostomatous Bryozoa. Mém Inst océanogr Monaco 14:1–134

Hicks JF, Obradovich JD, Tauxe L (1999) Magnetostratigraphy, isotopic age calibration and intercontinental correlation of the Red Bird section of the Pierre Shale, Niobrara County, Wyoming, USA. Cret Res 20:1–27

Hincks T (1880) On new Hydroida and Polyzoa from the Barents Sea. Ann Mag Nat Hist 5:277–286

Kaiser P, Voigt E (1983) Fossiler Schneckenlaich in Ammonitenwohnkammern. Lethaia 16:145–156

Kauffman EG (1977) Geological and biological overview: Western Interior Cretaceous basin. In: Kauffman EG (ed) Cretaceous facies, faunas, and paleoenvironments across the Western Interior basin. Mt Geol 14:75–99

Landman NH (1982) Embryonic shells of *Baculites*. J Paleontol 56:1235–1241

Larson NL, Jorgensen SD, Farrar RA, Larson PL (1997) Ammonites and the other cephalopods of the Pierre Seaway. Geoscience, Tucson

Larwood GP (1973) New species of *Pyripora* d'Orbigny from the Cretaceous and the Miocene. In: Larwood GP (ed) Living and fossil Bryozoa. Academic, London, pp 463–473

McKinney MJ, Dewel RA (2002) The ctenostome collar – an enigmatic structure. In: Wyse Jackson PN, Buttler CJ, Spencer Jones ME (eds) Bryozoan studies 2001. Swets and Zeitlinger, Lisse, pp 191–197

Meek FB, Hayden FV (1861) Descriptions of new Lower Silurian (Primordial), Jurassic, Cretaceous, and Tertiary fossils, collected in Nebraska, etc. Philadelphia Acad Nat Sci Proc 13:415–447

Molenaar CM, Rice DD (1988) Cretaceous rocks of the Western Interior Basin. In: Sloss LL (ed) Sedimentary cover–North American craton U.S., vol D-2. Geological Society of America, Boulder, pp 77–82

Palmer TJ, Taylor PD, Todd JA (1993) Epibiont shadowing: a hitherto unrecognized way of preserving soft-bodied fossils. Terra Nova 5:568–572

Roberts EM, Tapanila L, Mijal B (2008) Taphonomy and sedimentology of storm-generated continental shell beds: a case example from the Cretaceous Western Interior Basin. J Geol 116:462–479
Taylor PD (1988) Colony growth pattern and astogenetic gradients in the Cretaceous cheilostome bryozoan *Herpetopora*. Palaeontol 31:519–549
Taylor PD (1990) Bioimmured ctenostomes from the Jurassic and the origin of the cheilostome Bryozoa. Palaeontology 33:19–34
Taylor PD (2010) Bryozoans. In: Young JR, Gale AS, Knight RI, Smith AB (eds) The Palaeontological Association Field Guide to Fossils, Fossils of the Gault Clay, Wiley-Blackwell, London, pp 33–40
Taylor PD, Cuffey RJ (1992) Cheilostome bryozoans from the Upper Cretaceous of the Drumheller area, Alberta, Canada. Bull Br Mus (Nat Hist) Geol Ser 48:13–24
Taylor PD, James NP, Bone Y, Kuklinski P, Kyser TK (2009) Evolving mineralogy of cheilostome bryozoans. Palaios 24:440–452
Taylor PD, McKinney FK (2006) Cretaceous Bryozoa from the Campanian and Maastrichtian of the Atlantic and Gulf Coastal Plains, United States. Scr Geol 132:1–346
Taylor PD, Todd JA (2001) Bioimmuration. In: Briggs DEG, Crowther PR (eds) Palaeobiology II. Blackwell Science, Oxford, pp 285–289
Taylor PD, Wilson MA (1994) *Corynotrypa* from the Ordovician of North America: colony growth in a primitive stenolaemate bryozoan. J Paleontol 68:241–257
Todd JA (1993) Bioimmuration. Unpublished Ph.D. Thesis, University of Wales, pp 1–444
Todd JA, Taylor PD, Favorskaya TA (1997) A bioimmured ctenostome bryozoan from the early Cretaceous of the Crimea and the new genus *Simplicidium*. Geobios 30:205–213
Vialov OS (1961) Phenomena of vital immuration in nature. Dopovidi Akademi Nauk Ukrayin'skoi RSR 11:1510–1512 (In Russian)
Voigt E (1977) On grazing traces produced by the radula of fossil and Recent gastropods and chitons. In: Crimes TP, Harper JC (eds) Trace Fossils 2, Geological Journal Special Issues 9, Seel House Press, Liverpool, pp 335–346
Wilby PR, Briggs DEG, Riou B (1996) Mineralization of soft-bodied invertebrates in a Jurassic metalliferous deposit. Geol 24:847–850
Williams MY (1930) New species of marine invertebrate fossils from the Bearpaw Formation of southern Alberta. Nat Mus Canada Bull 63:1–6

# Chapter 29
# Krka River (Croatia): Case Study of Bryozoan Settlement from Source to Estuary

## Bryozoan Settlement in the Krka River

**Emmy R. Wöss and Maja Novosel**

**Abstract** Krka River, a river in Croatia's karstic Dalmatia region, has a length of 72 km and is famous for its seven tufa waterfalls of a total drop of 242 m. The source of the Krka River is at the base of the Dinaric Mountains, 3.5 km northeast of Knin, and it flows into the Adriatic Sea at Šibenik. The length of the freshwater section is 49.5 km; the brackish part is 23.5 km. In September 2008, 25 samples sites located between N 44°02.521′, E 16°14.087′ and N 43°49.121′, E 15°56.074′ were investigated for bryozoan occurrence combined with recording of environmental parameters. The specific situation with tufa deposition as well as heavy algae growth hampers colony settlement on the substrata. The species list contains phylactolaemate (*Cristatella mucedo, Fredericella sultana, Plumatella emarginata, P. fruticosa, P. fungosa, P. geimermassardi* and *P. repens)* and gymnolaemate (*Paludicella articulata* and *Conopeum seurati)* taxa. *P. repens* was the most abundant bryozoan, followed by *P. geimermassardi*, and *P. emarginata*. *C. seurati* in the Krka River estuary represents the first record of this species along the eastern (Croatian) coast of the Adriatic Sea.

**Keywords** Dalmatian karst region • Tufa deposition • Freshwater bryozoan distribution • *Conopeum seurati*

E.R. Wöss (✉)
Department of Limnology, University of Vienna, Althanstraße 14, Vienna A-1090, Austria
e-mail: emmy.woess@univie.ac.at

M. Novosel
Faculty of Science, Laboratory of Marine Biology, Rooseveltov trg 6, 10000 Zagreb, Croatia
e-mail: mnovosel@biol.pmf.hr

A. Ernst, P. Schäfer, and J. Scholz (eds.), *Bryozoan Studies 2010*,
Lecture Notes in Earth System Sciences 143, DOI 10.1007/978-3-642-16411-8_29,

## Introduction

In many countries, freshwater bryozoans are often one of the most poorly known faunal groups (Ricciardi and Reiswig 1994), although they are widely distributed in a broad range of freshwater habitats. They are, however, among the most important suspension-feeding animals, along with sponges and mussels (Wood et al. 2006). Freshwater bryozoans can be an abundant component in epibenthic and littoral communities in terms of biomass (Raddum and Johnsen 1983) and in regard of the recycling process of nutrients in small lentic habitats (Sørensen et al. 1986). They mostly belong to the class Phylactolaemata, which occurs exclusively in freshwater, while a few species can be assigned to the predominantly marine class of Gymnolaemata.

Freshwater bryozoans are characterized by complex life histories including the production of a high variety of propagules for reproduction and dispersal that is unique among freshwater invertebrates (Wöss 1996, 2002). Asexual reproduction includes formation of resting stages, in the case of the Phylactolaemata several kinds of statoblasts, while the Gymnolaemata produce hibernaculae. As colony morphology in many Phylactolaemate taxa does not offer enough reliable criteria for species distinction, taxonomic classification shifted to the analysis of the outer and recently also inner sclerotized surface of the statoblasts.

At the present stage, 44 species are recorded for the palaeoarctic region (Massard and Geimer 2008a) and 23 for Europe (Taticchi and Pieroni 2005; Taticchi et al. 2006, 2008; Taticchi 2010; Wood and Okamura 2005). Data on the occurrence and distribution of freshwater bryozoans in southeastern Europe are still fragmentary. Massard and Geimer (2008b) list 11 species for Bulgaria, 10 for Romania, and in Serbia 7 species are known (Simić and Ostojić 1997, and personal investigation of statoblasts of the Serbian Danube region). In Croatia, freshwater bryozoans have been neglected so far in faunal surveys and limnological studies. In the course of a joint Austrian-Croatian project (WTZ-HR 29/2008), a survey on this animal group started to take place in six areas (Wöss and Novosel unpubl data). The Krka River, a river of 73 km length situated in Dalmatia, was one of the areas chosen due to its special environment of tufa deposits as well as to trace the changes of species composition from a limnic to a brackish environment. The source of the river is at the base of the Dinaric Mountains, 3.5 km northeast of the base of Knin and 22 m below Topoljski Slap, Veliki Buk and Krčić Slap. The length of the freshwater section is 49.5 km and of the brackish section is 23.5 km. The Krka estuary runs in the Adriatic Sea in front of Šibenik. The Krka River is a typical groundwater-fed karstic river, which receives a considerable amount of water through diffuse subsurface recharge. The river has only a few tributaries, which usually dry out each summer (Lojen et al. 2009). The fish fauna of the freshwater part is characterized by a high degree of endemism, as 11 out of 20 fish species are only described for Croatia (Kerovec et al. 2007). With its seven tufa waterfalls and a total drop of 242 m, the Krka River is a natural and karstic phenomenon. Tufa deposits are formed by calcite precipitation from freshwater supersaturated with calcium carbonate (Horvatinčić et al. 2000). The deposition of tufa is a constant and dynamic process, involving physiochemical factors and organisms living in the water. In contrast to

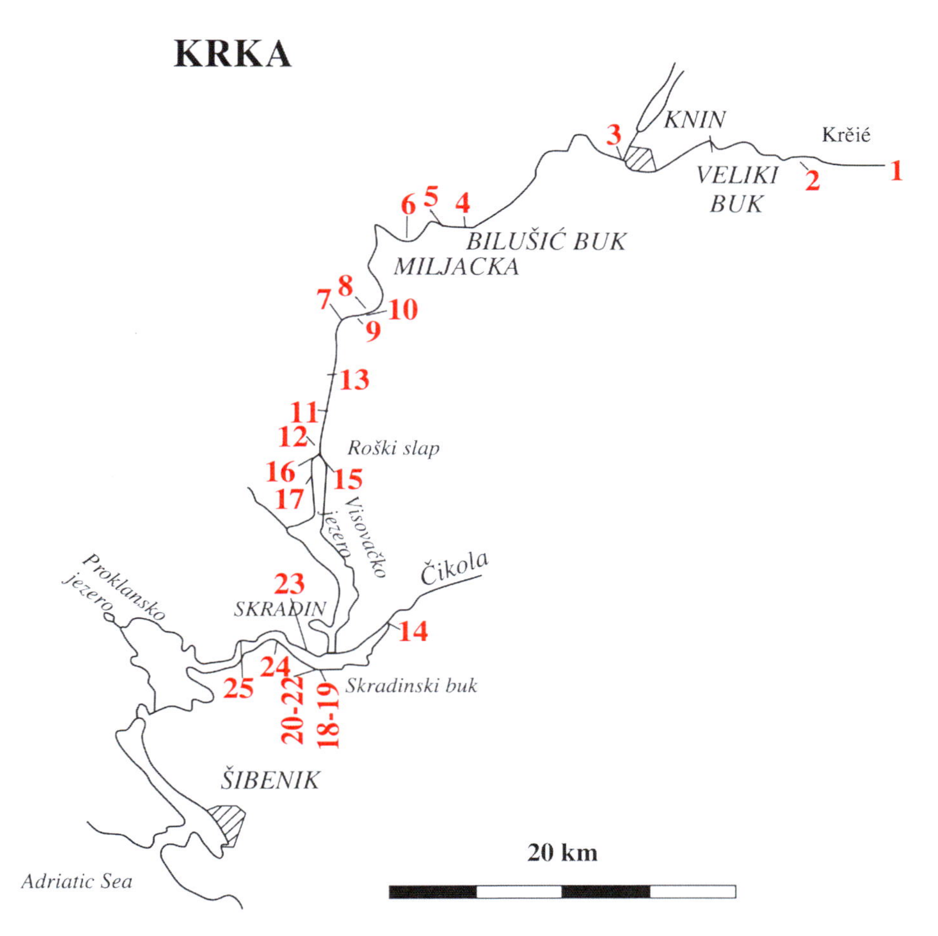


**Fig. 29.1** Krka River with 25 sampling sites

the Plitvice Lakes, one of the other areas chosen for the survey on freshwater bryozoans and characterized by deposition of tufa as well, deposition rate of calcium is lower. Juračić and Prohić (1991) documented 0.5 mm year$^{-1}$ in the estuary of the Krka River, while Emeis et al. (1987) reported an approximate deposition rate of 10 mm year$^{-1}$ at the Plitvice Lakes. At some locations there, e.g. at waterfalls, the tufa growth rate even exceeded 10 mm year$^{-1}$ (Horvatinčić et al. 2003).

## Investigation Area and Methods

In September 2008, 25 sites in the River Krka between the source around Knin and the estuary part at Skradin were examined for bryozoan species (Fig. 29.1). The locations were recorded through a GPS with Garminnüvi 760 instrument (Table 29.1).

**Table 29.1** Krka River: coordinates of investigated sites

| | | | |
|---|---|---|---|
| K1 | Topoljski Buk | N 44°02.521′ | E 16°14.087′ |
| K2 | Krčić Fishfarm | N 44°02.388′ | E 16°13.748′ |
| K3 | Sastavak | N 44°02.345′ | E 16°11.103′ |
| K4 | Marasovina (Liver) | N 44°00.620′ | E 16°05.557′ |
| K5 | Bilušića Buk | N 44°00.780′ | E 16°03.610′ |
| K6 | Brljan – Donje Lake | N 44°00.744′ | E 16°01.992′ |
| K7 | Orthodox Monastery | N 43°57.720′ | E 15°59.521′ |
| K8 | Nečven Brzak | N 43°58.504′ | E 16°01.411′ |
| K9 | Nečven-Orthodox Monastery | N 43°58.350′ | E 16°00.844′ |
| K10 | 200 m upflow Orthodox Monastery | N 43°57.839′ | E 15°59.526′ |
| K11 | Verovića Bare | N 43°57.836′ | E 15°59.532′ |
| K12 | Roški Waterfall | N 43°57.833′ | E 15°59.540′ |
| K13 | Ribarska Cave | N 43°56.540′ | E 15°59.484′ |
| K14 | Čikola River Delta | N 43°48.565′ | E 15°59.728′ |
| K15 | Roški Waterfall downflow | N 43°54.214′ | E 15°58.481′ |
| K16 | Među Gredama | N 43°51.861′ | E 15°58.351′ |
| K17 | Visovac Monastery | N 43°51.635′ | E 15°58.391′ |
| K18 | Skradinski Buk (harbour) | N 43°48.136′ | E 15°58.323′ |
| K19 | Skradinski Buk (old channel) | N 43°48.123′ | E 15°57.990′ |
| K20 | Skradinski Buk (under waterfall) | N 43°48.132′ | E 15°57.995′ |
| K21 | Skradinski Buk (shallow water body) | N 43°48.414′ | E 15°57.787′ |
| K22 | Skradinski Buk (touristic harbour) | N 43°48.473′ | E 15°57.675′ |
| K23 | Skradinski Buk harbour (right coast) | N 43°48.550′ | E 15°57.522′ |
| K24 | Skradinski Buk harbour (left coast) | N 43°48.703′ | E 15°57.205′ |
| K25 | Skradinski Bridge | N 43°49.121′ | E 15°56.074′ |

In shallow parts of the river, such as in the headwater or in some sections close to waterfalls, collecting of substrata took place by wading and was supported by using a rake. However, in most parts of the river, sampling was carried out from boats going along the shoreline. Substrata could be reached up to a water depth of approximately 1.8 m. Typical substrata investigated were submerged aquatic or helophytic plants, twigs, branches and roots from shrubs and small trees growing close to the waterside. In a provisional laboratory at Skradin subsamples of the collected material were analysed by stereomicroscope when bryozoans were still alive, while the rest dried or stored in 70% respectively 96% alcohol. The measurement of environmental parameters (temperature, conductivity and salinity) at each site was carried out by WTW instruments LF 330 and EC/Testr low+.

Subsequently, the exhaustive inspection of colonies included the preparation of different kind of statoblasts for SEM investigation. Resting stages were taken out of the colonies and transferred into a diluted solution of household bleach (approximately distilled water: bleach = 4:1) for 15–25 min, followed by 3x washing off in distilled water. After exposing them in an ultrasonic bath for 5–15 min, statoblasts were dried and mounted either directly on stubs with Tempfix Leit Cadhaesive glue and analysed with the VP SEM Leo 1455 at the Natural History Museum in London. Other samples were sputtered with gold in an Agar Sputtercoater 108 before mounting and the consecutive analysis with a SEM Philips XL 20 at the University of Vienna. The marine species were analysed with the stereomicroscope at the Faculty

**Table 29.2** Environmental data and occurrence of bryozoan species

| Date | Site | Temperature (°C) | Conductivity (μm) | Salinity (PSU) | Species |
|---|---|---|---|---|---|
| 3.9 | K1 | 10.6 | 441 | 0 | – |
| | K2 | 10.2 | 430 | 0 | – |
| | K3 | 15.6 | 855 | 0.2 | – |
| | K4 | 14.9 | 583 | 0 | – |
| | K5 | 15.6 | 789 | 0.1 | – |
| | K6 | 21.7 | 665 | 0.1 | *P. repens, P. fructicosa* |
| 4.9 | K7 | 18.5 | 633 | 0.1 | *P. repens* |
| | K8 | 19.0 (10 cm); 16.3 (1 m) | 645 | 0.1 | *P. emarginata* |
| | K9 | 16.4 (10 cm); 16.1 (1 m) | 645 | 0.1 | *P. geimermassardi, P. repens, P. emarginata* |
| | K10 | 18.3 (19 cm); 17.0 (1 m) | 640 | 0.1 | *P. emarginata* |
| | K11 | 19.7 (10 cm); 19.3 (1 m) | 630 | 0.1 | *P. repens, P. geimermassardi, P. fungosa, F. sultana, P. emarginata* |
| | K12 | 19.4 (10 cm); 18.9 (1 m) | 626 | 0.1 | *P. repens, P. geimermassardi, P. emarignata, P. articulata C. mucedo* |
| | K13 | 19.4 (10 cm); 19.2 (1 m) | 634 | 0.1 | *P. geimermassardi, P. repens, F. sultana* |
| 5.9 | K14 | 24.3 | 456 | 0 | *P. repens, P. fructicosa* |
| | K15 | 22.9 | 499 | 0 | – |
| | K16 | 23.5 | 484 | 0 | – |
| | K17 | 23.8 | 475 | 0 | – |
| | K18 | 24.3 | 460 | 0 | – |
| | K19 | 24.7 | 462 | 0 | *F. sultana, P. articulata* |
| 6.9 | K20 | 23.1 | 465 | 0 | – |
| | K21 | 17.3 | 510 | 0 | – |
| | K22 | 24.3 (10 cm); 26.0 (1 m) | 5.5 (10 cm); 35.5 (1 m) | 3.3 (10 cm); 11.6 (50 cm); 24.6 (1.5 m) | – |
| | K23 | 24.2 (10 cm); 26.3 (1 m) | 4.8 (10 cm); 40.8 (1 m) | 3.5 (10 cm); 24.0 (50 cm); 28.0 (1.5 m) | – |
| | K24 | 24.6 (10 cm); 26.0 (1 m) | 12.8 (10 cm); 33.8 (1 m) | 7.2 (10 cm); 19.2 (50 cm); 25.3 (1.5 m) | *Conopeum seurati* |
| | K25 | 24.9 (10 m); 26.1 (1 m) | 12.0 (10 cm); 30.0 (1 m) | 8.5 (10 cm); 9.5 (50 cm); 24.5 (1.5 m) | *Conopeum seurati* |

of Science University of Zagreb and with the SEM Tescan Vega TS5136LS at the Faculty of Mechanical Engineering and Naval Architecture University of Zagreb.

## Results

Table 29.2 displays the environmental data of all sampling sites scoring 12 of the 25 with bryozoan findings.

The bryozoans found belong to seven phylactolaemate and gymnolaemate species and include:

Gymnolaemata

Cheilostomata

Membraniporidae

*Conopeum seurati* (Canu, 1928)

Ctenostomata

Paludicellidae

*Paludicella articulata* (Ehrenberg, 1831)

Phylactolaemata

Fredericellidae

*Fredericella sultana* (Blumenbach, 1779)

Plumatellidae

*Plumatella emarginata* Allman, 1844

*Plumatella fruticosa* Allman, 1844

*Plumatella fungosa* (Pallas, 1768)

*Plumatella geimermassardi* (Wood and Okamura, 2004)

*Plumatella repens* (Linnaeus, 1758)

Cristatellidae

*Cristatella mucedo* (Cuvier, 1798)

Eight of the species were sampled in the freshwater section, and one species was found in the estuary part of the river (*Conopeum seurati*). Most frequent was *Plumatella repens* (7 sites), followed by *P. emarginata* (5 sites) and *P. geimermassardi* (4 sites). *F. sultana* was present at three sites, *P. fruticosa, P. articulata* and *C. seurati* at two, while *P. fungosa* as well as *C. mucedo* only occurred at one site.

The sites close to the source of the river (K1–K5) were lacking in bryozoans. At K1 (source) and K2 (section with fish farming), water temperature was extremely low (10.6° and 10.2°). K3 was situated downstream of the city of Knin close to a scrap yard. It was characterized by aberrant higher values of conductivity and also salinity.

Site K6 was the first site with considerably larger diversity of the benthic assemblage. Between filamentous algae intermingled with tufa precipitation, single sessoblasts of *P. repens* were found buried under calcareous deposits. However, living bryozoan colonies with mature floatoblasts were found on submerged twigs and small stems of aquatic plants, substrata typical also for the following sites with bryozoan aufwuchs.

Substrata at K7 were characterized by heavy fungi coverage, presumably caused by sewage of a monastery situated at the river bank. From K7 to K10 only a few living colonies were found. Although in K9 three different species could be sampled, only *P. geimermassardi* was present by a living colony growing on *Myriophyllum spicatum*. A remnant of a colony full with floatoblasts could be assigned to *P. repens,* as well as further floatoblasts of *P. emarginata* were identified on the stems of water lilies.

Sites K11 and K12, situated in the central part of the river, showed the most complex species composition as well as densest bryozoan growth. Substrata were nearly completely covered with colonies (e.g. honeycomb-fused tubes of *P. emarginata).* At each of both sites five different bryozoan species were found and, in total, seven of the eight freshwater bryozoan species were present in this section of the river.

From K13 onward, the first site after a large cascade, aufwuchs got less and freshwater bryozoans got rare. The substrata of sites K15–K 18 were characterized by high tufa coverage, however, filamentous algae and occasional sponges sometimes occurred in impressive sizes (Fig. 29.2a). In K19, situated in a backwater close to the main stream, two bryozoan species were present. *F. sultana* (>15 colonies, Fig. 29.2b) and *P. articulata* (3 colonies) were here more numerous than on the other sites where they were also found (K11 and K13 respectively K12).

Sites K20–K25 were situated below a chain of large cascades, and calcareous deposits reached the highest thickness. From site K22 onward, a significant rise in salinity could be observed. No bryozoans were found until the last two sites (K24 and K25) where *C. seurati* grew on stems of cane between 0.5 and 1.5 m water depth with a salinity ranging from 9.5 to 24.5 PSU.

In summary, except for the central part of the river, bryozoan aufwuchs was rarely dense and colonies were usually of small size (<8 cm length). Table 29.3 presents a ranking of total abundance of all colonies sampled in the Krka River. Most of the colonies could be assigned to three plumatellid species: *P. repens, P. emarginata,* and *P. geimermassardi. P. fungosa* was only found at K11 being next in abundance ("common") and followed by *F. sultana* and *C. seurati. P. articulata* and *P. fruticosa* were rare and *C. mucedo* was represented by a single finding (K12).

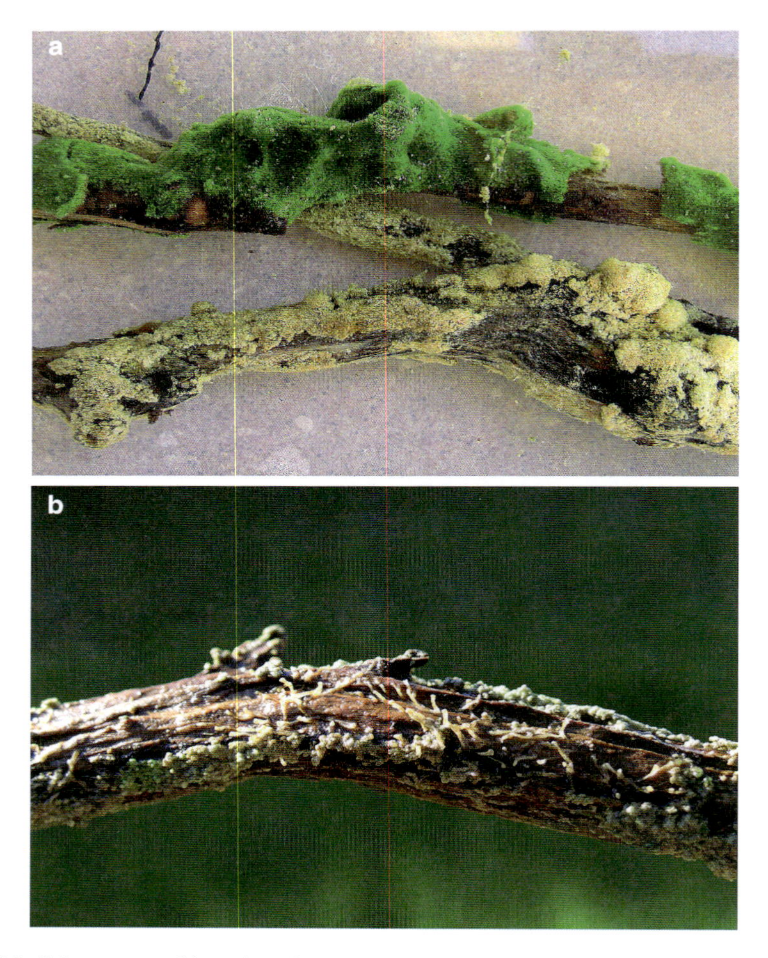

**Fig. 29.2** Substrata: wood branches, diameters 1.5–5 cm; (**a**) coverage with tufa and filamentous algae; (**b**) *Fredericella sultana* colonies between calcareous deposits

# Discussion

## *Tufa Deposition and Bryozoan Colonization*

The most striking character of the Krka River is the tufa coating found on substrata permanently submerged, an environmental condition that creates a special challenge for sessile benthic organisms. The mechanisms of calcite deposition are not completely explained, but it seems that tufa is a product of physico-chemical and biogenic precipitation associated with periphyton colonization (Frančišković-Bilinski et al. 2004). It has high porosity, and typically contains the remains of

**Table 29.3** Ranking abundance of species sampled in the Krka River

| Species | Total abundance | Sites |
|---|---|---|
| *P. repens* | Abundant (>26 colonies) | K6, K7, K9, K11, K12, K13, K14 |
| *P. emarginata* | Abundant (>26 colonies) | K8, K9, K10, K11, K12 |
| *P. geimermassardi* | Abundant (>26 colonies) | K9, K11, K12, K13 |
| *F. sultana* | Common (6–25 colonies) | K11, K13, K19 |
| *C. seurati* | Common (6–25 colonies) | K24, K25 |
| *P. fungosa* | Common (6–25 colonies) | K11 |
| *P. articulata* | Rare (2–5 colonies) | K12, K19 |
| *P. fruticosa* | Rare (2–5 colonies) | K6, K14 |
| *C. mucedo* | Presence (1 colony) | K12 |

micro- and macrophytes, bacteria, protozoa, fungi, invertebrates, and detritus (Primc-Habdija et al. 2001). The tufa growth rate at Plitvice Lakes follows closely the water temperature, with maximum deposition in August/September and minimum deposition in the winter months (Srdoč et al. 1985). This might cause a special handicap for bryozoan settlement as higher deposition rate coincides with the period of flourishing growth of bryozoan colonies (spring to autumn).

Deposition rates in the Krka River are an order of magnitude lower than in other carstic areas such as the Plitvice Lakes. Nevertheless, the impacts of tufa precipitation can be traced along the course of the whole Krka River, and especially rocks and stones were covered mostely by layers of tufa. Proliferate growth of bryozoan colonies seems to be overwhelmingly restricted to overhanging twigs of the vegetation nearby the riverbank and to ephemeral aquatic plants, although the situation was not as extreme as at several sites of Plitvice, where bryozoan colonies were sampled from decaying leaves having fallen from trees that grow along the lakeside. Rocks, old trunks, etc. are a less promising material for colonies. Tubes of bryozoan colonies attached to the substrata can be buried in the tufa, although to different degrees. However, this could also be explained by the different age of the colonies sampled. Furthermore, the overwintering stages of freshwater bryozoans deserve special interest. Attached bryozoan sessoblasts outside of colonies are also found to be covered by calcite precipitation. Karlson (1994) proclaimed that sessile propagules such as sessoblasts promote the local persistence of the bryozoan colony by fixation of resting stages to substrata. In this environment of heavy tufa sedimentation, the success of this strategy of asexual propagation seems to be questionable and further investigation of the germination rate of sessoblasts would be helpful.

## *Occurrence and Distribution of Bryozoan Species*

The census of the Krka River demonstrated an impressive list of eight freshwater species, although conclusions concerning the relevance of this bryozoan diversity have to remain fragmentary as data of comparable environments are still missing.

Investigations of the Danube River, for instance, reveal the same amount of species (Fesl et al. 2005; Wöss 2005). However, in the Krka River one family is clearly dominating: the Plumatellidae are present on all but one site (K19) where freshwater bryozoans were found. On the other hand, gelatinous species are nearly completely missing, as they could only be traced by one floatoblast of C. *mucedo* (in K12).

Bryozoan colonies are not equally distributed in the river and reach highest densities in the central part. In the upper section, the lack of bryozoans could be explained by low aestival water temperatures (K1 and K2) as well as by missing dissemination of statoblasts by animals, humans etc. at the source of the river. A few of the investigated sites showed anthropogenic influences on environmental conditions such as the influx of a scrap yard (K3) or sewage (K7–K10), which might have brought a deterioration for a number of benthic organisms. The decline (or complete lack) in bryozoan aufwuchs was obvious, however the phenomenon seemed to stay locally leaving no traces on sites further downstream.

In the lower section of the river, the lack of freshwater bryozoans on nearly all sites is striking (see also that K19 is not within the main river). Conductivity, water temperature, and heavy algae growth are distinctive in comparison to the upper section. The higher water temperature might be responsible for the pronounced tufa deposition as discussed above, but monitoring would be essential for all sites throughout the year. In terms of competition, a development of freshwater bryozoan colonies seems to be impeded by the thick coverage of filamentous algae on the substrata. The situation only changes towards the end of the river. The estuary is the only part of the river with a clear vertical zonation in the water column with *C. seurati* growing in deeper parts.

## Taxonomy and Zoogeography

*P. repens*, the most abundant species in the Krka River is a very common species found in many European countries (Massard and Geimer 2010). Records for southeastern Europe exist from Bulgaria, Romania, and Slovenia. Distribution data of *P. geimermassardi*, a species described first in 2004, are still rare and restricted to a few countries in Europe – Belgium, British Islands, France, Germany, and Italy – but it is highly probable that some former evidence for *P. repens* could be assigned to *P. geimermassardi*. *P. geimermassardi* is now also found in Austria and Serbia (Wöss unpubl data). SEM investigations exhibit that Croatian floatoblasts are more oval in shape than the Austrian ones (Fig. 29.3). In the course of a combined morphoplogical and genetical study (Wöss and Waeschenbach unpubl data), the validity of morphologically distinctive characters of statoblasts were tested by including molecular markers. Preliminary results of the differently shaped *P. geimermassardi* floatoblasts suggest that the different morphologies might reflect steps of a speciation process in progress but further confirmation is needed. *P. emarginata* is widely distributed in Europe including areas in Bulgaria, Greece, Romania, and Serbia in south-eastern Europe. The floatoblasts found at the

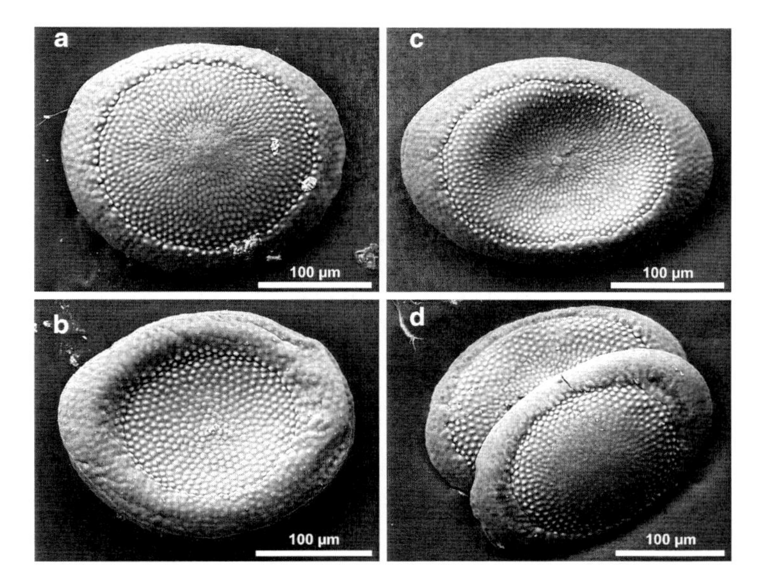


**Fig. 29.3** *Plumatella geimermassardi*, comparison of Austrian (*round-oval*) and Croatian (*long-oval*) floatoblasts; (**a**) ventral valve, Austrian sample; (**b**) dorsal valve, Austrian sample; (**c**) ventral valve, Croatian sample; (**d**) dorsal and ventral valves (Croatian sample)

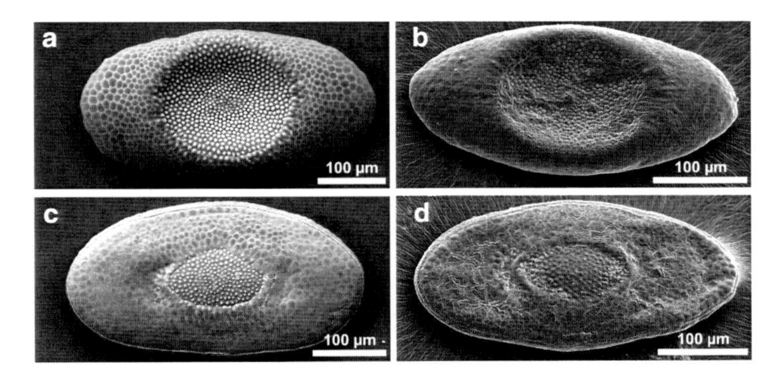


**Fig. 29.4** *Plumatella emarginata:* morphological plasticity of floatoblasts; (**a**) ventral valve; (**b**) dorsal valve; (**c**) ventral valve, tapered annulus; (**d**) dorsal valve, tapered annulus

Croatian sites showed a high morphological plasticity resulting in shapes with an unusual tapered annulus (Fig. 29.4). *P. fungosa*, *P. fruticosa*, *F. sultana*, *C. mucedo*, and *P. articulata* occur in many European countries as well, and are documented for some parts of south-eastern Europe (*P. fungosa:* Romania and Slovenia; *P. fruticosa:* Bulgaria, Romania and Serbia; *F. sultana:* Bulgaria, Greece, Romania

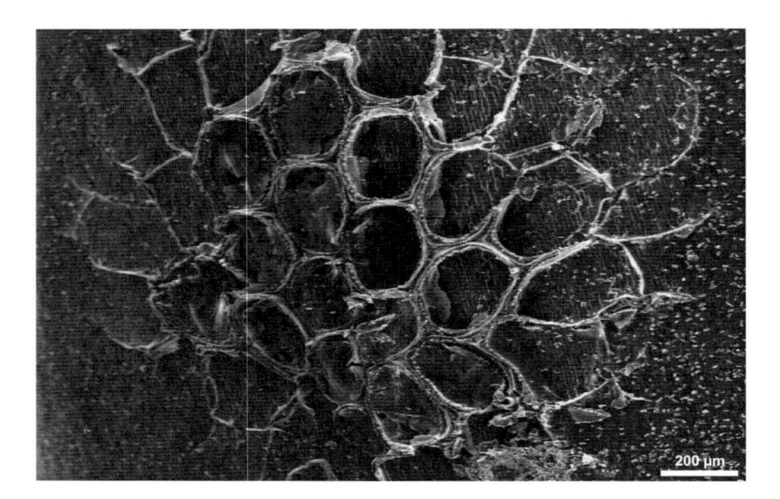


**Fig. 29.5** *Conopeum seurati*, colony

and Slovenia; *C. mucedo:* Bulgaria, Romania, Serbia and Slovenia; *P. articulata:* Romania and Slovenia).

*Conopeum seurati* (Fig. 29.5) is widely distributed in brackish lagoons of the Mediterranean Sea, such as those of the south of France, the coast of Tunisia and some localities all around Italy. It is a typical estuarine species, able to withstand very low salinities (<1) and tolerant to widely fluctuating concentrations. *C. seurati* was possibly introduced to the British Isles within historical times (Hayward and Ryland 1998) and appears to be a successful fouling organism. Occhipinti Ambrogi and Ambrogi (1987) described the population ecology of *C. seurati* in the Po Delta (Italy). Before 1983, when the invasion of the bryozoan *Tricellaria inopinata* started, it was one of the main bryozoan species in low salinity areas of the Venice Lagoon (Occhipinti Ambrogi 2000). The record of *C. seurati* from the Krka River is the first finding along the eastern (Croatian) coast of the Adriatic Sea.

**Acknowledgements** The authors wish to thank the personnel of the National Park Krka: Dr. Drago Marguš, Dr. Gordana Goreta and Branimir Pavelić for their outstanding support in organization of bryozoan sampling. Special thanks to Andrea Waeschenbach, Mary Spencer Jones and the team of the EMMA lab at the Natural History Museum London. We are also grateful to Daniela Gruber from the Institution of Cell Imaging and Ultrastructure Research of the University in Vienna for providing excellent working conditions.

## References

Allman GJ (1844) Synopsis of the genera and species of Zoophytes inhabiting the fresh waters of Ireland. Ann Mag Nat Hist 13(1):328–331

Blumenbach JF (1779) D. Joh. Fried. Blumenbachs Handbuch der Naturgeschichte. J.C. Dieterich, Göttingen

Canu F (1928) Trois nouveaux Bryozoaires d'eau douce. Bulletin de la Societe d'histoire naturelle de l'Afrique du Nord 19(7–8):262–264
Cuvier L (1798) Tableau élémentaire de l'Histoire naturelle des animaux. Baudouin, Paris
Ehrenberg CG (1831) Symbolae Physicae, seu Icones et descriptiones Corporum Naturalium novorum aut minus cognitorum, quae ex itineribus per Libyam, Aegyptum, Nubiam, Dongalam, Syriam, Arabiam et Habessiniam ... studio annis 1820–25 redirerunt Berolini, Berlin. Pars Zoologica, v. 4. Animalis Evertebrata exclusis Insectis
Emeis KC, Richnow HH, Kempe S (1987) Tufa formation in Plitvice National Park, Yugoslavia: chemical versus biological control. Sedimentol 34:595–609
Fesl C, Humpesch UH, Wöss ER (2005) Biodiversität des Makrozoobenthos der österreichischen Donau unter Berücksichtigung quantitativer Befunde der Freien Fließstrecke unterhalb Wiens. Denisia 16:139–158
Frančišković-Bilinski S, Barišić D, Vertačnik A, Bilinski H, Prohić E (2004) Characterization of tufa from the Dinaric Karst of Croatia: mineralogy, geochemistry and discussion of climate conditions. Facies 50:183–193
Hayward PJ, Ryland JS (1998) Cheilostomatous Bryozoa. I. Aeteoidea – Cribrilinoidea: notes for the identification of British species, vol 10, 2nd edn, Synopses of the British fauna (new series). Field Studies Council, Shrewsbury
Horvatinčić N, Čalić R, Geyh MA (2000) Interglacial growth of tufa in Croatia. Quaternary Res 53:185–195
Horvatinčić N, Krajcar Bronic I, Obelic B (2003) Differences in the 14C age, N13C and N18O of Holocene tufa and speleothem in the Dinaric Karst. Palaeogeogr Palaeclimatol Palaeoecol 193:139–157
Juračić M, Prohić E (1991) Mineralogy, sources of particles, and sedimentation in the Krka River Estuary (Croatia). Geol Vjesn 44:195–200
Karlson RH (1994) Recruitment and catastrophic mortality in *Plumatella emarginata* Allman. In: Hayward PH, Ryland JF, Taylor PD (eds) Biology and paleobiology of bryozoans. Olsen and Olsen, Fredensborg, pp 93–96
Kerovec M, Marguš D, Mrakovčić M, Kučinić M (2007) A review of research into the fauna of the Krka National Park. In: Marguš D (ed) Rijeka Krka i Nacionalni park "Krka" – prirodna i kulturna baština, zaštita i održivi razvitak. Javna ustanova "Nacionalni park Krka", Šibenik
Linnaeus C (1758) Systemae naturae per regna tria naturae, secundum classes, ordines, genera, species, cum characteribus, differentiis, synonymis, locis Ed.10. Laurentii Salvii, Holmiae
Lojen S, Trkov A, Ščančar J, Vázquez-Navarro JA, Cukrov N (2009) Continuous 60-year stable isotopic and earth-alkali element records in a modern laminated tufa (Jaruga, river Krka, Croatia): implications for climate reconstruction. Chem Geol 258:242–250
Massard JA, Geimer G (2008a) Global diversity of bryozoans (Bryozoa or Ectoprocta) in freshwater. Hydrobiol 595:93–99
Massard JA, Geimer G (2008b) Occurrence of *Plumatella emarginata* Allman, 1844 and *P. casmiana* Oka, 1908 (Bryozoa: Phylactolaemata) in Lake Pamvotis (Joannina, Greece). Bull Soc Nat luxemb 109:133–138
Massard JA, Geimer G (2010) Fauna Europaea. http://www.faunaeur.org. Last update 3 June 2010. Accessed 5 Nov 2010
Occhipinti Ambrogi A (2000) Biotic invasions in a Mediterranean Lagoon. Biol Invasions 2:165–176
Occhipinti Ambrogi A, Ambrogi R (1987) Short-term changes in a brackish water assemblage of Bryozoa with particular reference to *Conopeum seurati* (Canu). In: Ross JRP (ed) Bryozoa: present and past. Western Washington University, Bellingham, pp 183–190
Pallas PS (1768) Descripto Tubulariae fungosa propae Volodemirum mense Julio 1768 observatae. Novi Commentarii academiae scientiarum imperialis Petropolitanae 12:565–572
Primc-Habdija B, Habdija I, Plenković-Moraj A (2001) Tufa deposition and periphyton overgrowth as factors affecting the ciliate community on tufa barriers in different current velocity conditions. Hydrobiologia 457(1–3):87–96
Raddum GG, Johnsen TM (1983) Growth and feeding of *Fredericella sultana* (Bryozoa) in the outlet of a humic acid lake. Hydrobiologia 101:115–120

Ricciardi A, Reiswig HM (1994) Taxonomy, distribution, and ecology of the freshwater bryozoans (Ectoprocta) of eastern Canada. Can J Zool 72:339–359

Simić V, Ostojić A (1997) Ecological characteristics of benthic macrofauna of Vlasinkso jezero reservoir. In: Blaženčić J (ed) Vlasina lake – hydrobiological study, Faculty of Biology, Belgrade

Sørensen JP, Riber HH, Kowalczewski A (1986) Soluble reactive phosphorus release from bryozoan dominated periphyton. Hydrobiologia 132:145–148

Srdoč D, Horvatinčić N, Obelić B, Krajcar I, Sliepčević A (1985) Calcite deposition processes in karst waters with special emphasis on the Plitvice Lakes, Yugoslavia. (in Croatian with English abstract). Carsus Iugoslaviae 11:101–204

Taticchi MI (2010) *Plumatella viganoi*, a new freshwater bryozoan species (Phylactolaemata) from Lake Trasimeno (Umbria, Italy). Ital J Zool 77(3):316–322

Taticchi MI, Pieroni G (2005) Freshwater Bryozoa of Italy. A survey of some species from the Italian bryozoan collection of A. Viganò with new records. In: Moyano HIG, Cancino JM, Wyse Jackson PN (eds) Bryozoan studies 2004. A.A. Balkema Publishers, London, pp 317–327

Taticchi MI, Pieroni G, Elia AC (2006) A new species of the European freshwater bryozoan fauna: *Plumatella similirepens* Wood, 2001 (Bryozoa, Phylactolaemata) Linzer biol. Beitr 38 (1):47–54

Taticchi MI, Pieroni G, Elia AC (2008) First finding of *Plumatella vaihiriae* (Hastings, 1929) (Bryozoa, Phylactolaemata) in Europe. Ital J Zool 75:411–416

Wood TS, Okamura B (2004) *Plumatella geimermassardi*, a newly recognized freshwater bryozoan from Britain, Ireland, and continental Europe (Bryozoa: Phylactolaemata). Hydrobiologia 518(1–3):1–7

Wood TS, Okamura B (2005) A new key to the freshwater bryozoans of Britain, Ireland and Continental Europe, with notes on their ecology. Freshw Biol Assoc Sci Publ 63:1–113

Wood TS, Anurakpongsatorn P, Mahujchariyawong J (2006) Freshwater bryozoans of Thailand (Ectoprocta and Entoprocta). Nat Hist J Chulalongkorn Univ 6(2):83–119

Wöss ER (1996) Life-history variation in freshwater bryozoans. In: Gordon DP, Smith AM, Grant-Mackie JA (eds) Bryozoans in space and time. National Institute of Water and Atmospheric Research, Wellington, pp 391–399

Wöss ER (2002) The reproductive cycle of *Plumatella casmiana* (Phylactolaemata: Plumatellidae). In: Butler C, Spencer Jones M, Wyse Jackson PN (eds) Bryozoan studies 2001. A.A. Balkema, Rotterdam, pp 347–352

Wöss ER (2005) The distribution of freshwater bryozoans in Austria. In: Moyano HIG, Cancino JM, Wyse Jackson PN (eds) Bryozoan studies 2004. A.A. Balkema, London, pp 369–375

# Subject Index



A. Ernst, P. Schäfer, and J. Scholz (eds.), *Bryozoan Studies 2010*,
Lecture Notes in Earth System Sciences 143, DOI 10.1007/978-3-642-16411-8,

# Taxonomic Index




A. Ernst, P. Schäfer, and J. Scholz (eds.), *Bryozoan Studies 2010*,
Lecture Notes in Earth System Sciences 143, DOI 10.1007/978-3-642-16411-8,

## M



## N

## O



## P

**R**

MIX
Papier aus verantwortungsvollen Quellen
Paper from responsible sources
FSC® C105338

Printed by Books on Demand, Germany